Membrane Proteins in Energy Transduction

Membrane Proteins

EDITOR

RODERICK A. CAPALDI

Institute of Molecular Biology
University of Oregon
Eugene, Oregon

Volume 1 Membrane Proteins and their Interactions with Lipids
edited by Roderick A. Capaldi

Volume 2 Membrane Proteins in Energy Transduction
edited by Roderick A. Capaldi

Additional volumes in preparation

Membrane Proteins in Energy Transduction

edited by

RODERICK A. CAPALDI

Institute of Molecular Biology
University of Oregon
Eugene, Oregon

MARCEL DEKKER, INC. New York and Basel

Library of Congress Cataloging in Publication Data

Main entry under title:

Membrane proteins in energy transduction.

(Membrane proteins ; v. 2)
Includes bibliographies and indexes.
1. Membrane proteins. 2. Bioenergetics.
3. Enzymes. I. Capaldi, Roderick A. II. Series.
QP552.M44M46 574.1'9121 79-17584
ISBN 0-8247-6817-5

COPYRIGHT ©1979 by MARCEL DEKKER, INC. ALL RIGHTS RESERVED

Neither this book nor any part may be reproduced or transmitted in any form or by any means, electronic or mechanical, including photocopying, microfilming, and recording, or by any information storage and retrieval system, without permission in writing from the publisher.

MARCEL DEKKER, INC.
270 Madison Avenue, New York, New York 10016

Current printing (last digit):
10 9 8 7 6 5 4 3 2 1

PRINTED IN THE UNITED STATES OF AMERICA

Introduction to the Series

Membrane proteins are involved in many of the important functions of cells including energy transduction, transport of substances in and out of a cell or organelle, lipid metabolism, hormonal action, and cellular control. Procedures for isolating membrane proteins in an active form are continually being refined. As a result, more and more functionally important membrane proteins are being isolated and a vast body of literature on the structure and function of these proteins is accumulating. Studies of membrane proteins are published in journals of hematology, virology, bacteriology, and immunology as well as in journals of biology and biochemistry. Consequently, it becomes progressively more difficult to keep up with advances in this diverse area of research. The more diverse a field becomes, the more important it is to have available recent reviews on specific aspects of that area of research.

The aim of this series is to provide a forum for discussion of advances in our understanding of both the structure and function of membrane proteins. As far as possible, individual volumes will concentrate on areas in which significant progress is being made such as in protein-lipid interactions, energy transduction, and hormone-receptor interactions. In this way the volumes shall be an important reference source for advanced students interested in membrane-related phenomena as well as researchers in the selected areas.

Roderick A. Capaldi

Preface

Oxidative phorphorylation and photophosphorylation are the two major processes by which animal and plant cells produce the energy they require to sustain life. The seat of oxidative phosphorylation is mitochondria in both plants and animals, while photophosphorylation in plants occurs in specialized organelles called chloroplasts. Both energy-producing processes are used by different kinds of bacteria and, as in the case of mitochondria and chloroplasts, the components involved in energy transduction in bacteria are localized in membranes.

The pioneering work of David Green, E. Racker, Tsoo King, and their colleagues has lead to identification of most if not all of the components involved in oxidative phosphorylation and photophosphorylation. Most can now be purified, and as a result, considerable progress has been made toward understanding the structure and functioning of these membrane-bound components. It is this recent work which is reviewed in this book. Chapters on the flavin-containing dehydrogenases (Ohnishi) and on cytochromes (Trumpower and Katki, Capaldi) summarize the extent of our knowledge of the mitochondrial respiratory chain, while a chapter by Bragg discusses the progress that has been made in understanding the structure of the energy-transducing components in *Escherichia coli.* The molecular properties of the ATP-synthesizing enzyme from plant and animal cells is covered by Senior and there are chapters on energy-linked transport in *Halobacterium halobium* and on the structure of photoreaction centers from photosynthetic bacteria, systems on which much of our understanding of photophosphorylation has been obtained.

I am grateful to my colleagues in Oregon and to the personnel of Marcel Dekker, Inc. for their help in preparing this volume.

Roderick A. Capaldi

Contents

Contributors

Philip D. Bragg Department of Biochemistry, The University of British Columbia, Vancouver, British Columbia, Canada

Roderick A. Capaldi Institute of Molecular Biology, University of Oregon, Eugene, Oregon

Aspandiar G. Katki* Department of Biochemistry, Dartmouth Medical School, Hanover, New Hampshire

Janos K. Lanyi Extraterrestrial Research Division, NASA-Ames Research Center, Moffett Field, California

Tomoko Ohnishi Departments of Biochemistry and Biophysics, University of Pennsylvania, Philadelphia, Pennsylvania

John M. Olson Brookhaven National Laboratory, Upton, New York

A. E. Senior Department of Biochemistry, University of Rochester Medical Center, Rochester, New York

J. Philip Thornber Department of Biology, University of California, Los Angeles, California

Bernard L. Trumpower Established Investigator of the American Heart Association, and Department of Biochemistry, Dartmouth Medical School, Hanover, New Hampshire

*Current affiliation: Tufts-New England Medical Center, Boston, Massachusetts, and National Cancer Institute, Division of Cancer Treatment, Clinical Pharmacology Branch, Bethesda, Maryland.

Membrane Proteins
in Energy Transduction

1

Mitochondrial Iron-Sulfur Flavodehydrogenases

Tomoko Ohnishi

University of Pennsylvania
Philadelphia, Pennsylvania

I. Introduction

The mitochondrial inner membrane contains nonheme iron and acid-labile sulfur, each at about twice as much concentration as that of heme iron [1, 2, 3]. Important clues to the significance of this nonheme iron were provided by Beinert and Sands in 1960 [4] with the discovery of the unusual electron paramagnetic resonance (EPR) signals, generally known as the g = 1.94 signals, arising from iron-sulfur centers in the mitochondrial system. The function of these mitochondrial iron-sulfur centers, however, has for a long time been the least understood of all the components of the respiratory chain. This was mainly due to the fact that the commonly used cryogenic EPR techniques at temperatures above that of liquid nitrogen could detect resonance signals of iron-sulfur centers which correspond to only a small fraction of the total chemically measurable nonheme iron and acid-labile sulfide in the mitochondrial inner membrane.

New impetus was brought into the field when EPR spectra of reduced iron-sulfur centers, in yeast submitochondrial membrane preparations were examined in a lower range of temperature than hitherto, namely, between that of liquid nitrogen (77K) and that of liquid helium (4.2K) [5, 6].

To date, about six iron-sulfur centers in the NADH-ubiquinone reductase (UQ) segment, three in the succinate-UQ reductase segment, and one each in the electron-transferring flavoprotein (ETF) dehydrogenase and the cytochrome bc_1 region of the respiratory chain have been characterized, as summarized in Scheme I. Among these 11 iron-sulfur centers, EPR signals from only three centers are detectable above liquid nitrogen temperature.

The purpose of this chapter is to give the reader an up-to-date account of work on the mitochondrial iron-sulfur flavoprotein dehydrogenases, namely, succinate-, NADH-, and $ETFH_2$-UQ reductase segments of the respiratory chain. Many excellent reference sources are available on related subjects [9-13].

II. Succinate Dehydrogenase Segment

A. Background

Succinate dehydrogenase (SDH) has been extensively studied by many investigators since the first solubilization of SDH preparations by Morton [14] in 1950 from Keilin-Hartree heart muscle preparation. Singer et al. [15] and Wang et al. [16] independently purified SDH in 1955; preparations were found to contain flavin and iron (2-4 per flavin) and to catalyze electron transfer from succinate to nonphysiological electron acceptors, such as phenazine methosulfate (PMS) or ferricyanide. By improving the method of Wang et al., Keilin and King [17] first succeeded in isolating a more intact SDH preparation, which can transfer

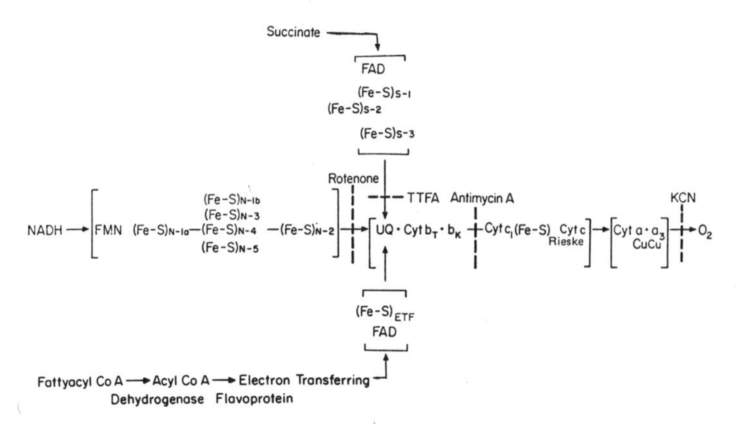

Scheme 1 Respiratory chain components present in the inner mitochondrial membrane, including newly detected iron-sulfur centers. Iron-sulfur centers associated with NADH-UQ reductase and succinate-UQ reductase segments are designated with suffixes N-X and S-X, respectively. The iron-sulfur center originally called center 5 [7] is now designated $(Fe\text{-}S)_{ETF}$ because of its identified function in the electron-transferring flavoprotein [8]. Reaction sites of the respiratory inhibitors are also shown. Other references are cited in relevant sections of the text.

electrons to the respiratory chain (reconstitutive activity) as well as to nonphysiological electron acceptors. It was established later that the SDH preparation contains eight iron and eight acid-labile sulfide atoms per flavin [18]. The flavin component of SDH was known to be acid-insoluble [3, 15, 16]. Subsequently, it was shown to be a unique FAD which is convalently bound to the polypeptide chain as 8α-[N(3)]-histidyl-FAD [19]. In 1971, Davis and Hatefi [20] isolated an essentially pure SDH preparation, by extracting the enzyme from purified particulate succinate-UQ reductase (complex II) [21, 22], using a chaotropic agent. From the analysis of this pure enzyme, it was shown that SDH has a molecular weight of approximately 100,000 and contains 8 g-atoms each of nonheme iron and acid-labile sulfur and 1 mol of flavin per molecule. It was further demonstrated that the SDH molecule is composed of two nonidentical subunits; namely ferroflavoprotein (Fp) and iron-sulfur protein (Ip), with molecular weights of 70,000 ± 7% and 27,000 ± 5%, respectively [20]. The Fp contains one covalently bound FAD and 4Fe·4S; Ip contains 4Fe·4S as illustrated in Scheme II. This molecular composition of SDH has been confirmed by most of the investigators in the field [23-25].

In spite of the multiplicity of nonheme iron and acid-labile sulfur in the SDH molecule (8Fe 8S), the EPR signal arising from one species of ferredoxin

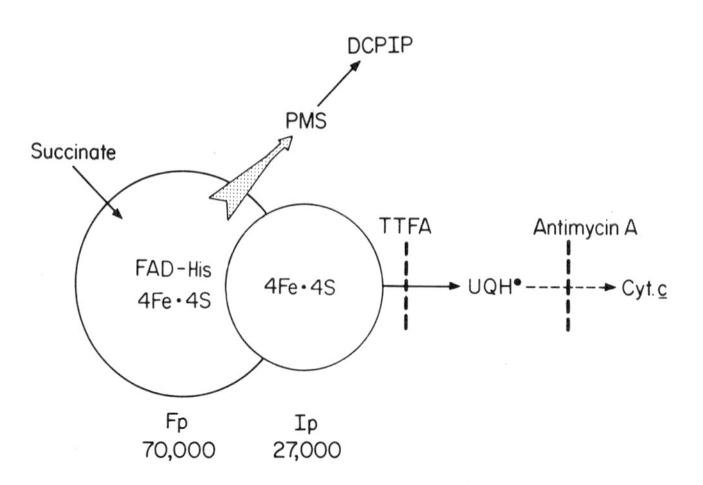

Scheme II Molecular composition of SDH subunits and schematic summary of electron transfer pathways associated with SDH. DCPIP, dichloroindophenol; TTFA, thenoyltrifluoroacetone.

(Fd)-type iron-sulfur center [4] was the only EPR-detected iron-sulfur component in this molecule. As will be described later, recent EPR studies of SDH at temperatures much lower than 77 K, have led to the identification of a second Fd-type iron-sulfur center [26, 27] and one HiPIP-type (paramagnetic in the oxidized state) center [28, 29] in the SDH molecule.

Neither direct analysis of redox components in the two subunits as isolated nor subunit reconstitution studies have been feasible, since the two subunits are tightly bound and only an extremely drastic method can separate them and such procedures render the isolated individual subunits enzymatically inactive (Fp even loses the ability to transfer electrons from succinate to redox dyes). However, several different soluble and particulate SDH preparations are available which are modified during the isolation process to different extents as described above. Comparative studies of structure-function relationship using these preparations have offered, as will be described later, a useful system for elucidating the distribution of redox components in two subunits and the possible mechanism of electron transfer within this molecule. Thus, representative SDH preparations are listed in Table 1 along with the codes used for them in this chapter. These preparations not only differ in their nonheme iron and acid-labile sulfur content but also in their enzymatic activities. Intact SDH has two different electron transfer pathways from succinate; one to artificial electron acceptors, the other to the respiratory chain after recombination with alkaline-treated submitochondrial particles (SMP) [30, 34, 35] or with purified cytochrome bc_1 complex [36], which is referred to as reconstitutive activity. The latter activity

Table 1 Soluble, Lipid-free SDH Preparations

Code of preparation	Solubilization methods	Prior succinate incubation	SDH-solubilized reconstitutive activity	Flavin:Fe:S	References
BS-SDH	Butanol extraction at pH 9.1 from HMP	+	+	1:8:8	18, 30
AS-SDH	Alkali (pH 10.6) extraction from complex II	+	+	1:8:8	18, 27
PS-SDH	Perchlorate (o.4-0.8 M) extraction from complex II	+[a]	+	1:8:8	20
B-SDH	Butanol extraction at pH 9.1 from HMP	–	–	1:8:8	18, 30
A-SDH	Alkali (pH 8) extraction of complex II	–	–	1:8:8	31
CN-SDH	Cyanide extraction of HMP	–	–	1:6:4	32
AA-SDH	Alkali extraction from acetone powder of mitochondria	–	–	1:4:4	33

[a]Dithiothreitol is present in addition to succinate.

is the most sensitive physiological parameter for examining the intactness of the isolated enzyme [26, 30]. Succinate dehydrogenase retains reconstitutive activity when extracted from mitochondrial membrane preparations in the presence of reducing agents, such as succinate, dithiothreithol, or dithionite, and purified anaerobically (e.g., BS-SDH, AS-SDH, and PS-SDH). If the reducing agents are omitted as in the case of B-SDH and A-SDH, no reconstitutive activity is preserved even though the chemically determined nonheme iron and acid-labile sulfur content is the same as in reconstitutively active enzymes (1:8:8 = flavin/Fe/S) [18, 31]. Enzyme which contains less than 8Fe 8S per flavin, such as CN-SDH or AA-SDH, exhibit no reconstitutive activity, in spite of reports purporting to demonstrate reconstitutive activity in AA-SDH [37]. All SDH preparations listed in the table show similar specific activity in transferring electrons to artificial electron acceptors, such as PMS and ferricyanide [27].

Recently Vinogradov et al. [38, 39] demonstrated that there are two different reaction sites for $K_3Fe(CN)_6$, one with low K_m (<200 μM) the other with high K_m (2 mM). The former activity is extremely labile (a characteristic of center S-3) while the latter activity is stable (a feature characteristic of other dye reductase activities), thus it can be argued that center S-3 is the site for the low-K_m, $K_3Fe(CN)_6$ reaction. It should also be pointed out that all soluble enzyme preparations listed contain both Fp and Ip subunits, as revealed by SDS electrophoresis or determination of the molecular weight [40, 41]. Particulate succinate-UQ reductase (complex II) [21] is also a useful preparation which contains 8Fe 8S per flavin. In this preparation, however, only about 50% of the protein arises from the SDH molecule; it also contains low-E_m cytochrome *b* (1 mol per flavin), trace amounts of cytochrome c_1, and 20-30% of phospholipid (deficient in UQ) [21].

B. Redox Components in the SDH Molecule

1. Iron-Sulfur Clusters

The EPR signal of the Fd-type iron-sulfur center (known as the g = 1.94 species) in purified SDH preparations was first detected by Beinert and Sands [4] in 1960 in EPR measurements above 77 K. It gives a spectrum of rhombic symmetry (g_z = 2.03, g_y = 1.93, g_x = 1.91). Relaxation times of this center are rather long and the EPR signals are readily saturated at temperatures below 25 K. This center is almost quantitatively reduced with a high concentration of succinate and its spin concentration was shown later to be approximately equivalent to that of flavin [42]. The presence of the second Fd-type iron-sulfur center with shorter relaxation times was recognized for the first time when Ohnishi et al. [26, 27] compared EPR spectra of reconstitutively active BS-SDH preparations reduced with succinate and with dithionite, respectively, at sample temperatures below 40 K.

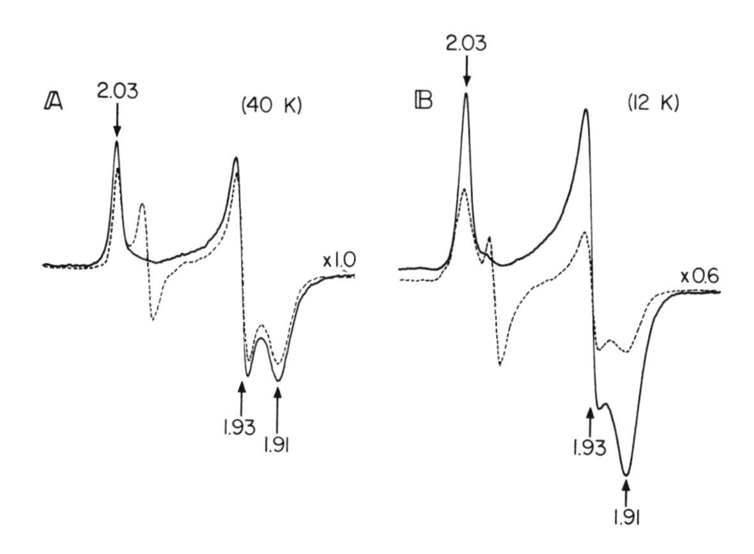

Figure 1 EPR spectra of succinate- or dithionite-reduced BS-SDH measured at two different temperatures. (The temperature reading was corrected later.) Reproduced with permission from Ohnishi et al. [27].

As presented in Figure 1, iron-sulfur spectra of succinate- and dithionite-reduced enzyme are not significantly different at 40 K (spectra A), while at lower temperatures (spectra B) dithionite-reduced enzymes give much more intense signals than those of succinate-reduced ones. Spin quantitation shows that dithionite-reduced BS-SDH gives spectra of 0.8-1.0 spins per flavin at temperatures above 40 K, while the spectra are equivalent to 1.6-1.8 spins per flavin at temperatures below 25 K [27, 43] (Fig. 2). This suggests that S-2 spins in the reconstitutively active enzyme have much shorter relaxation times (T_1 and T_2) than S-1 spins over a wide temperature range. Thus above 40 K, S-2 spins can not be detected because of extremely lifetime-broadened spectra. As will be described later, it has been established that S-1 and S-2 are spin coupled and the power saturation of center S-1 is relieved by the nearby S-2 spins. Therefore 1.6-1.8 spins per flavin can be obtained in quantitation of dithionite-reduced BS-SDH under EPR conditions where S-1 spins in the absence of S-2 spins are heavily saturated.

In addition, potentiometric titrations of centers S-1 and S-2 gave the $E_{m7.4}$ values of -5 ± 15 mV and -400 ± 15 mV, respectively, in all soluble SDH preparations measured [26, 27]. Both components titrate reversibly as one-electron transfer couples ($n = 1$). These observations indicated the presence of two distinct species of Fd-type iron-sulfur centers in the SDH with different relaxation

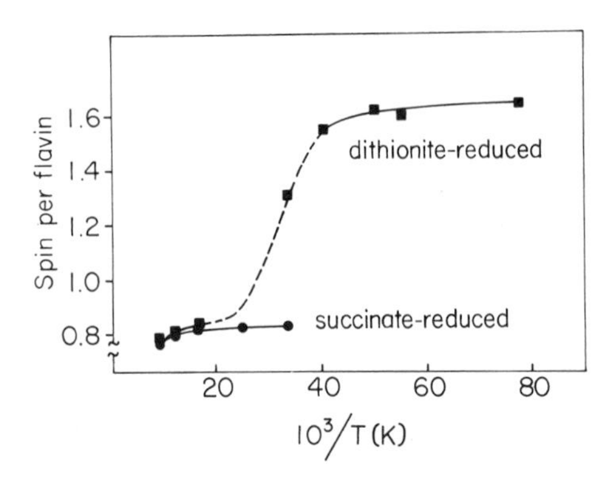

Figure 2 Spin concentration as a function of reciprocal temperature, which was determined with BS-SDH reduced with dithionite and succinate, respectively. Reproduced with permission from Salerno et al. [43].

rates and redox midpoint potentials; they were designated as centers S-1 and S-2. Although centers S-1 and S-2 show greatly different midpoint potentials their individual EPR spectra cannot be obtained by simply taking the difference spectra of dithionite-reduced (E_h = -500 mV) and succinate-reduced (E_h = 80 mV) BS-SDH preparations. This is because of the widely different relaxation behaviors of S-1 and S-2 spins and the effect of their spin-spin interactions, as will be described later. Fortunately it was found that the relaxation behavior of S-2 spins is very sensitive to its microenvironment, and S-2 spins can be converted to the less rapidly relaxing species without changing the spectral lineshape of both centers. For example, in the reconstitutively inactive enzymes such as B-SDH or AA-SDH, both center S-1 and center S-2 signals can be detected quantitatively at higher temperatures and they serve for a good approximation of the individual spectra.

Figure 3 shows the EPR spectra of B-SDH, reduced with dithionite (spectrum A) and succinate (spectrum B), respectively. Only center S-1 and flavin are reduced with succinate; addition of dithionite causes reduction of center S-2 and further reduction of flavin to a diamagnetic state. Under this EPR condition (27 K, 1 mW), spectra of both centers are not saturated (saturation begins at about 2-3 mW microwave power) and only a slight effect of the spin-spin interaction is discernible. Spectrum C, the computed difference between spectra A and B, gives the EPR spectrum of center S-2 alone. Spectra B and C indicate that both centers S-1 and S-2 exhibit similar EPR spectra of rhombic symmetry with essentially the same g values (g_z = 2.03, g_y = 1.93, g_x = 1.905) [27, 43]. In spite of the general tendency to obtain less spin concentration in reconstitutively in-

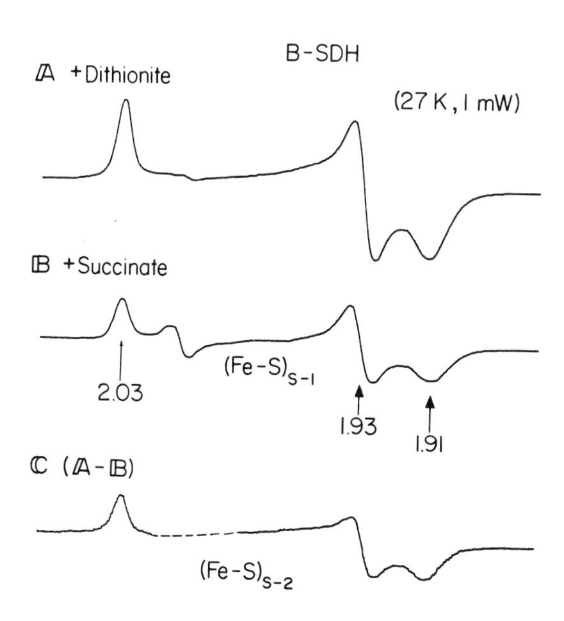

Figure 3 Individual EPR spectra of centers S-1 and S-2, which were obtained using B-SDH with a minimal contribution of spin-spin interaction. (A) 0.35 ml of B-SDH at the protein concentration of 40 mg/ml in 0.1 M phosphate buffer (pH 7.4) was reduced with an excess of dithionite added as a solid. (B) 3.5 μl of 1.0 M K-succinate was added to 0.35 ml of B-SDH, and incubated for 10 min at room temperature after being transferred into an EPR tube. EPR operating conditions: field modulation frequency, 100 KHz; modulation amplitude, 5 gauss; microwave frequency, 9.1 GHz; microwave power, 1 MW; time constant, 0.3 sec.; scanning rate, 200 gauss/min; sample temperature, 27 K. Under these EPR conditions, saturation of center S-1 starts around 2 mW. Subtraction of spectrum B (from A) was performed by the use of a Nicolette 1974 computer of average transients (Nicolette Instrument Corporation). Reproduced with permission from Salerno et al. [43].

active SDH preparations, quantitation of S-1 alone and S-1 plus S-2 in reconstitutively active BS-SDH supports the notion that both centers are present in the dehydrogenase in a concentration approximately equivalent to that of flavin, namely one S-1 and one S-2 per molecule.

A relatively symmetric highly temperature-sensitive EPR signal centered around g = 2.01 has been previously seen in various mitochondrial membrane preparations in the oxidized state [44-46]. However, it has received little attention because originally it was considered to be due only to ferric iron in a cubic environment and, in addition, the kinetic behavior of this component was reported to be too slow for a component of the respiratory chain [46]. However, purification of a protein exhibiting this signal [47] * has led to the recognition of a new type of iron-sulfur protein which is paramagnetic in the oxidized state, analogous to bacterial high-potential iron-sulfur protein [49] although its spectral lineshape and redox potential differ from those of the HiPIP isolated by Bartch [49]; thus it was classified as a HiPIP-type iron-sulfur center. A HiPIP-type iron-sulfur signal was identified in the particulate succinate-UQ reductase (complex II) by Beinert and his colleagues [28]. This center gives a rather symmetric signal around g = 2.01 with a peak-to-peak width of 23 gauss, which is readily detectable only below 30 K. Its spin concentration was found to be approximately equivalent to that of flavin in complex II; however, it was nearly undetectable in the soluble SDH preparation examined by Beinert and coworkers [28]. Ohnishi et al. subsequently demonstrated that this HiPIP-type center becomes extremely labile toward oxidizing agents upon removal of the dehydrogenase from the membrane and the EPR signal can be seen only in the reconsitutively active SDH preparations [29, 50]. Even in freshly isolated enzyme preparations, the minimal concentration of ferricyanide required in order to oxidize this HiPIP-type center causes modification of a large part of the enzyme, resulting partly in lineshape modification and partly in conversion to the EPR-undetectable form (Fig. 4) [29]. Since this center is easily modified even by a low concentration of ferricyanide, as seen in Figure 5, it is important to oxidize the enzyme preparation with a minimal concentration of ferricyanide, and only 0.3-0.5 equivalent center S-3 signals are maximally detected even in the fresh BS-SDH. The extrapolation to zero ferricyanide, however, gave above 0.7 spins per flavin [29]. Thus, this HiPIP-type center was considered as a constituent of the SDH present in unmodified enzyme in the amount of one center per molecule. It was designated as center S-3 following the nomenclature of centers S-1 and S-2 [26, 27]. Beinert et al. [53] recently confirmed these observations. They also estimated the spin concentration of center S-3 in soluble SDH after stabilizing this center by recombining the enzyme with alkaline-treated SMP, and showed that the increment of the flavin concentration was equivalent to the center S-3 concentration in the reconstituted succinate oxidase system. This provides additional support to the notion that center S-3 is an "integral" somponent of the SDH molecule, present as one center per molecule.

Since one cannot demonstrate biochemically that these three iron-sulfur centers are separate individual entities, the following question is a valid one:

*This HiPIP-type iron-sulfur protein was shown to exhibit EPR characteristics and redox behavior considerably different from that of center S-3 [48]. This protein is readily detached from the mitochondrial membrane during sonication [48] and its physiological function is still unknown. This center is not included in the respiratory chain components in Scheme I.

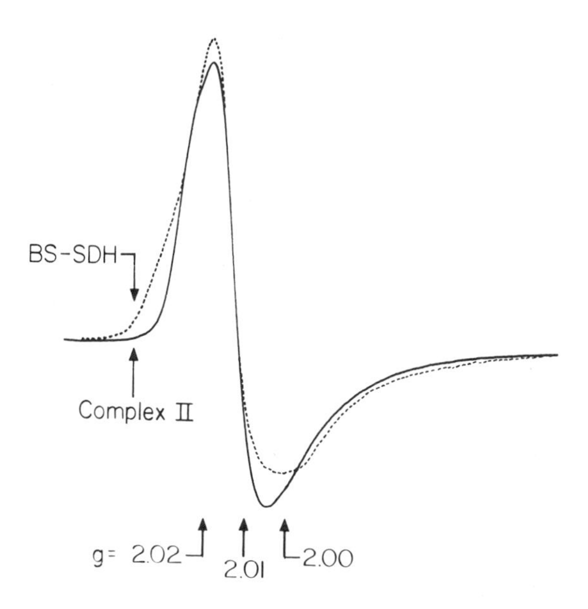

Figure 4 EPR spectra of center S-3 in particulate succinate-UQ reductase (complex II) and in soluble reconstitutively active SDH (BS-SDH). Concentrations of enzymes used were 19 mg protein/ml and 3.7 nmol flavin/mg protein (complex II) and 18.3 mg protein (BS-SDH). Complex II was oxidized with 150 μM ferricyanide in the presence of 50 μM PMS, while BS-SDH was oxidized with 100 μM ferricyanide and 10 μM PMS. Enzymes were rapidly frozen after 1 min incubation at room temperature. EPR operating conditions: modulation frequency, 100 kHz; modulation amplitude, 5 gauss; microwave frequency, 9.12 GHz; microwave power, 0.5 mW; time constant, 0.3 sec; scanning rate, 200 gauss/min; sample temperature, 10.1 K. The dotted-line spectrum was recorded at four times higher gain than that of the solid line. Reproduced with permission from Ohnishi et al. [29].

Do the three distinguishable iron-sulfur spectra arise from three distinct iron-sulfur clusters? So far no example of a binuclear iron-sulfur cluster is known which is paramagnetic in the oxidized state. Thus center S-3 is almost certainly a tetranuclear structure, consisting of 4Fe 4S similar to bacterial HiPIP [49] or *Bacillus polymixa* Fd [54]. Both center S-1 and Center S-2 spectra were detected not only in 8Fe SDH preparations but also in AA-SDH which contains only 4Fe 4S per molecule, in which no center S-3 signals were detected. Thus Ohnishi et al. [27] suggested that both S-1 and S-2 are binuclear centers similar to spinach ferredoxin or adrenodoxin [55]. Hence, three distinct iron-sulfur centers, namely two binuclear and one tetranuclear cluster, were proposed to be

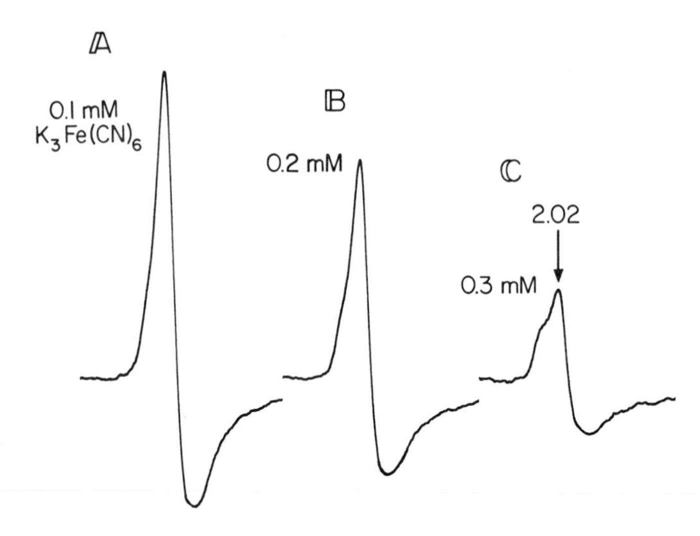

Figure 5 Destructive effect of ferricyanide on center S-3, revealed by the EPR spectra. The enzyme was oxidized with the concentration of ferricyanide as indicated in the presence of 10 μM PMS. EPR operating conditions: microwave frequency, 9.14 GHz; microwave power, 1 mW; scanning rate, 250 gauss/min; sample temperature, 10.6 K. Reproduced with permission from Ohnishi et al. [27].

present in the SDH molecule. The experimental evidence so far discussed, however, could not completely exclude an alternative possibility, i.e., only two tetranuclear clusters (responsible for S-1 and S-3 spectra, respectively) are present in the SDH molecule and the S-2 resonance spectra arise from a super-reduced state* of center S-3, because the spin concentration of centers S-1 and S-2 in the AA-SDH preparation examined was considerably lower than the flavin concentration. More recently, four independent lines of evidence [(1)-(4), below] showing directly the presence of at least two binuclear centers in the SDH molecule have been obtained and completely rule out an alternative two-tetranuclear hypothesis which has been discussed by Beinert and coworkers [28, 59, 60].

1. Salerno et al. [61] designed an iron-counting method in which iron-sulfur clusters were (gradually) converted to $(cysteine)_2$-Fe-$(NO)_2$ complex in which each iron atom yields a characteristic resonance spectrum equivalent to

*Carter et al. [56] proposed that tetranuclear clusters can undergo two sequential oxidation-reduction reactions with widely separated midpoint potentials. This was experimentally shown by the work of Cammack [57] and Sweeney et al. [58] for super-reduction and super-oxidation, respectively.

one spin with a peak at g = 2.035 [62]. This analysis quantitatively showed that destruction of each spin equivalent of g = 1.94 species gives rise to two spin equivalents of $(cysteine)_2$-Fe-$(NO)_2$ complex, showing that at least two binuclear clusters are present in the SDH molecule.

2. Cammack [51] devised an empirical procedure to distinguish iron-sulfur centers of binuclear and tetranuclear structure. Iron-sulfur centers in various proteins in the presence of a high concentration of dimethyl sulfoxide (DMSO) show EPR spectra having general features of lineshape and relaxation behavior depending only on binuclear or tetranuclear structure, but not on the individual protein structure. Binuclear iron-sulfur proteins exhibit rather isosymmetric spectra which are detectable above 77 K, while tetranuclear proteins exhibit spectra of axial symmetry detectable only at a much lower temperature. Fully reduced SDH in the presence of 70% DMSO exhibits a typical binuclear-type spectrum (Fig. 6b), which is almost identical with that of adrenodoxin-established binuclear iron-sulfur clusters (Fig. 6a) [57].

3. Exchange coupling constants J in various binuclear iron-sulfur clusters have been determined from the analysis of the temperature dependence of EPR linewidths and of integrated absorption intensities, and correlated with rhombic

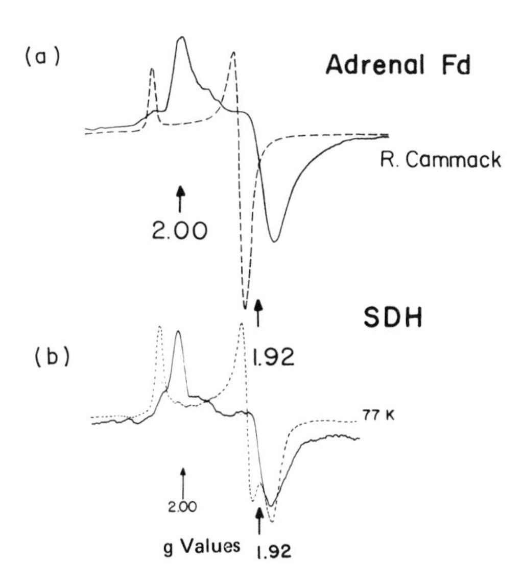

Figure 6 EPR spectra of dithionite-reduced adrenodoxin and BS-SDH in the presence (solid line) and absence (dashed line) of high DMSO concentration. Reproduced with permission from Cammack [51] (top spectra) and Salerno [52] (bottom spectra).

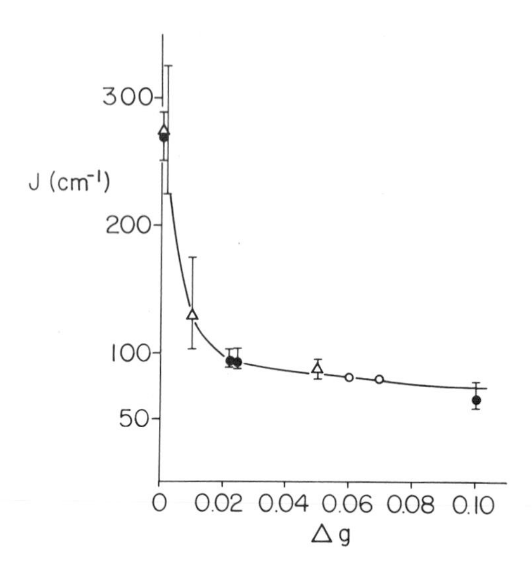

Figure 7 Variation of exchange coupling constant J as a function of the rhombic distortion of the EPR spectrum as measured by $\Delta g = g_y - g_x$ in various binuclear iron-sulfur clusters. In the order of decreasing J, the points represent putaredoxin (△), adrenodoxin (●), *Azotobacter vinelandii* protein I (△), *S. maximus* Fd (○), and Rieske's center (●) [centers N-1a and S-1 (●, ●) and *Azotobacter vinelandii* protein II (△) are also included]. Reproduced with permission from Salerno et al. [63].

distortion of their spectra [63]. The J value of about 90 cm^{-1} and distortion parameter, $g_y - g_x = 0.02$ of center S-1 offers independent supporting data for its binuclear structure (see Fig. 7). (Short relaxation times of center S-2 seem to arise from an additional relaxation mechanism [52].) The relaxation of J values of various HiPIP-type iron-sulfur clusters is also presently being analyzed [64] and strong indication of the tetranuclear structure of center S-3 has been obtained (Blum et al., unpublished result).

4. Albracht and Subramanian analyzed EPR spectra of iron-sulfur centers in *Candida utilis* SMP prepared from cells grown on a medium containing ^{57}Fe, and they demonstrated the binuclear structure of center S-1 from the g_z line broadening [65].

The next intriguing question is how are these three centers distributed in the two subunits of SDH, namely, Fp and Ip, each containing 4Fe 4S. As mentioned earlier, direct detection of S-1, S-2, and S-3 resonance signals in the isolated subunits has so far not been possible. However, the following three observations provide indirect evidence for their distribution in these two subunits.

1. All reconstitutively active SDH preparations contain 8Fe and 8S per molecule while any enzyme with less than 8Fe 8S per molecule cannot transfer electrons to the respiratory chain. In the reconstitutively active enzyme (BS-SDH or PS-SDH) electron transfer activity to the respiratory chain is very unstable. In contrast, electron transfer to nonphysiological redox dyes is rather stable. Center S-3 is very labile toward various reagents such as oxidants [29] or organic solvents [66], while flavin, and centers S-1 and S-2, are more stable to these reagents in an analogous way to the above two activities, respectively. As seen in Figure 8, center S-3 signals can be detected only in reconstitutively active SDH preparations, but not in inactive ones [28, 29]. Spectrum B shows that B-SDH, which contains 8Fe 8S per flavin as in the case of BS-SDH, exhibits no center S-3 signals. This phenomenon may arise either from modification of center S-3 to an EPR-undetectable form or to a molecular form which is more susceptible to destruction by $K_3Fe(CN)_6$. In any case, these observations indicate that even a subtle change of the molecular conformation around the S-3 cluster results in the loss of the reconstitutive activity.

2. A close relation between these two parameters is further demonstrated upon exposure of the freshly isolated BS-SDH to air. Rapid decay of the reconstitutive activity of the enzyme closely parallels the decrease of the resonance intensity of the center S-3 signal (decay half-life is approximately 35 min) (Fig. 9). On the other hand, about 70% of the dehydrogenase activity to artificial redox dyes and EPR signals of Fd-type iron-sulfur centers remains even after both center S-3 and reconstitutive activity is essentially lost [50, 28]. These observations have recently been confirmed by Beinert and coworkers [53], who also reported that the decay of low-K_m ferricyanide reductase activity parallels the decay of the center S-3 signal and reconstitutive activity.

3. Covalently bound FAD was shown to be located in the larger subunit (Fp) with the molecular weight of 70,000 [20]. As discussed in detail in Section II.E, the distance between electron transfer components in the SDH was estimated from the analysis of their spin-spin interactions. Center S-1 is about 10 Å distance either from center S-2 [27] or from flavin (Ohnishi et al., unpublished data). Likewise center S-3 is in close proximity to UQ which accepts electrons from SDH [67, 68]. From all this information, the most plausible distribution of flavin and iron-sulfur centers in the Fp and Ip subunits can be speculated as shown in Scheme III, although direct evidence for this assignment is still required. In this model, covalently bound FAD and two binuclear clusters (centers S-1 and S-2) are located in the Fp subunit, and a tetranuclear cluster (center S-3) is situated in the Ip subunit. Electrons are transferred to artificial dye (PMS) from either flavin or S-1, while electrons are transferred to the respiratory chain from S-3, first to UQ or UQH and then to the cytochrome chain.

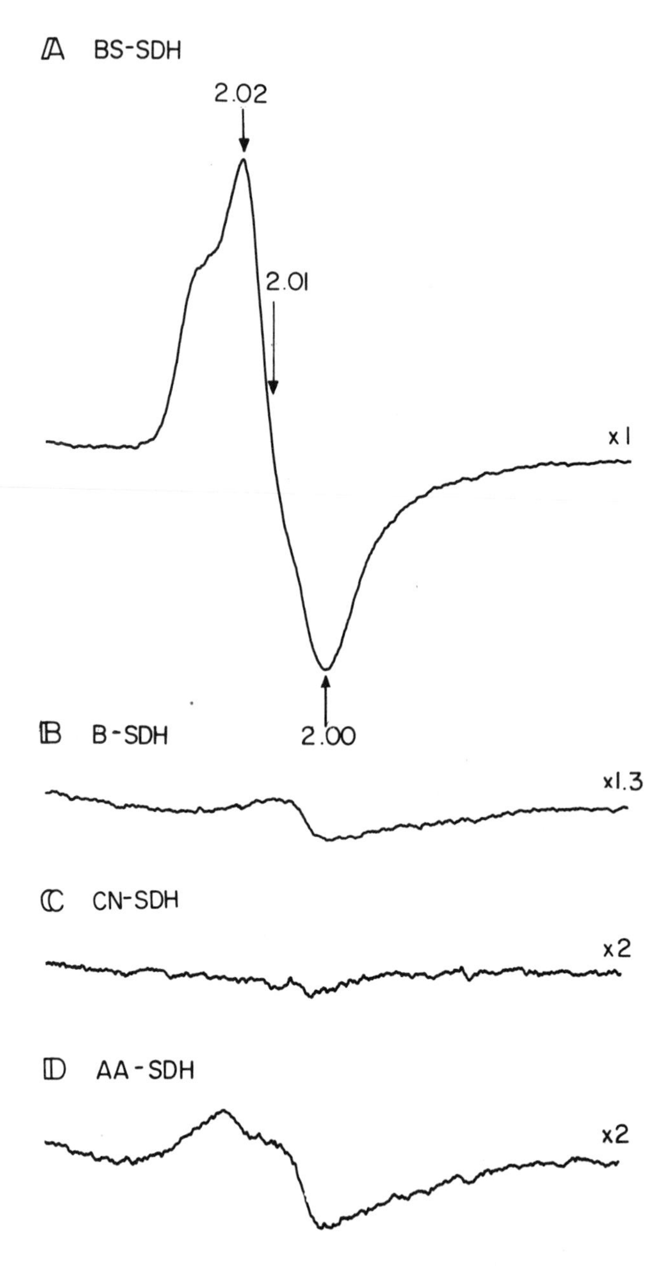

Figure 8 EPR spectra of iron-sulfur center S-3 of various succinate dehydrogenases. Enzymes were oxidized with 100 μM $K_3Fe(CN)_6$ in the presence of 43 μM PMS. Reproduced with permission from Ohnishi et al. [50].

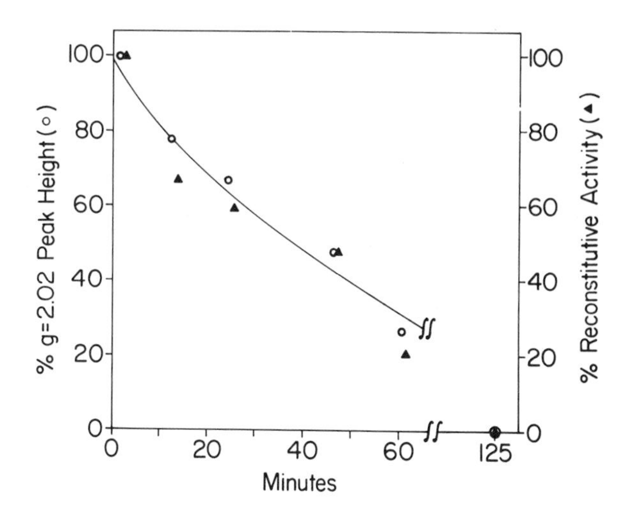

Figure 9 Decay of the reconstitutive activity of BS-SDH and concomitant loss of the center S-3 EPR spectra. Freshly isolated BS-SDH (10.0 mg protein/ml and 4.2 nmol flavin/mg protein) in 50 mM phosphate buffer (pH 7.4) was aged at 0°C in air. The dehydrogenase solution was withdrawn as indicated in the figure. Reconstituted succinate oxidase activity of fresh dehydrogenase was 14.5 nmol of succinate oxidized/min per mg protein at about 22°C. Samples for EPR measurement were simultaneously taken out and oxidized with 100 μM ferricyanide and 10 μM PMS and rapidly frozen. Peak heights at g = 2.02 from the baseline of the lower magnetic field were also plotted as a function of the aging time of the dehydrogenase. Reproduced with permission from Ohnishi et al. [27].

2. Flavin

Succinate dehydrogenase contains 1 mol of FAD per molecule, which is covalently linked to the larger subunit (Fp) [20]. The isoalloxazine structure was shown to be covalently linked to the histidyl residue of apoprotein polypeptide at the 8-methyl group (see Fig. 10). The flavin semiquinone signal at g = 2.00 was first detected by Beinert and Sands [4] in a purified SDH preparation at an intermediate redox state. Redox titration of flavin was conducted in 1961, poised with succinate/fumarate redox buffer, by Hollocher and Commoner [69] and by King et al. [70] by room temperature EPR analysis of the g = 2 signal. Both groups observed the maximum free radical signal (maximum radical; 20-30%) at a succinate/fumarate ratio of about 1.7 which corresponds to $E_{m7.4}$ = -7 mV. Potentiometric titration of flavin combined with optical absorption and fluorescence measurements in pigeon heart mitochondria gave a midpoint potential of $E_{m7.0}$ = -45 mV (pH 7.0, n = 2) [71]. More recently the midpoint potential of flavin in a SDH preparation was estimated indirectly by combining

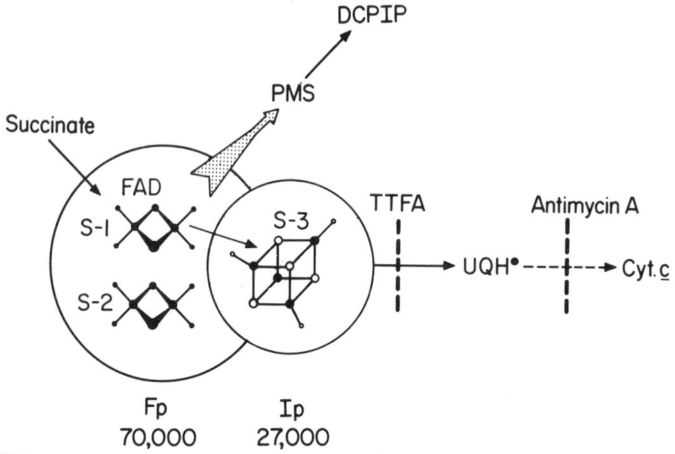

Scheme III Model structure of SDH and electron transfer pathways.

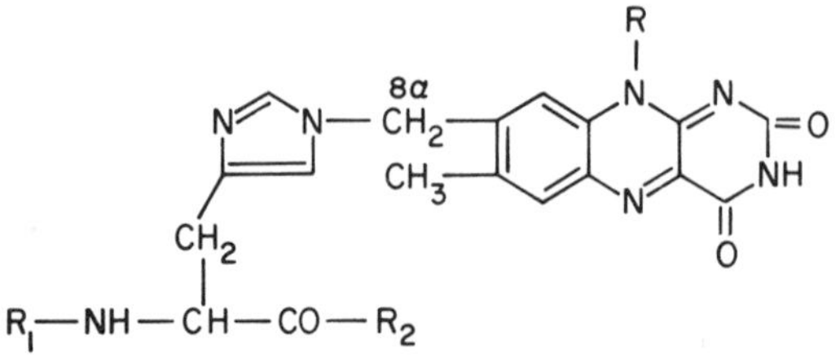

Figure 10 The structure of the flavin adenine nucleotide of SDH. The 8(α) methylene of the isoalloxazine ring is bonded to the inidazole-N(3) of the histidine residue of apoprotein.

potentiometric titration with the activation of the enzyme in the presence of the excess oxaloacetate [72]; this gave the midpoint potential value of $E_{m7.0}$ between -60 and -90 mV (n = 2). A direct potentiometric titration of the flavin free radical signal has been conducted in the author's laboratory, monitoring the g = 2.00 signal at -100°C. Redox mediators have been selected in order to prevent interference by dye signals. Preliminary results give an $E_{m7.4}$ value of -90 ± 20 mV (n = 2) [73] in agreement with the last indirect determination of midpoint potential [72]. It was pointed out previously that neutral flavin semiquinone (blue radical) shows spectra with linewidths of 19 gauss while anionic semiquinone exhibits spectra of 15 gauss linewidth [74]. The SDH free radical gives still narrower linewidths (12 gauss) and was interpreted as the anionic form being further narrowed because of a covalent bond at the 8α CH_3 [74].

C. *Thermodynamic Characteristics of the Redox Components in the SDH Molecule*

Midpoint potentials of the three iron-sulfur centers in SDH are summarized in Table 2. They were measured in soluble, particulate, purified preparations as well as in the intact mitochondrial membrane systems (namely in mitochondria or SMP). Center S-1 gives no significant variation of the $E_{m7.4}$ value in all systems studied; the E_m remains in the range of +30 mV. Center S-2 spectra cannot be clearly resolved in mitochondria or SMP from mammalian or avian systems due to multiple overlapping signals from the iron-sulfur centers in the NADH dehydrogenase region having similar midpoint potentials. A very low midpoint potential, -245 ± 15 mV was obtained for center S-2 in SMP using the *Saccharomyces* system in which no interferring iron-sulfur centers are present in the NADH dehydrogenase segment of the respiratory chain [27, 79]. A similar $E_{m7.4}$ value (-260 ± 15 mV) was obtained in detergent-treated bovine particulate preparations, such as complex II or succinate-cytochrome *c* reductase [27].

A further lowered midpoint potential (to approximately -400 mV) was obtained for center S-2 in all soluble SDH preparations examined, both in the reconstitutively active and inactive enzymes. This may reflect some environmental difference around center S-2 in soluble and membrane-bound SDH preparations and it also indicates that SDH binds with mitochondrial membrane more preferably when center S-2 is in the reduced form than in the oxidized form. It is also seen that after reconstituting soluble SDH with cytochrome bc_1 complex, the E_m of center S-2 reverts to the E_m value obtained in particulate preparations (-250 mV). To date, no functional role of center S-2 has been shown because this center is reduced only nonenzymatically by dithionite or borohydride together with a redox mediator. It should be noted, however, that in potentiometric titrations the midpoint potential of center S-2 is always measured after center S-1 becomes fully reduced (paramagnetic). It is shown that these two centers are located very close to each other (~10 Å distance). If centers S-1 and S-2 show not only spin-spin interaction, but also a strong redox interaction, namely, the redox state of S-1 affects the midpoint potential of S-2 (for example, paramagnetic S-1 makes S-2 more difficult to be reduced because of the coulombic interaction with its negative charge), the true midpoint potential of functioning S-2 can easily be much higher than the values listed in Table 2 (see Ref. 52 for the mathematical treatment of this argument).

Table 2 Thermodynamic Characteristics of Iron-Sulfur Centers in SDH in Membrane-bound and Soluble States

		E_m values (mV)			
Preparation	pH	S-1	S-2	S-3	References
Pigeon heart SMP		+30 ± 15	–[a]	>+150	75, 76
Beef heart SMP		–	–	>+120	75, 76
Beef heart SMP (ETP)		0 ± 15	–	70 ± 15	77
Mung bean mitochondria		10 ± 15	–	65 ± 10	78
Saccharomyces cerevisiae	7.2	0 ± 15	-245 ± 15	–	27
Succinate-cytochrome *c* reductase	7.4	0 ± 15	-260 ± 15	+65	29, 50
Complex II	7.4	0 ± 15	-260 ± 15	–	29, 50
Reconstituted succinate-cytochrome *c* reductase	7.4	–	-250 ± 20	–	27
BS-SDH, AS-SDH	7.4	-5 ± 15	-400 ± 15	–	27, 79
B-SDH	7.4	-5 ± 15	-400 ± 15	–	27, 29

[a] E_m values were not obtained because of multiple overlapping signals

The HiPIP-type center S-3 becomes extremely labile upon solubilization of the enzyme toward oxidizing agents, such as oxygen, ferricyanide, or redox mediating dyes [29, 50]. Thus E_m has been measured so far only in membrane-bound preparations.

In pigeon and beef heart SMP prepared by sonification procedures, midpoint potentials of center S-3 have been reported as approximately +120 mV [68, 80].

Recently, improved resolution of the center S-3 signal from the overlapping signals due to a spin-coupled QH• pair has been introduced to their potentiometric analysis using computer simulation [77]. One typical example is presented in Figure 11. On the right, EPR spectra in the g = 2 region at different redox potentials (E_h) are presented. At high E_h, center S-3 signal is seen which is centered aroung g = 2.01 (spectrum A). At intermediate E_h (spectra B and C), signals at g = 2.04 (on the left) and g = 1.96 (on the right) are observed, which disappear again at lower E_h (spectrum D). This indicates that these accompanying signals are associated with free radical states of the n = 2 electron transfer component such as flavin or UQ. Ruzicka and Beinert [81] reported that these multiple signals arise from a spin-spin interaction because their field position is independent of the frequency of the applied microwaves; ubisemiquinone (QH•) is one of the interacting species because these signals disappear upon depletion

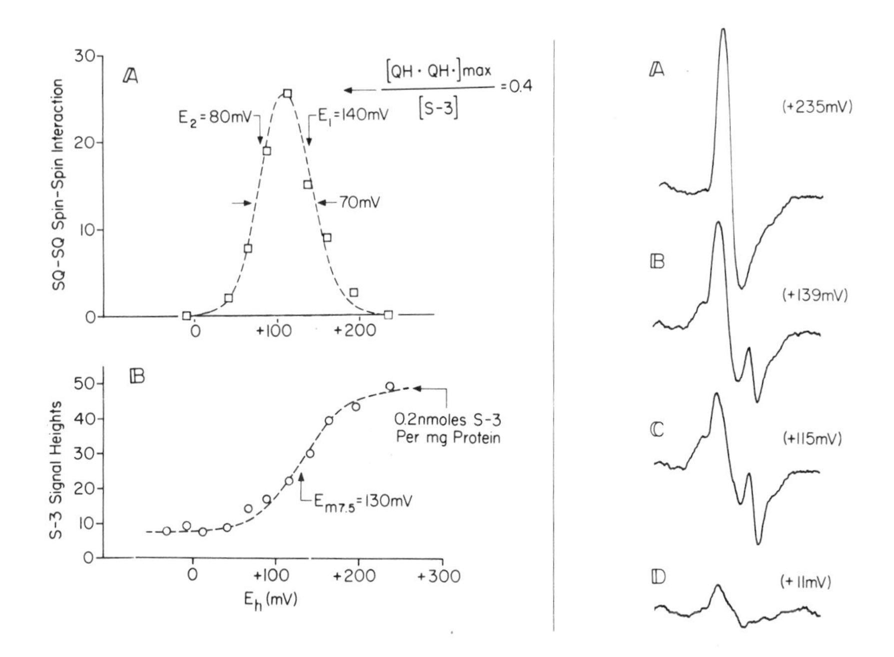

Figure 11 Potentiometric titration of EPR signals from dipolar coupled ubisemiquinone pair (SQ-SQ) and center S-3. Submitochondrial particles (18.5 mg protein/ml) were prepared by sonication of mitochondria in 0.25 M sucrose, 0.1 mM EDTA, and 50 mM HEPES buffer at pH 7.4. Potentiometric titration was performed as described previously [75, 82] in the presence of 1,4-naphthoquinone disulfonate, pyocyanine, Indigo tetrasulfonate, duroquinone, and 1,4-naphthoquinone (the last two were dissolved in DMSO) at concentrations between 20 and 50 μM. EPR operating conditions: microwave frequency, 9.18 GHz; modulation frequency, 100 KHz; modulation amplitude, 8 gauss/min; microwave power, 1 mW; time constant, 0.3 sec; scan rate, 250 gauss/min; sample temperature, 12 K. Reproduced with permission from Ohnishi et al. [77].

of UQ from the mitochondrial membrane and reappear upon replenishment of UQ. Initially the accompanying signals were suggested to arise from a spin-spin interaction between UQ and center S-3 spins, because of the concurrent appearance of their signals and because their short relaxation times were typical of transition metals [81]. The same workers in collaboration with Sands and his colleagues subsequently performed computer simulation studies of the EPR spectrum of complex II which was trapped kinetically at an intermediate redox state [72]. The best fit was obtained for spectra simulated as SQ-SQ spin-spin interaction overlapped with noninteracting S-3 signals rather than a spin-spin interaction between SQ and S-3.

Ingledew et al. [68] proposed spin-spin interactions between a ubisemiquinone pair from their thermodynamic analysis and a role of center S-3 as a magnetic relaxant to at least one of the ubisemiquinone pair. Figure 12A represents computer-simulated spectra of S-3 (dashed line) and dipolar coupled ubisemiquinone pair (solid line) using parameters reported by Ruzicka et al. [72]. Simulated S-3 and QH•QH• spectra were added in varying ratios to produce a series of spectra, exemplified by spectra B and C. Comparison of this series with experimental spectra taken at various redox potentials allowed the ratio of QH•QH• and S-3 and the percentage of maximum of each to be calculated. A good fit was obtained between simulated spectra and experimental data in that portion of the spectrum utilized. Potentiometric titrations of the spin-coupled UQ• pair and of center S-3, based on the simulation, were presented in Figure 11. Center S-3 shows an $n = 1$ titration curve with a midpoint potential of +130 mV (Fig. 11, bottom left). Potentiometric titration of EPR signals arising from the QH•QH• interaction gives a peak around +110 mV with about 70 mV width at half peak height. The maximum observed spin concentration of the dipolar coupled ubisemiquinone pair corresponds to approximately 40% of the center S-3 concentration. From the theoretical curve for QH•QH• interaction (Fig. 11, top left) the apparent midpoint potentials of the Q/QH• and QH•/QH_2 redox couples were obtained as $E_{1,7.4}$ = +140 mV and $E_{2,7.4}$ = +80 mV as illustrated in the figure. The QH•QH• titration gives a peak potential (E_m) which is about 30 mV higher than the E_m of the $g = 2$ signal measured at a much higher temperature (170 K) [77]. This shift may be caused by the indirect involvement of center S-3 as a QH• spin relaxant in its paramagnetic state.

More recently the midpoint potential of center S-3 was measured as a lower value, 70 ± 15 mV, in beef heart electron transport particles (ETP) which were prepared by alkaline treatment (pH 8.5 for 30 min) and homogenization according to Harmon and Crane [83]. Similarly, a midpoint potential of 65 ± 10 mV was obtained in mung bean mitochondria and SMP [78]. These E_m values are close to that obtained with bovine succinate-cytochrome *c* reductase [26, 50] in which no signals due to QH•QH• interaction are observed. Likewise, a lower E_m value for center S-3 was obtained in Q-extracted beef heart SMP [68]. Generally a higher E_m value ($>$100 mV) for S-3 is obtained where a strong QH•QH• interaction is present, while a lower E_m value ($<$100 mV) is obtained in the system in the presence of a weak QH•QH• interaction (or even in its absence). The controlling parameters, however, for these two widely different midpoint potentials of center S-3 in different systems still await further clarification.

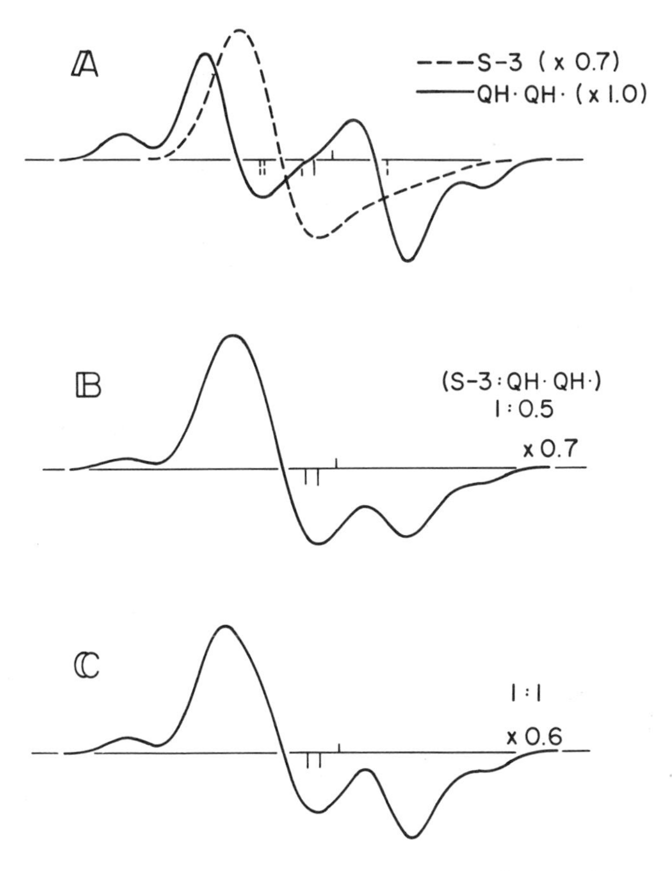

Figure 12 Computer simulation of individual spectra of center S-3 and dipolar coupled ubisemiquinone pair (A), and examples of spectra contributed from both species at varying proportions (B and C). The EPR spectra were simulated by numerical integration over the angles θ and ϕ, using a PDP-10 computer. Gaussian lineshape functions were used for all simulations. Spectra of center S-3 and dipolar coupled semiquinone pairs were simulated using the parameters of Ruzicka and Beinert [81]; small variations in simulation parameters were introduced to compensate for slight changes in the line shape in different preparations. Reproduced with permission from Ohnishi et al. [77].

The E_m values of centers S-1 and S-2 were shown to be independent of pH [84], while that of center S-3 was previously reported as being somewhat pH dependent [80]. This apparent pH dependence is explainable by the above-mentioned different E_m values depending on the kind of SMP preparations.

Analysis of the pH dependence using the same batch of particles established that center S-3 is not pH dependent (Ohnishi, unpublished data). Similarly, the pH independence of center S-3 was demonstrated in mung bean mitochondria [78].

The recent indirect or direct potentiometric titration of flavin gave an E_m value of $E_{m7.0} = -90 \pm 20$ mV (n = 2) as described in the preceding section [73]. The maximum free radical concentration during potentiometric titration is 10 (± 5)%, similar to the observation during the reductive titration with dithionite [28]. This indicates that the free radical state is rather unstable in this system, giving lower E_1 (Fl_{ox}/Fl_{sQ}) than E_2 (Fl_{sQ}/Fl_{red}). To understand the implications of these phenomena for the mechanism of hydrogen and electron transfer in SDH requires accurate measurements of E_1 and E_2 values and their pH dependence.

D. Kinetic Studies

Most kinetic studies of SDH preparations have been conducted by Beinert and his colleagues using rapid freeze-quenching techniques which were originally devised by Bray [85] and then developed in their laboratory [86]. The rate of reduction by substrate or reoxidation by ferricyanide, using partially [42] or fully activated [28, 53] enzyme preparations, showed that centers S-1 and S-3 are reduced or reoxidized within the turnover time of the enzyme (for example, within 6 msec at 14 K). Even in the fully activated complex II preparation, however, only 40-60% of center S-1 and less than 50% of center S-3 showed a rapid response consistent with the enzyme catalytic turnover.

In the complex II preparation, reduced with succinate, less than 10% of flavin appeared in the semiquinone form, while in soluble reconstitutively inactive SDH preparations 70-80% of flavin was accumulated in the semiquinone form. No significant difference was observed between the rate of semiquinone formation and reduction of centers S-1 and S-3 [53]. Very recently kinetic analysis of reconstitutively active enzyme which was extracted from ETP and purified by the procedure of King [18, 30] was repeated. A three-syringe two-chamber arrangement was used in order to assure minimal contact of the enzyme with oxidant, $K_3Fe(CN)_6$, before the addition of succinate to the enzyme [24].

In spite of all the efforts described above, no resolution has been attained of the kinetic sequence of the electron transfer reactions in SDH, which is the ultimate goal of the kinetic studies.

Is this due to an instantaneous equilibration of reducing equivalents within individual molecules (see Ref. 87) which is beyond the resolution limit of the rapid freeze techniques used so far? This problem remains open.

E. Spatial Relationship Among Redox Components in SDH Distance and Orientation

In order to approach the electron transfer mechanism from succinate to UQ and to the cytochrome chain, it is useful to obtain quantitative information about the distance and relative orientation among individual electron carriers in the molecule. It is especially important for a better understanding of the intramolecular electron transfer events in the dehydrogenase, where no kinetic or x-ray crystallographic resolutions are presently available.

Fortunately, both flavin and iron-sulfur centers are paramagentic in certain redox states. Thus detailed information can be obtained if one can detect magnetic interactions between neighboring components, such as cross relaxation or dipolar splitting phenomena of their EPR signals.

This approach is exemplified here with centers S-1 and S-2. Spin coupling of centers S-1 and S-2 can be demonstrated from the following phenomena: (1) the spin relaxation behavior of center S-1 is affected by S-2 spins [27, 43]; (2) the EPR lineshape of the fully reduced enzyme is modified at very low temperatures (both centers S-1 and S-2 are in the paramagnetic state); (3) a half-field signal is observed at g = 3.88 from the $m_s = 2$ transition of coupled S-1 and S-2 spins in the fully reduced enzyme [43, 88].

The first phenomenon can be demonstrated from the analysis of power saturation behavior of the S-1 signal in the presence and absence of S-2 spins. Simplifying the Bloch equations, the EPR signal amplitude for a homogeneous paramagnetic center can be expressed [89] as

$$\text{Signal amplitude} = K \frac{H_1 T_2}{(1 + \frac{1}{4} \gamma^2 H_1{}^2 T_1 T_2)^2}$$

where H_1 is the microwave magnetic field, γ is the magnetogyric ratio, T_1 is the spin-lattice relaxation time, T_2 is the spin-spin relaxation time, and K is a constant. At a given temperature, T_1 and T_2 are constant. Thus

$$\text{Signal amplitude} = K' \frac{P^{1/2}}{(1 + AP)^2}$$

where P is the input microwave power. When AP is much smaller than 1, log (signal amplitude) = 1/2 log P + K″. If the input power becomes too high, log (signal amplitude) deviates from the straight line with slope 1/2 and declines due to power saturation.

The dependence of the signal amplitude on the applied microwave power

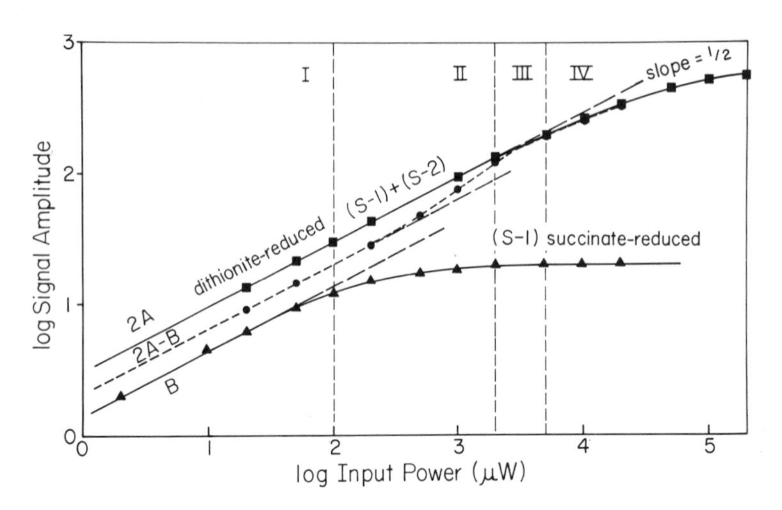

Figure 13 Saturation behavior of center S-1 and center S-2 EPR signals of BS-SDH at 10 K. Reproduced with permission from Salerno et al. [43].

is shown in Figure 13 using BS-SDH, reduced with succinate and dithionite, respectively. Center S-1 spins relax rather slowly, thus S-1 signals start to show power saturation at a microwave power of 50 μW at 13 K if center S-2 spins are in the diamagnetic state (succinate-reduced system). In the dithionite-reduced system (both S-1 and S-2 spins are paramagnetic), however, EPR signals of S-1 plus S-2 spins do not show power saturation up to a microwave power of 2 mW at the same temperature. Thus subtraction of the succinate-reduced spectra from the dithionite-reduced one gives a saturation curve with slope steeper than 1/2 which is not feasible for an independent spin [43]. This anomalous saturation behavior of S-1 spins can be readily illustrated by the following model system (Fig. 14).

There are two different species of spins which give at least partly overlapped spectra. Spin B relaxes more slowly than does spin A: spin B saturates at lower microwave power than spin A. If these spins A and B are not spin coupled (i.e., if they are independent spins), the saturation curve of the combined overall signal from A and B spins is just the A + B curve as shown in Figure 14 (top). Therefore the difference of signals from both centers and either center A or B gives the simple saturation curve of the other component. In this case, the saturation curve of A + B spins shows a shoulder at the comparable power to saturate spin B. What happens if spins A and B are spin coupled? If A is not paramagnetic, B is readily saturated as in the top case (Fig. 14).

If the rapidly relaxing (less saturable) A is also paramagnetic, both spins

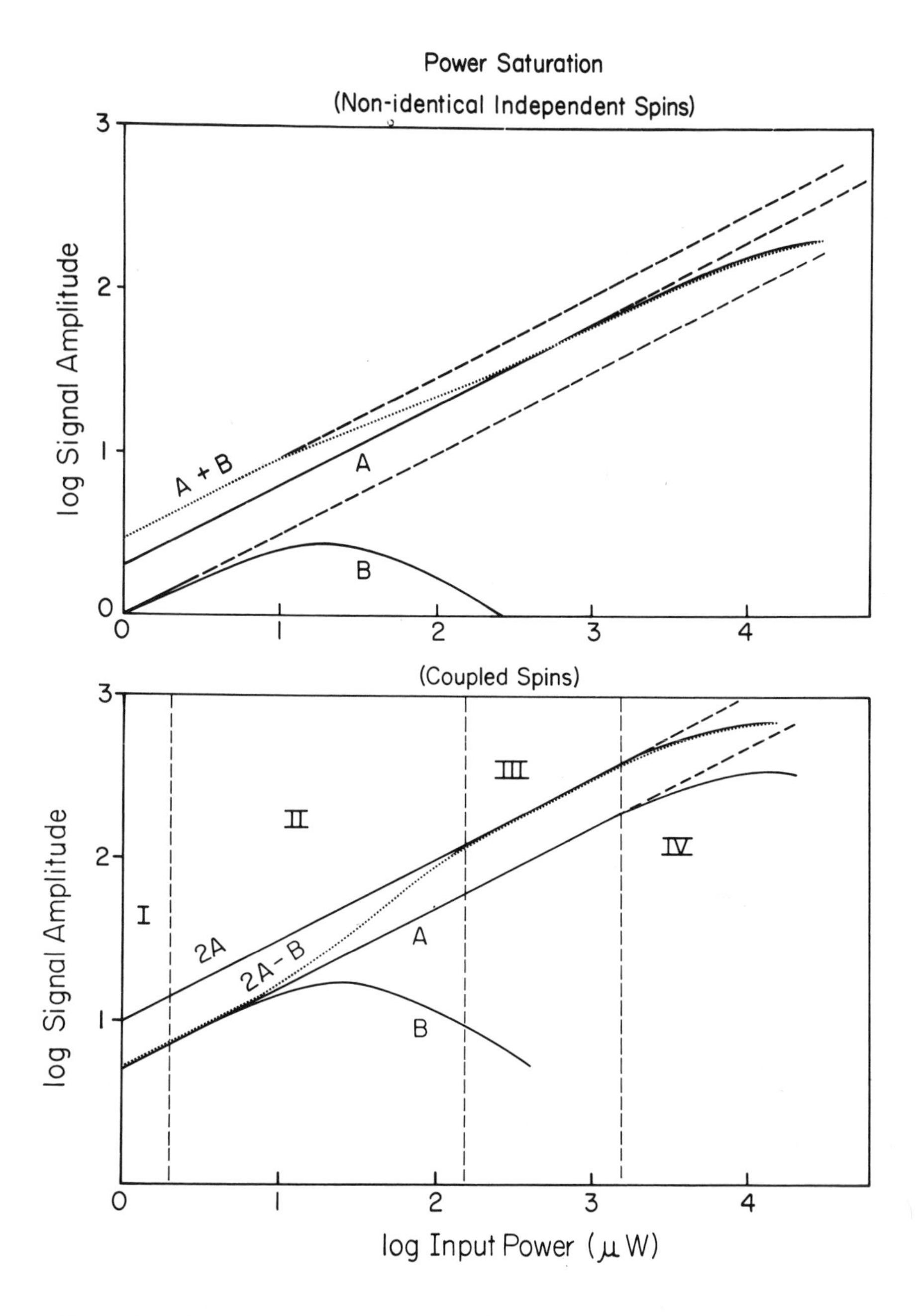

Figure 14 Model for power saturation of a system with two homogeneously broadened species. Top, independent species. Bottom, coupled species such that B spins are desaturated by A spins. Reproduced with permission from Salerno et al. [43].

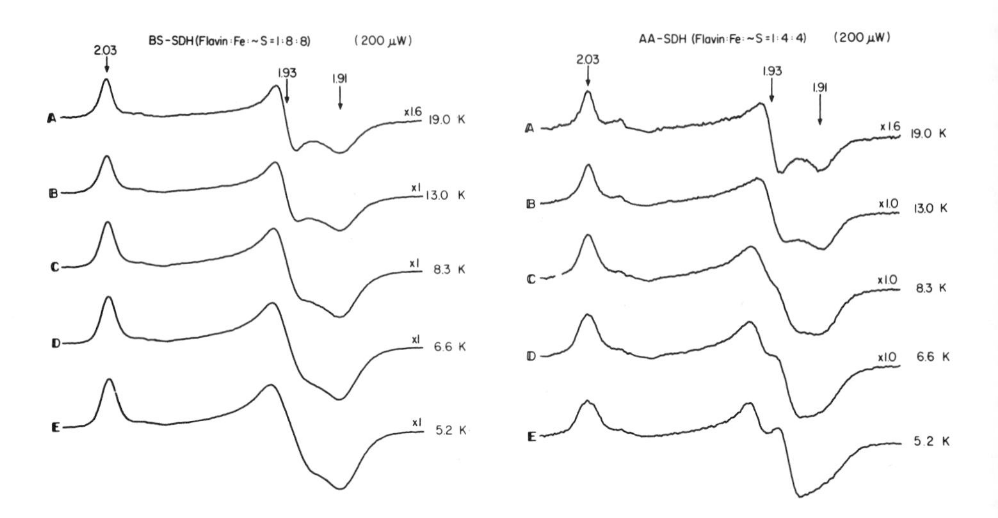

Figure 15 EPR spectra of dithionite-reduced BS-SDH and AA-SDH showing additional complexity as the temperature is lowered. Reproduced with permission from Ohnishi et al. [27].

A and B relax with a common rate equal to the rate of A alone, provided that the coupling between the spins is tight enough that energy transfer between A and B is not rate limiting [90]. Then signals from both centers become 2A instead of A + B. Therefore the difference of 2A – B gives an anomalous curve with a slope steeper than 1/2 in a certain power range.

The S-1 and S-2 spins behave similar to A and B in the latter model system (Fig. 14, bottom). This observation clearly demonstrates desaturation (relief of power saturation) of S-1 spins by S-2 spins. This also indicates that the distance between centers S-1 and S-2 is less than 10 Å. Similarly desaturation of S-1 by S-2 spins was observed in all reconstitutively inactive enzyme preparations, although S-2 spins saturate at somewhat lower microwave power than that of reconstitutively active enzyme [43].

More structural information can be obtained if one can observe dipolar splitting of fully reduced enzymes. As shown in Figure 15, the central EPR signal broadens upon lowering the temperature of dithionite-reduced BS-SDH, even under non-power-saturating conditions [27, 43]. An additional feature was observed with reconstitutively inactive enzymes, namely, splitting of the central resonance. Both phenomena arise from the spin coupling of centers S-1 and S-2, and the difference between these two enzyme systems (reconstitutively active or inactive) is mostly due to the conformational change around center S-2 which gives rise to, for example, different relaxation rates of center S-2 in a wide range of temperature.

The relationship between the splitting ΔH and the distance between the two spins (μ_{eff}) can be expressed in the form of

$$\Delta H = \pm \frac{\mu_{eff}}{r^3} (1 - 3 \cos^2 \Psi)$$

assuming dipolar coupling. Angle Ψ is the angle between the applied magnetic field and the line connecting S-1 and S-2. The absolute value of the angular term is at most 2, and from the lack of splitting along the z direction we expect the absolute value of $(1 - 3 \cos^2 \Psi)$ to be somewhat smaller than this along the y axis. Since the peak-to-peak distance of splitting is measured as $2 \Delta H = 23$ gauss, the distance between S-1 and S-2 is estimated as approximately 10 Å.

Additionally, structural information can be obtained from the computer simulation of the low-temperature spectra, assuming dipolar coupling of S-1 and S-2 spins.

The axes connecting the two iron atoms (g_z direction) are oriented similarly with respect to the line joining the clusters, with approximately 10 Å between them in agreement with the above estimate [43].

If centers S-1 and S-2 are spin coupled, these two spins are expected to behave as a spin 1 (S = 1) system. Thus the half-field signal due to the $\Delta m_s = 2$ transition ("forbidden" transition) will be observed when both centers are paramagnetic. Indeed rather symmetric g = 3.88 signals were observed in all dithionite-reduced enzymes. As expected, oxidized or succinate-reduced enzymes do not give rise to the half-field signal [43, 88]. Here again half-field signals in reconstitutively active and inactive enzymes are approximately of the same size, providing evidence for spin-coupling of similar strength in all SDH preparations.

Using the same approach, the distance between flavin and center S-1 was estimated as approximately 10 Å: in the presence of paramagnetic S-1, the SDH flavin g = 2 signal is more difficult to saturate than the flavodoxin free radical, and destruction of iron-sulfur centers converts SDH flavin to the flavodoxin type (Ohnishi et al., unpublished data).

The distance between S-3 and one of the spin-coupled QH•QH• pair was also estimated to be less than 10 Å. As shown in Figure 16, g = 2.04 and g = 1.99 signals exhibit microwave power dependence almost identical to that of the S-3 signal (g = 2.01). Ubisemiquinone free radical, observable in phospholipid vesicles in alkaline solution, is saturated at all microwave powers in excess of 1 mW, even at liquid nitrogen temperature, and is not even visible at the temperature shown in Figure 16 [68].

It has been generally accepted that SDH molecules are located on the

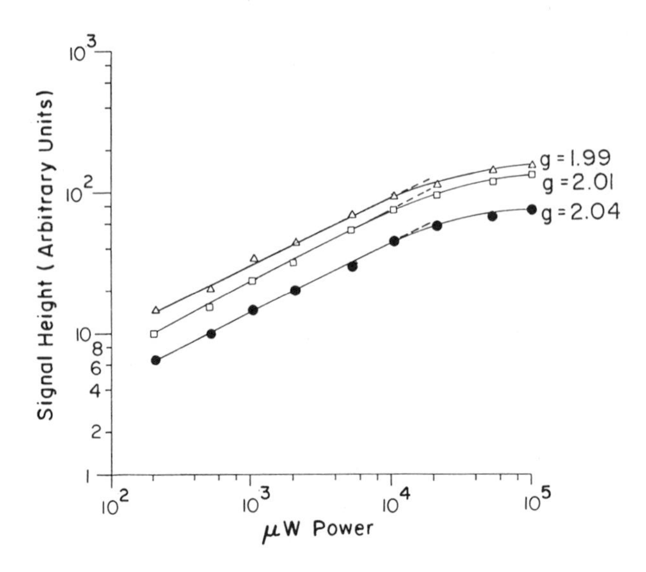

Figure 16 Microwave power dependence of the g = 2.01, g = 1.99, and g = 2.04 signals from beef heart SMP. The sample was poised at the E_h value of +105 mV. Reproduced with permission from Ingledew et al. [68].

matrix side of the mitochondrial inner membrane, based on the accessibility studies with membrane-impermeable probes such as substrates and redox agents [101-103]. Bathophenanthroline sulfonate, which was reported to be a membrane-impermeable iron-chelator, exerts an effect on HiPIP-type iron-sulfur centers after a long preincubation. The EPR signal of center S-3 is diminished in SMP, but not in mitochondria. On the other hand, another HiPIP-type center which was purified by Ruzicka and Beinert [47] was affected in mitochondria. These results indicate that center S-3 is located on the matrix side of the membrane osmotic barrier, while Ruzicka's center is localized in the cytoplasmic side [48]. As reported by Vinogradov et al. [38, 39], center S-3 detected as a low-K_m, $K_3Fe(CN)_6$ interaction site is not exposed on the surface of the inner membrane.

This information together with the above-described spatial relationships of the electron transfer components associated with SDH are summarized in Figure 17. Arrows are inserted to indicate the known distance between redox components. The precise position of center S-2 relative to flavin and to S-3 is an important piece of information which is still missing. Since it is difficult to poise these components simultaneously in the paramagnetic state, we cannot measure their distance from the analysis of spin-spin interaction. On interesting clue is the unusually short relaxation time(s) of center S-2 at low temperatures.

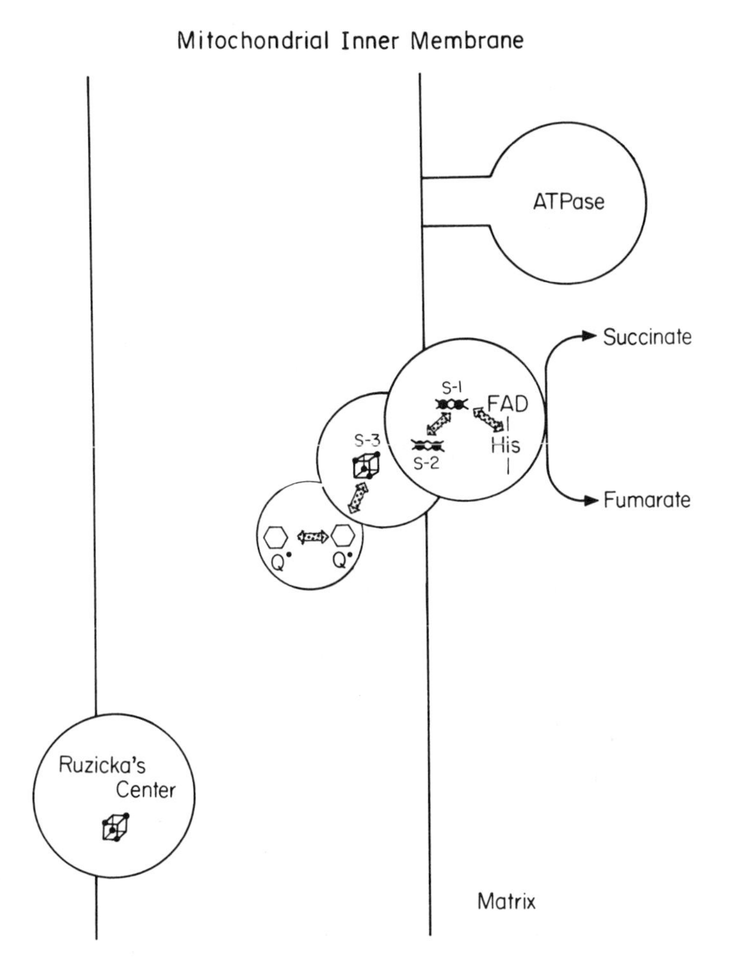

Figure 17 Spatial relationship of redox components associated with SDH.

One possible mechanism for this rapid relaxation would involve paramagnetic excited states of the reduced tetranuclear iron-sulfur cluster S-3. Our recent experiments showing depopulation of the ground state of this iron-sulfur cluster above 8 K give some credibility to this hypothesis (Salerno et al., unpublished results).

Center S-2 would then be close enough to interact magnetically with both S-1 and S-3. The more typical relaxation behavior of center S-1 would preclude direct interaction with center S-3, and center S-2 would be located between S-1 and S-3. This would in turn suggest the possibility of the involvement of S-2 as a metastable intermediate in electron transfer [43, 52].

The relaxation of center S-2 is indistinguishable from that of S-1 in 50% DMSO at low pH, conditions under which center S-3 is dissociated. However, in AA-SDH or CN-SDH preparations, where no EPR signals from center S-3 are detectable, S-2 relaxes more rapidly than S-1 over a wide range of temperature.

It is possible that the rapid low-temperature relaxation of the binuclear iron-sulfur clusters S-1 and S-2 in the fully reduced enzyme is caused by the interaction between these clusters themselves. The spin-coupled pairs of tetranuclear iron-sulfur clusters in 8Fe (Clostridial-type) [104] ferredoxins have been reported to relax more rapidly than analogous clusters in 4Fe ferredoxins [105]. In this case the biphasic power saturation of the fully reduced enzyme would be attributed to molecular inhomogeneity.

On the basis of the inhibition of electron transfer by hydrophilic metal chelators, Harmon and Crane [106] proposed a transmembrane location of iron-sulfur centers of SDH. Direct EPR examination, however, revealed that no iron-sulfur centers were affected by these inhibitors within the time range of the respiratory inhibition [48]. Only HiPIP-type iron-sulfur centers, but not Fd-type, responded to chelators after a long incubation. Center S-3 signals diminished in SMP but not in whole mitochondria, while the signal intensity of Ruzicka's center decreased in mitochondria, consistent with the allocation of these centers in Figure 17.

From the spectral simulation using complex II, Ruzicka et al. [67] indicated that spin-coupled ubisemiquinones orient mutually edge to edge, but not face to face, about 8 Å away from each other. Center S-3 is located in the vicinity of spin-coupled QH• as discussed in the preceding section [68]. Based on the EPR spectra of spin-coupled ubisemiquinones in the oriented multilayered beef heart mitochondria (following the procedure of Erecinska et al. [107]), the line joining coupled QH• was shown to be perpendicular to the mitochondrial membrane [108] as illustrated in Figure 17. These observations together with stable QH• formation, as described before, predict the existence of specific proteins which bind UQs and stabilize them in the free radical state. A timely isolation of Q-binding protein which may function between SDH and the Q pool was reported by Yu et al. [109].

F. Effect of Some Inhibitors, Thenoyltrifluoroacetone and Carboxin, on SDH

Thenoyltrifluoroacetone (TTFA) is a potential metal chelating agent which inhibits succinate oxidation at a concentration about two orders of magnitude lower than that required for similar inhibition of NADH oxidation [91]. It was shown that in micromolar concentration TTFA does not inhibit electron transfer from succinate to artificial redox dyes, but it inhibits almost completely succinate-UQ reductase [31] or succinate oxidase activity [92]. Singer et al. [93] reported that the reduction rate of center S-3 by succinate is not affected by TTFA, while the reoxidation rate of this center by a Q analog was several hundredfold suppressed. This inhibition site was assigned to be between center S-3 and the Q pool. More recently the same group proposed a modification of the microenvironment of center S-3 by TTFA and carboxin (the latter acts at the same site as TTFA but is effective at much lower concentrations [94]).

Recently Ingledew and coworkers reported that TTFA removes EPR signals arising from QH• QH• spin-spin interaction at comparable concentrations to that of inhibiting succinate oxidation [95, 96] (see Fig. 18). They also demonstrated that although the maximum signal intensity, lineshape, and relaxation behavior of center S-3 were not significantly altered by TTFA, cross-relaxation of the QH• spin by the rapidly relaxing center S-3 was completely perturbed by the low concentration of TTFA [96]. Ruuge and his coworkers previously detected anamalous saturation behavior of one of the QH• pools; i.e., its $g = 2$ signal can not be saturated up to 200 mW at 240 K [97]. They reported independently that this unsaturable $g = 2$ signal was lost in the presence of TTFA. Concomitantly, EPR signals arising from spin-spin interaction observed at $g = 2.04$ and $g = 1.99$ were completely suppressed [98, 99]. In this connection the following observation is very interesting. Succinate dehydrogenase is dissociated from mitochondrial membrane by KCN treatment. Prior reduction of SMP with succinate or dithionite prevents dissociation of this enzyme. Therefore KCN extraction methods can yield only reconstitutively inactive enzyme (but of very high purity). If mitochondrial particles are treated with TTFA prior to KCN treatment, SDH is completely resistant and is never detached from the membrane [100]. Thus further analysis of the interaction of TTFA and its analogs with the SDH system may lead to a clue for the binding mechanism of this enzyme to the membrane.

III. NADH-UQ Reductase Segment

A. Background

Studies on redox components in the NADH-UQ segment are more challenging than those on the SDH segment from the viewpoint of the bioenergetic membrane because (1) this segment functions not only for the electron and proton transfer but also serves for the energy coupling at site I, and (2) NADH-UQ reductase is an intrinsic membrane component and this enzyme can be solubilized only with the help of detergents. Thus, topics related to this segment include studies

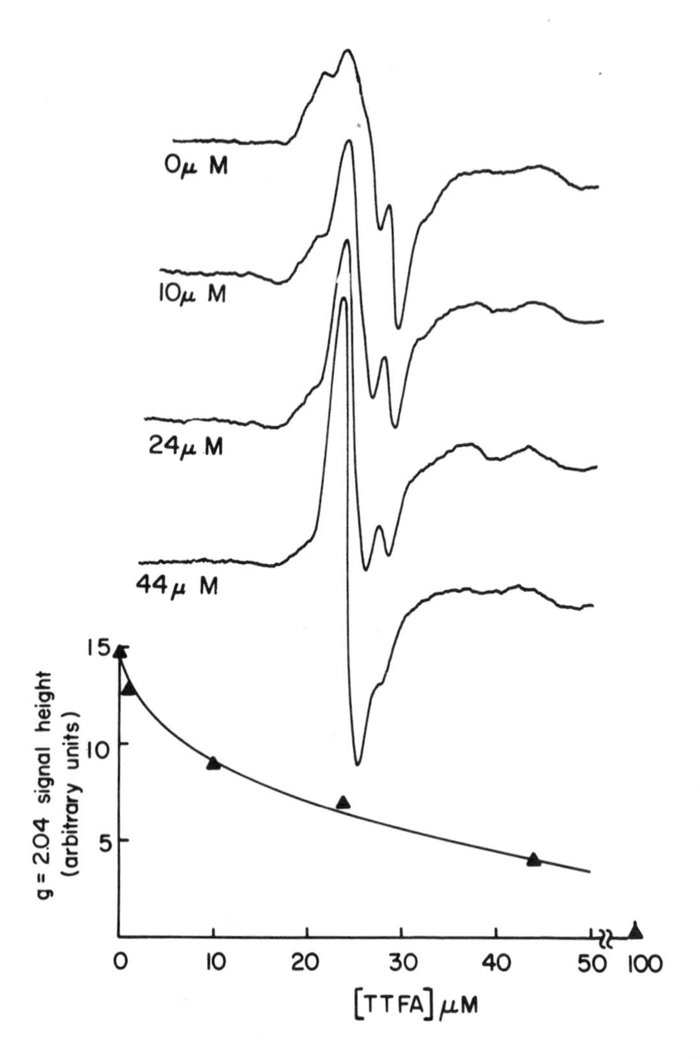

Figure 18 Effect of TTFA on the g = 2.04 and g = 1.99 signals and the HiPIP-center absorption (g = 2.02). Samples of pigeon heart SMP (10 mg protein/ml) were potentiometrically clamped at +100 mV (± 5 mV) and titrated with TTFA (see Section II.F). Samples were taken and freeze-clamped in EPR tubes approximately 10 sec after addition of TTFA. The EPR spectra shown (top) are those obtained in the presence of the indicated concentration of TTFA. The lower part of the figure shows the titration of the height of the g = 2.04 signal versus TTFA concentration. The experiment was conducted in a medium of 50 mM KCl / 50 mM 4-morpholine ethanesulphonic acid (pH 7.0). TTFA was dissolved in DMSO. EPR operating conditions: modulation amplitude, 0.5 mT (1 mT = 10 gauss), microwave frequency, 9.12 GHz; time constant, 1 sec; scan rate, 25 mT/min. Reproduced with permission from Ingledew and Ohnishi [95].

at widely different levels of structural integration including intact mitochondrial membrane systems, purified membrane fragments (complex I), and solubilized NADH dehydrogenase preparations of different molecular size (types I and II). Due to the complexity of its structure and function, information on this segment is still tentative in comparison with studies on the SDH segment, in spite of the large body of interesting information already available.

Complex I is a lipoprotein enzyme, NADH-UQ reductase, isolated by the use of deoxycholate and cholate together with ammonium sulfate fractionation [21, 110, 111]. It catalyses slow rotenone-sensitive NADH-UQ reduction and very rapid rotenone-insensitive NADH-$K_3Fe(CN)_6$ reduction, retaining properties present in the mitochondrial membrane. It contains FMN, nonheme iron, acid-labile sulfur, UQ, and phospholipid as presented in Table 3 [13, 112, 113]. Complex I is a much more useful system than complex II, because it can be prepared with a reasonably high yield and is contaminated only slightly with extraneous components. Complex I prepared by Ragan contains about 20% lower FMN and about 20% higher nonheme iron and acid-labile sulfur on the protein basis than the corresponding complex I prepared by Hatefi's group. Although it is not safe to base quantitation on FMN concentration (this acid-extractable FMN is rather readily removed from the membrane during the isolation process), 16-18 Fe and equivalent S per flavin is present in complex I according to Hatefi et al. [21, 111], while Ragan's complex I contains 23-24 Fe and equivalent S per flavin. In spite of such a high content of nonheme iron and acid-labile sulfur, the g = 1.94 EPR signal from only one species of iron-sulfur center is detectable above 77 K [4]. As will be detailed later, EPR spectra from several distinct species of iron-sulfur centers are observed at sample temperatures below 40 K as summarized in Scheme I (see Section I). Complex I contains at least 16 polypeptides according to Ragan [114] (see Fig. 19).

NADH dehydrogenase preparations with high molecular weight (type I) were isolated by Ringler et al. [115], Huang and Pharo [116], and more recently by Baugh and King [117]. These enzymes were solubilized using snake venom Naja Naja and detergents Lubrol and Triton X-100, respectively. All high molecular weight NADH dehydrogenase preparations are free of phospholipid and endogenous UQ, although they have a similar subunit composition to complex I. The former two preparations retain rapid NADH-$K_3Fe(CN)_6$ reductase activity while showing no rotenone-sensitive NADH-UQ reductase activity. In contrast, the last preparation was reported to retain both NADH-$K_3Fe(CN)_6$ and NADH-UQ reductase activity. Ragan and Racker [113], on the other hand, reported that phospholipid is required for physiological NADH-UQ reductase activity. Deletion of only about half of the phospholipid from complex I results in the complete loss of the activity of NADH-UQ reductase, but not of NADH-$K_3Fe(CN)_6$ reductase.

Table 3 Composition of Complex I

FMN (nmol/mg protein)	Nonheme iron (mg-atoms/mg protein)	Acid-labile sulfur (nmol/mg protein)	UQ (nmol/mg protein)	Lipid (mg dry weight/ (mg protein)	References
1.4–1.5	23–26	23–36	4.2–4.5	0.22	13
1.1–1.3	28–29	28–29	2.9	0.24	113

Source: Data from Hatefi and Stiggall [13] and Ragan and Racker [113].

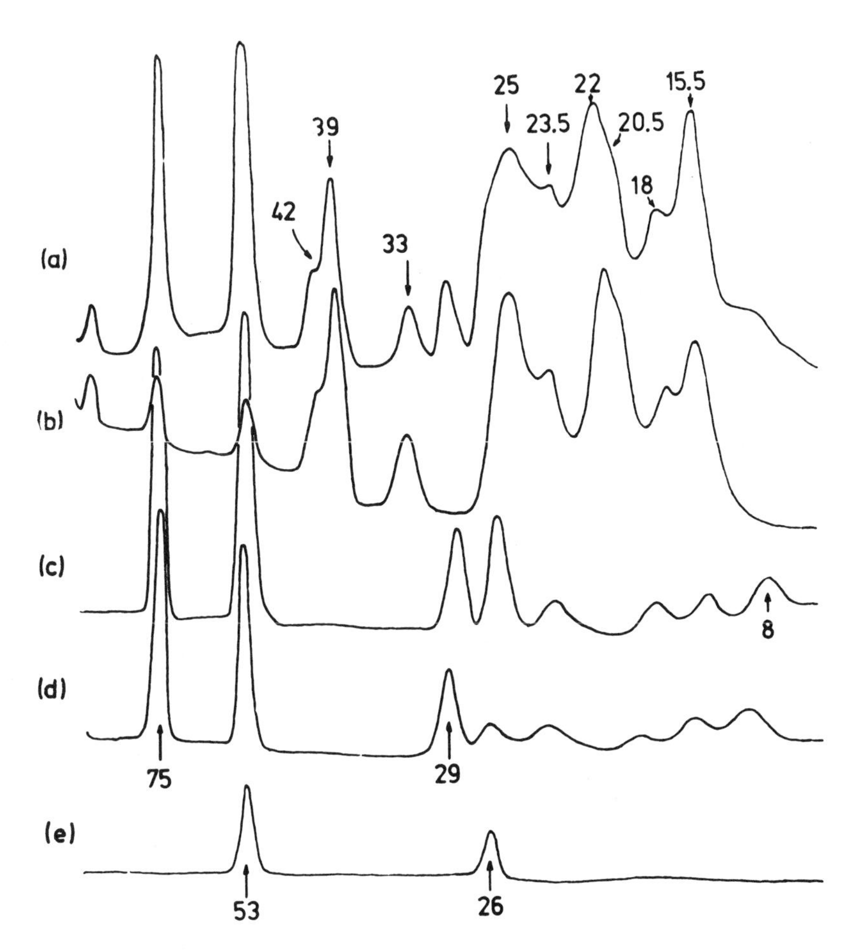

Figure 19 Subunit composition of complex I and fractions from chaotropic resolution. Electrophoresis in the presence of sodium dodecyl sulfate was performed on 10% acrylamide gels. (a) Complex I; (b) hydrophobic fraction of $NaClO_4$ resolution; (c) hydrophilic fraction; (d) iron protein fraction (Ip); (e) iron-sulfur flavoprotein fraction (Fp). Molecular weights ($\times 10^{-3}$) are indicated. Reproduced with permission from Ragan [112].

The low molecular weight form of NADH dehydrogenase preparation (type II) was reported as early as 1952 [118], which was followed by similar enzyme preparations. These enzymes were extracted by the treatment of heart muscle or SMP with 9-11% ethanol at pH 4.8-5.3 and 43-45°C [119-121], or by incubation with boiled snake venom at 37°C [122]. All these enzymes con-

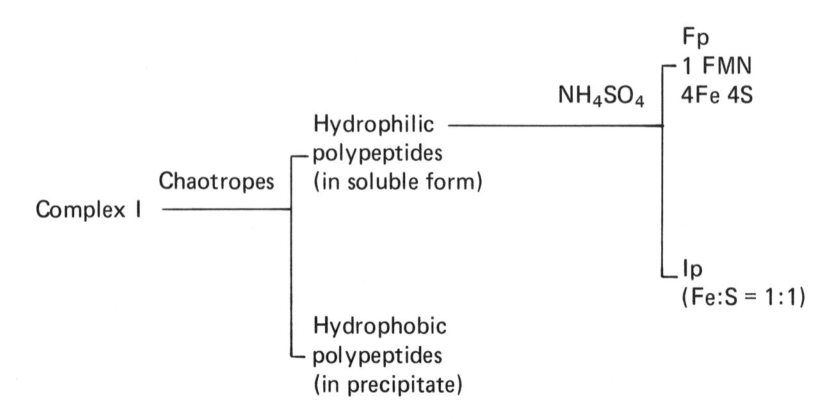

Scheme IV Chaotropic resolution of complex I. Reproduced with permission from Hatefi and Stempel [123] and Davis and Hatefi [124].

tain 1 FMN, 2-4 atoms of nonheme iron and equivalent acid-labile sulfur, and have a molecular weight of 70,000-120,000. They exhibit low NADH-$K_3Fe(CN)_6$ reductase together with various artifactual "rotenone-insensitive" NADH-UQ reductase and antimycin A-insensitive NADH-cytochrome *c* reductase activities.

The overall relationship of type I and type II NADH dehydrogenase preparations to complex I was clearly demonstrated by Hatefi and coworkers [123-125]. As summarized in Scheme IV, complex I was resolved by treatment with a chaotropic reagent, for example, 0.5 M perchlorate for 10 min at 35°C. Hydrophilic polypeptides were isolated in the soluble form which corresponds to about 20% of the total protein containing approximately 50% of nonheme iron and acid-labile sulfur (see Fig. 19). This soluble fraction is further resolved by ammonium sulfate fractionation to iron-sulfur flavoprotein (Fp) and iron-sulfur protein (Ip) fractions. The former contains 1 FMN, 4 nonheme iron, and 4 acid-labile sulfur atoms, and can be electrophoretically resolved into two polypeptide subunits with individual molecular weights of 53,000 and 26,000 [114]. The Ip fraction can be resolved into three major subunits by sodium dodecyl sulfate (SDS) gel electrophoresis with molecular weights of 75,000, 53,000, and 29,000 [114]. The Fp fraction corresponds to the unmodified type II NADH dehydrogenase and seems to be analogous to the Fp subunit of SDH. The precipitated fraction of chaotropic treatment is composed of hydrophobic polypeptides which contain nonheme iron and acid-labile sulfur in 1:1 ratio.

As will be described later, identification of EPR spectra of individual iron-sulfur centers in the NADH-UQ segment and their functional characterization have been conducted using mostly complex I and the intact mitochondrial membrane. Studies using isolated dehydrogenase preparations and subunits, however, have also given useful information.

B. Redox Components in the NADH-UQ Segment

1. Iron-Sulfur Centers

As discussed in detail in a review article by the author [126], a close correlation between site I energy conservation and EPR-detectable (measured at 77 K) iron-sulfur centers associated with the NADH-UQ segment of the respiratory chain has been suggested from comparative studies of phenotypical variants of yeast mitochondria. Some bacterial ferredoxins (iron-sulfur proteins) [127] and other enzymes containing iron-sulfur centers [128] were known to exhibit measurable EPR signals only below 15 K due to a very rapid spin relaxation. Thus, in order to establish rigorously the correlation between the presence or absence of EPR-detectable iron-sulfur centers and the occurrence of site I energy conservation, EPR spectra of reduced iron-sulfur centers in situ in *C. utilis* mitochondrial membrane preparations have been examined at temperatures between 77 and 5 K. This experiment has disclosed the presence of additional iron-sulfur center(s) in the NADH-UQ segment of the respiratory chain, although Ohnishi et al. [5, 6], at that time, attributed multiple signals to only one additional species of iron-sulfur center. The EPR spectra from hitherto unknown iron-sulfur centers were subsequently detected at temperatures below 20 K in mammalian systems in three laboratories, namely by Ohnish et al. [44], Orme-Johnson et al. [45], and Albracht and Slater [129].

Orme-Johnson et al. [45, 130] resolved EPR signals from four separate iron-sulfur centers (centers 1-4) in the anaerobic complex I titrating with measured quantities of reducing equivalents (NADH) (see Fig. 20). The g values of individual centers are summarized in Table 4 with peak positions illustrated in Figure 20.

These investigators have assigned relative oxidation-reduction potentials of the four iron-sulfur centers in the order of $2 > 3 > 4 > 1$, which was subsequently revised to $3 \geqslant 2 > 4 > 1$ [130]. From double integration of the EPR signals, each iron-sulfur center was estimated to take up a number of electrons approximately equal to the molarity of FMN in the NADH-UQ reductase preparation.

Ohnishi et al. subsequently have resolved EPR signals arising from multiple iron-sulfur centers in yeast [131] and pigeon heart [132] mitochondria and SMP and determined their redox midpoint potentials (see Scheme V) using the potentiometric titration method. Later the author designated iron-sulfur centers in the NADH dehydrogenase segment as N-1, N-2, . . . , N-4 in order to distinguish them from multiple iron-sulfur centers in SDH.

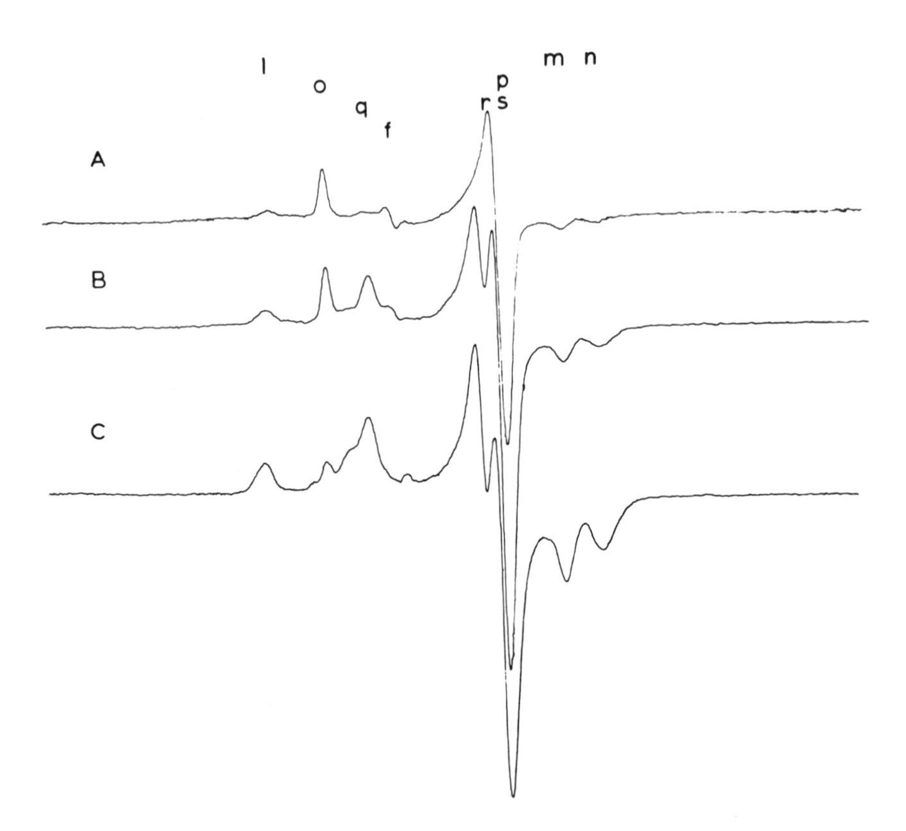

Figure 20 EPR spectra of NADH-reduced complex I showing iron-sulfur centers 1 (qrs), 2 (op), and 3 + 4 (lmn). A, reduced with 10.6 electron neq of NADH; B and C, reduced with 127 neq. EPR operating conditions: microwave frequency, ~9.2 GHz; power, 0.3 mW; modulation amplitude, 7.5 gauss; temperature, 13 K for A and B, 7.7 K for C. Reproduced with permission from Orme-Johnson et al. [45].

Important findings obtained in these early studies are (1) all iron-sulfur centers including center N-2 which has an E_m value close to that of cytochrome *b* are located on the substrate side of the rotenone inhibition site [44, 132]; (2) there is a large difference between the measured midpoint potentials (E_m) of low-potential iron-sulfur centers (N-1, N-3, N-4) and that of the high-potential iron-sulfur center (N-2). This indicates that energy transduction at site I is associated with the transfer of electrons across this potential gap.

In 1973, at the International Congress of Biochemistry in Stockholm, Albracht [133] and Ohnishi and Pring [134] independently reported EPR spectra

Table 4 Field Positions and Assignments of Resonances Observed in Complex I at 13 K

Iron-sulfur center	Field position[a]	Peaks
1	2.022, 1.938, 1.923	qrs
2	2.054, 1.922	op
3	2.100, 1.886, 1.862	lmn
4	2.103,[b] 1.864[c]	–

[a]The numbers are the measured field positions of prominent peaks given on the g-value scale.
[b]Since center 4 is only seen in the presence of center 3, i.e., the field position of the combined resonances is measured, these values may only be approximate. It is likely that the values differ somewhat more from those of center 3.
Source: Orme-Johnson et al. [45, 130].

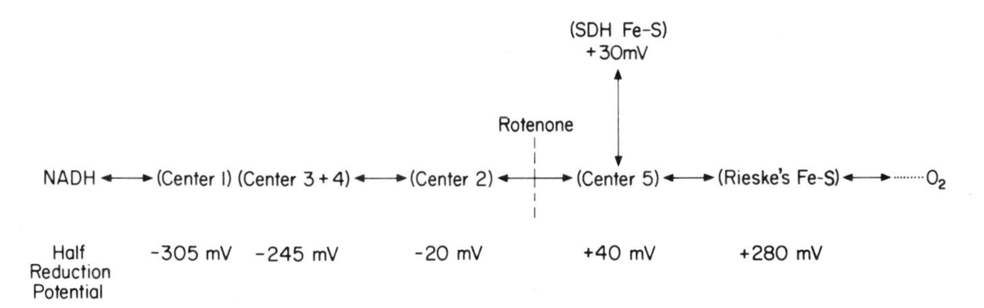

Scheme V Thermodynamic profile of iron-sulfur centers in pigeon heart SMP, as reported in 1972 by Ohnishi et al. [132]. Center 5 has been revised to $(Fe\text{-}S)_{ETF}$ [8] as described in Scheme I.

obtained at extremely low temperatures, namely 5 K and 4.2 K, using bovine heart complex I and pigeon heart SMP, respectively. Figure 21 gives EPR spectra of pigeon heart SMP reduced with NADH, measured at different sample temperatures [134]. As seen in spectrum D, multiple peaks at g values of 2.11, 2.07, 1.94, 1.93, 1.90, and 1.89 were detected at extremely low temperatures. Since the $g_x = 1.87$ component of centers N-3 and N-4, illustrated as n in Figure 20, is not discernible, at least two additional iron-sulfur centers were considered to be responsible for these multiple signals seen at extremely low temperatures; these were designated as centers N-5 and N-6.

Gradual introduction of more sophisticated experimental techniques, including refinement in potentiometric titration combined with computer analysis

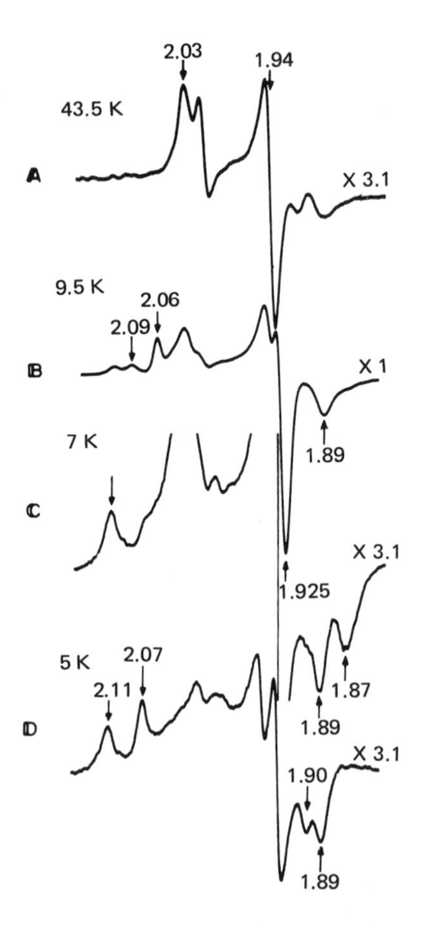

Figure 21 EPR spectra of multiple iron-sulfur centers in the uncoupled pigeon heart SMP, recorded at different sample temperatures. The SMP (56 mg protein/ml), pretreated with FCCP (0.2 nmol/mg protein), were incubated with 2.5 mM NADH and anaerobiosis was attained. EPR operating conditions: modulation amplitude, 12.5 gauss; microwave power, 20 mW. Reproduced with permission from Ohnishi and Pring [134].

[135] and simulation for the resolution of overlapped EPR spectra and for spin quantitation [136, 137], has led to revisions in the identification and assignment of EPR signals of all the above-described iron-sulfur centers except that of center N-2, as follows.

a. Center N-1. In the early stage the g = 1.94 signal was titrated as two unresolved (n = 1) components resulting in an n value of less than 1, hence the E_m

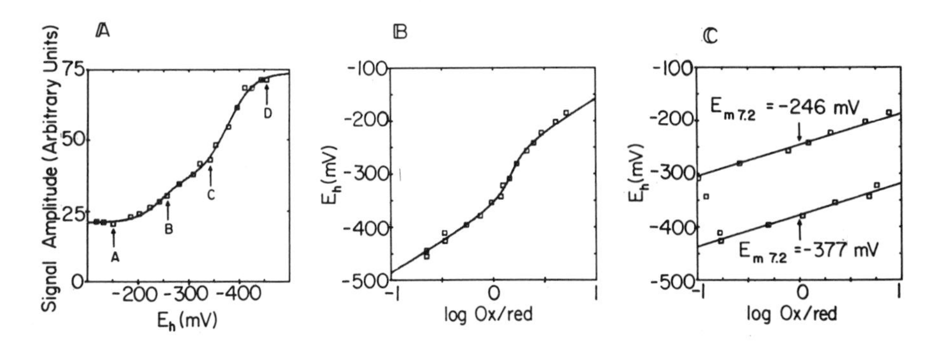

Figure 22 Computer analysis of the redox titration of iron-sulfur centers N-1a and N-1b in pigeon heart SMP. Pigeon heart SMP (30.2 mg protein/ml) were stirred under Ar in 0.2 M mannitol, 0.05 M sucrose, and 50 mM morpholinopropane sulfonate buffer (ph 7.2). Present as mediators were 31 μM duroquinone, 6 μM pyocyanine, 15 μM 2-hydroyl-1,4-naphthoquinone, 6 μM resorufin, 78 μM phenosafranine, 74 μM benzylviologen, and 133 μM methylviologen. EPR operating conditions: as in Figure 21, except that the temperature was 25 K. (A) Signal amplitude change as a function of E_h. The arrows at A, B, C, and D refer to the E_h at which the spectra of Figure 23 were taken. (B) Nernst plot of the data in (A). The abscissa scale is the logarithm of the ratio of oxidized to reduced iron-sulfur centers. (C) Resolution of curve B into two components. Reproduced with permission from Ohnishi and Pring [134].

value obtained was the mean value of the two [75, 132]. Contribution of two components to the g = 1.94 signal was first suggested from analysis of the ATP effect on center N-1 in pigeon heart mitochondria poised at very low redox potentials (E_h) [7]. The fraction of the g = 1.94 species which responded to the high phosphate potential was designated as center N-1a as distinguished from the unaffected part, N-1b. Subsequently resolution of the titration curve into two n = 1 curves was obtained by improved potentiometric titration [134, 135]. As presented in Figure 22 the center N-1 curve was titrated as two n = 1 curves with midpoint potentials of -240 ± 20 mV and -380 ± 20 mV, for center N-1b and center N-1a, respectively. From the difference spectra, N-1b and N-1a spectra were shown to have rhombic and axial symmetry, respectively, in the pigeon heart system (Fig. 23) [134]. Because of the high content of the respiratory components and availability of the tightly coupled mitochondria, Ohnishi had favored the use of the pigeon heart system. However, purified preparations, such as complex I or NADH dehydrogenases, are all prepared from beef heart; thus redox titrations of the iron-sulfur centers have also been conducted using beef heart SMP as the control for studies of purified systems. Titration curves for centers N-1a and N-1b were resolved in a way analogous to that of the pigeon

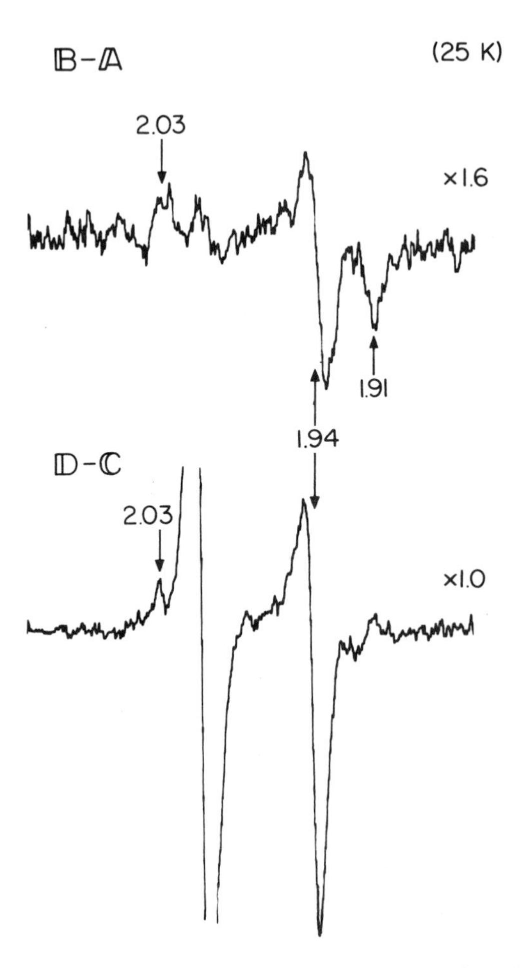

Figure 23 EPR spectra of the resolved iron-sulfur center N-1a or N-1b. The EPR spectrum of the component haveing an $E_{m7.2}$ value of -246 mV was taken by the difference spectra at -257 mV (B) and at -151 mV (A) as illustrated with arrows in Figure 22. Similarly the spectrum of the component having an $E_{m7.2}$ value of -377 mV was obtained as the difference between spectra at -455 mV (D) and at -343 mV (C). Subtraction of the EPR spectra was conducted using a Nocolet Signal Averager (NIC-1974). Reproduced with permission from Ohnishi and Pring [134].

system, although both centers give rise to rhombic spectra with somewhat different g_y and g_x values. In line with these findings, Albracht et al. also suggested earlier that the EPR spectrum of the g = 1.94 species arises from two centers in Complex I and showed that both centers give rhombic spectra. The resolution of the spectra was obtained utilizing the different lability of the two centers toward the added NADH [133].

Potentiometric titrations of various iron-sulfur centers in complex I are shown in Figure 24 (the samples of complex I were prepared by Widger and King). As shown, in curve I, center N-1a titrates as a typical n = 1 curve with $E_{m8.0}$ = -390 ± 20 mV, while center N-1b gives an anamalous titration curve with an apparent midpoint potential in the region of -150 mV. The spin concentration of N-1a is approximately equivalent to the FMN concentration, while the center N-1b content varies from preparation to preparation in the range of one-quarter to one-half of the FMN concentration. In addition, as shown in Figure 25, the spectrum of center N-1b is greatly broadened and is not evoked by reduction with NADH. Thus center N-1b seems to be labile, and is modified considerably during the complex I isolation process* [84]. There are additional inherent technical difficulties in connection with center N-1 analysis because of the lability of center N-1 in complex I toward NADH in the presence of trace O_2, as indicated by its lineshape modification [130, 133].

It should be mentioned here that the Fp subunit or type II NADH dehydrogenase contains 1 FMN and 4 atoms of nonheme iron and equivalent acid-labile sulfur analogous to the Fp subunit of SDH. As will be discussed later there are several independent lines of evidence showing that Fp contains two binuclear iron-sulfur clusters, probably center N-1a and N-1b or center N-1a and another N-1 type center, analogous to centers S-1 and S-2 in SDH. To make the situarion more complicated, there is a possibility that these two N-1 centers are redox coupled, as discussed previously for centers S-1 and S-2.

There is an additional ambiguity concerning center N-1; Orme-Johnson et al. reported that the center N-1 type signal is less saturable than center N-2 in the low-temperature range (13 K) [130]. We have also obtained preliminary data indicating an N-1 type center with a shorter relaxation time, and an $E_{m8.0}$ value of approximately -340 mV [52]. Whether the abnormal power saturation behavior of this N-1 type center is due to spin-spin interaction possibly with centers N-3 or N-4, or whether there is a distinct N-1 species with shorter relaxations requires further rigorous investigation.

More recently Albracht et al. have conducted computer simulation studies on the EPR characteristics of centers N-1a and N-1b [136, 137]. These investigators proposed that both centers N-1a and N-1b give rise to EPR spectra of axial symmetry rather than previously assigned rhombic character. Their g values were defined as g = 2.021 and g = 1.938 for center N-1a and g = 2.021

*One possibility which can not be excluded is that center N-1b is not an intrinsic component of NADH-UQ reductase; it may be a component of a different mitochondrial enzyme. It should be pointed out here that in beef heart SMP the spin concentration of center N-1a is equivalent to that of center N-2, a component which is widely accepted as present in 1:1 ratio to FMN, while center N-1b seems to be somewhat lower than center N-1a even in SMP.

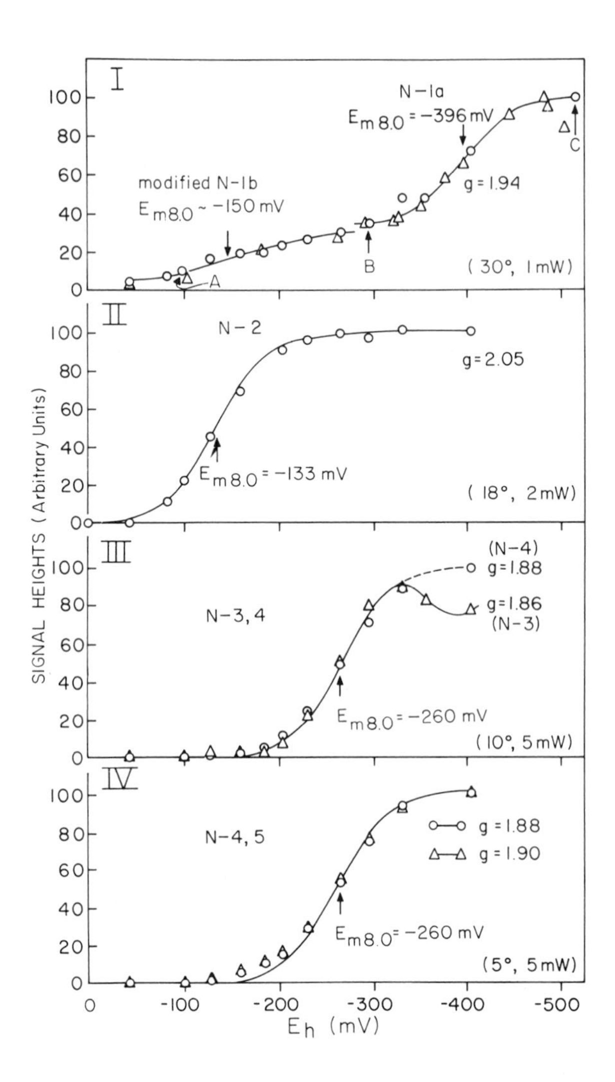

Figure 24 Potentiometric redox titration of iron-sulfur centers in complex I. Complex I was prepared by Widger and King and suspended in 50 mM Tris-HCl buffer at pH 8.0 and 0.6 M sucrose at a protein concentration of 35 mg/ml (flavin concentration, 1.2 nmol/mg protein) and titrated in the presence of several redox mediator dyes. Sample temperature and microwave power applied are shown in parentheses at the bottom right corner of each redox curve. Reproduced with permission from Ohnishi et al. [138].

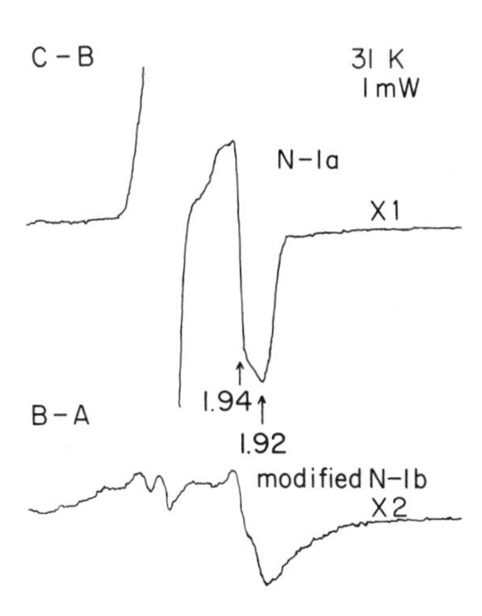

Figure 25 EPR spectra for the resolved center N-1a and N-1b in complex I. Redox potentials of samples A, B, and C [illustrated with arrows in Figure 24 (I)] are –98 mV, –295 mV, and –515 mV, respectively. Reproduced with permission from Ohnishi et al. [138].

and g = 1.928 for center N-1b. These investigators also reported that the spin concentration of the individual centers is only 25% of that of center N-2 or FMN. Their computer simulation was based on two assumptions: (1) center N-1 is reduced by NADH but not by dithionite, and (2) EPR spectra obtained by dithionite reduction arise from contaminants in complex I. It should be pointed out that the EPR lineshape obtained with their complex I which was reduced either with NADH or dithionite seems to be somewhat different from that commonly seen in other laboratories. Their centers N-1a are assigned differently from those designated by Ohnishi et al. [134, 135]; the two centers defined by Albracht et al. could well be center N-1a species in a different molecular environment.

b. Centers N-3, N-4, and N-5. From the potentiometric analysis of center N-3 and N-4, using troughs m and n (see Fig. 20) in pigeon heart SMP, where the center responsible for the m trough shifted the E_m value to -400 mV from -240 mV while the n trough remained as the -240 mV component, Ohnishi proposed that the g = 1.89 and g = 1.87 troughs (which correspond to m and n in Fig. 20) belong to separate centers having rhombic symmetry and assigned g components of center N-3 as g_z = 2.10, g_y = 1.93, g_x = 1.87, and of center N-4 as $g_z = 2.11$, $g_y = 1.93$, and $g_x = 1.89$ [135]. More recently Albracht et al. [136, 137] conducted simulations of spectral lineshape of centers N-3 and N-4, as cited above for center N-1. They demonstrated that g_z values of 2.10 for both centers fail to offer a good fit to experimental spectra, if each center is assumed to give one spin per FMN. The best fit was obtained with $g_{z,y,x}$ values of (2.036, 1.92-1.93, 1.863) and (2.103, 1.93-1.94, 1.884) for centers N-3 and N-4, respectively. Albracht et al. designated these centers with reversed number, but the present author intends to retain original names in order to keep the complexity to a minimum (see Fig. 26).

We have confirmed this revision of the g_z value, using an independent method [138]. Figure 27 represents spectra obtained with complex I poised at -292 mV, where no center N-1a signals (2.02, 1.94, 1.92) interfere with the measurement. Increase of the microwave power from 1 mW (spectrum in the solid line) to 20 mW (dotted line), at 9.3 and 5.6 K, respectively, results in the clear concomitant disappearance of the 2.04 and 1.86 signals of center N-3 and the 2.05 signal of center N-2. Decrease of the central signals in the 1.93-1.92 signal arise from both centers. Center N-4 peaks at 2.10, 1.94-1.93, and 1.88 are only partially saturated, and their signal height remains almost unchanged upon increase of the power. The previously reported shift of g_x component from g = 2.10 to somewhat larger g values in parallel with the titration of centers N-3 and N-4 in complex I [130] or in pigeon heart SMP [135] seem to arise from unidentified spin-spin interaction or from protein conformational change [138]. The establishment of the g values of centers N-3 and N-4 consequently affects the assignment of the multiple peaks observable at extremely low temperature. As described above, peaks at 2.10 and 1.88 can be explained by the partially saturated signals from center N-4. Thus only one center with extremely short relaxation time(s) can account for other peaks which appear at temperatures close to that of liquid helium, discarding center N-6.

Center N-5 peaks can be clearly discernible even under non-power-saturated conditions, (see g = 2.07 and g = 1.90 peaks in Figure 27), in spite of the skepticism voiced by Beinert and Ruzicka [59, 139]. The spin concentration of this center, however, is less than one-quarter of the FMN concentration. Again this leaves an ambiguity; this center may be very labile and readily modified during complex I preparation or may originally be a component of some other enzyme in equilibrium with the NADH dehydrogenase components. The EPR signals of this center were reported not to be detectable in *C. utilis* SMP [137] or in plant mitochondria [140]. Characteristics of all iron-sulfur centers in complex I are summarized in Table 5. As described so far, only Fd-type but not HiPIP-type iron-sulfur centers are present in the NADH-UQ segment of the respiratory chain.

The next question to consider is whether the structure of these iron-sulfur clusters is binuclear, similar to S-1, or tetranuclear as in the case of center S-3.

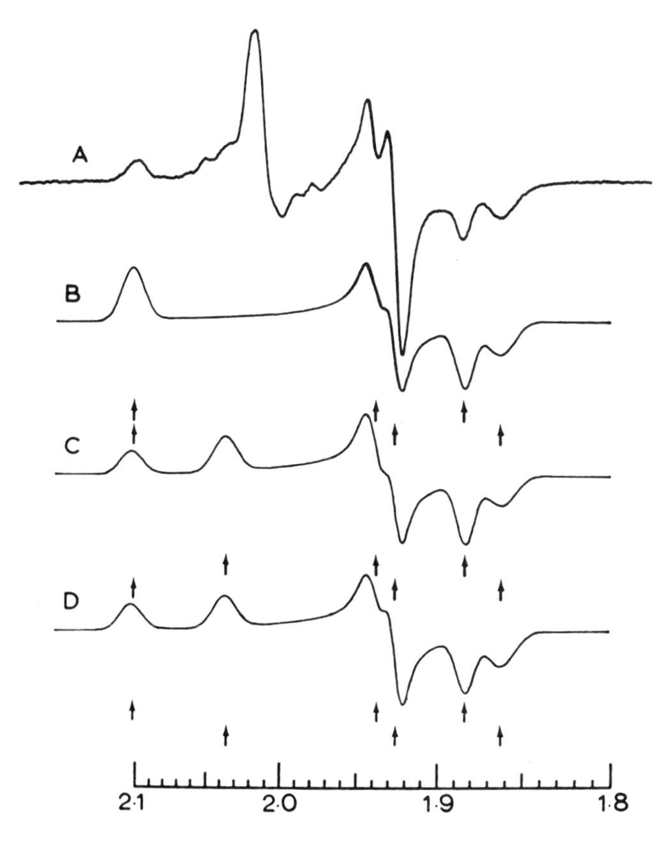

Figure 26 Determinations of the g values of centers N-3 and N-4 by simulation. (A) Experimental EPR spectrum at 12 K of complex I reduced with NADH. EPR conditions: sample temperature, 12 K; power, 2 mW; modulation amplitude, 0.63 mT. (B) Simulation of the sum of the centers N-3 and N-4 assuming two rhombic signals with a common g_z line at 2.10. Parameters: $g_{x,y,z}$ = 1.884, 1.938, 2.103, and widths (x, y, z) 1.9 mT, 1.9 mT, and 2.4 mT for the first signal; for the second signal $g_{x,y,z}$ = 1.863, 1.9263, 2.103, and widths (x, y, z) 3 mT, 1.4 mT, and 2.4 mT. (C) Simulation of the centers N-3 and N-4 assuming two rhombic signals with different g_z values. Parameters: for the first signal $g_{x,y,z}$ = 1.884, 1.928, 2.037, and widths (x, y, z) 1.9 mT, 1.9 mT, and 2.5 mT; for the second signal $g_{x,y,z}$ = 1.863, 1.9263, 2.103, and widths (x, y, z) 3 mT, 1.4 mT, and 2.4 mT. (D) As C but with the g_z values interchanged. Parameters: for center N-3, $g_{x,y,z}$ = 1.884, 1.938, 2.103, and widths (x, y, z) 1.9 mT, 1.9 mT, and 2.4 mT; for center N-4, $g_{x,y,z}$ = 1.863, 1.9263, 2.037, and widths (x, y, z) 3 mT, 1.4 mT, and 2.5 mT. The small arrows indicate the g values for the two signals (1 mT = 10 gauss). Reproduced with permission from Albracht et al. [137].

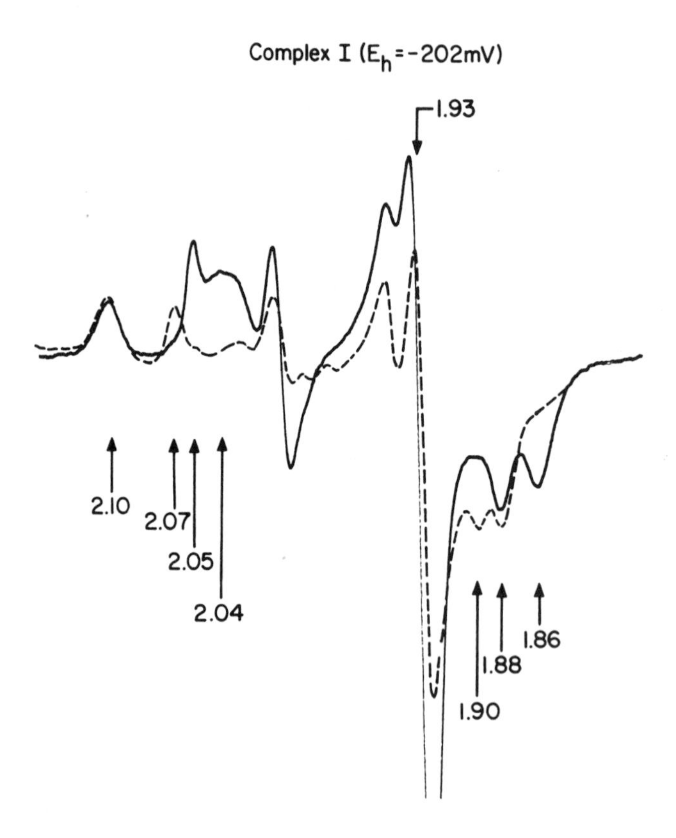

Figure 27 EPR spectra of complex I poised at −202 mV, measured at two different conditions. Solid line: microwave power, 1 mW; sample temperature, 9.3 K. Dashed line: power, 20 mW; temperature, 5.6 K. Center N-3 was power-saturated at a lowered temperature. Concomitant decrease of g_z and g_x signals are illustrated with arrows. Center N-4 signals are less power-saturated and an almost unchanged signal size of g_z = 2.10 and g_x = 1.88 is seen.

Albracht and Subramanian [65] reported that center N-1 in *C. utilis* SMP prepared from cells grown on ^{57}Fe showed typical broadening in g_z line which is indicative for a splitting by two nuclei with I = 1/2 with an A value of 12-14 gauss; ^{59}Fe also introduced a characteristic splitting in the xy direction. In contrast g = 2.05, g = 2.04, and g = 2.10 signals for centers N-2, N-3, and N-4 showed less broadening with an A value of 9 gauss, indicative of four equivalent nuclei contributing to the spectra. Thus, Albracht and Subramanian proposed N-1 as binuclear and other centers as tetranuclear centers.

In agreement with this proposed cluster structure, measurement of the ex-

Table 5 Summary of Iron-Sulfur Centers in Complex I

Center	[Spin]/[FMN]	Symmetry	g_x	g_y	g_z
N-1a	1	Rhombic	2.03	1.94	1.90
N-1b	1/4-1/2	Rhombic	–	–	–
N-2	1	Axial	2.05	~1.93	1.93
N-3	1	Rhombic	2.04	~1.93	1.86
N-4	1	Rhombic	2.10	~1.93	1.88
N-5	<1/4	Rhombic	(2.07	~1.93	1.90)

change coupling constant J of about 90 cm^{-1} and distortion parameter $g_y - g_x$ of 0.02 for center N-1a [63], similar to center S-1, gives independent supporting data for its binuclear structure. Our preliminary measurement of the energy gap between the ground state and first excited state of center N-2 also gives strong indication for its tetranuclear structure [141].

The multiplicity of N-1 type centers in complex I has not been established; however, the following observations provide evidence for the presence of at least two binuclear centers in the NADH-UQ segment; Widger and King [142] have recently isolated a small molecular weight NADH dehydrogenase which upon addition of NADH, elicits EPR signals from two distinct species of iron-sulfur centers and from the free radical of FMN, although the total spin concentration of iron-sulfur centers accounts for only approximately 5% of the FMN concentration. This enzyme contains 4 nonheme iron atoms and equivalent acid-labile sulfur, similar to the Fp subunit isolated by Hatefi and Stempel. The measured J value and the spectral distortion parameter of one species speak for its binuclear structure [142]. Thus the Fp subunit of NADH dehydrogenase appears to contain two binuclear centers (N-1 type centers) analogous to centers S-1 and S-2 in the Fp subunit of SDH.

Reductive titration with measured aliquots of NADH showed that complex I accepts about 20 n electron equivalents per mg protein [130], which corresponds to about 15 equivalents per FMN using the mean value of flavin content of Hatefi and Ragan's complex I preparations. Thus about 5 electron equivalents from iron-sulfur centers, 2 from flavin, plus 6-8 from UQ account for the total pool size reported by Orme-Johnson et al. [130].

2. Flavin

Flavin of NADH dehydrogenase is FMN [143] and it is noncovalently bound to the apoprotein. It is readily extractable from the dehydrogenase by acid treatment. There is no hard information about the kinetic and thermodynamic prop-

Table 6 Thermodynamic Characteristics of Iron-Sulfur Centers in the NADH-UQ Segment in the Membrane-bound and "Soluble" States

Center	PH Mt.[a,b]	BH SMP[a,b]	Complex I[c]	ADH[a]	IDH[a]	RDH[a]
N-1a	−380	−370	−385	−396	−396	−393
N-1b	−240	−220	−260	−253	−250	−226
N-2	−20	−80	−135	−205	−270	−135 (60%) −230 (40%)
N-3	−240	−240	−245	−245	−245	−252
N-4	−240[d]	−240				
N-5	−260	−275	−260	−260	−260	−270
References	135	138	144	144	144	144

[a]Abbreviations used: PH Mt., pigeon heart mitochondria; BH SMP, beef heart submitochondrial particles; ADH, reconstitutively active lipid-depleted complexI; IDH, reconstitutively inactive lipid-depleted complex I; RDH, ADH reconstituted with phospholipid.

[b]E_m was measured at pH 7.2 for PH Mt. while E_m values in BH SMP and other systems were measured at pH 8.0.

[c]Complex I used for this measurement was prepared by Ragan and Racker and was used as a starting material for the preparation of ADH and IDH. Center N-1b in this complex I seemed to be less modified than that in complex I of Widger and King.

[d]In PH Mt., the E_m value of center N-4 was shifted to −410 mV during the preparation process.

erties of this FMN, probably due to its rather unstable free radical state and lack of spectrophotometric resolution from iron-sulfur centers.

C. Thermodynamic Characteristics of the Redox Components in the NADH-UQ Segment

Midpoint potentials of iron-sulfur centers in the NADH-UQ segment in various systems, namely pigeon heart and beef heart SMP, complex I, reconstitutively active and inactive lipid-depleted complex I, and reconstituted NADH-UQ reductase [112, 113, 144], are summarized in Table 6. All these titrations except pigeon heart SMP were conducted at pH 8.0.

As discussed earlier, center N-2 has the highest value of the midpoint potential among iron-sulfur centers in this segment and is likely to be the component donating electrons to UQ. Ragan and Racker [113] reported that by depleting only 50% of phospholipid from complex I, using cholate and KCl, complex I completely loses its rotenone-sensitive NADH-UQ reductase activity although NADH-$K_3Fe(CN)_6$ reductase activity is fully preserved. They also reported that if this lipid depletion procedure is conducted under protective conditions, namely in the presence of dithionite and dithiothreitol, the resulting lipid-depleted complex I (for convenience they named this preparation NADH dehydrogenase in spite of the fact that about 50% of the phospholipid was retained) can be recombined with purified phospholipid and restores rotenone-sensitive NADH-UQ reductase activity. The resulting preparation combines with complex III to reconstitute antimycin A-sensitive NADH-cytochrome *c* reductase. If these reducing reagents are omitted during the lipid depletion, the resulting dehydrogenase gives a very low rate of rotenone-sensitive NADH-UQ reductase activity when it is combined with phospholipid. The former was thus called reconstitutively active and the latter reconstitutively inactive NADH dehydrogenase, although, strictly speaking, they are categorically different from the reconstitutively active and inactive SDH preparations.

Reconstitutively active and inactive NADH dehydrogenase preparations contain approximately the same concentration of nonheme iron and acid-labile sulfide, and spectrophotometric analysis does not reveal any differences [113]. However, as exemplified with centers N-1a and N-2 in Figures 28 and 29, the lineshapes of all iron-sulfur centers in the reconstitutively active preparations are almost identical to those found in complex I, but in inactive preparations all centers show broadened EPR spectra. Center N-2 especially showed the most dramatic modification in relaxation rate(s) as seen in the broad lineshape of Figure 18 [144].

As shown in Table 6, the midpoint potential of center N-2 in complex I is approximately -135 ± 20 mV, which is considerably lower than the E_m value obtained for center N-2 in beef heart SMP. During the lipid depletion, the E_m value of center N-2 becomes further electronegative, reaching -210 mV in active preparation and even lower, -265 mV, in the inactive preparation. In contrast, center N-1a (Table 6 and Fig. 30) as well as other centers do not show such a dramatic shift of the midpoint potential. In addition, the total signal amplitude of center N-2 gradually diminished during repeated redox titrations in the inactive preparation, while repeated titrations in the active preparation coincided well (reversible). Upon recombination of reconstitutively active NADH dehydrogenase with purified phospholipid, the E_m value of about 60% of center N-2 returned to -135 ± 15 mV, the original E_m value in complex I (see Fig. 31).

The fraction which is restored (60%) is in reasonable agreement with the extent of restoration of rotenone-sensitive NADH-UQ reductase.

These data suggest that iron-sulfur center N-2 may be surrounded by a phospholipid layer in the mitochondrial membrane, and exposure of the center to a more aqueous environment by removal of phospholipid with cholate tends to modify the molecular environment of center N-2. Upon recombination of the active dehydrogenase with phospholipids, the shift in midpoint potential is

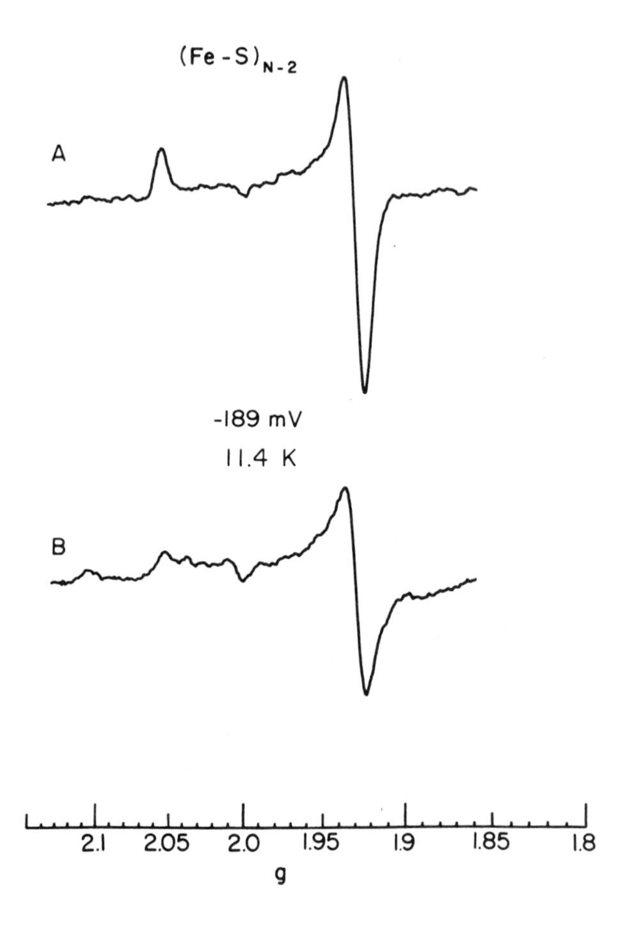

Figure 28 EPR spectra of center N-2 in reconstitutively active and inactive NADH dehydrogenases (ADH and IDH), respectively. Each preparation at a protein concentration of 10.6 mg/ml was poised at −189 mV in the presence of various redox mediating dyes and transferred to EPR tubes under anaerobic conditions. Reproduced with permission from Ohnishi et al. [144].

partially reversed, suggesting that during the reincorporation of phospholipid, part of the center N-2 is restored to its original environment. The location of center N-2 in relation to site I energy coupling will be discussed in more detail later.

Figure 32 summarizes the pH dependence of the midpoint potentials of some iron-sulfur centers. This accumulated data from various mitochondrial

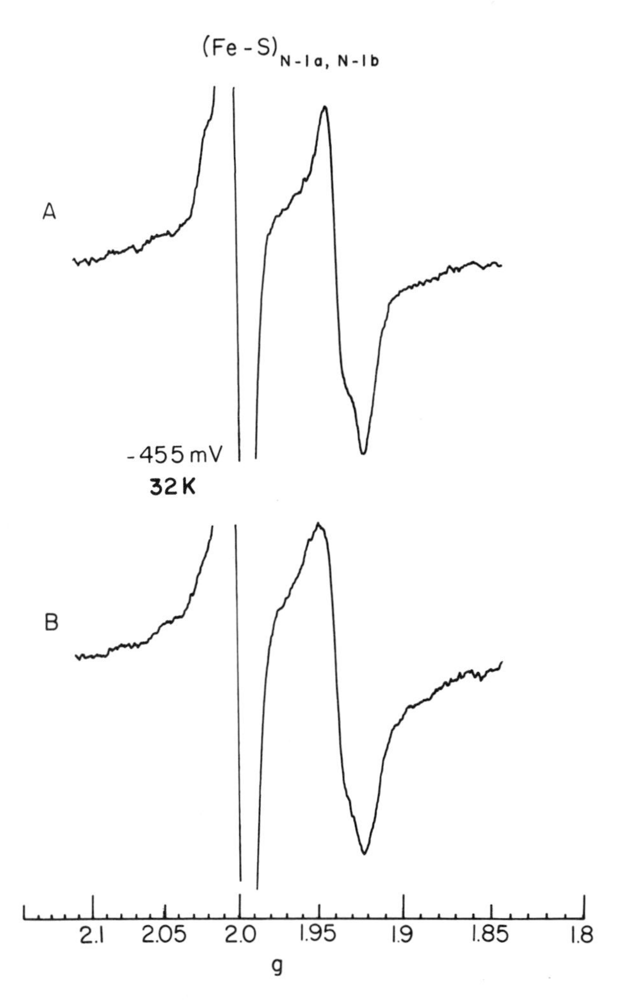

Figure 29 EPR spectra of centers N-1a plus N-1b in the reconstitutively active (A) and inactive (B) NADH dehydrogenase. Reproduced with permission from Ohnishi et al. [144].

systems show that centers N-1a and N-2 have pH-dependent apparent midpoint potentials which fall approximately 60 mV per pH unit, while the midpoint potentials of all other iron-sulfur centers are independent of pH in the range between pH 6 and 8.6 [145].

As will be described later, the apparent midpoint potentials of these two

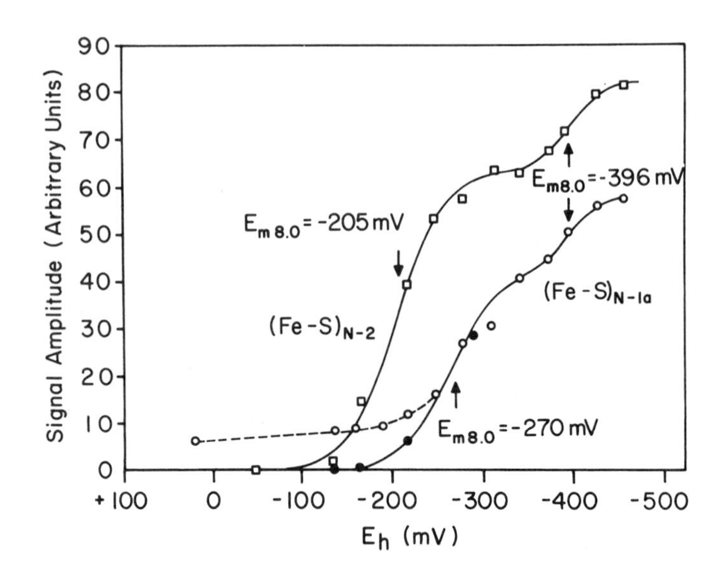

Figure 30 Potentiometric titration of centers N-2 and N-1a in reconstitutively active and inactive NADH dehydrogenase preparations. Signal amplitude between the negative peak at g = 1.92 and the baseline was plotted as a function of E_h. EPR conditions are the same as in Figure 28. Reproduced with permission from Ohnishi et al. [144] and Ohnishi and Ragan, unpublished data.

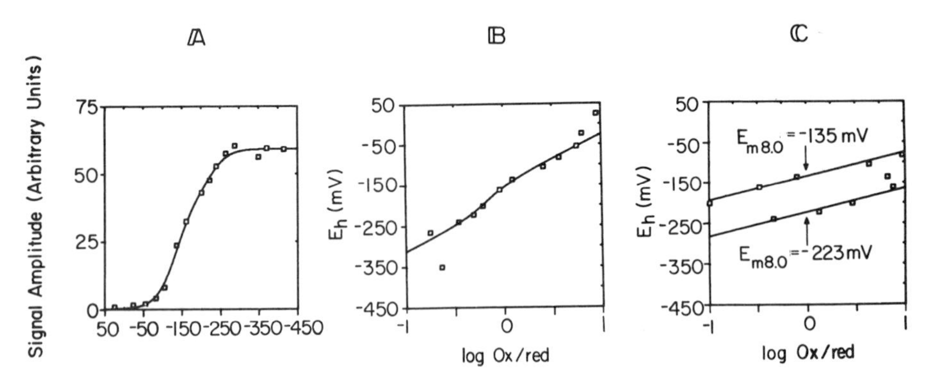

Figure 31 Computer analysis of the potentiometric titration of center N-2 in phospholipid vesicles reconstituted with reconstitutively active lipid-depleted complex I (final protein concentration 8.3 mg/ml). EPR conditions are the same as in Figure 29. Reproduced with permission from Ohnishi et al. [144].

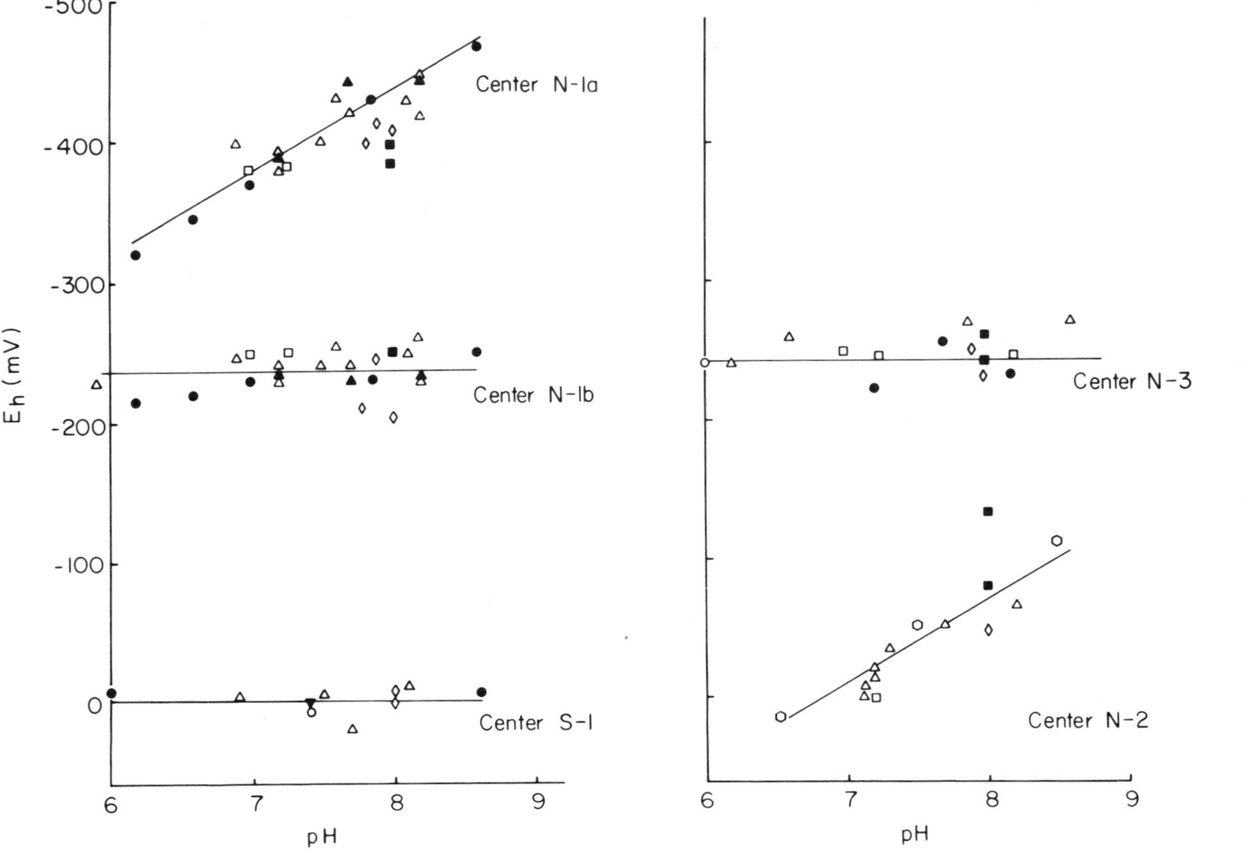

Figure 32 pH dependence of midpoint potentials of various iron-sulfur centers. Beef heart mitochondria (●—●) and SMP (◇—◇), pigeon heart mitochondria (□—□) and SMP (△—△), complex I (■—■), and SDH (▼—▼) were titrated potentiometrically, and pigeon heart SMP (⬡—⬡) and SDH (○—○) were titrated with the succinate-fumarate couple. Reproduced with permission from Ingledew and Ohnishi [145].

iron-sulfur centers are likewise dependent on phosphate potential, while none of the other centers either in the NADH or in the succinate-UQ reductase segment of the respiratory chain have been demonstrated to show this behavior.

D. Kinetic Studies

The earliest kinetic studies on type I NADH dehydrogenase were performed by Beinert et al. [146]. At the time, only the g = 1.94 signal visible at liquid nitrogen temperature had been observed, using freeze-quench techniques; it was found that NADH caused this signal to appear within 8 msec, which was too fast for the time resolution of the method. Use of the substrate analog acetylpyridine nicotinamide adenine dinucleotide (AcPy NADH) allowed the reduction kinetics of this component to be followed; the half-time at 1°C was 40 msec which corresponds to the one catalytic cycle in the NADH-$K_3Fe(CN)_6$ assay. This shows that the g = 1.94 iron-sulfur center N-1 is competent as an electron carrier in this segment of the respiratory chain. Beinert et al. concluded that the reaction of substrate with the enzyme was rate-limiting.

The discovery of the low-temperature EPR signals in NADH dehydrogenase gave rise to a second round of kinetic experiments. Orme-Johnson et al. [130] found that the EPR signals from centers 1, 2, and 3 + 4 in complex I could be elicited by NADH within 6 msec. Use of AcPy NADH slowed down the reaction enough to resolve the order of reduction, which was reported as 2, 3 + 4, and 1.

It seems clear that this "kinetic resolution" was in reality a thermodynamic resolution not necessarily reflecting the order of the centers in electron transfer. Assuming that intramolecular electron transfer is much faster than reaction with substrate (which is to be expected in a well-designed enzyme), rapid electron transfer within the dehydrogenase would maintain a quasi-equilibrium within the molecule at all stages of reduction. This would lead to the reduction of the highest potential center first (as is observed), even if it is last in the electron transfer sequence.

Although kinetic experiments have verified kinetic competence of iron-sulfur centers during redox reactions, they have contributed little to elucidating the sequence and mechanism of electron transfer in the dehydrogenase due to the rapidity of equilibration within the molecule compared with the reaction with substrate.

E. Energy Coupling at Site I

Evidence indicating direct involvement of iron-sulfur centers in energy coupling in the NADH-UQ segment of the respiratory chain has been discussed in detail in the reviews by Singer and Gutman [147], Garland et al. [148], and by the author [126]. Basically two different models concerning the mechanism of site I energy transduction have been reported from the available experimental data, namely, the chemiosmotic and chemical intermediate hypotheses.

Mitchell [149] proposed a chemiosmotic mechanism for site I energy coupling, mostly based on the following observations reported by Gutman et al. [150]: Beef heart SMP pretreated with a high concentration of piericidin A can still slowly oxidize NADH by leakage of electrons through the inhibition site. Upon exhaustion of the added low concentration of NADH, center N-2 remains reduced although all the low-potential (E_m) components in the site I segment are completely reoxidized. Center N-2 is reoxidized upon ATP addition [150]. Mitchell assigned FMN as the hydrogen carrier and centers N-1 and N-2 as electron carriers in the site I redox loop. This hypothesis was further elaborated by Garland et al. [148] and Lawford and Garland [151] based on the direct measurement of the H^+/site ratio of 2 coupled with electron transfer from NADH to UQ using both ox heart mitochondria and SMP. Recently, Singer and Gutman [152] and Gutman et al. [153] published a more quantitative analysis on the ATP-induced apparent midpoint potential shift of centers N-1 and N-2, and claimed that they had obtained experimental evidence supporting this chemiosmotic hypothesis (this is discussed later in detail). On the other hand, Ohnishi et al. [7] proposed earlier a possible involvement of center N-1a and subsequently both centers N-1a and N-2 [126, 134, 154], based on the phosphate potential dependence of their measured midpoint potentials according to the chemical intermediate hypothesis of the energy transduction developed by Wilson and his coworkers [155, 156]. Recently DeVault [157] discussed possible mechanisms for site I energy coupling based on available experimental data [150, 152, 154] and made quantitative predictions from the standpoint of both transductase (chemical intermediate) [157, 158] and chemiosmotic models [149, 159] for site I.

As shown in Figure 33, (spectra A and B) upon addition of ATP to a pigeon heart mitochondrial suspension which was poised at E_h values where center N-1a is partially or highly reduced, about 50% of center N-1a was oxidized (center N-1a contributes to about 45 ± 5% of the g = 1.94 signal amplitude under the EPR conditions used).* With the suspension poised at +50 mV, where center N-2 is mostly oxidized, ATP addition resulted in almost complete reduction of center N-2 (spectra C and D). The difference spectra represented at the bottom

*An excellent general criticism was raised by Jones et al. [160] on the application of the potentiometry to the redox components in the membrane-bound form, because of the selective permeability of redox mediators in the oxidized and reduced state. In the case of center N-1a, however, his prediction would cause the opposite effect, reduction of center N-1a by the addition of ATP, if center N-1a is located inside the osmotic barrier.

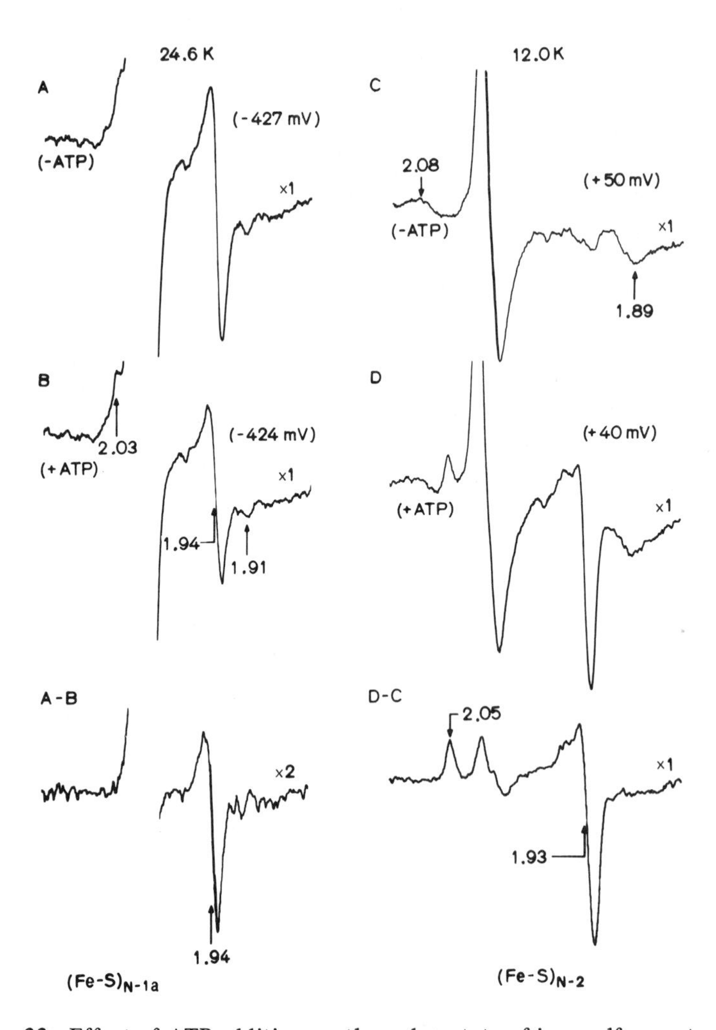

Figure 33 Effect of ATP addition on the redox state of iron-sulfur centers N-1a and N-2 in pigeon heart mitochondria equilibrated at fixed redox potentials. Mitochondria (12 mg protein/ml) were suspended in 0.2 M mannitol, 0.05 M sucrose, 0.2 mM EDTA, and 50 mM morpholinopropane sulfonate buffer (pH 7.2) and adjusted to redox potentials of −427 mV and +50 mV, respectively, in the presence of several redox dyes. Ten mM ATP was bubbled with Ar gas to remove dissolved oxygen and was added to the mitochondrial suspension anaerobically. The redox potential of the system was readjusted to an E_h value close to that prior to ATP addition. The suspension before and after ATP addition was transferred anaerobically to EPR tubes and rapidly frozen. EPR operating conditions: microwave frequency, 1.2 GHz; modulation amplitude, 12.5 gauss; microwave power, 20 mW. Difference spectra, A − B and D − C, were obtained using a Nicolet computer of average transients. Reproduced with permission from Ohnishi [154].

of Figure 33 identify centers N-1a and N-2 from their lineshape and signal position. These results show that the apparent midpoint potential of center N-1a decreases while that of center N-2 shifts toward a more positive value upon addition of ATP.* In contrast to these two iron-sulfur centers, the measured midpoint potentials of all other centers in the site I region, centers N-1b, N-3, N-4, and N-5, are not significantly affected by ATP addition [154].

In order to quantitate the midpoint potential shift ($E_{m7.2}$), a redox titration of center N-2 was performed with added ATP. As presented in Figure 34, the apparent midpoint potential of center N-2 shifts to +105 ± 20 mV in the presence of high phosphate potential, from the resting state midpoint potential of -20 ± 20 mV, giving a ΔE_m value of +125 ± 30 mV. The redox titration of center N-1a in the presence of added ATP is more difficult than that of center N-2 due to overlapping signals and to the much higher protein concentration required than for center N-2 to obtain a reasonable redox titration curve. Thus, $E_{m7.2}$ of center N-1a was calculated from experiments poised at two appropriate redox potentials (E_h). The addition of ATP to pigeon heart mitochondria poised at -510 ± 20 mV induced at most 10% oxidation of center N-1a. This result together with that of Figure 33 indicates that the apparent $E_{m7.2}$ of center N-1a shifts from -380 ± 20 mV to -440 ± 20 mV upon addition of ATP, giving a half-reduction potential shift ($E_{m7.2}$) of approximately 60 mV. Shift of the apparent E_m values of centers N-1a and N-2 was inhibited either by an uncoupler (FCCP) or by a phosphorylation inhibitor (oligomycin) or by both added together.

In pigeon heart SMP poised around -430 mV, ATP addition caused no significant change of the total g = 1.94 signal size. However if one assumes that about half of the change which is observed in the intact mitochondrial system can be seen in SMP [154, 161], only about 20% of the total g = 1.94 signal is expected to change. Thus, so far no definitive conclusion has been made on the apparent midpoint shift of center N-1a in the SMP system. The apparent midpoint shift of center N-2 was shown to be inhibited by rotenone, but not by antimycin A, suggesting that a dye interaction site for center N-1 is either at the UQ or at the cytochrome *b* levels. Gutman et al. [153] measured the apparent E_m shift of center N-2 in beef heart SMP using a physiological redox couple (succinate and fumarate). These investigators observed a positive shift of approximately 25 mV in the apparent $E_{m7.0}$ of center N-2. Although the E_m shift in their system is much smaller than the shift of about 125 mV described above, this quantitative difference can be rationalized by the use of different systems, intact pigeon heart mitochondria and beef heart SMP, because of their different

*For clarification, I want to point out that this information was cited in the Ref. 153 in two places, one as the oxidation (shift to a lower E_m value, page 55), and the other as the reduction (shift to a higher E_m value, page 58).

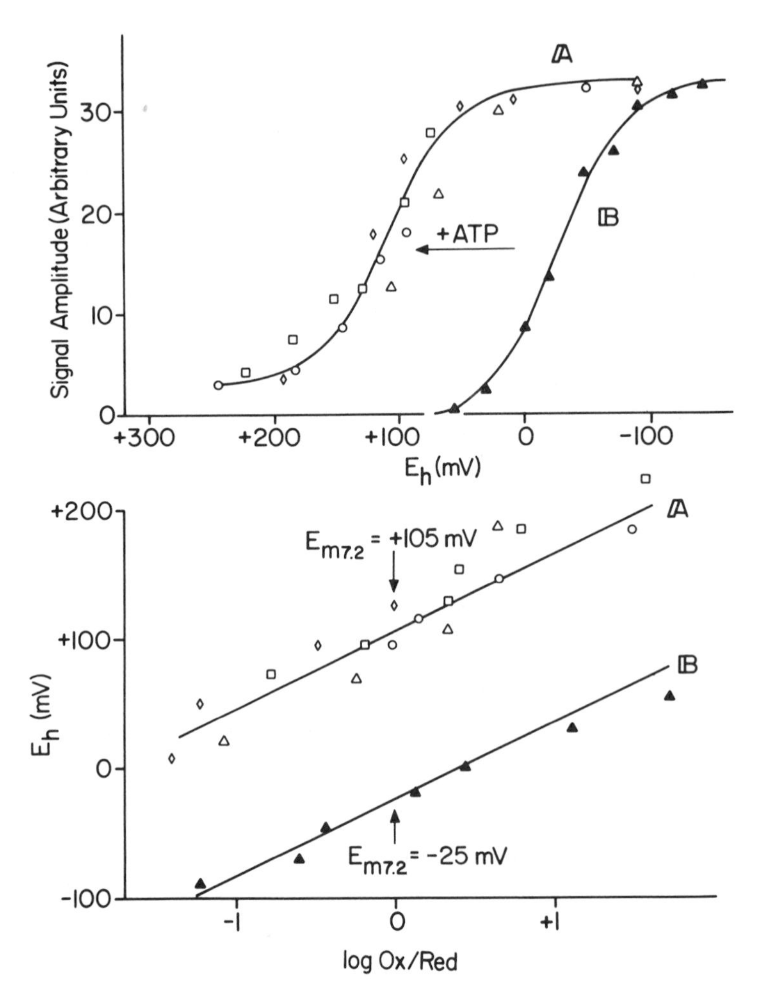

Figure 34 Redox titration of iron-sulfur center N-2 in the presence of added ATP. Titrations A and B were conducted with pigeon heart mitochondria (5.9 mg protein/ml) without and with prior treatment by carbonyl cyanide *p*-trifluoromethoxyphenyl hydrazone (0.5 nmol/mg protein), respectively. Oxidative (□, ◇) and reductive (○, △) titrations over a range of about 300 mV with coupled mitochondria were performed. Additions to the suspension were 8.3 mM ATP, 125 μM tetramethyl-*p*-phenylenediamine, 120 μM diaminodurol, 125 μM PMS, 125 μM phenazine ethosulfate, 75 μM duroquinone, 38 μM pyocyanine, 10 μM resorufin, and 50 μM 2-hydroxynaphthoquinone. The signal amplitude of center N-2 was obtained from the trough depth of the derivative spectrum measurement at g = 1.92 relative to the high magnetic field baseline. Reproduced with permission from Ohnishi [154].

inherent ATPase activity. Center N-2 responds to the actual phosphate potential of the system, but not directly to the added ATP concentration. Thus, in the author's view, experimental data from two laboratories on this point is, in principle, in good agreement. Indeed, ATP-dependent reduction of center N-2 was shown independently by the author when the suspension was poised on the higher redox potential side of center N-2 using succinate/fumarate as a redox buffer, in pigeon heart SMP inhibited by a high concentration of both antimycin A and KCN [154]. Additionally, Ohnishi repeated earlier experiments with beef heart SMP reported by Gutman et al. [150], using a physiological redox couple ($NADH/NAD^+$) which interacts at the low potential side of site I. Although the data obtained are different on a quantitative basis, as discussed in Ref. [154], oxidation of center N-2 upon ATP addition was confirmed when the redox potential of the system was suitably poised at the low-potential side of site I.

Assuming that a chemiosmotic mechanism is functioning, as illustrated in Figure 35, a phosphate potential (E_p) applied to site I in SMP is expressed as

$$E_p = E_2' - E_1' = -z\Delta pH + \Delta\Psi$$

using the expression of DeVault [157].

If we assume center N-2 is located at the M side of the inner mitochondrial membrane, as claimed by Gutman et al. [153], N-2 directly equilibrates with E_2' as illustrated in Figure 35. When the redox potential of the system is poised using the $NADH/NAD^+$ buffer at the E_h value of E_1', E_2' can be related to E_1' by

$$E_2' = E_1' - z\Delta pH + \Delta\Psi$$

Thus

$$\frac{(N\text{-}2)_{ox}}{(N\text{-}2)_{red}} = 10^{[E_1' - z\Delta pH + \Delta\Psi - E_{m(N\text{-}2)}]/60}$$

When $E_p = -z\Delta pH + \Delta\Psi$ is increased by added ATP, $(N\text{-}2)_{ox}$ should increase in agreement with experimental results. Even though $E_{m(N\text{-}2)}$ is pH dependent, center N-2 should still be oxidized. If E_2' is buffered using the succinate/fumarate couple, center N-2 should not change its redox level according to the chemiosmotic model of Figure 35. This prediction is not substantiated by the experimental results of either laboratory. Thus, use of a chemiosmotic model excludes the location of center N-2 on the M side of the mitochondrial inner membrane as shown in Figure 35, but is consistent with its location either in the middle of the membrane or on the C side, contrary to the conclusions reached by Singer and Gutman [152, 153]. The assignment of FMN as a transmembrane

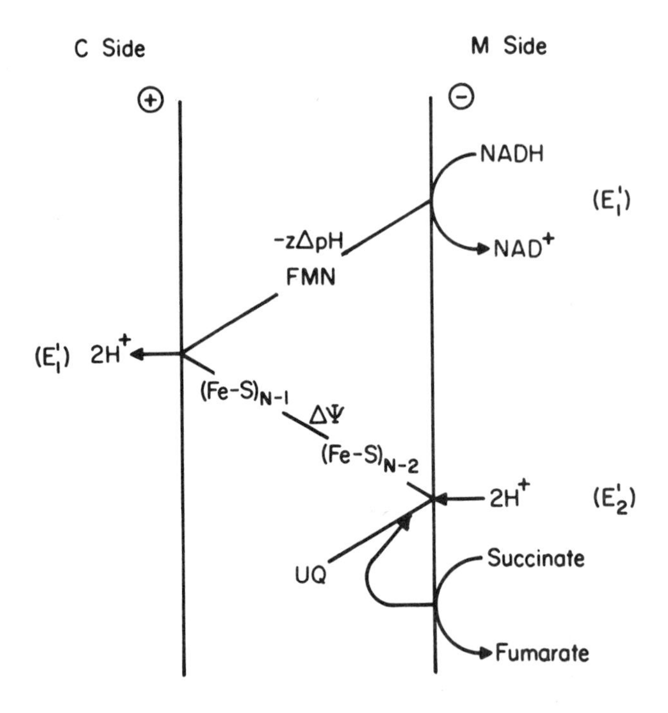

Figure 35 Diagrammatic arrangement of chemiosmotic "loop 1," after Mitchell and Garland.

hydrogen carrier as in this chemiosmotic loop is, however, very difficult to rationalize as will be discussed later (see Section III.F).

Regarding center N-1, Gutman et al. [153] reported the redox titration of an N-1 type center using an $NADH/NAD^+$ couple and obtained $E_{m7.0} = -344$ mV. Upon ATP addition, equilibrating at E_1' using a well-buffered $NADH/NAD^+$ couple (1:1 mixture each at the final concentration of 1 mM), an N-1 type center was reduced as presented in Figure 36. If we assume that the position of center N-1 is on the C side of the electron-carrying arm of the hypothetical site I loop, as proposed by Gutman et al. [153], Mitchell [149], and Garland et al. [148], the potential at center N-1 is related to E_1' as follows:

$$E_1 = E_1' - z\Delta pH$$

Thus

$$\frac{(N\text{-}1)_{ox}}{(N\text{-}1)_{red}} = 10^{[E_1' - z\Delta pH\ E_{m(N\text{-}1)}]/60}$$

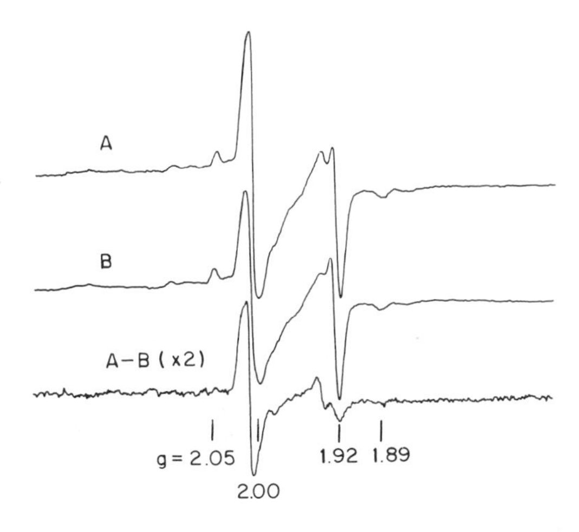

Figure 36 EPR spectra of energized and uncoupled ETP_H particles, inhibited by piericidin A, in the presence of equimolar concentrations of NAD and NADH. The membrane preparation was suspended and piericidin-treated, then mixed with the STM-BSA buffer, containing 1 mM each of NAD and NADH and 6 mM KCN, and frozen within 1 sec. In curve A, the NAD-NADH solution also contained 8 mM ATP; in curve B, the membrane preparation contained 20 μM CCP, the bottom curve is a difference spectrum (A − B), at twice the instrument sensitivity. EPR conditions: power, 0.9 mW; modulation amplitude, 6.3 gauss; modulation frequency, 100 KHz; temperature, 12.2 K; scanning rate, 400 gauss/min; time constant, 0.25 sec; microwave frequency, 9.2 GHz. Average of four scans, using a Nicolet signal averager. Reproduced with permission from Gutman et al. [153].

If the $E_{m(N\text{-}1)}$ value is independent of pH, addition of ATP will cause oxidation of center N-1. If its midpoint potential is pH dependent, the $-z\Delta pH$ effect will be cancelled out; in this case its redox state should remain unchanged upon ATP addition. Thus, in either case, proper analysis of their data based on the chemiosmotic model excludes the location of center N-1 on the C side of mitochondrial inner membrane, again contrary to the interpretation by Gutman et al. [153].

As can be seen from Figure 36, the redox state of center N-1 is based on a very small change in the lineshape at the top of the central $g = 1.94$ signal, using a computer-simulated spectral fit [130]. As discussed earlier (Section III.B), a correlation between redox level of center N-1 and lineshape was made before the contribution of centers N-3 and N-4 to the $g = 1.94$-1.92 signal was realized [136, 137] in addition to that of type N-1. Reduction of some component(s) among these iron-sulfur centers predicts (both from the chemiosmotic and transductase viewpoints) that at least one iron-sulfur center is functioning on the low-potential side of the NADH/NAD^+ couple, probably on loop 0 [162, 163].* In addition as discussed before, identification of this N-1 type center is another important problem; whether this center is distinct from centers N-1a or N-1b or equivalent to one of them under spin-coupled conditions is not yet established. In spite of the suggestion of the involvement of iron-sulfur center(s) in the loop 0, their observation can not completely exclude the following trivial cause for the reduction of N-1 type centers by ATP addition. If "uncoupler-insensitive" inhibition [167] of the NADH dehydrogenase (probably between NADH and FMN) by ATP is significantly effective under the experimental conditions, the site I loop becomes an isolated system blocked on both low- and high-potential sides. Thus ATP would cause internal electron transfer from N-2 to N-1, resulting in N-1 reduction which does not occur in the presence of uncoupler.

As described in the preceding section, center N-1a exhibits a pH-dependent apparent midpoint potential; thus partial oxidation of center N-1a (at most a 60 mV shift of apparent E_m to a lower value) is explicable from the chemiosmotic view by assigning its location somewhat inside the cytosolic surface in the electron-carrying arm of the site I loop.

Various transductase models described by DeVault [157, 158] can also be applied to the energy coupling at site I and can explain the reported redox behaviors of centers N-2 and N-1a as transducer components. Again the ATP induced reduction of center N-1 reported by Gutman et al. [153] excludes the involvement of this center in energy transductase at site I.

The transductase models of DeVault are based on the following assumptions: an energy-transducing site contains at least a single electron carrier with two molecular forms having low and high midpoint potentials $E_\ell{}^0$ and $E_h{}^0$, respectively), for example, under two different molecular conformations. These two forms cannot be distinguished spectrophotometrically and they are in equilibrium with isopotential pools at the low- and high-potential side of the site (E_1 and E_2, respectively). The potentials E_1 and E_2 differ by a phosphate potential E_p, and the potential difference across the transductase $E_2 - E_1 = E_p$ is used to synthesize ATP. The midpoint potential shift upon addition of ATP is apparent, because of the contribution from both forms of the energy-transducing component(s). The differences among the various transducer models used by De Vault et al. are mainly in the steps at which conversion to the high-energy form and release of the chemical intermediate occur.

The general trend of their reasoning will be illustrated using one simple

*This suggestion does not come as a great surprise because complex I is a heterogeneous multisubunit system and in fact Ragan and Widger reported that an ATP-dependent transhydrogenase reaction (reduction by $NADP^+$ by NADH) was observed in liposome reconstituted with complex I, hydrophobic protein, and F_1 [164]. Some experimental results, however, have indicated noninvolvement of iron-sulfur centers in "loop 0" [165, 167].

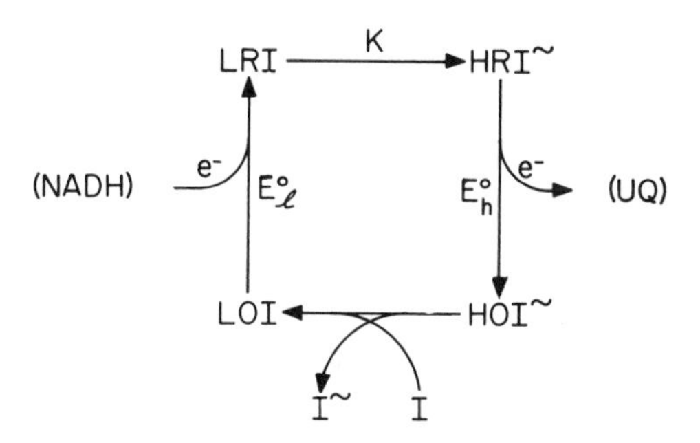

Figure 37 One example of the abridged transductase model of site I described by DeVault [157]. L and H represent the low- and high-potential forms; R and O, the reduced and oxidized forms; and I and $I^{\sim}$, the low- and high-energy forms of the intermediate. $E_{\ell}{}^{0}$ and $E_{h}{}^{0}$ are the midpoint potentials of the low- and high-energy forms, respectively.

case, model 1 in Ref. 157 (see Fig. 37). Equilibrium between the high- and low-potential reduced forms is given as $K = (HRI^{\sim})/(LRI)$. From the thermodynamic relationship illustrated in Figure 37, we can write Nernst's equations for two molecular forms:

$$E_1 = E_{\ell}{}^{0} + \frac{RT}{F} \ln \frac{(LOI)}{(LRI)}$$

$$E_2 = E_{h}{}^{0} + \frac{RT}{F} \ln \frac{(HOI^{\sim})}{(HRI^{\sim})}$$

where R is the gas constant, T is the absolute temperature, and F is the Faraday constant.

We define B_1 and B_2 as the concentration ratios of the oxidized to the reduced states of low and high midpoint potential forms, namely, (LOI)/(LRI) and $(HOI^{\sim})/(HRI^{\sim})$, respectively. Then the total concentration of these four redox forms is given as $(LOI)(1 + B_1 + K + KB_2)/B_1$. The fractional populations of the transductase in the four states are

$$f\,(LRI) = \frac{I}{D}$$

$$f\,(LOI) = \frac{B_1}{D}$$

$$f(\text{HRI}^{\sim}) = \frac{K}{D}$$

$$f(\text{HOI}^{\sim}) = \frac{B_2 K}{D}$$

where $D = 1 + B_1 + K + KB_2$.

If the high- and low-potential forms are spectrophotometrically indistinguishable, observation of the fraction of transductase in the oxidized and reduced forms will be $O = f(\text{LOI}) + f(\text{HOI}^{\sim})$ and $R = f(\text{LRI}) + f(\text{HRI}^{\sim})$. Using the fractional forms,

$$\frac{O}{R} = B_1 \frac{1 + KB_2/B_1}{1 + K} = B_1 \frac{1 + K \exp[(E_p - E_h{}^0 + E_\ell{}^0) F/RT]}{1 + K}$$

we can see that if E_1 is clamped, increase of phosphate potential E_p leads to the oxidation of the energy-transducing electron carrier. If O/R is rewritten in terms of B_2,

$$\frac{O}{R} = B_2 \frac{B_1/B_2 + K}{1 + K} = B_2 \frac{K + \exp[(E_h{}^0 - E_p - E_\ell{}^0) F/RT]}{1 + K}$$

It is evident that clamping of E_2 leads to the opposite effect, namely, reduction of the electron carrier upon addition of ATP. The other transductase models make qualitatively similar predictions.

Addition of ATP to SMP causes either oxidation or reduction of center N-2, depending on whether E_1 is clamped using $NADH/NAD^+$ redox buffer [150, 153] or E_2 is clamped using succinate/fumarate buffer or potentiometrically in the presence of redox dyes (Figs. 33 and 34). The intriguing redox behavior or center N-2 identifies it as the site I energy transducer or at least part of it.

According to this model, oxidation of center N-1a presented in Figure 33 (poised at -424 mV which corresponds to the $NADH/NAD^+$ pool) suggests that center N-1a is also a part of the site I energy transducing system. In order to accommodate the results of Gutman et al. [153] the center which was reduced on ATP addition must again be located in "site 0."

Regarding the identification of energy-transducing components in site I, Hatefi and his colleagues claimed that involvement of an N-1 type center is excluded by the observed phosphorylation efficiency i.e., P/O ratio of 3 in NADPH oxidation systems where center N-1 is not involved even as an electron carrier [13, 157, 169]. However, the experimental basis of their contention was not substantiated by their subsequent examination [111, 166] and was vulnerable to objections from other laboratories [112, 164].

In order to discriminate between these explanations based on chemiosmotic or chemical intermediate views that cannot be ruled out at the present stage, further quantitative examination of the predicted experiments has been proposed [157].

Concerning the topological distribution of the redox components in the mitochondrial inner membrane, Klingenberg and Bucholz [101] reported that membrane-impermeable $K_3Fe(CN)_6$ can interact with the NADH dehydrogenase from the M side but not from the C side on the substrate side of the cytochrome *c* level. Therefore it seems rather unlikely that any respiratory chain carrier in the NADH-UQ segment is located on the C side and can be directly in contact with the ambient medium outside of intact mitochondria. There is enough experimental evidence, however, to demonstrate that proton transfer actually occurs coupled with electron transfer from NADH to UQ, with $H^+/2e^-$ ratios of 2 in mitochondria and in SMP [148, 149], and of less than 2 in the reconstituted lipid vesicles with properly incorporated complex I [172].*

Thus an alternative scheme was proposed by Skulachev (see Fig. 38) in which all the redox components in the NADH-UQ segment are located on the matrix side of the mitochondrial inner membrane and proton translocation is mediated by an H^+-translocase (a specific proton channel or pore) analogous to the proton pump of the oligomycin-sensitive ATP synthesizing system [173, 174]. Skulachev also proposed as a simple model a proton pump in which reduction of a redox component changes the pK of one of the protolytic group of the protein; thus this group accepts H^+ from the matrix. This protonated carrier is oxidized by the next respiratory carrier only after a conformational change which accompanies transfer of protolytic groups through the hydrophobic barrier. Oxidation of the pump protein induces a pK change back to the initial value and hence releases H^+ to the outside of the mitochondrion. This proton pump model can be analyzed in terms of the transductase model of DeVault: the high-energy intermediate $I^{\sim}$ is replaced by an external proton and I is replaced by an internal proton. Related proposals include that of H^+ transfer based on the "membrane Bohr effect" of Chance et al. [175] as applied to site II energy coupling by Papa [176] or to site I by Slater [177]. Here we identify component(s) exhibiting a Bohr effect with electron transfer components which exhibit pH-dependent midpoint potentials. Thus, centers N-1a and N-2 are promising candidates for Bohr components in site I.

Although it is still very tentative, Ragan isolated a hydrophobic subunit (molecular weight 33,000) in complex I, which may play a role as a proton

*Recent reports of $H^+/2e^-$ ratio of 3-4 per site may also apply to site I [170, 171].

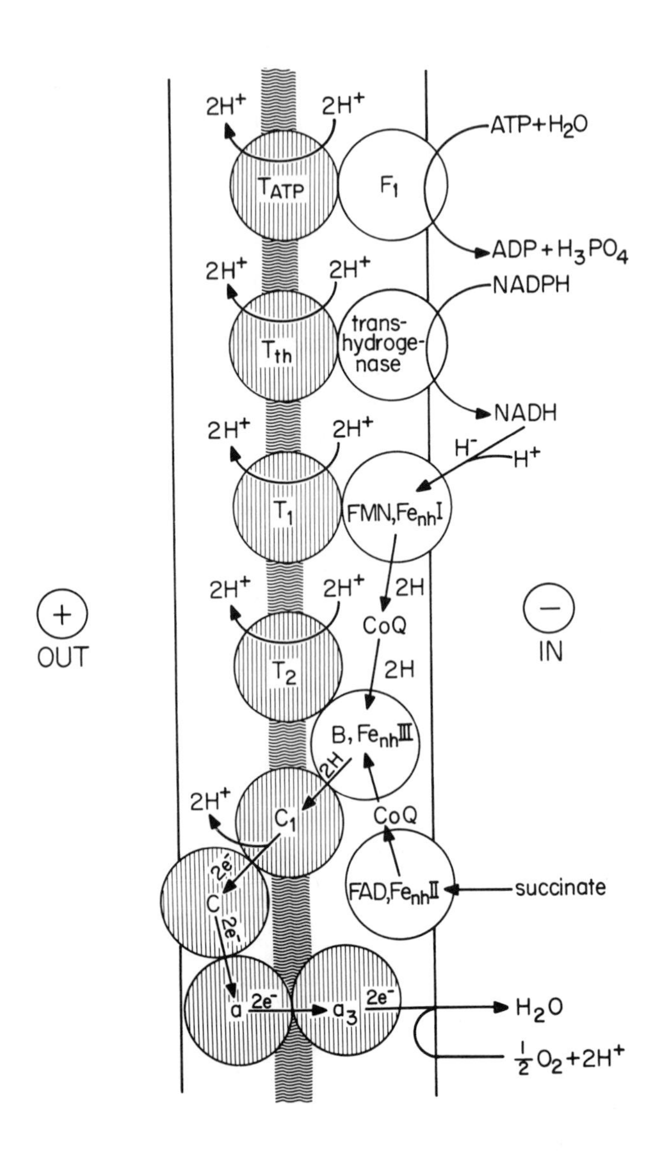

Figure 38 The general scheme of the membrane potential generators in mitochondrial membrane according to Skulachev [162]. The H^+ translocating components are designated as T_{ATP} (ATPase), T_{th} (transhydrogenase), T_1 (1st coupling site), and T_2 (2nd coupling site); Fe_{nh}, nonheme iron.

channel in site I energy coupling as will be discussed later. Very recently Mitchell also proposed the "indirect chemiosmotic" model where his original site I loop model was revised to a foreshortened non-transmembraneous loop shifted toward the M side of the inner mitochondrial membrane [178, 179]. Here each H^+ produced by a "local site I loop" is transferred to the C side via an H^+ channel, taking the above-described observations into consideration.

In order to formulate the mechanism of site I energy transduction, the most essential and missing experimental information at present is the detailed topology of all the redox components of the site I segment in the inner mitochondrial membrane and the actual electron and H^+ transfer mechanism in this segment of the respiratory chain.

F. Spatial Relationships Among Redox Components in NADH-UQ Segment

NADH-UQ reductase (complex I) is probably the largest component of the inner mitochondrial membrane. As already discussed in the preceding sections in this chapter, SDS gel electrophoresis of complex I showed that at least 16 subunits [114] are present in this segment of the respiratory chain. Chaotropic treatment of complex I resolved it into hydrophilic subunits in the soluble form [123, 124]. The hydrophilic fraction contains 20% of the protein and 50% of nonheme iron and acid-labile sulfur present in complex I, i.e., this part contains a fourfold higher concentration of Fe and S per mg protein than the hydrophobic fraction. The hydrophilic part consists of iron-sulfur flavoprotein (Fp) containing 1 FMN and 4Fe 4S, and iron-sulfur protein (Ip) containing Fe and S in 1:1 ratio.

The Fp can be resolved into two subunits with molecular weights of 53,000 and 26,000 and the Ip into three major subunits with molecular weights of 75,000, 53,000, and 29,000 and traces of other low molecular weight subunits. Surface labeling of complex I with ^{125}I with lactoperoxidase [114] according to Phillips and Morrison [180] indicated that the water-soluble polypeptides including Fp and Ip were largely buried in complex I. Hydrophobic proteins were extensively labeled with ^{125}I, suggesting that they are exposed to the ambient medium. The presence of a hydrophilic cleft in complex I was thus speculated to allow a hydrophilic substrate, NADH, to interact with FMN in the Fp subunit [114].

Ragan also examined the effects of gradual proteolytic treatment using various enzymes and correlated polypeptide degradation and modification of various enzymatic activities of complex I. For example, transhydrogenase activity of complex I was rapidly inactivated by trypsin with the loss of a polypeptide of molecular weight of 20,500 [112]. Orme-Johnson et al. [130] reported earlier that the Fp subunit gives EPR signals arising from at least three species of iron-sulfur centers, which was suggested from the power and temperature dependence of individual lines (See Fig. 39, spectrum B). More recently, Widger and King isolated a type II NADH dehydrogenase which contains 4Fe 4S and gives EPR spectra from two species of iron-sulfur center [142]. EPR studies conducted in the author's laboratory on this enzyme indicated that one iron-sulfur species is more labile than the other, and the measured J value and spectra distortion parameter indicate that the latter species is binuclear. Thus the Fp subunit in the NADH-UQ segment seems to be analogous to the Fp subunit of SDH, containing one flavin and two binuclear iron-sulfur centers. In the type II dehydrogenase of Widger and King [142], both centers can be reduced similarly either with NADH or with dithionite, although the spin concentration of the EPR spectra accounts for only about 5% of the chemically measured content of nonheme iron and acid-labile sulfur. Salerno et al. [63], from the J value of 90 cm^{-1} and the spectra distortion parameter, indicated that center N-1a examined in complex is binuclear like center S-1. Albracht and Subramanian also reported that center N-1 has a binuclear structure but all other iron-sulfur centers in the NADH-UQ reductase segment have a tetranuclear structure, based on the analysis of EPR lineshape of these iron-sulfur centers in the SMP preparation of *C. utilis* cells grown in ^{57}Fe culture medium [65]. Thus, most probably centers N-1a and N-1b or another N-1 type center are two binuclear clusters in the Fp subunit.

Orme-Johnson et al. [130] reported that the Ip subunit gives a spectrum arising largely from a single species, similar to a somewhat broadened center N-2 spectra (see Fig. 39, spectrum C). Cammack recently pointed out, however, that upon removing the constraints of the apoprotein around the iron-sulfur cluster (for example, iron-sulfur proteins in the presence of a high concentration of DMSO), all tetranuclear iron-sulfur centers exhibit a common EPR spectra of axial symmetry ($g_{/\!/} = 2.05$, $g_{\perp} = 1.93$) with a lineshape very similar to this Ip spectrum [51]. Thus it is impossible to assign the Ip spectrum to center N-2 without any additional information, but it can only be ascribed to at least one species of tetranuclear iron-sulfur center.

The hydrophobic fraction appears to contain tetranuclear iron-sulfur center(s). The dramatic lineshape and midpoint potential change in center N-2 upon depletion of phospholipid from complex I, as discussed in Section III.C suggest the possibility that center N-2 is located close to the binding site of complex I to the mitochondrial inner membrane and is the component close to the H^+ channel. Ragan also reported that the hydrophobic subunit with molecular weight of 33,000 seen in complex I or in the hydrophobic fraction of chaotic resolution (see Fig. 19a,b) is iodinated only in the isolated complex I, but it is not iodinated

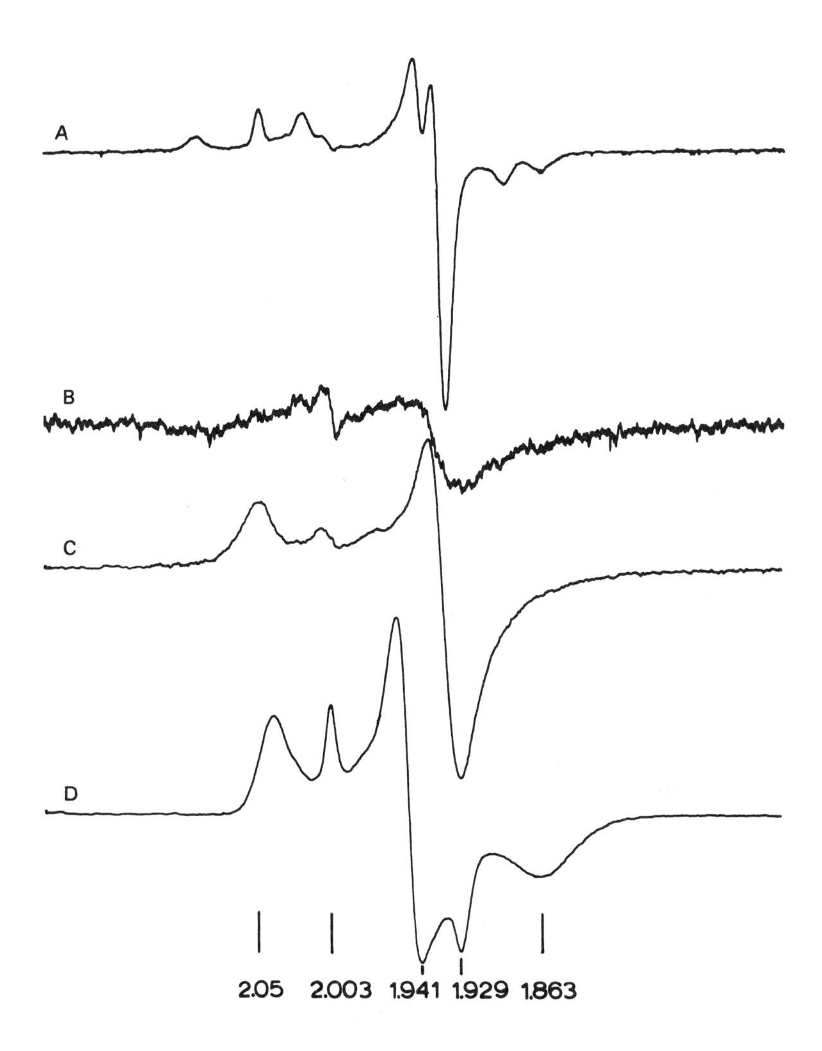

Figure 39 ERP spectra of complex I and of proteins obtained by treatment of complex I according to Orme-Johnson et al. [130]. A, Complex I reduced with an excess of DPNH. B, Sediment obtained by low-speed centrifugation of complex I that had been treated with 2.05 M urea for 10 min at 37°C. C and D, Ammonium sulfate fractions of the supernatant from B obtained at 27.5% and 36.4–52.9% saturation, respectively. The protein concentrations were 13.0, 19.8, and 10.1 mg/ml for B, C, and D, respectively. The samples for B, C, and D were reduced with an excess of solid dithionite.

when complex I is incorporated into phospholipid vesicles. Thus he speculated that this 33,000-dalton subunit may play a role as the proton channel protein in the site I segment [114]. These interesting observations, however, offer only circumstantial or complementary information on the actual spatial distribution of redox components of the site I segment in the mitochondrial membrane. Direct information on the distance and orientation of flavin and iron-sulfur centers in situ relative to neighboring electron carriers and to the inner mitochondrial membrane is essential in order to define the mechanism of energy transduction and electron transfer in the site I segment. Two different strategies seem to be promising: one is to examine spin-spin interaction between redox components, and the other is to make use of spin-spin interactions between respiratory components and externally added membrane-impermeable paramagentic probes. The former approach was described in detail for the assignment of the spatial distribution of redox components in Section II; the latter was devised by Case and Leigh [181] and has been applied to studies of localization of cytochromes, UQ, and iron-sulfur centers in mitochondrial [181, 182] and photosynthetic systems [183].

Spin-spin interaction between intrinsic redox components in the NADH-UQ segment was first detected between center N-3 and (most probably) flavin [84, 88]. Center N-3 exhibits a rhombic EPR spectrum with g values of g_z = 2.035, g_y = 1.93-1.92, and g_x = 1.86. The potentiometric titration of the g = 1.86 peak amplitude gives an n = 1 curve with a midpoint potential of approximately -240 mV, as shown in dashed lines in Figure 40. Further lowering of the potential (E_h) gives an interesting anomaly; the signal decreases in amplitude, reaches a minimum at about -380 mV, and then increases again to its maximum value. In contrast to the center N-3 titration, the N-4 titration measured with either g_x = 1.88 or g_z = 2.10 peak amplitudes gives rise to a typical n = 1 curve with the same midpoint potential (Fig. 40, solid line). This anomaly is also seen in the lineshape change in the g_x region as illustrated in spectra A, B, and C on the right of Figure 40.

Concomitantly, a g = 3.93 signal in the half-field with low anisotropy appears as the potential is lowered, reaches a maximum at -380 mV, and disappears with further lowering of the potential (see Fig. 41).

The transient nature of these phenomena suggest the possible involvement of an n = 2 electron carrier with a paramagnetic intermediate redox state. An attractive and plausible candidate for the interaction partner of center N-3 is FMN, although the possibility of two strongly coupled iron-sulfur centers cannot be excluded. Preliminary results [145] from my laboratory indicate that the potential of the dip point of the center N-3 titration curve and of the maximum of the half-field titration curve are pH dependent (60 mV per pH unit), as anticipated for FMN. Although not yet well established, this approach would esti-

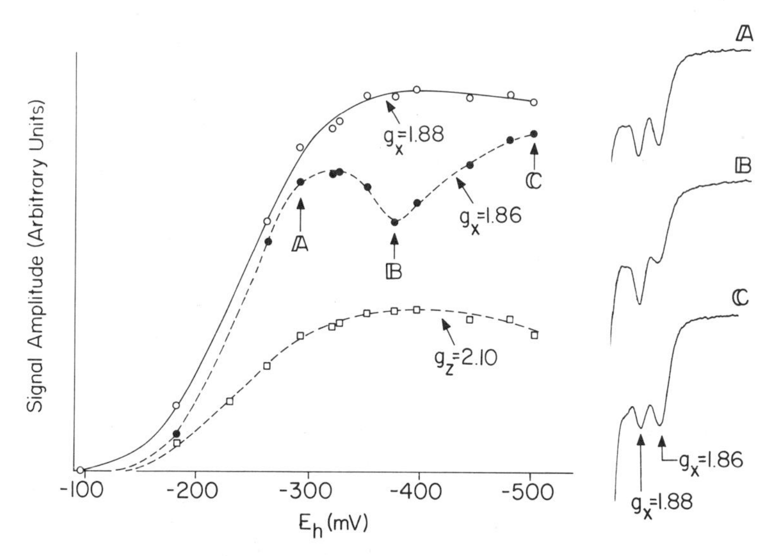

Figure 40 Potentiometric titration of centers N-3 and N-4, using the g_X = 1.86 signal amplitude for N-3 and g_X = 1.88 or g_Z = 2.10 for N-4. The lineshapes in the g_X range at A)–292 mV), B (–377), and C)–503) are shown on the right. EPR recording conditions were the same as in Figure 24 (III).

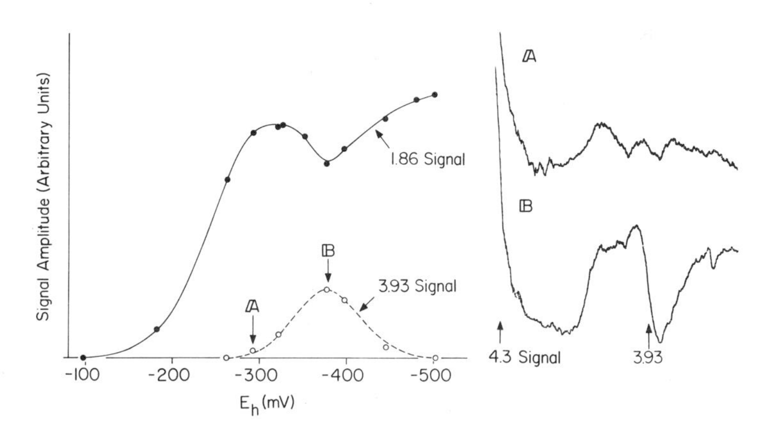

Figure 41 Potentiometric titration of the half-field signal (Δm_s = 2 transition) at g = 3.93 and titration curve of the g = 1.86 signal of center N-3. EPR spectra of the half-field signal of samples A and B are shown on the right of titration curves. EPR conditions: microwave power, 20 mW; modulation amplitude, 12.5 gauss; sample temperature, 7.4 K.

mate the distance between center N-3 and its interacting partner as about 10 Å or less. This approach will be extended to interactions between other components, as exemplified with SDH components.

Introduction of a paramagnetic species capable of close approach to an electron carrier in the mitochondrial membrane will cause magnetic dipole-dipole interactions which can result in perturbation of the EPR spectrum of the electron carriers; dipole interaction can appear as an EPR line "splitting" (or shift), as an EPR line broadening, or as relief of power saturations, depending on the strength of the interaction, the relaxation rate and magnetic moment of the probe, and the linewidth of the EPR spectrum of the electron carrier. Paramagnetic probes should be inert and should not exhibit EPR spectra which interfere with spectra of the electron carriers; in addition they should not be transported through the mitochondrial osmotic barrier before freezing the sample for EPR examination. To meet these qualifications, Gd^{3+} and Ni^{2+} have been used so far as paramagnetic probes [181-183].

Depending on the relaxation rate of the probe, splitting or broadening of the signal by interaction with the probe can be observed. If the strength of the interaction (in Hz) is much greater than $1/T_1$ (where T_1 is the spin-lattice relaxation time), well-resolved splitting is observed which is proportional to r^{-3} assuming dipolar coupling (where r is the distance between the interacting probe and redox component). At the other extreme, a very rapidly relaxing probe should broaden the signals of nearby electron carriers with an r^{-6} dependence, again assuming a dipolar mechanism. Variation of the relaxation rate through the intermediate region gives rise to transition from broadening. One can also expect the probe to increase the relaxation rate (decrease the relaxation time) of the electron carrier if it is in close proximity to it. The effect on the EPR spectrum would be to render the power saturation more difficult; dipolar relaxation processes follow r^{-6} [184, 185]. This approach has been applied successfully to the topographical studies of electron carriers in the photosynthetic chromatophores [183] and to cytochrome hemes in mitochondria [181]. It has also been applied to iron-sulfate centers (Fd-type and HiPIP-type) [182]. The EPR spectra of Ruzicka's iron-sulfur center became more difficult to saturate upon treatment with Ni in the intact mitochondria. This indicates that the location of Ruzicka's center is on the C side of the mitochondrial membrane in agreement with the results obtained with the bathophenanthroline sulfonate method as described in the preceding section [48]. Neither Ni^{2+} nor Gd^{3+} gave a significant effect on the spectra of other iron-sulfur centers either in mitochondria or in SMP. However, extension of this approach to different paramagnetic probes or partially modified membrane systems may enable us to estimate the distances of electron carriers from the exterior surface of the membrane.

IV. ETF Dehydrogenase

The Fd-type EPR signal with g_z = 2.08 and g_x = 1.89 was recognized first by Ohnishi et al. at temperatures below 40 K in pigeon heart mitochondria or SMP and its midpoint potential was determined to be approximately +40 mV [132]. It was tentatively designated as center 5, simply because at that time numbers 1-4 were used to name centers on the substrate side of this center. Because of its midpoint potential value and the detection of the signal in the cytochrome bc_1 complex prepared in King's laboratory, its location in the cytochrome bc_1 region in the respiratory chain was suggested [126]. It was found, however, that EPR signals from Rieske's center, but not from center 5, were detectable in pure preparations of bc_1 complex III. Thus for some time, the location and function of this center remained unknown.

Subsequently, Ruzicka and Beinert purified an iron-sulfur flavoprotein from beef heart mitochondria, following the EPR signals through the purification steps [186]. It is a membrane-bound iron-sulfur flavoprotein and was solubilized by K-cholate and purified by ammonium sulfate fractionation and DEAE-Sepharose chromatography. Although it is very difficult to isolate a water-insoluble enzyme to a high degree of purity, they were able to obtain the enzyme about 25-30% pure. The molecular weight is approximately 70,000; it contains 1 FAD, and 4-5 atoms each of nonheme iron and acid-labile sulfur. Acid treatment of the enzyme released FAD, Fe, and S in the ratio of 1:4:4. As presented in Figure 42, this enzyme gives an EPR spectrum of rhombic symmetry at g_z = 2.086, g_y = 1.939, and g_x = 1.886 and a flavin free radical signal at g = 2. Spin quantitation of the iron-sulfur signal showed approximately one spin per flavin. Thus the enzyme was proposed to contain a single tetranuclear iron-sulfur center [186].

The iron-sulfur group is reduced with a half-time of 5 msec when the protein is mixed with an equivalent amount of ETF of the β-oxidation cycle, prereduced with an acyl-CoA dehydrogenase and a saturated fatty acyl CoA. Omission of any component of this hydrogen-donating system prevents significant reduction of the iron-sulfur group. Neither iron-sulfur nor flavin is reduced by NADH, NADPH, succinate, glycerol-3-phosphate, or dihydroorotate. Since its function in the respiratory chain is now known, the name of this iron-sulfur center should be revised to ETF dehydrogenase iron-sulfur center, abbreviated as $(Fe\text{-}S)_{ETF}$ as presented in Scheme I (in Section I).

Reductive titration of the flavin free radical signal gives a typical bell-shaped curve, with the maximum free radical size about 0.3 equivalent of the total flavin, indicating a rather stable free radical intermediate state in this enzyme [186, 187].

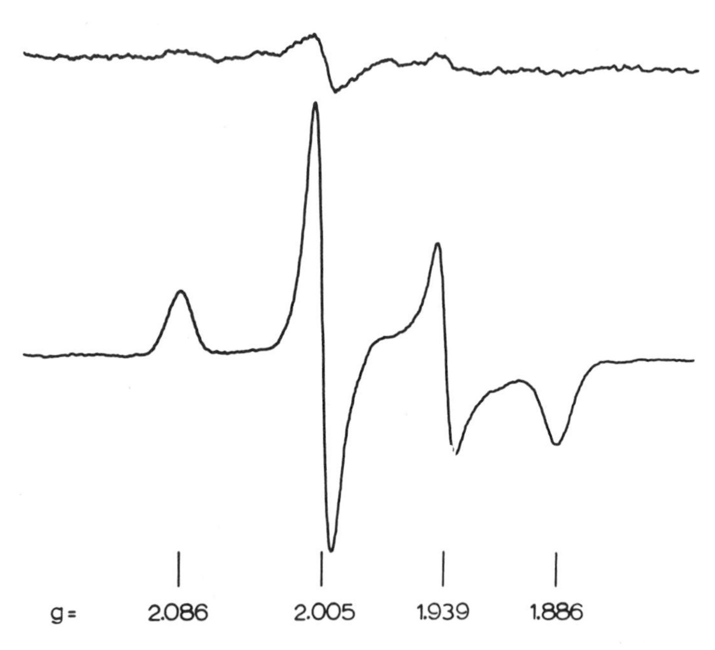

Figure 42 EPR spectrum (first derivative) of iron-sulfur flavoprotein from mitochondria. The protein, 5 mg/ml, was dissolved in 10 mM Tris-HCl (pH 8). Upper trace, protein as isolated; lower trace, reduced with solid dithionite. EPR operating conditions: microwave power, 270 mW; modulation amplitude and frequency, 8 gauss and 9.2 GHz, respectively; temperature 13 K; scanning rate 400 gauss/min. The radical signal at g = 2 is strongly saturated under these conditions. Reproduced with permission from Ruzicka and Beinert [186].

The iron-sulfur flavo enzyme $ETFH_2$-UQ reductase appears to function as a link between the fatty acyl dehydrogenation pathway (n = 2 transfer steps) and the (n = 1) electron transfer system.

V. Conclusion

This chapter has described the author's present view on some functional and topological features of iron-sulfur centers and other components associated with mitochondrial iron-sulfur flavoproteins. From various independent evidence it can be said that SDH contains two binuclear and one tetranuclear, NADH dehydrogenase two (or three) binuclear and three (or four) tetranuclear, and $ETFH_2$ dehydrogenase one tetranuclear iron-sulfur center(s), and that all contain in addition one flavin component. All these three dehydrogenase systems function to transfer electrons from substrate to UQ via flavin and iron-sulfur centers. It would be nice to be able to resolve a specific sequence of electron transfer but, unfortunately, to date the rapid freeze-quenching technique is not able to resolve any intramolecular electron transfer kinetic sequences because of rapid intramolecular redox equilibration.

It is very interesting that only two iron-sulfur centers (centers N-1a and N-2) exhibit phosphate potential dependent and pH-dependent apparent midpoint potentials, which distinguish them from all other centers in the site I segment. Analysis of the reported redox behavior of these two centers suggests their possible involvement in energy transduction at site I, either from chemiosmotic or energy transductase models. Proton pump mechanisms seem to be more plausible models, taking into account the suggested topological distribution of the redox carriers in the site I segment of the respiratory chain.

Structural studies using spin-spin interaction phenomena among respiratory carriers and added probes seem to be a promising avenue for further progress.

VI. Recent Developments

Ruzicka and Beinert [188] purified the soluble HiPIP-type iron-sulfur protein from beef heart mitochondria to homogeneity and showed that this protein exhibits aconitase activity when activated by reducing agents in the presence of ferrous iron. This HiPIP-type iron-sulfur protein and aconitase show identical light absorption, EPR spectra, isoelectric point, and are inseparable by SDS-polyacrylamide electrophoresis. This is the first example of mitochondrial iron-sulfur protein functioning as a non-electron-carrier.

Transhydrogenase was also purified to homogeneity by Höjeberg and Rydström [189]. The enzyme was shown to be a single polypeptide with molecular weight of 97,000 dalton which is devoid of flavin and any other respiratory redox activities. Using transhydrogenase reconstituted with synthetic liposomes, they demonstrated uncoupler sensitive reduction of NAD^+ by NADH and uptake of lipophilic anions [189]. Rydström et al. [190] provided experimental evidence showing that rotenone-sensitive NADPH oxidation in beef heart submitochondrial particles is via the same pathway as NADH oxidation.

Acknowledgments

I would like to express my deep gratitude to Dr. B. Chance for his continuing interest and encouragement, and to my colleague Dr. J. C. Salerno for his penetrating discussions and critical reading of the manuscript. I also would like to thank Dr. T. E. King and his group who permitted me to cite our unpublished data. Thoughtful comments by other colleagues in the department, especially Drs. W. J. Ingledew, P. R. Rich, H. Blum, D. DeVault, and J. S. Leigh, are also gratefully acknowledged. I am indebted to Drs. H. Beinert and R. Sands for their stimulating criticism and helpful advice to our EPR studies on iron-sulfur proteins. Preparation of this chapter was supported by Grants GM 12202-14 from the National Institutes of Health and PCM 75-13459 A02 from the National Science Foundation.

References

1. B. Mackler, R. Repaske, P. M. Kohout, D. E. Green, *Biochim. Biophys. Acta 15:* 437 (1954).
2. F. L. Crane, J. L. Glenn, and D. E. Green, *Biochim. Biophys. Acta 22:* 475 (1956).
3. T. E. King, *Adv. Enzymol. 28:* 155 (1966).
4. H. Beinert and R. H. Sands, *Biochem. Biophys. Res. Commun. 3:* 41 (1960).
5. T. Ohnishi and B. Chance, in *Flavins and Flavoproteins* (H. Kamin, ed.), University Park Press, Baltimore, Md., 1971, pp. 681-693.
6. T. Ohnishi, T. Asakura, H. Wohlrab, T. Yonetani, and B. Chance, *J. Biol. Chem. 245:* 901 (1970).
7. T. Ohnishi, D. F. Wilson, T. Asakura, and B. Chance, *Biochem. Biophys. Res. Commun. 49:* 1087 (1972).
8. F. J. Ruzicka and H. Beinert, *Biochem. Biophys. Res. Commun. 66:* 622 (1974).
9. W. Lovenberg (ed.), *Iron-Sulfur Proteins,* Vols. 1 and 2 (1973), Vol. 3 (1977), Academic, New York.
10. H. Kamin (ed.), *Flavins and Flavoproteins,* University Park Press, Baltimore, Md., 1971.
11. T. Singer (ed.), *Flavins and Flavoproteins,* Elsevier, Amsterdam, 1976.
12. T. E. King, H. S. Mason, and M. Morrison (eds.), *Oxidases and Related Redox Systems,* University Park Press, Baltimore, Md., 1973.
13. Y. Hatefi and D. L. Stiggall, in *The Enzyme* 3rd Ed. (P. Boyer, ed.), Academic, New York, 1977, pp. 175-295.
14. R. K. Morton, *Nature 166:* 1092 (1950).
15. T. P. Singer, E. B. Kearney, and P. Bernath, *J. Biol. Chem. 223:* 599 (1956).
16. T. Y. Wang, C. L. Tsou, and Y. L. Wang, *Sci. Sinica 5:* 73 (1956).
17. D. Keilin and T. E. King, *Nature 181:* 1520 (1958).
18. T. E. King, *J. Biol. Chem. 238:* 4032 (1963).

19. P. Hemmerich, A. Ehrenberg, W. H. Walker, L. E. G. Erikson, J. Salach, P. Bader, and T. P. Singer, *FEBS Lett. 3:* 37 (1969).
20. K. A. Davis and Y. Hatefi, *Biochemistry 10:* 2509 (1971).
21. Y. Hatefi, A. G. Haavik, and D. E. Griffiths, *J. Biol. Chem. 237:* 1676 (1962).
22. D. M. Ziegler and K. A. Doeg, *Arch. Biochem. Biophys. 97:* 41 (1962).
23. G. Zanetti, P. G. Righetti, and P. Cerletti, in *Biochemistry and Biophysics of Mitochondrial Membrane* (G. F. Azzone, E. Carafoli, and A. L. Lehninger, eds.), Academic, New York, 1972, pp. 33-39.
24. C. J. Coles, H. D. Tisdalf, W. C. Kenny, and T. P. Singer, *Physiol. Chem. Phys. 4:* 301 (1972).
25. D. B. Winter and T. E. King, *Biochem. Biophys. Res. Commun. 56:* 290 (1974).
26. T. Ohnishi, D. B. Winter, J. Lim, and T. E. King, *Biochem. Biophys. Res. Commun. 53:* 231 (1973).
27. T. Ohnishi, J. C. Salerno, D. B. Winter, J. Lim, C. A. Yu, L. Yu, and T. E. King, *J. Biol. Chem. 251:* 2094 (1976).
28. H. Beinert, B. A. C. Ackrell, E. B. Kearney, T. P. Singer, *Eur. J. Biochem. 54:* 185 (1975).
29. T. Ohnishi, J. Lim, D. B. Winter, and T. E. King, *J. Biol. Chem. 251:* 2105 (1976).
30. T. E. King, in *Methods in Enzymology,* Vol. 10 (R. W. Estabrook and M. E. Pullman, eds.), Academic, New York, 1967, pp. 322-332.
31. M. L. Baginsky and Y. Hatefi, *J. Biol. Chem. 244:* 5313 (1969).
32. T. E. King, D. B. Winter, and W. Steele, in *Structure and Function of Oxidation-Reduction Enzymes* (A. Akerson and A. Ehrenberg, eds.), Pergamon, New York, 1972, pp. 519-532.
33. P. Bernath and T. P. Singer, in *Methods in Enzymology* Vol. 5 (S. P. Colowick and N. O. Kaplan, eds.), Academic, New York, 1962, pp. 597-614.
34. T. P. Singer, *Comprehensive Biochemistry 14:* 127 (1966).
35. W. G. Hanstein, K. A. Davis, M. A. Ghalambor, and Y. Hatefi, *Biochemistry 10:* 2517 (1971).
36. C. A. Yu, L. Yu, and T. E. King, *J. Biol. Chem. 249:* 4905 (1974).
37. T. P. Singer, M. Gutman and V. Massey, in *Iron-Sulfur Proteins* (W. Lovenberg, ed.), Vol. 1, Academic, New York, 1973, pp. 225-300.
38. A. D. Vinogradov, E. V. Gavrikova, and V. G. Goloveshkina, *Biochem. Biophys. Res. Commun. 65:* 1264 (1975).
39. A. D. Vinogradov, B. A. C. Ackrell, and T. P. Singer, *Biochem. Biophys. Res. Commun. 67:* 803 (1975).
40. D. B. Winter, Ph.D. Thesis, State University of New York at Albany, 1975.
41. C. J. Coles, H. D. Tisdale, W. C. Kearney, and T. P. Singer, *Physiol. Chem. Phys. 4:* 301 (1972).
42. D. V. Dervartanian, C. Veeger, W. H. Orme-Johnson, and H. Beinert, *Biochim. Biophys. Acta 191:* 22 (1969).

43. J. C. Salerno, J. Lim, T. E. King, H. Blum, and T. Ohnishi, *J. Biol. Chem.* (in press).
44. T. Ohnishi, T. Asakura, T. Yonetani, and B. Chance, *J. Biol. Chem. 246:* 5960 (1971).
45. N. R. Orme-Johnson, W. H. Orme-Johnson, R. E. Hansen, H. Beinert, and Y. Hatefi, *Biochem. Biophys. Res. Commun. 44:* 446 (1971).
46. N. R. Orme-Johnson, R. E. Hansen, and H. Beinert, *J. Biol. Chem. 249:* 1928 (1974).
47. F. J. Ruzicka and H. Beinert, *Biochem. Biophys. Res. Commun. 58:* 556 (1974).
48. T. Ohnishi, W. J. Ingledew, and S. Shiraishi, *Biochem. J. 153:* 39 (1976).
49. R. G. Bartch, in *Bacterial Photosynthesis* (H. Guest, A. San Pietro, and L. P. Vernon, eds.), Antioch Press, Yellow Springs, 1963, pp. 315-326.
50. T. Ohnishi, D. B. Winter, J. Lim, and T. E. King, *Biochem. Biophys. Res. Commun. 61:* 1017 (1974).
51. R. Cammack, *Biochem. Soc. Trans. 3:* 482 (1975).
52. J. C. Salerno, Ph.D. Thesis, University of Pennsylvania, 1977.
53. H. Beinert, B. A. C. Ackrell, A. D. Vinogradov, E. B. Kearney, and T. P. Singer, *Arch. Biochem. Biophys. 182:* 95 (1977).
54. N. A. Stombaugh, R. H. Burris, and W. H. Orme-Johnson, *J. Biol. Chem. 248:* 7951 (1973).
55. W. H. Orme-Johnson and R. H. Sands, in *Iron-Sulfur Proteins* (W. Lovenberg, ed.), Vol. 2, Academic, New York, 1973, pp. 195-238.
56. C. W. Carter, J. Kraut, S. T. Freer, R. A. Alden, L. C. Sicker, E. Adman, and L. H. Jensen, *Proc. Nat. Acad. Sci. U. S. 69:* 3526 (1972).
57. R. Cammack, *Biochem. Biophys. Res. Commun. 54:* 548 (1973).
58. W. V. Sweeney, A. J. Bearden, and J. C. Rabinowitz, *Biochem. Biophys. Res. Commun. 59:* 188 (1974).
59. H. Beinert and F. J. Ruzicka, in *Electron Transfer Chains and Oxidative Phosphorylation* (E. Quagliariello, S. Papa, F. Palmieri, E. Slater, and N. Siliprandi, eds.), North Holland, Amsterdam, 1975, pp. 37-42.
60. H. Beinert, in *Iron Sulfur Proteins* (W. Lovenberg, ed.), Vol. 3, Academic, New York, 1977, pp. 61-100.
61. J. C. Salerno, T. Ohnishi, J. Lim, and T. E. King, *Biochem. Biophys. Res. Commun. 73:* 833 (1976).
62. J. C. Woolum, E. Tiezzi, and B. Commoner, *Biochim. Biophys. Acta 160:* 311 (1968).
63. J. C. Salerno, T. Ohnishi, H. Blum, and J. S. Leigh, *Biochim. Biophys. Acta 494:* 191 (1977).
64. H. Blum, J. C. Salerno, R. Prince, J. S. Leigh, and T. Ohnishi, *Biophys. J. 20:* 23 (1977).
65. S. P. J. Albracht and J. Subramanian, *Biochim. Biophys. Acta 462:* 36 (1977).
66. J. C. Salerno and T. Ohnishi, *Arch. Biochem. Biophys. 176:* 757 (1976).
67. F. J. Ruzicka, H. Beinert, K. L. Schepler, W. R. Dunham, and R. H. Sands, *Proc. Nat. Acad. Sci. U. S. 72:* 2886 (1975).

68. W. J. Ingledew, J. C. Salerno, and T. Ohnishi, *Arch. Biochem. Biophys. 177:* 176 (1976).
69. T. C. Hollocher and B. Commoner, *Proc. Nat. Acad. Sci. U. S. 47:* 1355 (1961).
70. T. E. King, R. L. Howard, and H. S. Mason, *Biochem. Biophys. Res. Commun. 5:* 329 (1961).
71. M. Erecinska, D. F. Wilson, Y. Mukai, and B. Chance, *Biochem. Biophys. Res. Commun. 41:* 386 (1970).
72. B. A. C. Ackrell, E. B. Kearney, and D. Edmondson, *J. Biol. Chem. 250:* 7114 (1975).
73. T. E. King, T. Ohnishi, C. A. Yu, and L. Yu, in *Structure and Function of Energy-Transducing Membranes* (K. Van Dam and B. F. Van Gelder, eds.), Elsevier, Amsterdam, 1977, pp. 49-60.
74. G. Palmer, F. Muller, and V. Massey, in *Flavins and Flavoproteins* (H. Kamin, ed.), University Park Press, Baltimore, Md., 1971, pp. 123-140.
75. D. F. Wilson, M. Erecinska, P. L. Dutton, and T. Tsuzuki, *Biochem. Biophys. Res. Commun. 41:* 1273 (1970).
76. T. Ohnishi, D. F. Wilson, T. Asakura, and B. Chance, *Biochem. Biophys. Res. Commun. 49:* 1087 (1972).
77. T. Ohnishi, J. C. Salerno, H. Blum, J. S. Leigh, and W. J. Ingledew, in *Bioenergetics of Membranes* (L. Packor, G. C. Papageorgiou, and A. Trebst, eds.), Elsevier/North Holland, Amsterdam, 1977, pp. 209-216.
78. P. R. Rich, A. L. Moore, W. J. Ingledew, W. D. Bonner, Jr., *Biochim. Biophys. Acta 462:* 501 (1977).
79. T. Ohnishi, J. S. Leigh, D. B. Winter, J. Lim, and T. E. King, *Biochem. Biophys. Res. Commun. 61:* 1026 (1974).
80. W. J. Ingledew and T. Ohnishi, *FEBS Lett. 54:* 167 (1975).
81. F. J. Ruzicka and H. Beinert, *Fed. Proc. 34:* 579 (1975).
82. P. L. Dutton, *Biochim. Biophys. Acta 226:* 63 (1971).
83. H. J. Harmon and F. L. Crane, *Biochem. Biophys. Res. Commun. 59:* 326 (1974).
84. T. Ohnishi, W. J. Ingledew, B. Widger, C. I. Ragan, J. Lim, and T. E. King, presented in the Symposium on Bioenergetics in the Xth International Congress of Biochemistry, Hamburg, 1976.
85. R. C. Bray, *Biochem. J. 81:* 189 (1961).
86. H. Beinert, R. E. Hansen, and C. R. Hartzell, *Biochim. Biophys. Acta 423:* 339 (1976).
87. J. S. Olson, D. P. Ballou, G. Palmer, and V. Massey, *J. Biol. Chem. 249:* 4363 (1974).
88. J. C. Salerno, T. Ohnishi, J. Lim, and T. E. King, *Biochem. Biophys. Res. Commun. 75:* 618 (1977).
89. F. Bloch, *Phys. Rev. 70:* 460 (1946).
90. G. E. Pake, *Paramagnetic Resonance,* Benjamin, New York, 1962, pp. 129-134.
91. A. L. Tappell, *Biochem. Pharmacol. 3:* 289 (1960).
92. T. E. King, *Adv. Enzymol. 28:* 155 (1966).

93. T. P. Singer, H. Beinert, B. A. C. Ackrell, and E. B. Kearney, *Proceedings of the 10th FEBS Meeting, Paris* (Y. Raoul, ed.), Associated Scientific, Amsterdam, 1975, pp. 173-185.
94. P. C. Mowery, D. J. Steenkamp, B. A. C. Ackrell, T. P. Singer, and G. A. White, *Arch. Biochem. Biophys. 178:* 495 (1977).
95. W. J. Ingledew and T. Ohnishi, *Biochem. J. 164:* 617 (1977).
96. W. J. Ingledew, J. C. Salerno, and T. Ohnishi, *Fed. Proc. 36:* 901 (1977).
97. E. K. Ruuge and A. A. Konstantinov, *Biofizika 21:* 586 (1976).
98. A. N. Tikkonov, D. S. Burbaw, I. V. Grigolawa, A. A. Konstantinov, M. Y. Ksenzenko, and E. K. Ruuge, *Biofizika 22:* 734 (1977).
99. A. A. Konstantinov and E. K. Ruuge, *Bioorganic Chem. 3:* 787 (1977).
100. T. E. King, T. Ohnishi, D. B. Winter, and J. T. Wu, in *Iron and Copper Proteins* (K. T. Yasunobu, H. F. Mower, and O. Hayaishi, eds.), Plenum, New York, 1976, pp. 182-227.
101. M. Klingenberg and M. Bucholz, *Eur. J. Biochem. 13:* 247 (1970).
102. D. D. Tyler and J. Newton, *FEBS Lett. 8:* 325 (1970).
103. C. P. Lee, B. Johansson, and T. E. King, in *Probes of Structure and Function of Macromolecules and Membranes* Vol. 1 (C. P. Lee and J. K. Blasie, eds.), Academic, New York, 1971, pp. 401-406.
104. R. Mathews, S. Charlton, R. H. Sands, and G. Palmer, *J. Biol. Chem. 249:* 4326 (1974).
105. R. N. Mullinger, R. Cammack, K. K. Rao, D. O. Hall, D. D. E. Dickson, C. E. Johnson, J. D. Rush, and A. Simopoulos, *Biochem. J. 151:* 75 (1975).
106. H. J. Harmon and F. L. Crane, *Biochim. Biophys. Acta 440:* 45 (1976).
107. M. Erecinska, J. K. Blasie, and D. F. Wilson, *FEBS Lett. 76:* 235 (1977).
108. J. C. Salerno, H. J. Harmon, H. Blum, J. S. Leigh, and T. Ohnishi, *FEBS Lett. 82:* 179 (1977).
109. C. A. Yu, L. Yu, and T. E. King, *Fed. Proc. 36:* 815 (1977).
110. Y. Hatefi, W. G. Hanstein, K. A. Davis, and K. S. You, *Ann. N. Y. Acad. Sci. 227:* 504 (1974).
111. Y. Hatefi, in *Iron and Copper Proteins* (K. T. Yasunobu, H. F. Mower, and O. Hayaishi, eds.), Plenum, New York, 1976, pp. 150-160.
112. C. I. Ragan, *Biochim. Biophys. Acta 456:* 249 (1976).
113. C. I. Ragan and E. Racker, *J. Biol. Chem. 248:* 6876 (1973).
114. C. I. Ragan, *Biochem. J. 154:* 295 (1976).
115. R. L. Ringler, S. Minakami, and T. P. Singer, *J. Biol. Chem. 238:* 801 (1963).
116. P. K. Huang and R. L. Pharo, *Biochim. Biophys. Acta 245:* 240 (1971).
117. R. F. Baugh and T. E. King, *Biochem. Biophys. Res. Commun. 49:* 1165 (1972).
118. H. R. Mahler, N. K. Sarkar, L. P. Vernon, and R. A. Alberty, *J. Biol. Chem. 199:* 585 (1952).
119. B. Mackler, *Biochem. Biophys. Acta 50:* 141 (1961).
120. R. L. Pharo, L. A. Sordahl, S. R. Vyas, and D. R. Sanadi, *Biol. Chem. 241:* 4771 (1966).

121. S. A. Kumar, N. A. Rao, S. P. Felton, F. M. Huennekens, and B. Mackler, *Arch. Biochem. Biophys. 125:* 436 (1968).
122. T. E. King and R. L. Howard, *F. Biol. Chem. 237:* 1686 (1962).
123. Y. Hatefi and K. E. Stempel, *J. Biol. Chem. 244:* 2350 (1969).
124. K. A. Davis and Y. Hatefi, *Biochemistry 8:* 3355 (1969).
125. Y. Hatefi, K. E. Stempel, and W. G. Hanstein, *J. Biol. Chem. 244:* 2358 (1969).
126. T. Ohnishi, *Biochim. Biophys. Acta 301:* 105 (1973).
127. G. Palmer, R. H. Sands, and L. E. Mortenson, *Biochem. Biophys. Res. Commun. 23:* 357 (1966).
128. W. H. Orme-Johnson and H. Beinert, *Biochem. Biophys. Res. Commun. 36:* 337 (1969).
129. S. P. J. Albracht and E. C. Slater, *Biochim. Biophys. Acta 245:* 503 (1971).
130. N. R. Orme-Johnson, R. E. Hansen, and H. Beinert, *J. Biol. Chem. 249:* 1922 (1974).
131. T. Ohnishi, T. Asakura, D. F. Wilson, and B. Chance, *FEBS Lett. 21:* 59 (1972).
132. T. Ohnishi, D. F. Wilson, T. Asakura, and B. Chance, *Biochem. Biophys. Res. Commun. 46:* 1631 (1972).
133. S. P. J. Albracht, *Biochim. Biophys. Acta 347:* 183 (1974).
134. T. Ohnishi and M. Pring, in *Dynamics of Energy Tranducing Membrane* (L. Ernester, R. W. Estabrook, and E. C. Slater, eds.), Elsevier, Amsterdam, 1974, pp. 169-180.
135. T. Ohnishi, *Biochim. Biophys. Acta 387:* 475 (1975).
136. S. P. J. Albracht and G. Dooijewaard, in *Electron Transfer Chains and Oxidative Phosphorylation* (E. Quagliariello, S. Papa, F. Palmieri, E. Slater, and N. Siliprandi, eds.), North Holland, Amsterdam, 1975, pp. 49-54.
137. S. P. J. Albracht, G. Dooijewaard, F. J. Leenwerik, and B. Van Swol, *Biochim. Biophys. Acta 459:* 300 (1977).
138. T. Ohnishi, J. C. Salerno, W. Widger, and T. E. King, manuscript in preparation.
139. H. Beinert, in *Iron and Copper Proteins* (K. T. Yasunobu, H. F. Mower, and O. Hayaishi, eds.), Plenum, New York, 1976, pp. 137-149.
140. P. Rich and W. D. Bonner, *Functions of Alternative Respiratory Oxidases* (D. Lloyd, H. Degn, and G. C. Hill, eds.), Pergamon, New York, 1978, pp. 61-68.
141. H. Blum, J. C. Salerno, P. R. Rich, and T. Ohnishi, submitted for publication.
142. W. Widger and T. E. King, submitted for publication.
143. N. A. Rao, S. P. Felton, F. M. Huennekens, and B. Mackler, *J. Biol. Chem. 238:* 449 (1963).
144. T. Ohnishi, J. S. Leigh, C. I. Ragan, and E. Racker, *Biochem. Biophys. Res. Commun. 56:* 775 (1974).
145. W. J. Ingledew and T. Ohnishi, Hamburg Poster presentation, 1976.

146. H. Beinert, G. Palmer, T. Cremona, and T. P. Singer, *J. Biol. Chem. 240:* 475 (1965).
147. T. P. Singer and M. Gutman, *Adv. Enzymol. 34:* 79 (1971).
148. B. P. Garland, R. A. Clegg, J. A. Downie, T. A. Gray, H. G. Lawford, and J. Skyrme, in *Mitochondria/Biomembranes* (S. G. van der Bergh, P. Borst, L. L. M. Van Deenen, J. C. Riemersma, E. C. Slater, and J. M. Tager, eds.), Vol. 28, North Holland, Amsterdam, 1972, pp. 105-117.
149. P. Mitchell, in *Mitochondria/Biomembranes* (S. G. van der Bergh, P. Borst, L. L. M. Van Deenen, J. C. Riemersma, E. C. Slater, and J. M. Tager, Eds.), Vol. 28, North Holland, Amsterdam, 1972, pp. 358-370.
150. M. Gutman, T. P. Singer, and H. Beinert, *Biochemistry 11:* 556 (1972).
151. H. G. Lawford and P. B. Garland, *Biochem. J. 130:* 1029 (1971).
152. T. P. Singer and M. Gutman, *Horizons Biochem. Biophys. 1:* 261 (1974).
153. M. Gutman, H. Beinert, and T. P. Singer, in *Electron Transfer Chains and Oxidative Phosphorylation* (E. Quagliariello, S. Papa, F. Palmieri, E. Slater, and N. Siliprandi, eds.), North Holland, New York, 1975, pp. 55-62.
154. T. Ohnishi, *Eur. J. Biochem. 64:* 91 (1976).
155. D. F. Wilson, M. Erecinska, and P. L. Dutton, *Ann. Rev. Biophys. Bioeng. 3:* 203 (1973).
156. P. L. Dutton and D. F. Wilson, *Biochim. Biophys. Acta 346:* 165 (1974).
157. D. DeVault, *J. Theor. Biol. 62:* 115 (1976).
158. D. DeVault, *Biochim. Biophys. Acta 226:* 193 (1971).
159. P. Mitchell, *Chemiosmotic Coupling and Energy Transduction,* Glynn Research Ltd., Bodmin, England, 1968.
160. R. W. Jones, T. A. Gray, and P. Garland, *Biochem. Soc. Trans. 4:* 671 (1976).
161. P. L. Dutton, D. F. Wilson, and C. P. Lee, *Biochemistry 9:* 5077 (1971).
162. V. P. Skulachev, in *Current Topics in Bioenergetics* Vol. 4 (D. R. Sanadi, ed.), Academic, New York, 1971, pp. 127-185.
163. J. Moyle and P. Mitchell, *Biochem. J. 132:* 571 (1973).
164. C. I. Ragan and W. R. Widger, *Biochem. Biophys. Res. Commun. 62:* 744 (1975).
165. J. Rydström, J. B. Hoek, and L. Ernster, in *The Enzyme* Vol. 13, (P. D. Boyer, ed.), Academic, New York, 1976, pp. 51-79.
166. Y. M. Galante and Y. Hatefi, *Fed. Proc. 35:* 1435, Abstr. 409 (1976).
167. M. Gutman and N. Silman, *FEBS Lett. 47:* 241 (1974).
168. L. D. Ohaniance and Y. Hatefi, *J. Biol. Chem. 250:* 9397 (1975).
169. Y. Hatefi and W. G. Hanstein, *Biochemistry 12:* 3515 (1973).
170. M. D. Brand, B. Raynafarje, A. L. Lehninger, *J. Biol. Chem. 251:* 5670 (1976).
171. B. Raynafarje, M. D. Brand, A. L. Lehninger, *J. Biol. Chem. 251:* 7442 (1976).
172. C. I. Ragan and P. C. Hinkle, *J. Biol. Chem. 250:* 8472 (1975).

173. V. P. Skurachev, in *Energy Transducing Mechanism* (E. Racker, ed.), University Park Press, Baltimore, Md., 1975, pp. 31-73.
174. V. P. Skurachev, *Proceedings of the 10th FEBS Meeting, Paris* (Y. Raoul, ed.), Associated Scientific, Amsterdam, 1975, pp. 225-238.
175. B. Chance, A. R. Crofts, M. Nishimura, and B. Price, *Eur. J. Biochem. 13:* 364 (1970).
176. S. Papa, *Biochim. Biophys. Acta 456:* 39 (1976).
177. E. C. Slater, in *Electron Transfer Chains and Oxidative Phosphorylation* (E. Quagliariello, S. Papa, F. Palmieri, E. Slater, and N. Siliprandi, eds.), North Holland, Amsterdam, 1975, pp. 3-14.
178. P. Mitchell, *Biochem. Soc. Trans. 4:* 399 (1976).
179. P. Mitchell, *FEBS Lett. 78:* 1 (1977).
180. D. R. Phillips and M. Morrison, *Biochemistry 10:* 1766 (1971).
181. G. D. Case and J. S. Leigh, *Biochem. J. 160:* 769 (1976).
182. G. D. Case, T. Ohnishi, and J. S. Leigh, *Biochem. J. 160:* 785 (1976).
183. G. D. Case and J. S. Leigh, *Proceedings of the 11th Rare Earth Research Conference* (J. M. Haschke and H. D. Eick, eds.), Vol. 2, U.S. Atomic Energy Commission, Oak Ridge, Tenn., 1974, pp. 706-715.
184. C. P. Slichter, *Principles of Magnetic Resonance,* Harper and Row, New York, 1963.
185. A. Carrington and A. D. McLachlan, *Introduction to Electron Spin Resonance,* Harper and Row, New York, 1967.
186. F. J. Ruzicka and H. Beinert, *Biochem. Biophys. Res. Commun. 66:* 622 (1975).
187. H. Beinert, in *Structure and Function of Energy-Transducing Membranes* (K. Van Dam and B. F. Van Gelder, eds.), Elsevier, Amsterdam, 1977, pp. 11-22.
188. F. J. Ruzicka and H. Beinert, *J. Biol. Chem. 253:* 2514 (1978).
189. B. Höjeberg and J. Rydström, *Biochem. Biophys. Res. Commun. 78:* 1183 (1977).
190. J. Rydström, J. Montelius, D. Bäckström, and L. Ernster, *Biochim. Biophys. Acta 501:* 370 (1978).

2

Succinate-Cytochrome *c* Reductase Complex of the Mitochondrial Electron Transport Chain

Bernard L. Trumpower*
Aspandiar G. Katki†

Dartmouth Medical School
Hanover, New Hampshire

*Established Investigator of the American Heart Association.
†Current affiliation: Tufts-New England Medical Center, Boston, Massachusetts, and National Cancer Institute, Division of Cancer Treatment, Bethesda, Maryland.

I. Introduction

A. *An Overview of the Succinate-Cytochrome* c *Reductase Complex*

Succinate-cytochrome *c* reductase complex is the portion of the mitochondrial electron transport chain which catalyzes electron transfer from succinate to cytochrome *c* according to Equation 1:

$$\text{Succinate} + 2c^{3+} \rightarrow \text{fumarate} + 2c^{2+} + 2\text{H}^+ \qquad (1)$$

A schematic representation of the electron transport chain, including the succinate-cytochrome *c* reductase reaction, is shown in Figure 1a. Electron transfer from succinate to cytochrome *c* spans the second coupling site of oxidative phosphorylation, and the free energy available from the succinate-cytochrome *c* reductase reaction can be conserved and utilized for the synthesis of ATP.

The stoichiometry of the succinate-cytochrome *c* reductase reaction is such that as one molecule of succinate is oxidized to fumarate two electrons are transferred to two molecules of cytochrome *c* and two H^+ are liberated (Equation 1). The equilibrium of this reaction lies extensively in the direction of fumarate and reduced cytochrome *c*. The reaction is driven to this equilibrium as a result of the increment in oxidation-reduction potential between the succinate/fumarate couple and the ferrocytochrome *c*/ferricytochrome *c* couple as shown in Equations 2-4:

$$\text{Succinate} \rightarrow \text{fumarate} + 2H^+ + 2e^- \quad E_0' = -24 \tag{2}$$

$$2c^{3+} + 2e^- \rightarrow 2c^{2+} \quad E_0' = +254 \tag{3}$$

$$\text{Succinate} + 2c^{3+} \rightarrow \text{fumarate} + 2c^{2+} + 2H^+ \quad \Delta E_0' = +230 \tag{4}$$

In mitochondria respiring under conditions in which the rate of electron transport is limited by the availability of ADP (state 4), the ratio succinate/fumarate is such that the apparent potential (E_h) of this couple may be as low as 0 to -30 mV and the ratio of ferri- to ferrocytochrome *c* couple is +250 to +300 mV [1, 2]. Thus, under conditions in which respiration is coupled to the synthesis of ATP, succinate-cytochrome *c* reductase complex catalyzes an electron transfer process with an available free energy change of 11-15 kcal/mol succinate.

Electron transfer from succinate to cytochrome *c* is sufficiently exergonic to allow the synthesis of one ATP per electron pair transferred. However, as we shall discuss below in attempting to write a mechanism in which electron transfer is accompanied by H^+ translocation, it may be somewhat oversimplified and inaccurate to think of electron transfer through the second coupling site as generating in 1:1 stoichiometry the molecular intermediates involved in ATP synthesis.

B. *Scope and Content of this Review*

In this chapter we have described what is understood about the mechanism whereby succinate-cytochrome *c* reductase complex transfers electrons from succinate to cytochrome *c* and how this electron transfer affects the outward translocation of H^+ across the inner mitochondrial membrane. Consequently we have empha-

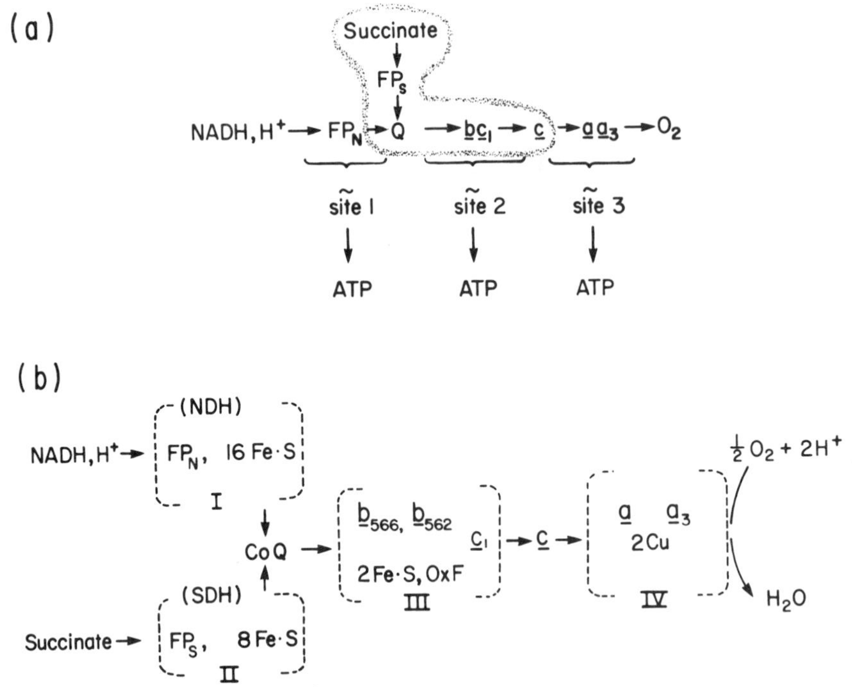

(a)

Figure 1 (a) Diagram of the mitochondrial electron transport chain. The components of the electron transport chain are arranged in a classical linear sequence as commonly represented in textbooks. The three brackets delineate the three segments of the electron transport chain corresponding to the "coupling sites" of oxidative phosphorylation. The free energy from electron transfer through these three segments is conserved in the form of a common high-energy intermediate (~), which in turn is used for synthesis of ATP or other endergonic reactions. The stippled area includes the components of the succinate-cytochrome *c* reductase complex. (b) Diagram of the electron transport chain in which functionally related components are arranged in four complexes (I-IV). In addition to a distinct flavoprotein, the NADH dehydrogenase complex (NDH) and succinate dehydrogenase complex (SDH) contain 16 g-atoms and 8 g-atoms of nonheme iron, respectively, and equivalent amounts of acid-labile sulfide. The cytochrome bc_1 complex contains two *b* cytochromes, cytochrome c_1, an iron-sulfur protein of the 2Fe 2S type, and a newly discovered protein known as oxidation factor (OxF).

sized functional aspects of the reductase complex and have included structural information which is relevant to that purpose.

We have attempted to write for an audience somewhat broader than usual and thus have tried to describe a complex subject in language which is understandable to students, young investigators in the field of bioenergetics, and interested colleagues in related fields. Established investigators, who have made such notable contributions to this field and who have given freely of their time in training young people, will surely recognize the usefulness of this approach.

Current understanding of the electron transport chain complexes has evolved through extensive experimentation. Our understanding of the reductase complex is incomplete and some current views will likely change. Under such circumstances it is neither possible nor desirable to write a strictly factual review. Where there is appreciable disagreement we have tried to describe those experimental results which, to us, decide the issue. In some instances we have postulated solutions to currently unsolved problems in the hope that these might receive additional experimental support.

C. Relationship of Succinate-Cytochrome c *Reductase Complex to the Inner Mitochondrial Membrane*

Succinate-cytochrome *c* reductase complex is composed of succinate dehydrogenase, ubiquinone, the *b* cytochromes, cytochrome c_1, the iron-sulfur protein of the bc_1 segment, a newly discovered protein named oxidation factor, and phospholipids. In addition, the reductase complex contains several polypeptides, such as those of core protein, for which a function in electron transport is currently not known. The components of the reductase complex are integrated into the inner mitochondrial membrane in such a manner that they contribute to the structure of the membrane in addition to fulfilling a functional role in electron transport. Certain components, such as succinate dehydrogenase and oxidation factor, appear to be tightly associated with the inner membrane, but in a manner such that they can be reversibly removed from the membrane leaving the essential membrane structure unchanged in their absence. Other components, such as the *b* cytochromes, appear to be integral membrane proteins which cannot be isolated without destruction of the membrane.

As first shown by Hatefi and coworkers, the mitochondrial inner membrane can be fractionated with detergent and salt into four lipoprotein complexes in which functionally related components of the electron transport chain remain structurally associated [3]. This elegant work has provided the conceptual basis upon which all subsequent fractionations of the membranous electron transport chain have been based. The four complexes, which are referred to as the primary enzyme complexes of the electron transport chain [3], are identified by Roman numerals I-IV and include NADH-ubiquinone reductase (I), succinate-ubiquinone reductase (II), ubiquinol-cytochrome *c* reductase (III), and cytochrome *c* oxidase (IV). A diagram of the electron transport chain, in which functionally related components are segregated into these four complexes, is shown in Figure 1b.

In addition to complexes I-IV, it is also possible to isolate what appear to be hybrid complexes such as succinate-cytochrome *c* reductase complex (an apparent combination of complexes II + III) and NADH-cytochrome *c* reductase complex (I + III). The difficulties associated with purifying individual components of the electron transport chain by further resolving these lipoprotein complexes is evidence that the complexes represent structural units by which the electron transport chain is integrated into the mitochondrial membrane.

D. Succinate-Cytochrome c *Reductase Complex as a Structural Entity in the Inner Mitochondrial Membrane*

Although the electron transport chain probably consists of several lipoprotein complexes, it is questionable whether each of the complexes I-IV is a discrete structural entity in the inner membrane. Whether succinate dehydrogenase is structurally separated from the bc_1 segment and whether there is such a structure as complex II is particularly questionable.

To illustrate this, several possible structural relationships between the complexes are represented in Figure 2. The model in case A is one in which each primary electron transport complex is a segregated structural unit in the inner membrane. Ubiquinone would function as a mobile carrier which shuttles reducing equivalents between the dehydrogenase complexes and the bc_1 complex by diffusion in the lipid phase of the membrane continuum. In this model succinate dehydrogenase is a component of complex II, which contains proteins such as a unique *b* cytochrome and tightly bound phospholipid which are separate from those of complex III. According to this model, resolved reductase complex is a fusion product created by the coincident isolation of two separate primary complexes.

The model in case B is one in which the reductase complex is composed of succinate dehydrogenase in fixed association with the bc_1 segment. In such a model, complex II and complex III are not unique and distinct complexes separate from each other but rather are cleavage products derived from a common parent reductase complex. This model predicts that many of the polypeptides found in complex II would also be found in complex III and vice versa. Ubiqui-

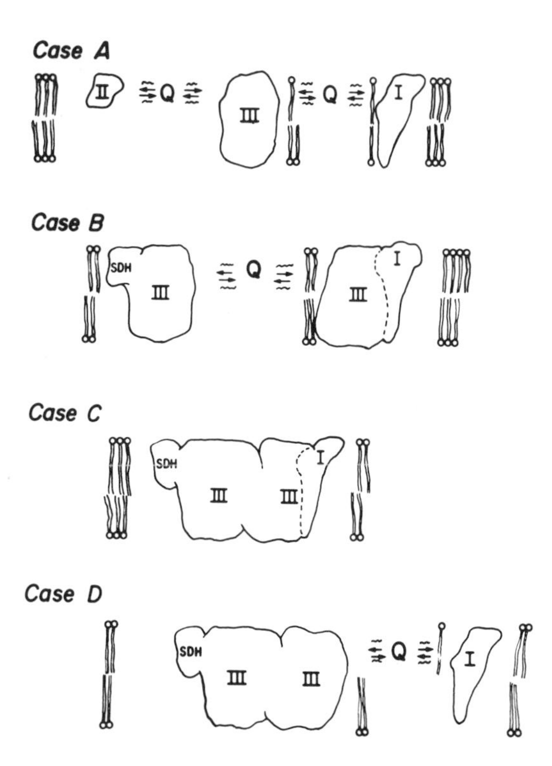

Figure 2 Diagram showing several possible structural relationships among the complexes of the electron transport chain. Complexes I and III are represented as spanning the inner mitochondrial membrane, shown as a phospholipid bilayer. For simplicity the cytochrome *c* oxidase complex (IV) is omitted. As discussed in the text, it is not known whether complexes II and III exist as discrete lipoprotein complexes as represented in case A, or whether they are operational entities derived by cleavage of succinate-cytochrome *c* reductase complex, which is composed of succinate dehydrogenase in permanent association with the bc_1 segment (case B). The diagram also illustrates the possible function of ubiquinone (Q) as a mobile carrier which shuttles reducing equivalents between the complexes by free diffusion in the lipid bilayer of the membrane continuum.

none would function as a bound carrier within the reductase complex and also as a mobile carrier in the bulk phase of the membrane.

Case C is an extension of case B except that it represents a fusion of two reductase complexes at the level of the bc_1 segment. Such a model would predict a stoichiometry of one succinate dehydrogenase per two c_1 cytochromes. Ubiquinone would function as a carrier whose mobility is restricted to the combined reductase complex and thus would not necessarily enter the bulk phase of the membrane continuum.

Case D differs from case B only in that one succinate dehydrogenase is bound to two bc_1 segments. Such a structure might arise as a partial cleavage product derived from case C if there is a loose association of complex I with the bc_1 segment or alternatively such a structure may exist independent of any association with complex I. The point which is illustrated by cases C and D is that the inner mitochondrial membrane contains a molar excess of bc_1 segment relative to succinate dehydrogenase [4, 5]. Depending on the tissue and species source, the ratio of c_1 to covalent flavin may vary from 2:1 to 10:1. This raises the possibility that one dehydrogenase may service more than one bc_1 segment. In addition, other substrate dehydrogenases may associate with one or more bc_1 segments.

One might argue that there is a dynamic relationship in which hybrid complexes are formed by rapid reversible association and that the most accurate representation would be a composite of the models shown in Figure 2. However, this view circumvents questions such as whether there is a *b* cytochrome uniquely associated with succinate dehydrogenase in complex II and separate from the *b* cytochromes of the bc_1 segment, and whether ubiquinone freely diffuses in the lipid bilayer of the membrane continuum.

Having been given the choice of writing on complex III or succinate-cytochrome *c* reductase complex, we chose the latter to emphasize that the reductase complex may represent the simplest membrane unit in which both native structure and function are preserved. In particular, we suggest that succinate dehydrogenase is tightly bound to the bc_1 segment and that complex II and complex III may be operational entities which arise as cleavage products from a larger lipoprotein complex as represented by cases B, C, or D. At the same time we recognize that resolved reductase complex as usually prepared may include populations of the bc_1 segment which are not otherwise associated with succinate dehydrogenase in situ. This possibility does not detract from the evidence suggesting a tight association of the dehydrogenase with the bc_1 segment.

The analogy of succinate-cytochrome *c* reductase complex may not extend to include an NADH-cytochrome *c* reductase complex as a discrete structural entity. For instance, unlike complex II, complex I is readily obtained free of cytochromes *b* and c_1, and the core protein polypeptides of the bc_1 segment appear to be absent when complex I is analyzed by electrophoresis [6]. Thus, it seems likely that complex I is a separate lipoprotein complex. Whether complex I is loosely associated with the bc_1 segment in situ cannot be decided at present.

Our view that succinate-cytochrome *c* reductase complex is an elementary structural unit composed of succinate dehydrogenase plus the components of the bc_1 segment and that complex II is a cleavage product derived from the reductase complex is supported by circumstantial evidence. Hatefi first pointed out that "complex IV can be separated readily from a fraction containing complexes I, II, and III, but the resolution of the latter fraction into individual complexes is not so easily achieved" [3]. In addition, complex II can be prepared from succinate-cytochrome *c* reductase complex [3], and there is extensive coidentity of components in complex II and complex III as discussed below.

The tight association of succinate dehydrogenase with the bc_1 segment is illustrated by the fact that it is quite difficult to isolate complex III free of succinate dehydrogenase by fractionation with anionic detergents and salt, whereas the reductase complex is readily separated from the cytochrome oxidase complex. This distinction would not be expected if complex II has the same independent status as a lipoprotein unit, as appears to be the case with cytochrome oxidase. Small quantities of succinate dehydrogenase remaining associated with the bc_1 segment in the preparations of complex III can frequently be detected by acrylamide gel electrophoresis in dodecyl sulfate [7-10]. The flavin-containing subunit of the dehydrogenase [apparent molecular weight (M_r) = 70,000] is recognizable as a high molecular weight polypeptide distinct from the two core protein polypeptides of the bc_1 segment (see Fig. 7, Section II.B). Although the staining intensity of the M_r = 70,000 polypeptide of succinate dehydrogenase relative to the M_r = 30,600 polypeptide of cytochrome c_1 is less than expected from equivalent amounts of succinate dehydrogenase and cytochrome c_1, this may simply reflect a situation in situ in which every dehydrogenase is tightly bound to a bc_1 segment but not the converse (see Fig. 2).

Succinate-cytochrome *c* reductase activity is an inadequate criterion for establishing the absence of succinate dehydrogenase in preparations of complex III. Preparations of complex III which on analysis by electrophoresis appear to contain small amounts of succinate dehydrogenase (see Fig. 7) commonly show no appreciable succinate-cytochrome *c* reductase activity. However, if such preparations are reconstituted with phospholipid and ubiquinone, succinate-cytochrome *c* reductase activity appears as shown in Table 1. This result illustrates that functional interaction of succinate dehydrogenase with the bc_1 segment requires phospholipid and ubiquinone, as discussed in Section III.B, and because this activity is latent, succinate dehydrogenase may go undetected when activity is measured. It follows that many preparations of complex III would more accurately be described as succinate-cytochrome *c* reductase complex having latent or inactivated succinate dehydrogenase.

There are two procedures by which succinate dehydrogenase can be dissociated from the bc_1 segment. Yu and coworkers [11] removed the dehydrogenase from the bc_1 segment by extraction with alkali followed by further fractionation of the bc_1 segment with detergent. The bc_1 segment retained its

Table 1 Effect of Phospholipid and Ubiquinone on Succinate-Cytochrome *c* Reductase Activity of Complex III

Additions to complex III[a]	Cytochrome *c* reductase activity (units/mg)
None	0.20
Phospholipid	2.80
Phospholipid + Q-10	3.40
Phospholipid + Q-10 + antimycin	0.05

[a]Complex III [10] was reconstituted with soybean phospholipids and ubiquinone-10 (Q-10) where indicated by incubating 300 μg of complex III with 900 nmol of phospholipid and 22.8 nmol of Q-10 in 200 μl of buffer containing 20 mM Na-succinate / 40 mM Na-P_i (pH 7.4) / 0.5 mM Na-EDTA for 60 min at 37°C. This mixture was then diluted to 2 ml by addition of the same buffer, and 1.5 μg of complex III was withdrawn for assay of succinate-cytochrome *c* reductase activity [25].

ability to rebind succinate dehydrogenase, and succinate-cytochrome *c* reductase activity could thus be restored. Riccio and coworkers [12] dissociated succinate dehydrogenase from the bc_1 segment by extraction of mitochondria with Triton X-100 followed by hydroxyapatite chromatography. However, this procedure also removed phospholipid, ubiquinone, and the iron-sulfur protein of the bc_1 segment in addition to the dehydrogenase, and the resulting bc_1 segment was devoid of electron transfer activity.

The coidentity of components in complex II and complex III is evidence that these preparations are overlapping cleavage products derived from a common origin. All preparations of complex II contain cytochrome *b* and many also contain cytochrome c_1 [13-19]. A *b* cytochrome which has spectral properties virtually identical to those of complex II can be prepared by cleavage of complex III [18]; consequently, an obvious conclusion one might draw is that at least some portion of the cytochrome *b* in complex II is coincident with that in complex III. Hatefi and coworkers, using fourth-derivative analysis of low-temperature difference spectra, have shown that the cytochrome *b* of complex II is heterogeneous and have advanced the view that a portion of this cytochrome *b* is unique to complex II [18, 19]. Although further experimentation is required to resolve the question of whether there is a *b* cytochrome unique to complex II, the common occurrence of c_1 in preparations of complex II suggests that succinate dehydrogenase has remained associated with the integral membrane proteins of the bc_1 segment. It seems to us that the difficulty of preparing complex II free of c_1 reflects the coincident difficulty of splitting the bc_1 segment as would be required to cleave succinate dehydrogenase in association with cytochrome *b* from the reductase complex. Another way of looking at this problem is to ask why it is so difficult to prepare complex II free of c_1, while resolved cytochrome oxidase free of c_1 is readily obtained. The answer, it seems to us, is that cytochrome oxidase is a bona fide lipoprotein complex and complex II is not.

To further test the relationship between complex II and complex III it would be especially useful if their polypeptide compositions were compared by acrylamide gel electrophoresis in dodecyl sulfate. The polypeptide profile of complex II is not so commonly documented in the literature. However, an indirect comparison of our preparation of complex III (see Fig. 5, Section II.B.3) with an electrophoresis profile of complex II (see Fig. 6 in Ref. 15) leads us to speculate that complex II contains the two core protein polypeptides from the bc_1 segment in addition to the heme-containing polypeptide of c_1 (M_r = 30,600) and a low molecular weight polypeptide ($M_r \simeq$ 14,000 in Fig. 5a, Section II.B) whose identity in the bc_1 segment is unknown but which we suspect may be apocytochrome *b* (see Sections V.B and VII.A).

Preparations of complex II which contain cytochrome c_1 in addition to *b* appear to differ from reductase complex only in that they lack oxidation factor and the iron-sulfur protein of the bc_1 segment. The primary functional difference between complex II containing cytochromes *b* and c_1 and succinate-cytochrome *c* reductase complex appears to be that in such preparations of complex II, succinate can reduce cytochrome *b* but not c_1 [13, 20], while in the reductase complex both cytochromes *b* and c_1 are rapidly reduced by succinate. Such preparations of complex II resemble reductase complex which has been depleted of oxidation factor, whose presence is required for reduction of cytochrome c_1, but not for reduction of the *b* cytochromes [21]. The similarity between these preparations of complex II and reductase complex depleted of oxidation factor is extended by the observation that antimycin inhibits the reduction of *b* in complex II by succinate [20], as has been found with reductase complex depleted of oxidation factor [21].

In addition to oxidation factor, complex II also lacks the iron-sulfur protein of the bc_1 segment. This conclusion is arrived at from the fact that, based on flavin content, the nonheme iron and acid-labile sulfide of complex II are accounted for by the content of succinate dehydrogenase [15]. In addition, complex II lacks the g = 1.90 electron paramagnetic resonance (EPR) signal [22] and the polypeptide (M_r = 25,000) expected for the iron-sulfur protein of the bc_1 segment [15].

If the above evaluation is correct, it follows that complex II might be reconstituted with oxidation factor plus the iron-sulfur protein from the bc_1 segment to form functionally active reductase complex. Thus, even though the weight of evidence suggests that complex II is an operational entity, it may turn out to be an extremely useful one, as it might be used in combination with oxidation factor to develop a reconstitution assay capable of measuring functional activity of the iron-sulfur protein of the bc_1 segment.

II. Preparation and Properties of Resolved Succinate-Cytochrome *c* Reductase Complex

A. Preparation of Resolved Reductase Complex

Succinate-cytochrome *c* reductase complex can be obtained as a highly resolved functional segment of the electron transport chain by extraction and purification from the inner mitochondrial membrane. With one exception [23, 24], all methods for isolation of the reductase complex involve the use of detergent and salt to disperse and purify the complex from the mitochondrial membrane; thus, the conceptual basis for such preparations was devised more than 15 years ago (see Ref. 3). The fact that alternative methods for isolating the reductase complex have not been devised is both a testimony to the merits of the classical approach and a contributing factor to certain inadequacies in our knowledge of the complex.

The purification schemes involve a protocol in which mitochondria are stored in a frozen state and then thawed and washed with buffer of moderately high ionic strength to remove peripheral and soluble matrix proteins. The washed mitochondria are then dispersed with detergent in combination with salts such as ammonium sulfate. By increasing the concentration of salt in controlled fashion, it is possible to achieve a "red-green split," in which a sedimentable pellet is separated from a dispersed supernatant by centrifugation. The pellet contains cytochrome oxidase (green) and residual membrane proteins, while the supernatant is enriched in NADH-cytochrome *c* reductase and succinate-cytochrome *c* reductase. The dispersed reductase complex is then precipitated by higher concentrations of salt.

The initial split is usually less than completely effective and small amounts of cytochrome oxidase may be dispersed along with the red supernatant. By repeating the fractionation with detergent and salt, one obtains reductase which is bright red in color and transparent due to bound detergent. When the detergent is removed by dialysis, the reductase complex becomes visibly turbid and can be sedimented by low-speed centrifugation. A scheme outlining such a purification procedure has been discussed by Hatefi [3, 19].

From the above outline and published procedures, one might expect, in the absence of experience to the contrary, that isolation of the reductase is a simple straightforward procedure. However, from noting the variations in activity reported from various laboratories (see Table 2), it is evident that difficulties are commonly encountered in attempting to prepare reductase complex of high purity and high activity.

In the past 5 years we have prepared reductase complex from bovine heart mitochondria on more than 100 occasions. Although the number of such preparations does not necessarily bear on their quality, this has allowed us to en-

Table 2 Succinate-Cytochrome *c* Reductase Activity of Preparations of Succinate-Cytochrome *c* Reductase Complex Reported in the Literature

Activity[a]	Comments	References
25	From beef heart mitochondria with *tert*-amyl alcohol supplemented by cholate and salt.	23, 24
0.95	From beef heart mitochondria with cholate and ammonium sulfate.	25
1.00	From beef heart mitochondria. Reported identical properties with preparations according to Yamashita and Racker [25]	53
0.08-0.30	Prepared according to Yamashita and Racker [25]. Reported fourfold increase of activity in presence of uncoupler.	205
0.30	From pigeon breast muscle with Triton and deoxycholate. As used for measuring midpoint potentials of cytochromes in resolved complexes and determining stoichiometry of redox components in bc_1 complex [134]	27, 149
2.0	Referred to as a partially purified preparation of complex III.	28
0.34	Prepared according to Yamashita and Racker [25]. Used to measure rate constants of "active" and "sluggish" forms of *b*.	155
2.4-6.6	From beef heart mitochondria with cholate and ammonium sulfate. Used to prepare bc_1 complex free of succinate dehydrogenase.	11
3.5-6.0	As prepared in this laboratory by the procedure of Yamashita and Racker [25] with minor modifications.	See text

[a]Activity is expressed as μmoles of cytochrome *c* reduced per minute per milligram. The conditions by which the activities of various preparations were measured are frequently not completely described. However, it seems unlikely that reasonable variations in assay procedures, such as temperature, concentration of cytochrome *c*, ionic strength, or pH of buffers, could account for the wide variation of activities reported.

counter many of the vagaries of the preparation. As a guide to obtaining a reproducible preparation of reductase complex having high activity, we would offer the following comments.

All published procedures for isolating the reductase should be expected to require some modification in order to be successfully implemented in a laboratory other than that in which they were developed. From first-hand experience we can state that the procedures described by Yamashita and Racker [25] and Yu and coworkers [11] are capable of yielding reductase with succinate-cytochrome c reductase activity of 3-6 units/mg. We routinely use a procedure modified from that of Yamashita and Racker [25].

As reported by Yu and coworkers [26] and discussed below, reductase complex of low activity can frequently be reactivated by addition of phospholipid and ubiquinone. In addition, detergents can alter the spectra and midpoint potentials of the b cytochromes in resolved reductase complex [27]. These observations indicate that it is advisable to perform the red-green split with as low a concentration of detergent as possible, and repeated fractionations with detergent and salt, which remove phospholipid from the reductase [26], should be avoided.

B. Characterization of Resolved Reductase Complex

The usefulness of resolved reductase complex depends on recognizing the extent to which it retains native functional properties. Thus, it is important to characterize the complex by methods which measure function and selected physical properties relevant to function. Measurements of electron transport activities, content and spectral properties of the cytochromes, and polypeptide composition meet these criteria.

1. Electron Transport Activity*

Electron transport activity is a quantitative index of function. Measurement of activity is a convenient and sensitive means of detecting damage to the complex during isolation and of localizing the site of damage which may have occurred. Resolved reductase complex catalyzes four electron transfer reactions which are conveniently measured. These are succinate-cytochrome c reductase, ubiquinol-cytochrome c reductase, succinate-ubiquinone reductase, and succinate dehydrogenase.

Succinate-cytochrome c reductase activity is measured under conditions in which the rate of cytochrome c reduction is zero order in cytochrome c concentration and first order in reductase complex [25] and thus measures the activity of some rate-limiting component within the complex. The component which is rate limiting may be different in preparations obtained by different procedures. For instance, the results in Table 1 illustrate that interaction of succinate dehydrogenase with the bc_1 segment may be rate limiting, apparently as a secondary consequence of alteration of the phospholipid and ubiquinone in the preparation

*Throughout the text and in tables we have expressed rates of electron transfer as units per milligram. For comparison of different electron transfer reactions all rates have been converted to one electron equivalent values.

Table 3 Cytochrome Content and Succinate-Cytochrome *c* Reductase Activity of Washed Beef Heart Mitochondria and Resolved Succinate-Cytochrome *c* Reductase Complex

	Washed mitochondria	Reductase complex
Cytochrome *b* content (nmol/mg)	0.295	2.79
Cytochrome c_1 content (nmol/mg)	0.156	1.55
Ratio b/c_1	1.9:1	1.8:1
Succinate-cytochrome *c* reductase activity		
μmol/min per mg	0.345	4.08
μmol/min per nmol c_1	2.21	2.63

of complex III. In some preparations of reductase complex, the activity of oxidation factor appears to be limiting, possibly due to depletion of ubiquinone, as discussed below (see Table 4 and Section VIII.C).

Succinate-cytochrome *c* reductase activity of the resolved complex should be equal to or greater than that of mitochondria when expressed as rate of *c* reduction per mole of complex. The activity and cytochrome content of washed mitochondria and reductase complex derived therefrom for several of our recent preparations are summarized in Table 3. The resolved complex is enriched 10-fold in cytochrome c_1 content and exhibits a slight increase in activity relative to that of the mitochondria when expressed on the basis of c_1 content. The activity of 4 units/mg shown here is somewhat lower than the value of 6-7 units/mg which we feel is the maximum to be expected.

The expected activity of 6-7 units/mg is based on the fact that we occasionally obtain such active preparations, although our most typical preparations have activity of 5 units/mg. Preparations of reductase complex with activity under 3 units/mg can be activated more than twofold by addition of phospholipid and ubiquinone as shown in Table 4. In addition, reductase complex which has been intentionally depleted of oxidation factor can be reconstituted to catalyze succinate-cytochrome *c* reductase activity with rates of 6-7 units/mg (see Fig. 18, Section VIII.C). The activity of 6-7 units/mg is in agreement with the results of Yu and coworkers [11].

Most preparations of reductase described in the literature have low succinate-cytochrome *c* reductase activity, even allowing for variations in assay conditions (Table 2). Although the low activity of such preparations may not be detrimental to the experiments for which they were used, the functional integrity of such preparations has clearly been altered. In regard to the difficulty of obtaining complex III devoid of succinate dehydrogenase, it is interesting to note that a

Table 4 Effect of Phospholipid, Ubiquinone, Succinate Dehydrogenase, and Oxidation Factor on Succinate-Cytochrome *c* Reductase Activity of Succinate-Cytochrome *c* Reductase Complex

Additions to reductase complex[a]	Cytochrome *c* reductase activity (units/mg)
None	2.30
P'lipid	2.29
P'lipid + Q-10	4.81
SDH	3.26
OxF	4.34
SDH + OxF	5.58
P'lipid + Q-10 + SDH + OxF	5.26

[a]Succinate-cytochrome *c* reductase complex was reconstituted with the indicated components by incubating 300 μg of reductase complex, 750 nmol of soybean phospholipids (P'lipid), 19 nmol of ubiquinone-10 (Q-10), 60 μg of succinate dehydrogenase (SDH), and 50 μg of oxidation factor (OxF) in 200 μl of buffer containing 20 mM Na-succinate / 40 mM Na-P_i (pH 7.4) / 0.5 mM Na-EDTA for 60 min at 37°C. The mixture was then diluted to 2 ml by addition of the same buffer and 1.5 μg of reductase complex was withdrawn for assay of succinate-cytochrome *c* reductase activity.

preparation described as partially purified complex III [28] has greater activity than many preparations of reductase complex (Table 2).

Ubiquinol-cytochrome *c* reductase activity measures the rate of electron transfer through the bc_1 segment to cytochrome *c* in a reaction which is inhibited by antimycin [29]. This activity should be equal to or greater than the succinate-cytochrome *c* reductase activity, and because it does not require succinate dehydrogenase activity it is useful for characterizing reductase complex in which the dehydrogenase is suspected as being rate limiting.

One difficulty in measuring ubiquinol-cytochrome *c* reductase activity is obtaining a suitable substrate. The hydrophilic-lipophilic balance of the reduced quinone substrate must be such that it preferentially donates reducing equivalents to the membranous bc_1 segment rather than directly to cytochrome *c* under assay conditions in which the molar ratio cytochrome *c*/complex typically exceeds 1×10^4. The concentration of cytochrome *c* must be high enough (>10 μM) to insure that the reaction is zero order in cytochrome *c*, and the solubility of the hydroquinone in the assay medium must be sufficient to generate saturating concentrations of substrate. Thus, hydrophilic hydroquinones such as durohydroquinone react directly with cytochrome *c* due to the mass action effect, and the normal biological substrate, ubiquinol-10, is too insoluble to obtain maximum initial rates.

Ubiquinol-2 (Q-2), a synthetic analog of the naturally occurring ubiquinones, has been used for measuring the ubiquinol-cytochrome *c* reductase activity of complex III [29], and with this substrate the reduction of cytochrome *c* is more than 90% sensitive to antimycin. Hatefi has reported that with complex III the rate of cytochrome *c* reduction by Q-2 is close to 1000 units/mg [19], although rates of 100 units/mg are more typical [30]. However, the low yield in synthesis and lability of Q-2 have contributed to a shortage which has limited experimentation.

Wan and coworkers recently synthesized analogs of ubiquinone containing short, saturated side chains and showed that these are suitable substrates for energy-linked reactions in mitochondria [31]. We find that the decyl analog of Q-2 is well suited for measuring ubiquinol-cytochrome *c* reductase activity with the resolved complex. The Q-2 analog gives rates greater than 25 units/mg at 30°C and the reduction of cytochrome *c* is more than 90% sensitive to antimycin. By comparison, when durohydroquinone was used to measure ubiquinol-cytochrome *c* reductase activity of the resolved complex, the reduction of cytochrome *c* was completely insensitive to antimycin. These observations are consistent with the results of Hare and Crane [32], who detected no antimycin-sensitive ubiquinol-cytochrome *c* reductase activity in the isolated reductase complex using durohydroquinone as substrate. The use of the decyl analog of Q-2 in measuring activity of reductase complex reconsititued with oxidation factor is described below (see Fig. 18, Section VIII.C).

Succinate-ubiquinone reductase activity is uniquely suited for measuring the activity of those components of the reductase complex which are required for electron transfer from succinate dehydrogenase to ubiquinone. This activity is commonly used to characterize complex II [33] and to document reconstitution of succinate dehydrogenase to membranous segments derived from the reductase complex [20, 34]. Succinate-ubiquinone reductase activity is measured in an assay in which reduction of ubiquinone is followed indirectly by including 2,6-dichloroindophenol as a terminal electron acceptor [33]. The details of this assay procedure are such that it should more accurately be referred to as a succinate-ubiquinone-dichloroindophenol reductase assay. This activity requires ubiquinone and is maximal when measured with Q-2 [33]. The distinguishing feature of succinate-ubiquinone reductase activity is that it is inhibited by thenoyltrifluoroacetone (TTB) [33], whereas succinate dehydrogenase activity is not [35]. In measuring this activity with the resolved reductase complex and in reconstitution experiments, it seems probable that the synthetic analog of Q-2 can be substituted for Q-2 in the succinate-ubiquinone-dichloroindophenol assay.

Succinate dehydrogenase activity is measured by following the rate of reduction of dichloroindophenol in the presence of phenazine methosulfate (PMS), which functions as a mediator to transfer reducing equivalents from the flavoprotein to the dye [36, 37]. Activity is determined at infinite concentrations of

PMS by extrapolation [36, 37]. Preparations of reductase complex having succinate-cytochrome *c* reductase activity of 6 units/mg typically have succinate dehydrogenase activities of 8-10 units/mg. This assay is applicable to both soluble and membrane-bound dehydrogenase and can be used in combination with the succinate-ubiquinone-dichloroindophenol reductase assay to measure the extent to which the dehydrogenase is functionally incorporated into the reductase complex. Vinogradov and coworkers have discussed the relationship between the multiple activities expressed by succinate dehydrogenase and reconstitution of the enzyme to the membrane [38].

2. Content and Spectral Properties of Cytochromes

Cytochrome content is a quantitative index of the extent to which the reductase complex has been purified, and certain spectral properties of the cytochromes are related to function. Reductase complex contains cytochromes b and c_1 and available evidence indicates these are present in a 2:1 molar ratio. The cytochrome content can be measured by difference spectroscopy using a procedure in which separate spectra of cytochromes b and c_1 are obtained by selective reduction as shown in Figure 3.

Cytochrome c_1 content is measured from a difference spectrum in which c_1 is reduced by ascorbate (Fig. 3a). Measurement of c_1 content is generally more accurate and reliable than that of b for several reasons. There is only one species of c_1 in mitochondria, and the heme group of c_1, unlike that of b, is covalently linked to the protein [39-41] and thus not susceptible to disruption during isolation of the reductase complex. The absorbance coefficient of c_1 has been determined with the purified protein [42] and is probably identical to that of c_1 in the membrane since the spectrum is unchanged [42] and there is only a small change, if any, in midpoint potential [43, 44], during isolation of this cytochrome.

Measurement of the c_1 content is subject to several sources of error. Underestimation of c_1 content may result from incomplete reduction by ascorbate, and oxidation by cytochrome *c* oxidase, especially with mitochondria, can contribute to this problem. The latter is prevented by inclusion of KCN. Overestimation of c_1 content may result from reduction of high-potential b, which contributes to the absorbance difference, $\Delta A_{553-539}$, but can be recognized (compare Figs. 3a and 4d) and corrected for. Measurement of c_1 in mitochondria is complicated by extensive overlapping absorbance of cytochrome c, which can be removed by extraction with KCl [45].

Cytochrome b can be measured from difference spectra of the b cytochromes (Fig. 3b,c). Since substrates and artificial reductants which reduce b will also reduce c_1, difference spectra of the b cytochromes are measured against a reference in which c_1 is selectively reduced by ascorbate. Thus one can obtain

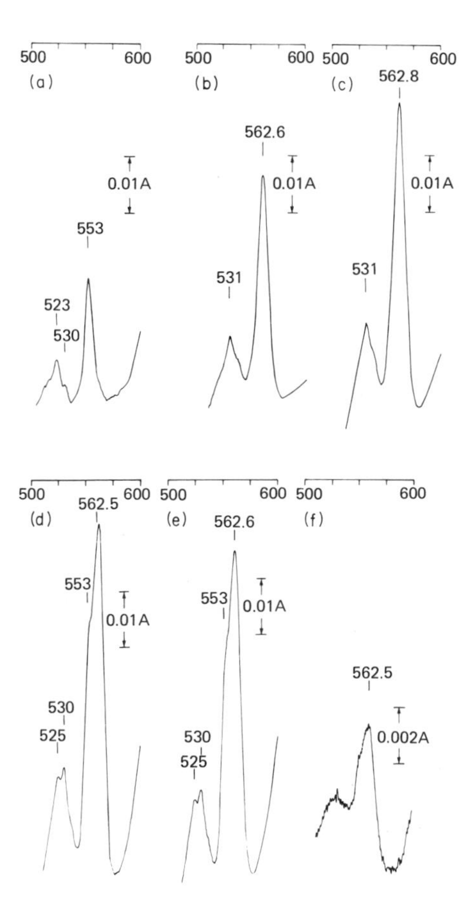

Figure 3 Absorption difference spectra of the cytochromes in resolved reductase complex. The reductase complex was suspended at 1mg/ml in 100 mM Na-P_i / 0.1 mM EDTA, pH 7.4, for obtaining the spectra. (a) Difference spectrum of cytochrome c_1. The spectrum was obtained from a sample reduced by ascorbate versus a reference oxidized by ferricyanide. The cytochrome c_1 content calculated from the coefficient $\Delta A_{553\text{-}539} = 17.5\ mM^{-1}$ is 1.20 nmol/mg. (b) Difference spectrum of the cytochrome *b* reduced by succinate recorded against a reference reduced by ascorbate. The content of succinate-reducible *b* calculated from the coefficient $\Delta A_{562\text{-}577} = 25.6\ mM^{-1}$ is 1.59 nmol/mg. (c) Difference spectrum of the cytochrome *b* reduced by dithionite recorded against a reference reduced by ascorbate. The calculated content of dithionite-reducible *b* is 2.13 nmol/mg. (d)–(f) Spectra showing the measurement of CO-reactive cytochrome *b*. The spectrum in (d) is of the cytochromes *b* and c_1 reduced by dithionite recorded against a reference oxidized by ferricyanide. The spectrum in (e) is identical to that in (d) except after addition of dithionite CO was bubbled through the sample. The decreased absorbance results from reaction of a portion of the *b* with CO. The spectrum in (f) is of the CO-reactive *b*, obtained as a difference of the spectrum in (d) versus (e). The calculated content of CO-reactive *b* is 0.21 nmol/mg.

separate spectra for the total *b* cytochromes, reducible by dithionite, or the portion thereof which is reduced by substrate. The spectrum in Figure 3b is of the cytochrome *b* reduced by succinate and that in Figure 3c is of the *b* reduced by dithionite.

Measurement of cytochrome *b* content by difference spectroscopy is more susceptible to error than is measurement of c_1. Underestimation of *b* may result from reduction of high-potential *b* by ascorbate in the reference cuvette. A reliable absorbance coefficient from a purified *b* protein is not yet available, and the absorbance spectrum results from two *b* species whose ability to exhibit multiple absorbance maxima and midpoint potentials is well documented (see Ref. 46). Finally, the extinction coefficient(s) of the *b* cytochromes may change during isolation of the reductase complex, since the *b* heme group is not covalently linked to the protein [48-51] and the midpoint potential is known to change [52].

The absorbance coefficient of 25.6 mM^{-1} which we employ was obtained by Berden and Slater [53] from measuring the fluorescence quenching of antimycin upon binding to the bc_1 segment. This value should be considered tentative, since it includes a correction factor (see Ref. 53) and these same measurements gave an absorbance coefficient $\Delta A_{533\text{-}539} = 20.1\ mM^{-1}$ for cytochrome c_1 as compared to the value of 17.5 mM^{-1} obtained with pure c_1 [42].

There is a natural tendency to state that the ratio of *b* to c_1 in resolved preparations is "approximately 2:1." Using the procedures described above, we commonly obtain a value of b/c_1 of 1.80:1 to 1.85:1. Careful reading of the literature and recalculation of some reported values indicates that this lower than expected ratio is not unique to our preparations. In addition to possible errors as discussed above, this lower ratio might result from loss of heme or of cytochrome *b* protein during preparation of the reductase complex.

Assuming a single absorbance coefficient for the *b* cytochromes, 75-80% of the total *b* in the reductase complex is reducible by succinate (Fig. 3). This value is not increased by addition of antimycin or KCN nor by anaerobic conditions. Less than 10%, and frequently none, of the dithionite-reducible *b* is reactive to CO as shown in Figure 3d-f.

Although the different spectral forms of *b* can be damaged by detergent [27], the presence of spectrally distinct forms of *b* in the resolved reductase complex [14, 27] and in complex III [18, 54] has been established. The spectrum of the low-potential *b*, here identified as that not reduced by succinate, is shown in Figure 4c. The spectrum exhibits a maximum at 564.1 nm, with a low-wavelength shoulder at ~558 nm, suggesting that the low-potential *b* consists primarily of b_{566} with some contribution from b_{562}, in agreement with previous reports [14, 18]. The low-wavelength shoulder is probably due to a split α band of b_{566}, but the possibility remains open that there may be a spectral form of *b*

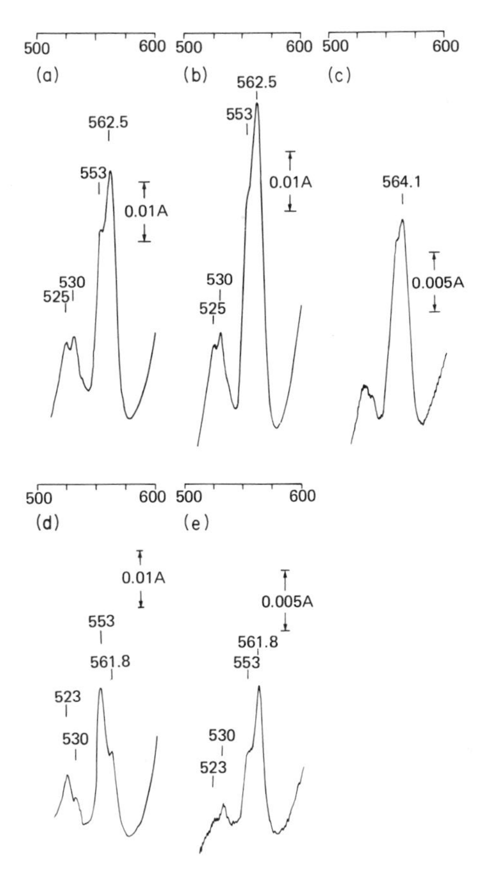

Figure 4 (a)-(c) Demonstration of low-potential cytochrome *b* in resolved reductase complex. The spectrum in (a) is of the cytochromes *b* and c_1 reduced by succinate and that in (b) is of the cytochromes reduced by dithionite. Both spectra were recorded versus a reference oxidized by ferricyanide. The spectrum in (c) is of the low-potential cytochrome *b*, obtained as a difference spectrum of the dithionite-reduced versus the succinate-reduced cytochromes. (d) and (e) Demonstration of high-potential cytochrome *b* in resolved reductase complex. The spectrum in (d) shows the reduction of c_1 and high-potential *b* by 1 mM ascorbate plus 0.1 mM dithiothreitol recorded versus a reference oxidized by ferricyanide. The high-potential *b* appears as a small peak or shoulder at 561.8 nm in the c_1 spectrum. The spectrum in (e) shows the high-potential *b*. The sample was reduced with ascorbate plus dithiothreitol as in (d) but the spectrum was recorded versus a reference in which most of the c_1 was reduced by addition of 0.1 mM ascorbate. The shoulder at 553 nm in (e) is from the difference spectrum of the small amount of c_1 which is not reduced in the reference.

with a separate absorbance maximum at 558 nm as discussed in Section V.D. Evidence for the presence in the isolated reductase complex of high-potential *b*, having a maximum at 561.8 nm, is provided by the spectrum shown in Figure 4e.

3. Polypeptide Composition

The usefulness of characterizing resolved reductase complex by polypeptide composition is illustrated by considering the ideal situation in which one could list the functional components of the complex, separate each of the polypeptides of the complex, and recognize each of the separated polypeptides as one of the functional components. This information would help establish the molecular stoichiometry within the complex, facilitate reconstitution of the complex from individual purified proteins, and provide a basis for investigating the relationship between structure and function at the molecular level.

Such an ideal situation has not yet been realized. The resolved reductase complex must include at least those polypeptides which are present in the bc_1 segment plus succinate dehydrogenase. A profile showing polypeptides of complex III separated by acrylamide gel electrophoresis in the presence of dodecyl sulfate is shown in Figure 5a. The profile reveals seven major polypeptides, indicated by arrows, having apparent molecular weights of 48,200, 44,300, 30,600, 24,500, 21,000, 15,500, and 13,000-14,000. The polypeptide of M_r = 13,000-14,000 is resolved into equal amounts of two polypeptides if electrophoresis is performed according to Laemmli [55]. An additional low molecular weight polypeptide (M_r = 4,000-6,000, see Ref. 8), barely visible in the photograph, is also present. The flavin-containing subunit of succinate dehydrogenase (M_r = 70,000) is visible in the upper region of the gel. The smaller (IP) subunit of succinate dehydrogenase (M_r = 27,000) is observable as a species distinct from the above polypeptides if larger amounts of sample are applied to the gel.

With the exception of the subunits of succinate dehydrogenase, the only polypeptide which can unequivocally be identified as a functional component of the complex is the heme-containing polypeptide of c_1. This identification is shown in Figure 5b. The apparent molecular weight of the c_1 polypeptide (30,600) has been obtained by comparison to standards including carbonic anhydrase (Fig. 5a) and validated by Ferguson-type analysis [10]. When purified c_1 is mixed with complex III, the c_1 polypeptide migrates as a single symmetrical species (Fig. 5b).

Because of confusion which is arising in the literature in assigning polypeptides to cytochrome *b*, it is important to establish whether the M_r = 30,600 polypeptide of the complex is composed solely of c_1. With purified c_1 the staining of the M_r = 30,600 heme-containing polypeptide by Coomassie brilliant blue is proportional to the amount of c_1 applied to the electrophoresis gel as shown in Fig. 6. Using this standard curve to measure the c_1 content of complex III, an average c_1 content of 3.69 nmol/mg is obtained as shown in Table 5. Difference spectroscopy indicates a c_1 content of 3.80 nmol/mg for this same preparation. This experiment provides strong evidence that the M_r = 30,600 polypeptide of complex III resolved in this electrophoresis system is composed solely of c_1.

To summarize what is known about the requisite polypeptides of the reductase complex, we have compiled in Table 6 the polypeptide composition of complex III reported from various laboratories. To simplify comparison, we have eliminated from the tabulation the flavoprotein subunit of succinate dehydrogenase (M_r = 70,000) and have aligned apparently identical polypeptides. In addition, we have designated those polypeptides which have been assigned to cytochromes *b* and c_1.

Although the limitations on comparing molecular weights reported from various laboratories are obvious, several conclusions can be drawn. All preparations of the bc_1 segment contain two polypeptides corresponding to those whose molecular weight we report as 48,200 and 44,300. These two polypeptides are known as core protein I and core protein II [56, 57]. Their function is not known. The reported molecular weights for core protein I range from 45,000 to 55,000 and for core protein II from 41,500 to 50,000.

Values of M_r = 26,700-37,000 have been reported for the heme-containing polypeptide of c_1. We have reported M_r = 30,600 for the c_1 polypeptide [10], and this analysis included comparison to carbonic anhydrase (29,000) and glyceraldehyde-3-phosphate dehydrogenase (36,000).

Resolving and identifying polypeptides of $M_r < 25,000$ is especially difficult and is reflected in the literature (see Table 6). The accuracy of molecular weight determinations and resolving capabilities of most electrophoresis systems decline in the low molecular weight range [58-60] and, to the extent that staining intensity is proportional to amount of protein, low molecular weight polypeptides will stain less intensely.

Various laboratories have reported that a polypeptide in complex III having M_r = 38,000 [8], 28,600 [8], 34,000 [9], or 37,000 [11, 61] is a subunit of cytochrome *b* (see Table 6). This identity is open to question, since the heme of *b* is not retained during denaturation and cytochrome *b* from mammalian mitochondria has not been purified. Our preparations of complex III appear to contain no polypeptide of M_r = 24,500-44,300 which might be attributed to *b* (see Figure 5a), and evidence cited above indicates that the reputed *b* polypeptide does not comigrate with the heme-containing c_1 polypeptide. In addition, comparison of our preparation of complex III with a preparation of bc_1 complex known to contain this polypeptide [11] establishes that this polypeptide is resolvable from c_1 and absent from our preparation as shown in Figure 7. The reputed *b* polypeptide also appears to be missing from complex III prepared in other laboratories [57, 63]. However, the resolution and detection of this re-

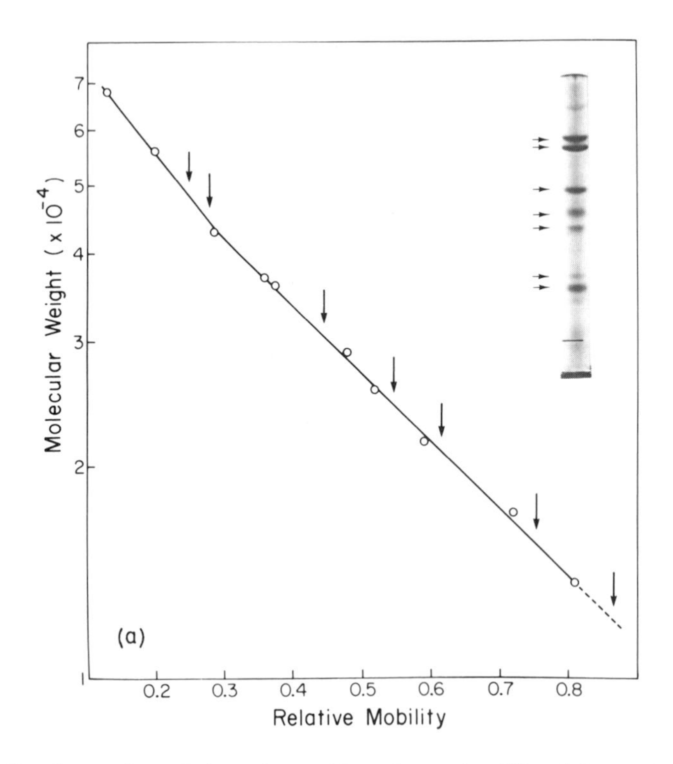

Figure 5a Separation of the polypeptides of complex III and determination of their molecular weights by acrylamide gel electrophoresis in dodecyl sulfate [204]. Proteins used as molecular weight standards, whose relative mobilities are indicated by the open circles, include serum albumin (68,000), glutamate dehydrogenase (53,000), yeast alcohol dehydrogenase (37,000), glyceraldehyde-3-phosphate dehydrogenase (36,000), carbonic anhydrase (29,000), chymotrypsinogen (25,700), myokinase (21,700), myoglobin (17,200), and ribonuclease (13,700). The relative mobilities of the seven major polypeptides of complex III are indicated by arrows. The molecular weights and identities of the polypeptides of complex III are discussed in the text.

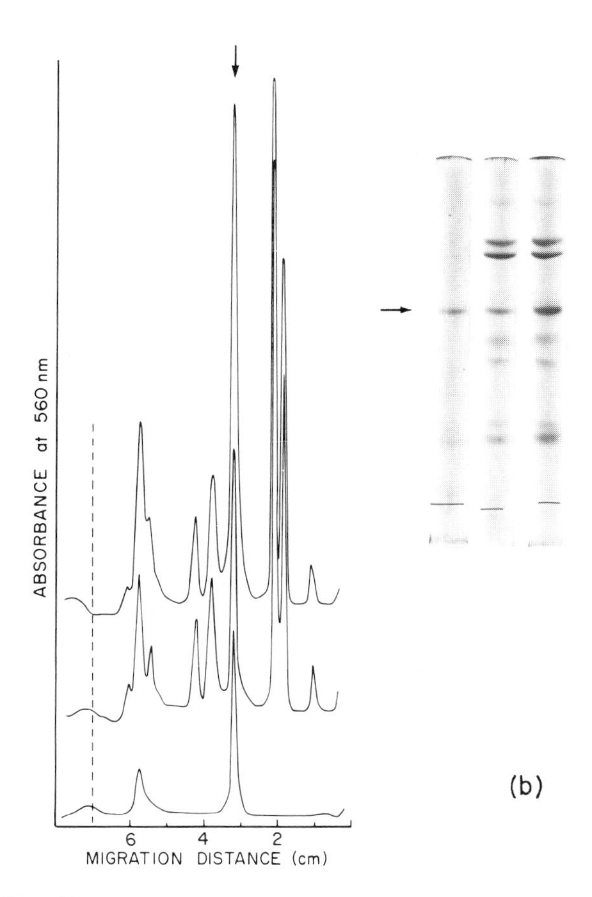

Figure 5b Identification of the cytochrome c_1 heme-containing polypeptide in complex III. The electrophoresis gel on the left and the bottom densitometer tracing are of purified cytochrome c_1, having a heme content of 23-25 nmol/mg. The gel and scan in the middle are of complex III, having a heme c_1 content of 3.8 nmol/mg. The gel on the right and the upper tracing were obtained by electrophoresis of a sample containing a mixture of complex III plus purified cytochrome c_1. The arrow designates the heme-containing polypeptide of cytochrome c_1 (M_r = 30,600). The low molecular weight c_1 associated polypeptide is evident in the bottom region of the gel. The possibility that this polypeptide may be an impurity in the purified c_1 preparation is discussed in the text.

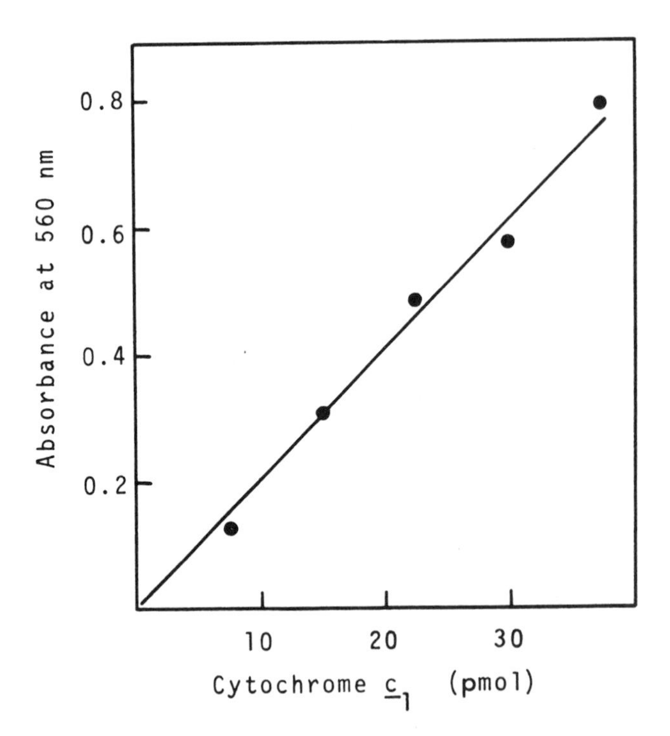

Figure 6 Proportional staining of cytochrome c_1 in electrophoresis gels by Coomassie brilliant blue. Purified cytochrome c_1 was applied to electrophoresis gels in known amounts as indicated. After electrophoresis, fixing, staining, and destaining the intensity of the stained c_1 heme-containing polypeptide was determined by scanning the gels at 560 nm.

puted *b* polypeptide by electrophoresis is not as simple as one might wish. Capaldi and coworkers have shown that the reputed *b* polypeptide exhibits anomalous behavior on electrophoresis and migrates with M_r = 41,500, 38,000, or 32,000 in the Weber-Osborn system, depending on conditions of denaturation [8]. In view of these results it seems premature to assign this anomalous polypeptide to cytochrome *b*.

If succinate-cytochrome *c* reductase complex consists of the bc_1 segment plus succinate dehydrogenase, then our electrophoresis analysis of complex III suggests that purified reductase complex would consist of 11 polypeptides. These would include the flavoprotein subunit of succinate dehydrogenase (M_r = 70,000), the core protein polypeptides (M_r = 48,200 and 44,300), the heme-

Table 5 Measurement of Cytochrome c_1 Content in Complex III by Polyacrylamide Gel Electrophoresis in Dodecyl Sulfate[a]

Complex III applied to gel (μg)	Cytochrome c_1 content (nmol/mg)
1.50	3.73
2.25	3.64
3.00	3.83
4.50	3.69
5.25	3.66
6.00	3.50
6.75	3.78
	Average = 3.69

[a]Cytochrome c_1 content of complex III was determined by applying varying amounts of complex III to electrophoresis gels and measuring the intensity of staining by Coomassie blue of the M_r = 30,600 heme-containing polypeptide. The amount of cytochrome c_1 in each gel was then calculated from the curve shown in Figure 6, which was obtained with the purified cytochrome.

containing c_1 polypeptide (M_r = 30,600), the smaller (IP) subunit of succinate dehydrogenase (M_r = 27,000), a polypeptide of M_r = 24,500 which may be the iron-sulfur protein of the bc_1 segment (see Ref. 63), and five lower molecular weight polypeptides of M_r = 21,000, 15,500, 13,000-14,000 (two polypeptides), and 4,000-6,000.

Resolved reductase complex contains the major polypeptides present in complex III, the subunits of succinate dehydrogenase, and three "extra" polypeptides as shown in Figure 8. Two of these extra polypeptides may correspond to subunits of F_1-ATPase [7, 9], and the 33,000 polypeptide appears to be identical with the reputed b polypeptide whose identity we questioned above. The extra polypeptides are digested when the reductase is treated with trypsin (Fig. 8). The trypsin digestion causes a parallel increase of succinate-cytochrome c reductase activity and content of cytochromes b and c_1. Thus it seems likely that these extra polypeptides are impurities. After trypsin treatment the composition of the reductase complex is very similar to that of complex III plus succinate dehydrogenase (Fig. 8), although fragments generated by the trypsin treatment obscure differences which might exist in the low molecular weight polypeptides.

Table 6 Polypeptide Composition Reported for Complex III

	Reference					
(See text)[a]	[8][b]	[8][c]	[9][d]	[62, 63]	[11]	[7]
–	–	–	59,000	–	–	62,000
–	–	–	55,000	–	–	–
48,200	46,600	45,500	50,000	46,000	53,000	55,000
44,300	41,500	44,500	47,000	43,000	50,000 (b)	50,000
–	38,000 (b)	28,600 (b)	34,000 (b)	–	37,000 (b)	37,000 (c_1)
30,600 (c_1)	30,000 (c_1)	26,700 (c_1)	29,000 (c_1)	29,000 (c_1)	30,000 (c_1)	30,000
24,500	23,300	24,600	25,000	25,000	28,000	26,000 (b)
21,000	–	–	18,000 (b)	15,000 (b)	17,000 (b)	19,000
15,500	12,700 (c_1)	15,000 (c_1)	14,500 (b)	14,000 (c_1)	15,000 (c_1)	16,000
13,000-14,000	14,300 (b)	9,000 (b)	13,000 (c_1)	11,500	–	12,000
4,000-6,000	5,600	4,800	10,000	–	–	10,000

[a]In the electrophoresis system of Weber and Osborn [204] there appears to be a single polypeptide of molecular weight 13,000-14,000 (see Fig. 5a). In the Laemmli system [55] two polypeptides are resolved as discussed in the text. The value of M_r = 4,000-6,000 is adopted here from the work of Capaldi and coworkers [8].

[b]Apparent molecular weights as measured in the Weber-Osborn [204] electrophoresis system.

[c]Apparent molecular weights as measured in the Swank-Munkres [60] electrophoresis system.

[d]In a subsequent report [57], Gellerfors and coworkers have shown an electrophoresis profile of complex III which lacks the M_r = 59,000 and M_r = 55,000 polypeptides attributed to the ATPase contamination and which lacks the M_r = 34,000 polypeptide which is presumably identical to the reputed apocytochrome b whose identity is discussed in the text.

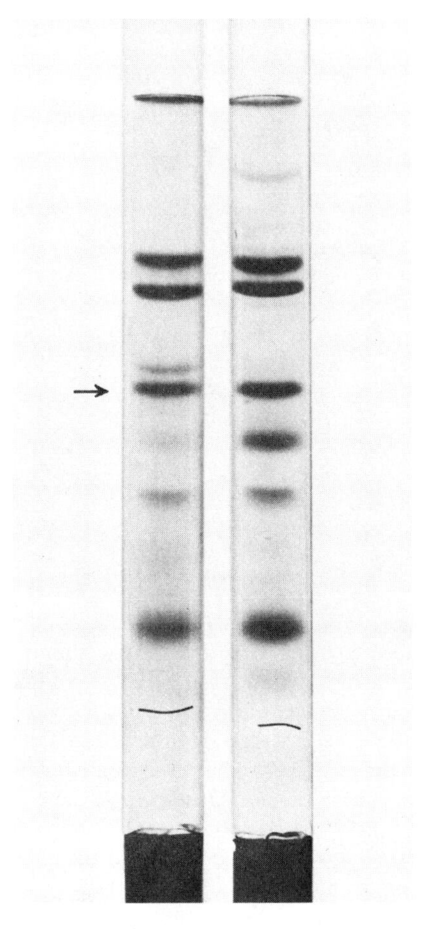

Figure 7 Acrylamide gel electrophoresis in dodecyl sulfate of complex III (right) and a preparation of resolved cytochrome bc_1 complex (left). The bc_1 complex was a gift of Drs. C. A. Yu and T. King. The migration of the c_1 heme-containing polypeptide is indicated by the arrow. Note the presence of trace amounts of the flavoprotein subunit from succinate dehydrogenase which is visible in the upper region of the electrophoresis gel of complex III and which is absent in the preparation of bc_1 complex. The polypeptide immediately above c_1 in the gel of the bc_1 complex has an apparent molecular weight of 33,000 in this electrophoresis system (see Ref. 8) and appears to be missing in complex III. The question of whether this polypeptide is apocytochrome *b* is discussed in the text.

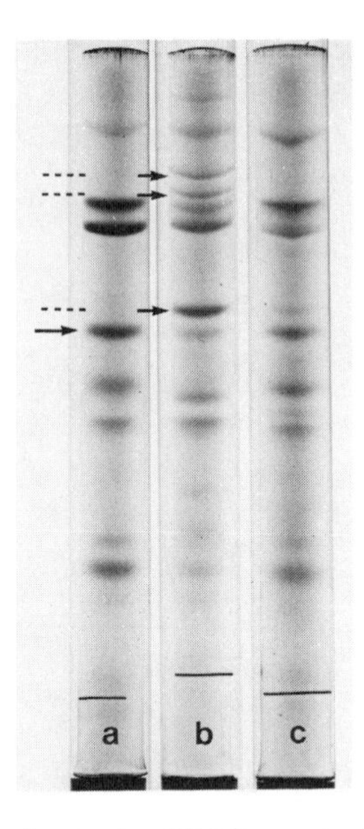

Figure 8 Electrophoresis of complex III (a), succinate-cytochrome *c* reductase complex (b), and succinate-cytochrome *c* reductase complex (c) after treatment with trypsin. The solid arrow indicates the c_1 heme-containing polypeptide. The dashed arrows indicate three polypeptides in the reductase complex which are absent from complex III and which are digested by treatment of the reductase complex with trypsin. Several additional polypeptides, of lower molecular weights, are also observable in the reductase complex and absent from complex III if larger amounts of protein are applied to the electrophoresis gel. The reductase complex was treated with trypsin by incubating for 60 min at 30°C with 100 μg trypsin/mg complex and the reaction stopped by addition of trypsin inhibitor. Reductase complex which was incubated as a control, minus trypsin, had succinate-cytochrome *c* reductase activity of 4.32 units/mg and a cytochrome *b* content of 1.70 nmol/mg. The reductase complex after treatment with trypsin had succinate-cytochrome *c* reductase activity of 7.03 units/mg and a cytochrome *b* content of 2.67 nmol/mg. Note that the three "extra" polypeptides of the reductase complex are extensively digested by the trypsin treatment.

III. Succinate Dehydrogenase

A. Properties of Purified Succinate Dehydrogenase

Succinate dehydrogenase is a flavoprotein which contains nonheme iron and acid-labile sulfide. The solubilized enzyme catalyzes electron transfer from succinate to artificial acceptors such as ferricyanide and, if not damaged during isolation, will rebind to the bc_1 segment and catalyze electron transfer from succinate to the membranous components of the electron transport chain. This latter property of the dehydrogenase is known as reconstitutive activity. The enzyme from beef heart has a molecular weight of 97,000 and contains 1 mol flavin, 8 g-atoms iron, and 8 g-atoms acid-labile sulfide per mol of enzyme [15]. The enzyme consists of two subunits whose molecular weights are 70,000 and 27,000 [15]. The larger, flavin-containing subunit (FP subunit) contains 1 molecule of FAD, which is covalently bound through the imidazole group of a histidine residue [64], and 4 atoms of iron and 4 atoms of acid-labile sulfide. The smaller subunit (IP subunit) contains 4 atoms of iron and 4 atoms of acid-labile sulfide [15].

Succinate dehydrogenase from bovine heart has been purified by Davis and Hatefi by extraction from complex II with sodium perchlorate, a chaotropic agent [15]. The purity of the soluble enzyme was demonstrated by activity measurements (T_n = 10,000), chemical analysis (flavin, nonheme iron, and acid-labile sulfide), and several physical measurements including acrylamide gel electrophoresis in dodecyl sulfate. The purified enzyme is capable of rebinding to the membranous bc_1 segment and has high reconstitutive activity [65].

The iron and sulfur atoms of succinate dehydrogenase are structurally clustered into iron-sulfur centers giving rise to three EPR signals. The FP subunit apparently contains two ferredoxin-type centers, each presumably composed of a 2Fe 2S grouping, giving rise to EPR signals S-1 and S-2 [16]. The IP subunit contains a high-potential type iron-sulfur center, presumably composed of a 4Fe 4S grouping, giving rise to EPR signal S-3 [66]. Since the iron-sulfur centers and related aspects of succinate dehydrogenase are discussed by Ohnishi in Chapter 1, we will focus our discussion on the integration of the dehydrogenase into the reductase complex.

B. Succinate Dehydrogenase as a Functional Component of the Reductase Complex

Succinate dehydrogenase is an allotopic enzyme [67]. Certain catalytic activities and physical properties of the soluble dehydrogenase are reversibly changed from those of the membranous enzyme. The mechanism by which succinate dehydrogenase transfers reducing equivalents to the bc_1 segment is not understood; thus, as suggested by Racker [68], it seems especially appropriate to focus attention on the interaction of this protein with its allotopic site in the membrane.

In the absence of PMS mediator, soluble succinate dehydrogenase catalyzes electron transfer from succinate to ferricyanide in a reaction whose biphasic kinetics suggest that there are two sites on the enzyme at which ferricyanide reacts. These two sites are different in that one has high affinity (K_m = 0.25 mM) while the other has low affinity (K_m = 3 mM) for ferricyanide [38, 69]. When succinate dehydrogenase is bound to the membrane, the high-affinity ferricyanide site is masked or latent, since activity associated with this site cannot be measured in the membranous enzyme but appears when the enzyme is solubilized [38, 69]. The low-affinity ferricyanide site is accessible and measurable in the membrane-bound dehydrogenase as well as in the soluble enzyme.

The reconstitutive activity of soluble succinate dehydrogenase is labile, such that at 4°C there is a first-order decay of reconstitutive activity with a half-life ($t_{1/2}$) = 1.6 hr [70]. Membranous dehydrogenase retains inherent reconstitutive activity for several days at 4°C. The activity of the high-affinity ferricyanide site in the soluble enzyme is also labile. Reconstitutive activity of the soluble enzyme and the activity of the high-affinity ferricyanide site are lost in parallel [38, 69], which indicates a relationship between these two activities.

Succinate dehydrogenase which is functionally integrated into complex II [71] or the reductase complex and soluble dehydrogenase which is reconstitutively active both exhibit the S-3 EPR signal in addition to the S-1 and S-2 signals [66]. Soluble dehydrogenase loses the S-3 signal in parallel with loss of reconstitutive activity [66]. These and related findings (see Chapter 1 and Refs. 38 and 66) are convincing circumstantial evidence that the iron-sulfur center responsible for the S-3 signal is the high-affinity ferricyanide site, that this site is located in the IP subunit (M_r = 27,000), and that reconstitutive activity is associated with the functional integrity of this labile iron-sulfur center.

It appears therefore that the IP subunit is responsible for binding of succinate dehydrogenase to the bc_1 segment and that this binding stabilizes the labile S-3 center and permits electron transfer from the dehydrogenase to an endogenous acceptor in the bc_1 segment. The proposed sequence of electron transfer can be written as succinate → FP (FAD, S-1, and S-2) → IP (S-3) → bc_1 segment acceptor.

An observation whose significance has perhaps been overlooked is that binding of succinate dehydrogenase to the bc_1 segment and stabilization of reconstitutive activity can occur under conditions which do not restore electron transport activity from the dehydrogenase to the bc_1 segment [70]. Protein-protein interaction appears to be of primary importance in preserving latent reconstitutive activity and phospholipids are additionally required for electron transfer from the dehydrogenase to the acceptor in the bc_1 segment [70]. This phenomenon is evident in the results shown in Table 1, in which latent reconstitutive activity is retained in the absence of electron transport activity which can be restored by addition of lipid.

Although there is evidence indicating that binding of the dehydrogenase to the bc_1 segment involves the IP subunit of the dehydrogenase, less is known about the corresponding polypeptide of the bc_1 segment which might be involved in such protein-protein interaction. It is of particular interest to know if the *b* cytochromes exhibit any spectral or potentiometric changes upon reconstitution with succinate dehydrogenase. Such information might be obtainable from the bc_1 complex of Yu and coworkers [11]. Also, it seems important to establish whether differences exist in the *b* cytochromes of the bc_1 complex of Yu and coworkers, which can be reconstituted with succinate dehydrogenase [11], compared with complex III, which cannot be thus reconstituted [11, 65].

The acceptor of reducing equivalents from the dehydrogenase within the bc_1 segment may be either (or both) the quinone or semiquinone form of ubiquinone. The troublesome experimental fact which is central to the postulated acceptor role of ubiquinone is that in the soluble form, reconstitutively active succinate dehydrogenase does not catalyze reduction of ubiquinone by succinate. Membranous dehydrogenase does catalyze electron transfer from succinate to ubiquinone and this activity is inhibited by TTB [33]. The succinate-dye reductase activities of the soluble dehydrogenase are not inhibited by TTB [35]. Thus the membranous segment restores succinate-ubiquinone reductase activity to succinate dehydrogenase and this allotopic property of the dehydrogenase can be recognized by its sensitivity to TTB. The question is, how does the bc_1 segment accomplish this?

There are three general mechanisms by which the bc_1 segment might confer TTB-sensitive succinate-ubiquinone reductase activity to the dehydrogenase. One possibility is that the bc_1 segment contains a redox component (other than ubiquinone) which is an obligatory carrier of reducing equivalents between the dehydrogenase and ubiquinone. A second possibility is that the bc_1 segment, on binding the dehydrogenase, induces a conformational change in the dehydrogenase such that the membranous enzyme can bind and reduce ubiquinone while the soluble protein cannot. The third possibility is that the bc_1 segment induces a physical or chemical change in ubiquinone which is prerequisite to its being an effective acceptor.

To evaluate these three possibilities it is useful to identify which components of the bc_1 segment are required for TTB-sensitive succinate-ubiquinone reductase activity. Complex II [33] and a reconstituted complex formed from a partially purified preparation of cytochrome *b*, phospholipids, and ubiquinone [72] exhibit TTB-sensitive activity. From these results it can be deduced that cytochrome c_1 and the iron-sulfur protein of the bc_1 segment are not required. We would speculate that oxidation factor is not required since it seems unlikely that oxidation factor is retained in complex II or in the partially purified *b* preparation.

Thus it appears that cytochrome *b*, ubiquinone, and phospholipid are required for TTB-sensitive ubiquinone reductase activity. It is also possible that the two core protein polypeptides are required since they appear to be present in complex II (see figure 6 in Ref. 15), and it has been reported that reaction of a core protein polypeptide in complex III with *p*-hydroxymercuribenzoate causes a coincidental loss of ubiquinone reductase activity [73].

Several studies have led to the conclusion that cytochrome *b* fulfills a structural role but does not undergo oxidation-reduction in the succinate-ubiquinone reductase reaction. Bruni and Racker found that reconstitutively active succinate dehydrogenase formed a stable complex with a partially purified preparation of cytochrome *b* and the succinate-ubiquinone-dichloroindophenol reductase activity of the reconstituted complex was sensitive to TTB [72]. This activity was absolutely dependent on the presence of cytochrome *b*, and certain preparations of *b* were apparently damaged in a manner such that they were not capable of forming an active complex. However, it was also observed that the cytochrome *b* which formed an active complex with succinate dehydrogenase did not appear to undergo oxidation-reduction [72]. It was suggested that cytochrome *b* fulfills a structural role in organizing the components of the electron transport chain in addition to its functional role as a redox component. The possibility that this structural activity may provide the basis for an assay for cytochrome *b* is discussed in Section V.C.

Although cytochrome *b* most likely does fulfill a structural role as postulated, it also seems possible that oxidation-reduction of *b* is obligatory in the succinate-ubiquinone reductase reaction as it occurs in the intact succinate-cytochrome *c* reductase complex but that the turnover of *b* is not required in the succinate-ubiquinone-dichloroindophenol reductase assay. In highly active preparations of complex II, having TTB-sensitive quinone reductase activity of 100 units/mg, addition of succinate causes a slow and partial reduction of *b* [20]. This observation has been interpreted as evidence that redox turnover of *b* is not required, and this interpretation is strengthened by the fact that lack of *b* reducibility in these experiments cannot be attributed to low ubiquinone reductase activity [20]. However, it was also found that addition of Q-2 increased the rate and extent of *b* reduction by succinate (see figure 1 in Ref. 20). Since addition of Q-2 is also required for TTB-sensitive ubiquinone reductase activity, the latter result can be interpreted as evidence that oxidation-reduction of *b* is associated with activity. In addition, TTB inhibits the reduction of *b* by succinate both in the absence and presence of Q-2 [20]. Although only 40% of the *b* is rapidly reduced by succinate in the presence of Q-2, this extent of reduction may be sufficient to support the observed ubiquinone reductase activity. Incomplete reduction of *b* by succinate is also observed in highly active reductase complex (see Fig. 3).

Several results lead us to suggest that TTB may inhibit succinate-ubiquinone reductase by blocking electron transfer between iron-sulfur center S-3 and the semiquinone form of ubiquinone. The amount of TTB required to inhibit succinate-ubiquinone-dichloroindophenol reductase activity in a reconstituted system has been shown to be proportional to the amount of succinate dehydrogenase added [20]. Mowery and coworkers [74] have observed that TTB causes some broadening and loss of the S-3 EPR signal but noted that there is no direct evidence that the inhibitor chelates the iron of the iron-sulfur center. On the expectation that chelation by TTB would cause a more dramatic change in the EPR signal than observed, it was proposed that the inhibitor may exert an indirect effect on center S-3 by interacting vicinal to this iron-sulfur center. We would point out that TTB normally forms a trivalent chelate with iron [75] and that coordination of the iron by sulfur in an iron-sulfur center may sequester the metal in a manner such that a monovalent or otherwise weak chelate is formed. EPR studies have detected resonance signals whose properties indicate there is a spin-spin interaction between center S-3 and ubisemiquinone (see Section IV.C and Refs. 76-78). More recent findings indicate that TTB disrupts a spin-spin interaction between two ubisemiquinone radicals and also shifts the midpoint potential of center S-3 [79]. To account for all of these results, we suggest that TTB forms a weak chelate with the iron of center S-3 and thus displaces ubisemiquinone which is otherwise bound in proximity to this iron-sulfur center.

To illustrate the dilemma currently challenging investigators, we have outlined in Figure 9 a mechanism of electron transfer which is consistent with available information and which draws attention to several troublesome unanswered questions. Suppose for a moment that electron transfer from the dehydrogenase to the bc_1 segment starts with a pre-existing, catalytic quantity of semiquinone, whose initial formation is discussed below. According to this mechanism, an electron is then transferred from iron-sulfur center S-3 to the semiquinone form of ubiquinone (QH•) along with H^+ to form hydroquinone (QH_2). This reaction is inhibited by TTB. Hydroquinone is oxidized to semiquinone and quinone by an acceptor (A) such as dichloroindophenol. An important point to note is that dichloroindophenol functions both as a terminal acceptor and as a semiquinone

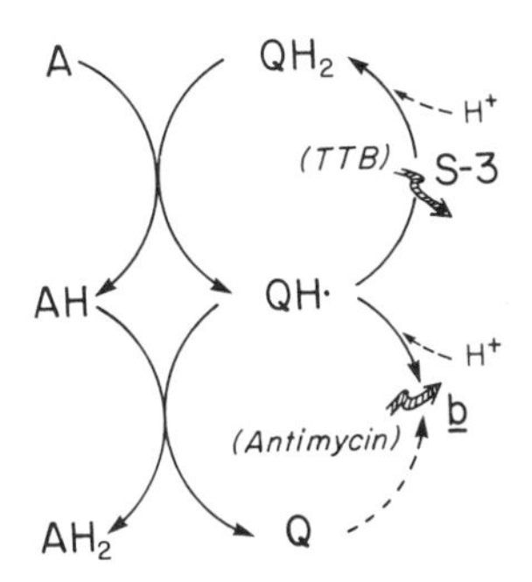

Figure 9 Proposed mechanism of electron transfer in the succinate-ubiquinone-dichloroindophenol reductase reaction. The diagram describes the sequence of electron transfer which may occur under steady-state conditions as electrons are transferred from iron-sulfur center S-3 of succinate dehydrogenase to the acceptor dye (A), dichloroindophenol.

regenerating system. Additional quinone can then enter the hydroquinone-semiquinone cycle via the dismutation reaction $2QH\cdot \rightarrow Q + QH_2$. In the functionally intact reductase complex, reduced *b* might function as a semiquinone regenerator by transferring an electron and H^+ to quinone. This reaction would be lacking in complex II (as indicated by the broken arrow) due to loss of redox components such as c_1 which are otherwise required for regeneration of reduced *b*.

In other words, turnover of *b* could be obligatory in the functionally intact succinate-ubiquinone reductase reaction. But in preparations of complex II in which *b* is damaged, the required turnover of *b* may be circumvented by assay conditions in which an artificial acceptor generates semiquinone in sufficiently high concentrations to accommodate the influx of reducing equivalents from the dehydrogenase.

Reduction of *b* by succinate, as observed with some preparations of complex II [20], might occur via semiquinone whose formation is inhibited by TTB. Reduction of *b* by succinate would be accelerated by added Q-2 and inhibited by antimycin as observed [20]. Although antimycin would inhibit reduction of *b* by succinate and its oxidation by fumarate, it would not inhibit the succinate-quinone reductase reaction as routinely measured because of the short circuit introduced by dichloroindophenol (Fig. 9).

Herein then arises the troublesome question. How does membranous succinate dehydrogenase form ubisemiquinone, as required to initiate the hydroquinone-semiquinone cycle, if soluble dehydrogenase does not catalyze reduction of Q-2 by succinate? One is confronted by this same problem in postulating several otherwise feasible mechanisms, and the answer is clearly not yet at hand.

We would suggest that there may be separate donor sites on succinate dehydrogenase for ubiquinone and ubisemiquinone. Specifically, it seems possible that iron-sulfur center S-2 may catalyze the reaction $Q \rightarrow QH\cdot$ and that center S-3 may catalyze the reaction $QH\cdot \rightarrow QH_2$. The observed stabilization of reconstitutive activity in the absence of succinate-ubiquinone reductase activity [70] could result from protein-protein interaction between center S-3 of the IP subunit and a polypeptide of the bc_1 segment, which would stabilize center S-3 and thus protect reconstitutive activity, but which would leave center S-2 incapable of carrying out the reaction $Q \rightarrow QH\cdot$ and succinate-quinone reductase activity would be absent.

The additional requirement for phospholipid which allows succinate-quinone reductase activity to appear [70, 72] may be due to one of several effects on center S-2. One possibility is that phospholipid facilitates the transfer of $QH\cdot$ from center S-2 to center S-3, thus allowing the reaction $Q \rightarrow QH_2$ to proceed. If this is the case, center S-2 in the soluble dehydrogenase may participate in a suicide reaction in which a single turnover of $Q \rightarrow QH\cdot$ renders center S-2 inactive or, alternatively, QH· may be sequestered in a protected environment at center S-2. Otherwise one would expect that QH· would dismutate at center S-2 and complete the conversion $Q \rightarrow QH_2$. Another possibility is that the lack of succinate-ubiquinone reductase activity in the reconstitutively active enzyme may be due to the anomalously low midpoint potential of center S-2 in the soluble enzyme [16], in which case the reaction Q → QH· would be precluded by lack of reduction of center S-2 by succinate.

The mechanism we suggest is consistent with the midpoint potentials obtainable by the QH·/Q and QH_2/QH· couples (see Section IV.C). In this regard it is interesting to note that the midpoint potential of center S-2 is raised 150 mV when the dehydrogenase is functionally integrated into the membrane [16]. In view of evidence that stabilization of S-3 can occur without reconstitution of electron transfer activity [70], it seems important to learn if the midpoint potential of S-2 changes coincident with stabilization of S-3 by the membrane or only under conditions in which electron transfer activity is restored.

IV. Ubiquinone

A. An Overview of Ubiquinone

Ubiquinone was first postulated to be a member of the electron transport chain simply because this ubiquitous, neutral lipid is capable of oxidation-reduction and is present in mitochondria [80]. Some 20 years after this proposal, ubiquinone is now established as an obligatory redox component of the electron transport chain. In addition, ubiquinone might mediate the active transport of H^+ outward across the mitochondrial membrane as predicted by the chemiosmotic hypothesis [81].

Evidence that ubiquinone is an obligatory participant in electron transport is as follows. Ubiquinone is present in mitochondria of tissues from all mammalian species tested [82, 83] and is concentrated in the inner mitochondrial membrane [84]. With either succinate [85, 86] or NADH [87] as substrate, electron transport activity is lost when Q is extracted from mitochondria and restored when the quinone is reincorporated. Succinate-cytochrome *c* reductase complex reconstituted from resolved membrane components requires addition of Q [25] and mutants of *Escherichia coli* lacking Q show a loss of dicoumarol-sensitive respiration, which is fully restored by addition of Q [88].

The steady-state redox poise of Q changes in parallel with that of the cytochromes in coupled mitochondria [89], and Q is reduced in an antimycin-sensitive reaction during reversed electron transport [90]. NADH and succinate oxidation are inhibited by competitive structural analogs of Q [91] and piericidin A, a potent inhibitor of NADH oxidase [92], bears a striking structural resemblance to Q. When antimycin is added to mitochondria in the aerobic steady state, Q and the *b* cytochromes become more reduced, while c_1 is oxidized [4]. Further, in O_2 pulse experiments the kinetics of Q oxidation are comparable to those of other redox components of the bc_1 segment, and there is a kinetic and stoichiometric correlation between Q oxidation and cytochrome *b* reduction [93].

The possibility that ubiquinone mediates the active transport of H^+ across the mitochondrial membrane as predicted by the chemiosmotic hypothesis [81] has not received universal acceptance. The mechanism by which Q might function as a transporter of H^+ operating within the electron transport chain is as follows: if Q is reduced by an electron donor with uptake of H^+ from the medium on the matrix side of the membrane and oxidized by an electron acceptor with release of H^+ to the medium on the cytosol side, the redox turnover of Q would accomplish net active transport of H^+ outward across the membrane. If such is the case, two predictions follow. First, the electron donor and acceptor for Q must be asymmetrically disposed across the inner membrane. Second, the oxidation-reduction of Q must include a transmembrane movement of ubiquinone and ubiquinol or the equivalent of such movement. These two predictions are open to experimental support.

B. The Mobile Carrier Hypothesis and the Pool Function of Ubiquinone

The concept that ubiquinone might function as a mobile carrier of reducing equivalents within the mitochondrial membrane was first proposed by Green and coworkers [94-96]. As evidence evolved that the electron transport chain consisted of a group of lipoprotein complexes, it was proposed that Q acts as a mobile carrier which collects and shuttles reducing equivalents from the dehydrogenase complexes to the bc_1 complex (see Figs. 1 and 2) and that cytochrome *c* fulfills an analogous mobile carrier function between the bc_1 and cytochrome oxidase complexes.

The mobile carrier hypothesis implies that there is a single, homogeneous pool of Q which diffuses between the primary electron transport complexes in the phospholipid continuum of the membrane and that there are specific sites for oxidation-reduction of Q located in the lipoprotein domains of the dehydrogenase complexes and the bc_1 complex. Evidence consistent with the mobile carrier hypothesis includes the following. Ubiquinone is generally present in molar excess, relative to the bc_1 complex, and there does not appear to be a fixed ratio of ubiquinone per bc_1 complex [35]. Appreciable quantities of Q are found in preparations of complex I, II, and III [3], and since ubiquinone is soluble in hydrocarbon solvents, it might be expected to freely diffuse within the hydrophobic interior of the phospholipid bilayer of the membrane continuum. Ubiquinone is a common acceptor of reducing equivalents from the various dehydrogenases, including NADH dehydrogenase, succinate dehydrogenase, and glycerol phosphate dehydrogenase [35] and appears to function as a confluence point through which reducing equivalents from one dehydrogenase can equilibrate with another [4, 97].

Kroger and Klingenberg have adopted the mobile carrier hypothesis and have shown that the kinetics of oxidation-reduction and the steady-state redox poise of Q are consistent with a model in which 80-90% of the Q exists as a functionally homogeneous pool and there are not separate compartments of Q for the different substrates such as NADH and succinate [98]. These authors have stressed that the redox behavior of Q appears to be determined solely by the activity of the donor which reduces the quinone and the acceptor which oxidizes the hydroquinone [98]. Thus the redox poise of Q can be mathematically predicted by a set of equations which include a single donor function and a single-acceptor function, and the equations are equally valid when the rate of electron flow through the Q pool is limited by antimycin [99]. In other words, the pool function of Q implies that only a single donor site is required for the reduction of Q to QH_2 and likewise a single acceptor site for the oxidation of QH_2 to Q.

The reversible conversion of quinone to hydroquinone must proceed through a semiquinone or equivalent intermediate or else must occur as a concerted reaction. Because of the properties of the ubiquinone molecule (see Section IV.D), and because ubisemiquinone has been detected in mitochondria [100], it seems most likely that the former type of reaction mechanism applies. However, whether significant concentrations of ubisemiquinone exist under steady-state conditions of electron transfer and whether there are redox donors and acceptors in the reductase complex which specifically participate in electron transfer to and from ubisemiquinone has been questioned.

Kroger and Klingenberg [98] have proposed that ubisemiquinone occurs only in minute concentrations as a transient intermediate in electron transfer and that conversion of ubisemiquinone to ubiquinone and ubiquinol occurs spontaneously, without participation of donor or acceptor for ubisemiquinone, by the dismutation reaction $2QH\cdot \rightarrow Q + QH_2$. Involvement of a rapid dismutation reaction in the redox turnover of Q is consistent with the results of kinetics measurements [98]. To support this view it has been argued that the stability constant of durosemiquinone in water is so low ($K_{eq} = 10^{-24}$) that when applied to Q in the mitochondrial membrane, one can calculate that less than 0.001% of the total Q would exist as ubisemiquinone [101]. Thus the molar ratio of ubisemiquinone to any redox donor, such as succinate dehydrogenase, would be less than 10^{-3}:1, which would make it unlikely that the semiquinone would be effective as an acceptor of reducing equivalents.

C. Binding Sites for Ubiquinone, Ubisemiquinone, and Ubiquinol

While the kinetics measurements are consistent with the pool function proposed for Q [98, 99], they obviously do not exclude the involvement of an acceptor for semiquinone. It seems unlikely that the stability constant of durosemiquinone in water [102] is applicable to ubisemiquinone in the mitochondrial membrane. Although semiquinone does dissociate a proton with pK = 6 [103], as noted [101], this merely indicates that such dissociation will occur if semiquinone is in an aqueous environment; it does not establish that ubisemiquinone in situ would concentrate at the aqueous membrane interface. In addition to being a hydrophobic environment, the inner mitochondrial membrane contains proteins which may specifically contribute to the stability of ubisemiquinone, and a donor or acceptor for ubisemiquinone might fulfill such a role. Furthermore, as discussed below (Section IV.D), the ubiquinone molecule includes several structural features which may also contribute to the stability of ubisemiquinone in situ, and these are distinctly lacking in the structure of duroquinone.

Experimental evidence indicates that ubisemiquinone exists as a detectable intermediate in electron transport [100]. Measurements by EPR spectroscopy have recently identified a paramagnetic species attributable to ubisemiquinone and power saturation and temperature dependence measurements suggest there is an interaction between ubisemiquinone and iron-sulfur center S-3 [76-78]. In addition, Boveris and coworkers have shown that generation of superoxide anion by mitochondria is significantly enhanced by antimycin and is dependent on the presence of ubiquinone [104]. This important finding suggests that antimycin might promote the reaction of ubisemiquinone with O_2, either by causing an increase in the steady state concentration of semiquinone or by displacing semiquinone from an acceptor site, or both. In oxygenated buffers the autooxidation of ubiquinol results in production of superoxide anion [105].

Identifying the components of the electron transport chain which transfer reducing equivalents to and from the various redox forms of Q and establishing their topographical location in the membrane is central to understanding the mechanism of electron transfer within the succinate-cytochrome *c* reductase complex. In this regard it is useful to recognize that the physicochemical properties of ubisemiquinone are markedly different from those of ubiquinone or ubiquinol. In addition, as pointed out by Mitchell [106], the midpoint potential of the $QH\cdot/Q$ couple may be 300-600 mV more negative than that of the $QH_2/QH\cdot$ couple, depending on the stability constant of ubisemiquinone. Thus, unless one accepts the dismutation of ubisemiquinone as an obligatory step in electron transport, it seems obvious that the reactions $Q \rightarrow QH\cdot$ and $QH\cdot \rightarrow QH_2$ would require separate donor sites. Similar considerations apply to the acceptor sites for oxidation of QH_2.

In discussing the mechanism of the succinate-ubiquinone-dichloroindophenol reductase reaction, we suggested that succinate dehydrogenase may have two dissimilar donor sites, iron-sulfur centers S-2 and S-3, which catalyze the low-potential $Q \rightarrow QH\cdot$ and high-potential $QH\cdot \rightarrow QH_2$ reactions, respectively. According to this proposal, iron-sulfur center S-3 would function as an acceptor/donor site for oxidation-reduction of ubisemiquinone (see Fig. 9) and TTB would disrupt the interaction between the semiquinone and its acceptor/donor site. Further indirect evidence implicating the TTB-sensitive site as an acceptor/donor for ubisemiquinone comes from the finding that ubiquinone antagonizes binding of TTB [107]. The asymmetric location of the S-3 iron-sulfur center of succinate dehydrogenase on the matrix side of the inner mitochondrial membrane [108] and evidence that reduction of Q by NADH dehydrogenase occurs on the matrix side of the membrane [109] are consistent with the chemiosmotic postulate that reduction of Q by the dehydrogenase donors results in uptake of H^+ from the matrix medium.

Cytochrome *b* may be a second acceptor/donor site for Q. Nelson and coworkers have shown that the rate of antimycin binding to the bc_1 segment is increased several-fold in submitochondrial particles depleted of Q, and the rate of antimycin binding decreases when Q is replenished [107]. This suggests that Q competes for the high-affinity antimycin binding site and, if interpreted with evidence that antimycin binds to one of the *b* cytochromes [110], leads to the postulate that this *b* cytochrome is an acceptor/donor for Q. Recalling that the Q-dependent production of superoxide anion is enhanced by antimycin [104], it seems possible that antimycin prevents electron transfer from ubiquinol or ubisemiquinone to cytochrome b_{562} by binding to the latter.

D. Size and Properties of the Ubiquinone Molecule

In anticipation of a mechanism by which ubiquinone might transport hydrogen through the mitochondrial membrane, it is worthwhile considering whether the naturally occurring ubiquinones are capable of transmembrane movement. Although it is not universally accepted that ubiquinone operates as a H^+ carrier as predicted by the chemiosmotic hypothesis, it does seem to be commonly assumed that ubiquinone is inherently capable of diffusion-dependent transmembrane movement. This assumption is without experimental support.

In this regard we wish to draw attention to the size of the ubiquinone molecule relative to the cross-sectional dimension of a phospholipid bilayer, as shown in Figure 10, and to certain other properties of the ubiquinone molecule which may be relevant to its function. In vertebrates the most commonly occurring homologue of ubiquinone is Q-10, having a 50-carbon isoprenoid sidechain [83]. One exception to this generality is the occurrence of Q-9 in the rat [111].

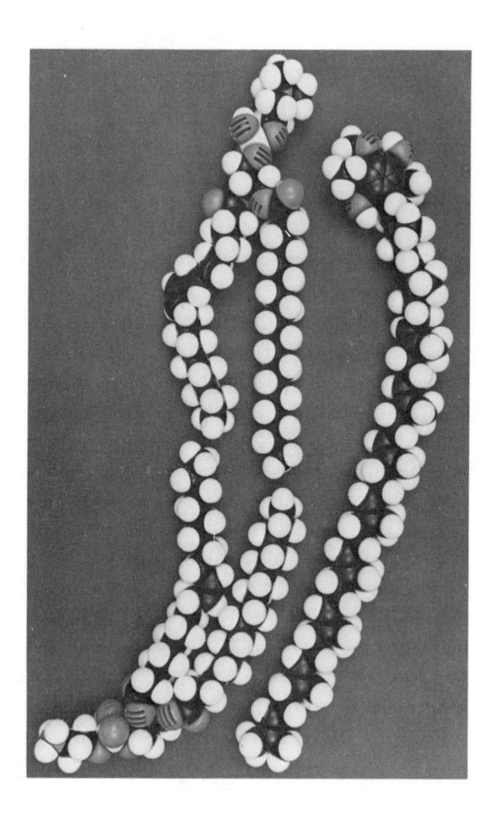

Figure 10 Dimensions of the ubiquinone-10 molecule relative to a phospholipid bilayer as illustrated by CPK models. The ubiquinone-10 molecule, located on the right, is oriented to show eight of the 10 isoprenoid double bonds. The vicinal methyl groups extend to the right beside each double bond. The substituted hydroquinone ring is at the upper end of the molecule. The OH groups of the hydroquinone which are *ortho* and *meta* to the isoprenoid sidechain are visible in the 7 o'clock and 1 o'clock positions on the ring respectively. The phospholipid molecule to the upper left is 18:0, 18:2 phosphatidylcholine. The pronounced bending of the fatty acyl sidechain resulting from two *cis* double bonds is evident. The phospholipid molecule to the lower left is 16:0, 18:1 phosphatidylethanolamine.

Ubiquinone-6, which occurs in certain yeasts, is the shortest sidechain homologue found in nature [83].

The extended dimension of the 50-carbon isoprenoid sidechain of Q-10 is 48 Å, and that of the entire molecule is 56 Å. The Q-10 molecule clearly can extend through most of a simple phospholipid bilayer of dimensions 40-50 Å (Fig.

10). One might expect that the effective size of the Q molecule is reduced by free rotation around single bonds within the sidechain and that random folding of the sidechain facilitates free diffusion of Q within the hydrophobic interior of a bilayer. Although some folding is certainly possible, the structure of the isoprenoid sidechain places strict limits on internal rotation around single bonds. The 10 double bonds introduce a periodic rigidity into the sidechain and the van der Waals radii of the 10 vicinal methyl groups further limit rotation around adjacent single bonds by steric hindrance. The rigidity of this molecule relative to a fatty acid can be readily appreciated by handling the CPK models.

The isoprenoid sidechain contributes to molecular rigidity rather than to fluidity of the local membrane environment. Unlike the *cis* double bonds in the unsaturated fatty acyl groups of membrane phospholipids, which increase membrane fluidity [112, 113], the *trans* double bond of the isoprenyl group does not give rise to an increased precessional radius as the distal region of the sidechain is precessed around a proximal single bond. Consequently the *trans* double bonds of ubiquinone, even when repeated 10 times within the molecule, probably make a smaller contribution to local membrane fluidity than would a *cis* double bond. Because lateral mobility of a lipid molecule in a membrane is inversely proportional to the square of its length [114], the isoprenoid sidechain additionally limits, rather than enhances, diffusion. If ubiquinone manifests its apparent size and extends into both halves of the membrane bilayer, the hydrophobic association of the sidechain with adjacent phospholipids and the associated ordering requirement of the otherwise independently mobile monolayers would very likely make an even more pronounced contribution to the rigidity and immobility of ubiquinone.

The substituents of the benzoquinone ring, and possibly the isoprenoid sidechain, contribute to the stability of ubisemiquinone. The two methoxyl groups increase semiquinone stability by electron delocalization [115]. It is conceivable that the isoprenoid sidechain contributes to semiquinone stability by forming a six-membered ring involving the quinone oxygen adjacent to the sidechain and the first sidechain double bond. Such a structure would be analogous to ubichromenol, which is formed from ubiquinone under alkaline conditions [116]. Interestingly enough, molecular orbital calculations indicate that the preferred conformation of ubiquinone is one in which the first *trans* double bond of the sidechain forms a chroman-type ring structure with the quinoid oxygen [117]. In addition, by comparison to the synthetic *cis* analog, it was found that the *trans* double bond in the first isoprenyl unit of the sidechain was essential for restoration of succinate oxidase activity to rat liver mitochondria depleted of ubiquinone [118]. An important consequence of such a conformation is that the ring methyl group acquires methine character. If ubisemiquinone adopts a conformation analogous to a chroman, either spontaneously or by virtue of its binding to a protein, the semiquinone would be stabilized by resonance delocalization into the quinone ring, two methoxyl groups, a methine methyl, and a six-membered chroman-type ring. We would speculate that the trace amounts of ubichromenol which occur naturally [116], and whose function is unknown, may be coincident covalent products which result from such a conformation of ubisemiquinone.

From rotating the quinone ring around the first single bond of the sidechain in the CPK model, it is evident that the quinone oxygen adjacent to the isoprenoid sidechain may reside in a more hydrophobic environment than does the opposite oxygen. This asymmetry, under optimal conditions, could stabilize ubisemiquinone by localizing the quinone and semiquinone oxygens in preferred hydrophobic and hydrophilic environments, respectively.

What emerges from these considerations is the realization that whereas ubiquinone is sufficiently hydrophobic to reside in a membrane bilayer, the structure of this redox component is highly specialized. To the extent that structure is selected for biological function, it seems likely that the function of Q may entail phenomena more complex than simple diffusion. If ubiquinone transports hydrogen across the mitochondrial membrane during electron transfer, this process may involve proteins which optimize the unique structural properties of Q. Such proteins might mediate the movement of hydrogen atoms between Q binding sites juxtaposed across the membrane or facilitate the transmembrane movement of the quinone functional group.

V. The *b* Cytochromes

A. *An Overview of the* b *Cytochromes*

There are two molecules of cytochrome *b* per molecule of c_1 in mitochondria and succinate-cytochrome *c* reductase complex. This stoichiometry is established by chemical analysis, in which the *b* heme is extracted into organic solvent and measured as the pyridine hemochromogen [50], whose quantitation is based on analysis of iron content [119]. The heme prosthetic group of the *b* cytochromes is ferroprotoporphyrin IX, which is noncovalently bound to the apoprotein [35]. In the native cytochrome the heme is not reactive to external ligands such as CO or O_2.

Cytochrome *b* was discovered in 1925 by Keilin with the use of a hand spectroscope [120]. To an interested observer not working in the field this may appear to have been the last straightforward observation to be made on the *b* cytochromes. For example, it is still not known whether the two *b* cytochromes in mammalian mitochondria are identical, having the same primary amino acid sequence derived from a common gene, or whether they are different protein species. In mitochondria and resolved electron transfer complexes, the *b* cytochromes give rise to multiple absorption spectra, have multiple midpoint potentials, and exhibit complex, nonclassical kinetics of oxidation-reduction. While some of the diverse properties of the *b* cytochromes are probably iatrogenic, it seems clear that the two *b* cytochromes differ markedly from each other, both in the physical properties they manifest and in their functional behavior. Some of these differences could be explained by there being two apoproteins of different genetic origin, if such turns out to be the case. However, other differences, such as the opposite redox behavior of the two *b* cytochromes under certain metabolic conditions (see Section V.C) are not readily attributable to genetic factors only. The explanation of these phenomena, to state the obvious, is that two cytochrome molecules can experience different environments and both physical properties, such as spectra and midpoint potentials, and functional behavior reflect these different environments.

To establish perspective in the midst of what appears to be a discouragingly complex field, one might consider the following. Eventually cytochrome *b* from a mammalian source such as beef heart mitochondria will be purified. At that time, if cytochrome *b* is a single genetic entity as appears to be the case in yeast [121], we will most likely find ourselves in the position of trying to reconstruct a membrane segment containing a heterogeneous mixture of cytochrome *b* from a disturbingly homogeneous protein.

B. *Past Attempts and Future Prospects for the Purification of Cytochrome* b

Cytochrome *b* from mammalian mitochondria has not been purified. Although there have been reports of extensively purified preparations from heart mitochondria using detergents such as dodecyl sulfate [51] and protease [122], all of these have appeared to be denatured and in no instance has their homogeneity been convincingly demonstrated. This early work has been reviewed by Wainio [35].

The current status of cytochrome *b* purification is such that it is not possible to identify with certainty which of the polypeptides of the bc_1 segment is cytochrome *b* and, more importantly, a functional assay for cytochrome *b* is not yet available. The notion that apocytochrome *b* might have a molecular weight of approximately 30,000 appears to originate with the recent work on biosynthesis of mitochondrial proteins in *Neurospora crassa* [123]. The *b* cytochrome from this fungus has been purified and its purity has been documented by several methods, including heme analysis and acrylamide gel electrophoresis in dodecyl sulfate [124]. The *b* apoprotein has a minimum molecular weight of 26,000-30,000 based on heme content and behaves as a dimeric heme protein of molecular weight 51,000 on sedimentation equilibrium [124]. On hydroxyapatite chromatography this *b* cytochrome is separated into two polypeptides having virtually identical amino acid composition (see table 3 in Ref. 124). On electrophoresis in dodecyl sulfate these two polypeptides migrate with an apparent molecular weight of 30,000 and their mobilities are so similar that they can be distinguished but not resolved. Whether this *b* apoprotein is a single gene product which has been slightly modified subsequent to translation or whether it is two genetic species is not yet established [124].

The possibility that there may be a mammalian *b* cytochrome whose molecular weight is in the same range as that reported for the *b* from *Neurospora* is based on a recent report by Yu and coworkers, who isolated a highly purified polypeptide containing ferroprotoporphyrin IX [61]. The *b*-type polypeptide was purified by chromatography in dodecyl sulfate to yield an apparently pure polypeptide having a heme content of 26 nmol/mg, which agreed well with the M_r = 37,000 determined by electrophoresis [61]. The typical *b*-type spectrum was slightly shifted to a maximum of 560 nm and the purified heme protein appeared to be CO-reactive, since prior to exposure to dodecyl sulfate the partially purified *b* was 35% reactive to CO [61].

However, as discussed above, there is reason to question whether the *b* apoprotein of mammalian mitochondria has a molecular weight in the 25,000-40,000 range. In this regard it might be noted that the preparation of *b* originally described by Goldberger et al. [51] had a heme content of 36.1 μmol/g protein and this value was confirmed by iron analysis. This preparation did not employ proteases, and thus it seems that the molecular weight of *b* must be no greater than 27,700. In the absence of evidence to the contrary, we would speculate that the *b* preparation of Goldberger was less than 100% pure and that the true minimum molecular weight is correspondingly lower.

Riccio and coworkers have recently taken an important step toward purification of cytochrome *b* from heart mitochondria in devising a simple method for preparation of highly resolved bc_1 complex [12]. This preparation of bc_1 complex appears to consist of as few as five major polypeptides, has a *b* content of 8 nmol/mg protein, and is obtained in 30% yield. The *b* cytochromes show absorption spectra of b_{562} and b_{566} and are not reactive to CO. It appears that this highly purified bc_1 complex does not contain the polypeptide attributed to *b* by Yu and coworkers [61]. The resolved complex is virtually free of nonheme iron and Q-10 and no phospholipid could be detected by extraction with organic solvent, although the possible presence of tightly bound phospholipids, analogous to those occurring in cytochrome oxidase [125, 126], has not been ruled out.

The bc_1 complex has no succinate-cytochrome *c* reductase or ubiquinol-cytochrome *c* reductase activities [12]. The lack of these activities is attributable to the removal of several required components, including succinate dehydrogenase, ubiquinone, phospholipid, the iron-sulfur protein of the bc_1 segment, and oxidation factor. In addition, the complex may retain antimycin which was used to stabilize the complex during its isolation. Further experimentation is required to establish whether an active complex can be obtained by adding back the components noted above. The high yield, ease of preparation, and physical properties of the *b* cytochromes in this bc_1 complex are most promising for the purification of cytochrome *b*. What is now required is an assay for cytochrome *b*.

C. *An Assay for Cytochrome* b

The lack of a suitable assay, capable of measuring biological function, is perhaps the most significant impediment currently delaying the purification of cytochrome *b*. The purification of any protein is generally facilitated by a functional assay, and convincing evidence of purification requires that the protein whose homogeneity is documented by various physicochemical measurements also must be shown to be capable of performing its ascribed biological function. This is no less true for cytochrome *b* than for any other protein, notwithstanding the fact that *b* is only one of several functional components whose native domain is an intricately organized lipoprotein complex.

One might expect a functional assay for cytochrome *b* to involve reconstitution of antimycin-sensitive electron transfer through the bc_1 segment. Thus, one would have in hand a suitable assay for *b* if it were possible to show that partially purified *b*, dispersed from the membrane, could be recombined with reductase complex which was specifically depleted of *b* and thus restore electron transport activity originally manifested by the native complex.

Such an assay is not available because it is not possible to selectively deplete either reductase complex or the bc_1 segment of cytochrome *b*. Methods capable of dispersing *b* cause apparently irreversible denaturation of the residual complex. The reason for this denaturation is probably that cytochrome *b* is an integral membrane protein whose presence is a prerequisite to further assembly of the lipoprotein complex. Because of this structural role, cytochrome *b* is fundamentally different from succinate dehydrogenase and oxidation factor, whose reversible removal can be achieved while the primary structural features and latent activity are preserved in the depleted lipoprotein complex.

What follows from this line of reasoning is the realization that an assay which measures the complete function of *b* might require that one reassemble the entire reductase complex from individual resolved components. While such a de novo reconstitution will probably be accomplished in the foreseeable future, this is currently not possible since it requires an equally problematic purification and assay for each of the remaining functional components of the reductase complex. Obviously, complete reconstitution of the complex is not a practical or realistic approach to assaying cytochrome *b*.

An assay for cytochrome *b* might be available but not widely recognized. For the reasons discussed above (Section III.B) it seems possible that reduction of *b* by succinate is an obligatory step in the succinate-ubiquinone reductase reaction, but that the requirement for oxidation-reduction of *b* is circumvented in the succinate-ubiquinone-dichloroindophenol reductase assay. Thus it might be possible to develop a succinate-ubiquinone-cytochrome *b* reductase assay in which liposomes containing Q-10 are reconstituted with partially purified *b* protein followed by addition of reconstitutively active succinate dehydrogenase. One might expect that the *b* protein would preserve the reconstitutive activity of succinate dehydrogenase and the *b* would be reduced by succinate in a reaction whose rate is accelerated by Q-2 and inhibited by TTB (see Ref. 20). If this rationale is correct, such an assay would measure several of the biological functions ascribed to cytochrome *b*, including its ability to bind and stabilize succinate dehydrogenase [72], its reducibility by substrate [20], and its ability to bind antimycin with high affinity [110].

The limitations to developing such an assay are as follows. First, if the two *b* cytochromes are genetically different, only one might reconstitute with the dehydrogenase. Second, if one accepts the view that succinate-cytochrome *c* reductase complex is a primary unit of structure and function and extends this view to its extreme, it might be expected that the *b* cytochromes would show specificity toward the various dehydrogenases. Third, the hypothetical succinate-ubiquinone-cytochrome *b* reductase assay may require additional proteins, such as the core protein polypeptides (see Ref. 73), either for binding and stabilization of the dehydrogenase or for electron transfer from the dehydrogenase to *b*.

And finally, two separate laboratories have reported that reconstitutively active succinate dehydrogenase and complex III cannot be recombined to form an active succinate-cytochrome *c* reductase complex [11, 65]. In one instance this observation was cited to support the argument that complex II has functional capabilities beyond those of succinate dehydrogenase [65]; in the other instance this observation was cited to support the claim that a highly resolved bc_1 complex has functional capabilities beyond those of complex III [11]. Both of these arguments may be correct and, if so, the question arises, what is missing?

These limitations might seem to preclude the possibility of developing a succinate-ubiquinone-cytochrome *b* reductase assay of the type we have proposed for cytochrome *b*. Alternatively, the solution to certain of these problems may well provide the necessary breakthrough leading to the purification of active *b*. Only the latter assumption will support experimentation and thus it might be worthwhile attempting to develop such an assay.

D. *Spectroscopic and Thermodynamic Properties of the* b *Cytochromes*

Two molecules of cytochrome *b* give rise to a single absorption spectrum, having an α-band maximum at 562-563 nm at room temperature, if the cytochromes are fully reduced (see Fig. 3). This spectrum is actually a composite, consisting of an absorption maximum at 562 nm, a second maximum at 566 nm, and a shoulder at 558 nm. The simplest explanation consistent with available information is that one molecule of *b* gives rise to the spectral form b_{562}, and the second *b* molecule gives rise to the spectral form b_{566}, which has a split α band due to an asymmetric heme ligand, giving rise to an absorption maximum at 566 nm and a shoulder at 558 nm.

The two spectral forms of *b* are present in intact mitochondria [128], submitochondrial particles [129], isolated reductase complex (see Fig. 4 and Refs. 14 and 27) and complex III [19, 54]. There is no convincing evidence to indicate that the spectral forms of *b* observed in isolated reductase complex are anything other than a statistical representation of the total cytochrome *b* of the parent mitochondria.

The suggestion that there are multiple spectral forms of *b* was first made in 1958 by Chance on the basis of inhibitor studies [127]. Subsequently Chance and coworkers reported that a novel short-wavelength form of *b* was functionally active and characterizable by its response as the mitochondria assume an energized state [128-130]. Slater and coworkers [131] then showed that a long-wavelength form of *b* (566 nm) likewise responded when mitochondria are energized by addition of ATP, and subsequent work established the coincident response of the short (558 nm) and long (566 nm) wavelength forms of *b* [46, 132] (see however Ref. 133). This work initiated an intensive period of experimentation, still ongoing, whose central objective has been to establish the relationship between the multiple spectral and potentiometric forms of cytochrome *b* and the role of *b* in electron transport and energy conservation.

The absorption spectrum of b_{562} and a composite spectrum of b_{566} plus b_{558} can be demonstrated by artificially controlling the oxidation-reduction potential of the cytochromes in uncoupled mitochondria or resolved reductase complex by virtue of the fact that b_{562} has a higher midpoint potential than b_{566}. Thus, one would expect that, in a potentiometric titration in which the potential is varied from high potentials (+300 mV) to low potentials (-200 mV), if spectra were recorded at various applied potentials, one would obtain spectra composed of reduced c_1 plus b_{562} at high potentials and spectra of b_{562} and b_{566} plus c_1 at low potentials. Subtraction of the c_1 spectrum would yield a spectrum of b_{562} at high potentials and a composite spectrum of b_{562} and b_{566} at low potentials. This experiment, in which difference spectra of the *b* cytochromes are repeatedly recorded with high resolution at various potentials during a potentiometric titration, is not documented in the literature as frequently as might be expected.

Erecinska and coworkers have reported an experiment with resolved reductase complex [134] in which difference spectra were recorded at intervals during an oxidative titration, using a fully reduced sample as a reference. The resulting difference spectra show b_{566} at low potentials and progressive contributions from b_{562} at higher potentials (see figure 3 in Ref. 134).

The resolution of b_{562} from b_{566} plus b_{558} is commonly achieved by a potentiometric titration in a dual wavelength spectrophotometer according to the technique developed by Dutton and Wilson [135]. In such an experiment the oxidation-reduction potential of the cytochromes is equilibrated to an electrode potential by the use of appropriate mediators under anaerobic conditions. As the applied potential is then varied, the redox status of the *b* cytochromes is monitored at a fixed wavelength, such as 562 nm, versus a fixed reference, such as 575 nm. One thus measures the relative amount of reduced cytochrome as a function of applied potential and these are plotted as log ox/red versus E_h (mV). If the plotted values of log ox/red versus E_h obey a linear relationship, one can use the Nernst equation [$E_h = E_m + (RT/nF)\ln$ ox/red] to calculate the midpoint potential (E_m) and to determine the number (n) of electron equivalents required for an oxidation-reduction event.

In a typical experiment, when this procedure is applied to the *b* cytochromes of pigeon heart submitochondrial particles the experimental data yield a nonlinear plot of log ox/red versus E_h [136]. Such a nonlinear plot can be arithmetically resolved into three linear components, having midpoint potentials of -30, +40, and +120 mV [136].

The relationship between these three potentiometric forms of *b* and the spectral forms of *b* is deduced by recording difference spectra at various potentials (see figure 7 in Ref. 136). The spectrum recorded in the potential range +190 to +115 mV has an absorption maximum at approximately 560 nm and corresponds to a portion of the b_{562}, having E_{m7} = +120 mV. The spectrum recorded in the potential range +115 to 0 mV has a maximum at 562 nm and coresponds to the major portion of b_{562}, having E_{m7} = +40 mV. The spectrum recorded between 0 and -150 mV has a maximum at approximately 566 nm and a shoulder at 558 nm, corresponding to b_{566}, having E_{m7} = -30 mV.

What is now obvious is that one of the spectral forms of cytochrome *b* (b_{562}) can give rise to two potentiometric forms, having E_{m7} values of +120 mV and +40 mV. There is some question whether the +120 mV form might be a damaged product derived from b_{562} (see Ref. 46), since the high-potential b_{562} is demonstrable in resolved reductase complex (see Fig. 4) and submitochondrial particles [136], but is either absent or difficult to detect in intact mitochondria. However, it is still applicable to interpret these results in terms of two *b* molecules in a heterogeneous environment. Accordingly, it appears that one molecule of *b* exists as a single spectral form, b_{562}, which either has two potentiometric forms with E_{m7} = +120 mV and +40 mV or is especially susceptible to slight modification causing a change in potential. The second *b* molecule gives rise to the spectral form b_{566}, having a low-wavelength shoulder (558 nm). If the low-wavelength shoulder is a spectral form distinct from b_{566}, then the midpoint potential (-30 mV) of these is either identical or so similar that they cannot be resolved potentiometrically.

The most important difference between the two *b* molecules in relation to their function is that they exhibit characteristic differences in redox behavior under a variety of simulated metabolic states and in the presence of inhibitors. The oxidation-reduction behavior of the two *b* molecules can be monitored separately by using the wavelength pair 561 versus 575 nm to monitor b_{562} and 566 versus 575 nm to monitor b_{566} [137].

That these two wavelength pairs report the redox behavior of the two separate *b* molecules requires several assumptions. If one of the *b* molecules exists as two spectral forms, b_{558} and b_{566}, it must be assumed that the absorbance changes of b_{566}, as measured at 566-575 nm, are always representative of this *b*. In other words, if the spectral forms b_{558} and b_{566} are interconvertible forms of the same molecule, the b_{566} portion must always reflect the redox behavior of both spectral forms. This assumption seems valid since the oxidation-reduction behavior of b_{558} and b_{566} is similar under a wide variety of conditions and the b_{558} absorbance can be explained as a shoulder in the spectrum of b_{566}, possibly resulting from a strained heme ligand [138].

A similar assumption must be made in regard to the homogeneity of b_{562}, which may be a mixture of two potentiometric forms. For the examples discussed below, this assumption is apparently valid since b_{562} is composed of only one potentiometric form in pigeon heart mitochondria [139]. And finally, since the spectra of b_{562} and b_{566} overlap, the absorbance changes measured at 561-575 nm and 566-575 nm must be appropriately corrected using two simultaneous equations [137].

When pigeon heart mitochondria are oxidizing malate plus glutamate and respiration is limited by lack of ADP (state 4), cytochrome b_{566} is 42% reduced and b_{562} is only 10% reduced [137]. If one assumes that the midpoint potentials in state 4 are equal to those in uncoupled mitochondria, it can be calculated that the apparent potentials of b_{566} and b_{562} in state 4 are substantially different: ΔE_h (b_{566} minus b_{562}) = -109 mV. Upon addition of ADP (state 3) b_{566} becomes *less* reduced to a level of 19%, and b_{562} becomes *more* reduced to a level of 20%. Under these conditions, the increment in apparent potentials is less than in state 4, such that ΔE_h (b_{566} minus b_{562}) = -57 mV.

A similar response occurs with succinate plus glutamate as substrates [137]. When ADP is limiting, b_{566} is 56% reduced and b_{562} is 19% reduced. Upon addition of ADP, b_{566} declines to 26% reduced and b_{562} remains essentially unchanged. In this experiment ΔE_h (b_{566} minus b_{562}) in state 4 is -105 mV and in state 3 is -72 mV. If an uncoupler is added in experiments such as these the increment in apparent potentials between b_{566} and b_{562} is essentially abolished [137].

In view of the extensive evidence that state 4 respiration establishes an electrochemical H^+ potential across the mitochondrial membrane, positive outside, the above results illustrate that in the presence of the membrane potential there is a large increment in the apparent potentials of b_{562} and b_{566}, such that b_{566} is poised more than 100 mV more negative than b_{562}. As the membrane potential is lowered by addition of ADP, the potential increment between the two cytochromes decreases. Addition of uncoupler brings the cytochromes to an isopotential state [137].

A related phenomenon can be demonstrated in an experiment in which the *b* cytochromes are initially in equilibrium and then segregated by addition of ATP [137]. When freshly prepared mitochondria are supplemented with KCN, followed by ascorbate plus tetramethylphenylene diamine (TMPD) to reduce the *c* cytochromes and oxaloacetate to oxidize pyridine nucleotides, the *b* cytochromes assume an isopotential poise in which b_{566} is 17% reduced, b_{562} is 57% reduced, and ΔE_h (b_{566} minus b_{562}) = -12 mV. Subsequent addition of ATP causes b_{566} and b_{562} to change redox poise in opposite directions, b_{566} becoming more *reduced* than initially and b_{562} becoming more *oxidized.* The resulting increment in potentials is 145 mV, b_{566} more negative than b_{562} [137]. Since ATP hydrolysis results in H^+ extrusion via the proton translocating ATPase [140], it again appears that the redox poise of the two *b* molecules responds to the electrochemical potential, such that b_{566} becomes more highly reduced while b_{562} becomes more oxidized as the membrane potential is generated, positive outside.

Changes in the redox poise of the *b* cytochromes such as those described above can be treated as a change in apparent potential (E_h), assuming the midpoint potential (E_m) does not change, or one can assume the change in redox poise results solely from a change in the midpoint potentials of the two *b* molecules. Either assumption supports the same interpretation of how two *b* molecules might show different responses as the mitochondrial membrane is "energized" by formation of an electrochemical potential across the membrane.

As proposed by Mitchell [141] and Wikstrom [46], these results are consistent with a model in which the two *b* molecules are asymmetrically disposed across the membrane. As the membrane potential is formed, the heme group of the *b* molecule located toward the cytosol side of the membrane would reside in an environment which is more electropositive than the heme environment of the second *b* molecule located toward the electronegative matrix side. The result of such an arrangement is that the membrane potential which is formed during state 4 respiration or by ATP hydrolysis would oppose electron transfer from the *b* molecule in the electropositive cytosol region to the *b* molecule oriented toward the matrix. The simplest version of this model is one in which b_{566} is located toward the cytosol side of the membrane and b_{562} is located more toward the matrix region.

A second consideration arises from the chemical potential associated with the concentration gradient of H^+ across the membrane. To the extent that either molecule of *b* is exposed to the aqueous medium at the membrane surface, a local change of pH would perturb the *b* cytochrome by mass action. What emerges from this discussion is the realization that the observed changes in redox poise of the *b* cytochromes probably reflect changes in both E_h and E_m. Under nonequilibrium conditions of electron transfer the membrane potential would act as a kinetic barrier, favoring reduction of b_{566}. Likewise a membrane potential would very likely alter the intrinsic properties which determine the midpoint potentials of the *b* cytochromes. However, the measurement of E_m requires that the potential of the cytochrome must reach equilibrium with the electrode system, and whether such an equilibrium can be obtained under conditions in which the membrane potential is retained is a debatable point (see Refs. 46 and 47).

Throughout the above discussion we have referred to b_{558} and b_{566} as absorption bands of a single spectral form of the same *b* molecule, possibly due to an asymmetric or strained ligand in b_{566} [138]. Under most conditions used to resolve the different spectral forms, one half of the total *b* of the mitochondria exhibits a spectrum with an α-band maximum at 566 nm and a shoulder at 558 nm. Although there have been several reports that b_{558} and b_{566} are separate spectral forms of *b*, these have been based on observed variations in $\Delta A_{558}/\Delta A_{566}$, which may be misleading due to redox changes in b_{562} (see Ref. 46).

Higuti and coworkers [133] recently obtained spectra with intact rat liver mitochondria which they attributed to b_{558} and b_{566} as separate spectral forms of *b*. When mitochondria, metabolizing endogenous substrates, were supplemented with rotenone and KCN, b_{562} became reduced along with cytochromes *c*, c_1, *a*, and a_3 [133]. Subsequent addition of uncoupler caused reoxidation of b_{562}, whose difference spectrum was thus demonstrable. Under similar conditions, in which endogenous substrate reduced b_{562}, subsequent addition of ATP to the sample cuvette caused reduction of b_{566}, whose spectrum showed an α-band at 565.5 nm. Higuti and coworkers described this spectrum as being b_{566} free of b_{558}. However, the published spectrum does seem to show a small, but less than normal, shoulder on the low-wavelength side of the peak (see figure 4B in Ref. 133). Consequently, it seems important to establish that addition of ATP to the sample cuvette in this experiment has not caused some oxidation of b_{562} coincident with reduction of b_{566}, since oxidation of b_{562} would effectively subtract an "inverted" b_{562} spectrum from the difference spectrum of b_{566} plus b_{558}.

A difference spectrum attributed to b_{558} was obtained by adding succinate to mitochondria supplemented with rotenone and ATP and then adding a controlled amount of KCN to the reference cuvette while the sample cuvette was maintained anaerobic [133]. It was concluded that b_{558} undergoes rapid reduction and is a spectral form of *b* distinct from b_{566}, but that b_{558} may be autooxidizable. It is unlikely that an autooxidizable *b* cytochrome is part of the mitochondrial respiratory chain, and since these mitochondria were apparently coupled, such autooxidation would not be attributable to abusive treatment and associated damage to the native heme-protein. To exclude the possibility that this cytochrome might be peripheral to the respiratory chain, perhaps associated with the outer membrane, this experiment might be repeated with mitochondria from another source, such as pigeon heart. Also, it might be useful to examine the effects of CO on the composite spectrum of b_{566} plus b_{558} in rat liver mitochondria.

As noted above, there appears to be no spectral form of *b* unique to succinate-cytochrome *c* reductase complex. This important point is central to the question of whether the reductase complex is a primary structural unit of the electron transport chain. In addition, this bears directly on the mechanism of electron transport at the dehydrogenase-cytochrome *b* junction. If complex III is an isolated entity which services a homogeneous mobile pool of ubiquinone and does not physically interact with the different dehydrogenases, there would be no spectral form of *b* unique to the reductase complex. If, on the other hand, there is a constant association of succinate dehydrogenase with a portion of the bc_1 segment as we have proposed, or if a homogeneous pool of bc_1 complex services the dehydrogenases in a time-sharing model in which there is reversible association between the dehydrogenase and the bc_1 complex, one might expect to see spectral forms of *b* attributable to its interaction with the dehydrogenase.

The binding of succinate dehydrogenase to the bc_1 segment causes pronounced changes in two of the iron-sulfur centers of the dehydrogenase [16, 66]. A comparable change might be expected in the spectrum or midpoint potential of any *b* cytochrome involved in such an interaction. Although further experimentation is required to establish that some undetected differences might not exist, it appears that the existence of b_{562} and b_{566} is independent of any association with the dehydrogenase, since these spectral forms are present in bc_1 complex lacking succinate dehydrogenase [12]. Of course it is possible that succinate dehydrogenase may bind to a polypeptide other than *b*, such as a core protein polypeptide [73], and it is also possible that binding of the dehydrogenase to a *b* polypeptide might not alter the spectrum or midpoint potential. However, in light of detailed information now available on the spectroscopic and thermodynamic properties of the *b* cytochromes in intact mitochondria and recent advances in obtaining the bc_1 segment free of succinate dehydrogenase [11, 12], it should now be possible to investigate what contribution, if any, succinate dehydrogenase makes to the environment of cytochrome *b*.

E. Kinetics of Oxidation and Reduction of the b *Cytochromes*

The kinetics of cytochrome *b* oxidation and reduction are not easily explained from a classical view of the electron transport chain (see Fig. 1a) in which there is a linear sequence of electron carriers. If a pulse of O_2 is added to anaerobic mitochondria in the presence of antimycin, there occurs a rapid, transient *reduction* of cytochrome *b*, after which *b* goes oxidized as the O_2 is consumed [132, 142-144]. This phenomenon is known as the oxidant-induced reduction of cytochrome *b*. The oxidant-induced reduction of *b* is not predictable from a classical view of the electron transport chain. Thus, the commonly depicted sequence $Q \rightarrow b \rightarrow c_1$ is either an incorrect or oversimplified representation of electron transfer to and from the *b* cytochromes.

The oxidant-induced reduction of *b* does not require electron transfer through cytochrome oxidase to O_2. In intact mitochondria, supplemented with antimycin, rapid oxidation of cytochrome *c* by cytochrome *c* peroxidase induces a transient reduction of *b* [145] and resolved bc_1 complex manifests a pronounced oxidant-induced reduction of *b* if ferricyanide is used as oxidant [146, 147]. Although the oxidant-induced reduction of *b* is generally demonstrated in the presence of antimycin, Erecinska [148] and Wilson and coworkers [149] demonstrated the induced reduction of *b* in resolved reductase complex in the absence of antimycin at low temperatures, which suggests that the inhibitor merely stabilizes or amplifies the effect. An experiment which illustrates the relationship between the oxidant-induced reduction of *b* and the kinetics of cytochrome c_1 and ubiquinone oxidation-reduction is described in Section VII.B.

As noted by Chance [93], the oxidant-induced reduction of *b* is maximal under conditions in which the *b* cytochromes are initially highly oxidized and the *c* cytochromes are reduced. This redox poise is difficult to maintain and the resulting induced reduction of *b* is transient; thus the oxidant-induced reduction of *b* is especially unsuited to quantitative experimentation.

The solution to this difficulty comes from recognizing that the oxidant-induced reduction of *b* is a reversible phenomenon. Thus, if a high-potential oxidant, such as ferricyanide, causes an increase in the reducibility of *b*, one would predict that a high-potential reductant, such as ascorbate, would decrease the reducibility of *b*. This prediction is confirmed with resolved succinate-cytochrome *c* reductase complex, in which addition of ascorbate, in the presence of antimycin, causes a dramatic inhibition of *b* reduction by succinate [150]. A similar phenomenon, unexplained at the time, was noted by Davis and coworkers with resolved NADH-cytochrome *c* reductase complex (complex I + III), in which addition of ascorbate plus TMPD in the presence of antimycin appeared to prevent reduction of *b* by NADH [18]. Likewise, Eisenbach and Gutman have shown that the biphasic reduction of *b* by succinate in submitochondrial particles includes a slow reaction whose occurrence is coincident with reduction of the *c* cytochromes by high-potential reductants in the presence of antimycin [151, 152]. In addition, Wikstrom and Berden [154] showed that the oxidant-induced reduction of *b* was relaxed by the reduced form of certain mediators.

In resolved reductase complex the reduction of *b* by succinate is normally a fast reaction, having a half-time of less than 100 msec [15]. However, in the presence of ascorbate and antimycin, the reduction of *b* by succinate is inhibited so markedly that a spectrum recorded 2 min after succinate addition shows almost no *b* reduction [150]. Subsequent addition of ferricyanide causes a rapid reduction of *b*, as shown in Figure 11, demonstrating that this controlled reduction is reversible and is the apparent inverse of the oxidant-induced reduction of *b* [150].

The controlled reduction of *b* is a pseudo first-order reaction ($k_1 = 3 \times 10^{-2}\ min^{-1}$) involving all of the succinate-reducible *b* (Fig. 11). This experiment shows clearly that the controlled reduction involves both b_{562} and b_{566}. Likewise, the dynamic control of *b* reduction in submitochondrial particles involves both b_{562} and b_{566} [152]. Thus the experiments on controlled *b* reduction [150] agree with earlier reports [147, 153, 154] indicating that, under appropriate conditions (see Ref. 46), the oxidant-induced reduction of *b* involves both b_{562} and b_{566}. Furthermore, the monotonic nature of the controlled *b* reduction in resolved reductase complex demonstrates that, in the presence of antimycin, the reduction of both *b* molecules is most likely controlled by the same rate-limiting step [150].

Eisenbach and Gutman confirmed the occurrence of the controlled reduction of *b* in resolved reductase complex and correctly noted that the rate constant we obtain (typically $3 \times 10^{-2}\ min^{-1}$) is 60-fold lower than that which they have measured [155]. It was speculated [155], incorrectly, that this difference might be due to omission of cyanide in our reaction mixture. Although we routinely include 50 μM KCN, we subsequently tested and found that inclusion of 2 mM KCN or complete omission of KCN resulted in no change in the rate constant. A more likely reason for the enhanced rate measured by Eisenbach and Gutman is the use of mediators in attempting to measure the controlled reduction of *b* [151, 152, 155]. The ability of mediators to reverse the oxidant-induced reduction of *b* is well documented [46, 154, 156]. In this regard it is useful to note that the controlled reduction of *b* is *pseudo* first-order in *b*. For instance, if one compares succinate and durohydroquinone as substrates, the controlled *b* reduction proceeds as a psuedo first-order reaction with k_1 (succinate) $= 2.7 \times 10^{-2}\ min^{-1}$ and k_1 (durohydroquinone) $= 1.1 \times 10^{-1}\ min^{-1}$. The importance of this consideration is that the rate constant measured in the pres-

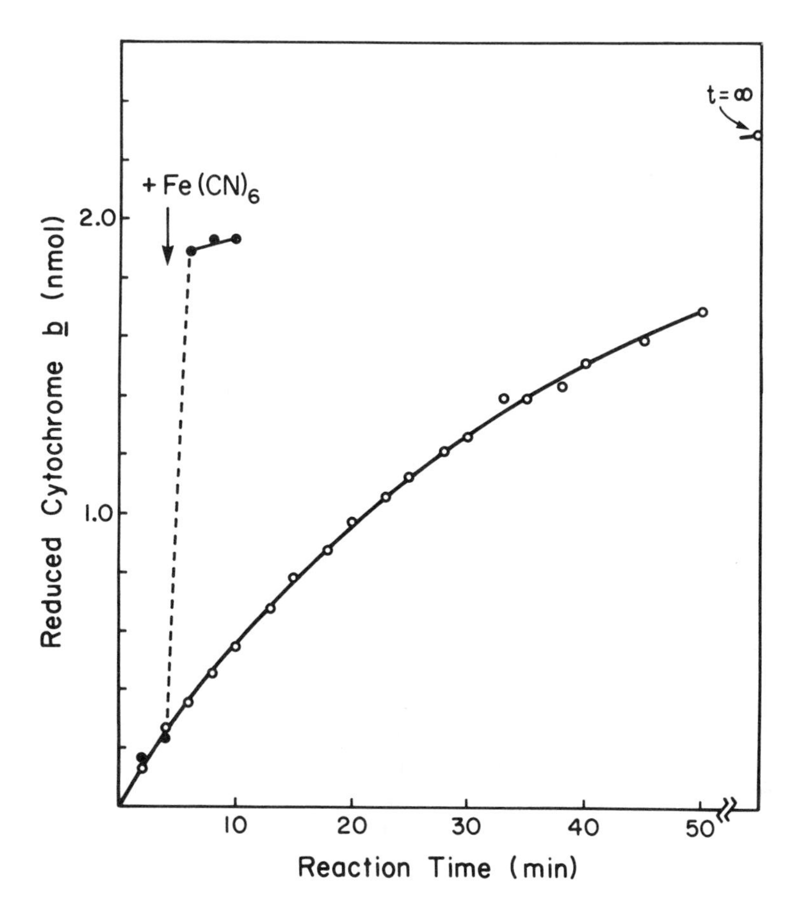

Figure 11 The controlled reduction of cytochrome *b* in resolved reductase complex. The open circles record the time-course of the reduction of cytochrome *b* by succinate after preincubation of the reductase complex with antimycin plus ascorbate. The dashed line shows the rapid, oxidant-induced reduction of *b* which results from addition of ferricyanide. Reproduced from Trumpower and Katki [150] with permission of Academic Press.

ence of mediators may be that of a bypass reaction, whose rate is governed by the mediator, and thus irrevelant to the rate-limiting step within the bc_1 segment controlling the reduction of *b*.

The controlled reduction of *b* requires addition of both ascorbate and antimycin prior to succinate [150]. If a limited amount of either ascorbate or antimycin is added in the presence of an excess of the other, reduction of *b* occurs as

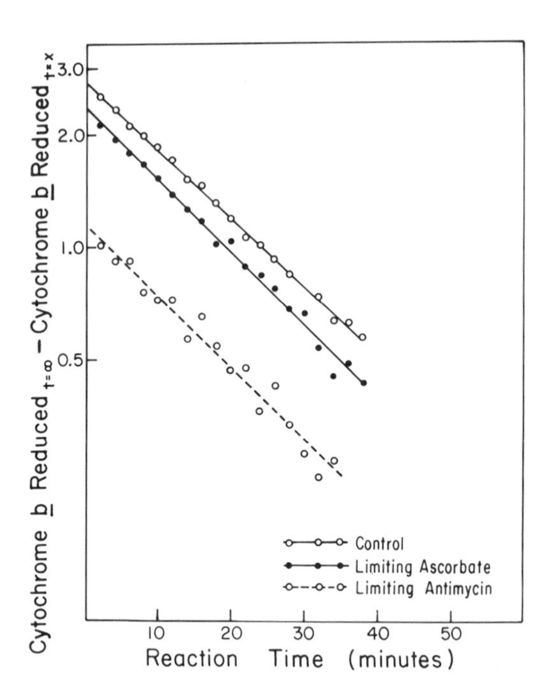

Figure 12 Semilogarithmic plots showing the first-order nature of the controlled reduction of cytochrome *b*. In the presence of excess amounts of ascorbate and antimycin (control) all of the succinate-reducible *b* (2.60 nmol) is reduced in controlled fashion. If a limiting amount of ascorbate (see Fig. 14) or a limiting amount of antimycin (see Ref. 150) is used, the reduction of *b* is biphasic. The second slow phase of the reaction under these conditions proceeds as a first-order reaction with a rate constant identical to that in the fully controlled state as shown by the three curves.

a biphasic reaction, in which the rate constant of the slow second phase is identical to that of the fully controlled reduction as shown in Figure 12. The amount of antimycin required for the controlled reduction of *b* is identical to that required for inhibition of electron transfer through the bc_1 complex [150], suggesting that both processes involve the same antimycin binding site [110].

The possibility that mediators might introduce a bypass reaction in the controlled reduction of *b* also raises the question of what reaction rate is measured in the absence of mediators and, in particular, the relationship between the measured rate of controlled *b* reduction and the presence of ascorbate. If addition of succinate is delayed up to 45 min after addition of ascorbate and antimycin, there is no change in the rate constant as shown in Figure 13. This

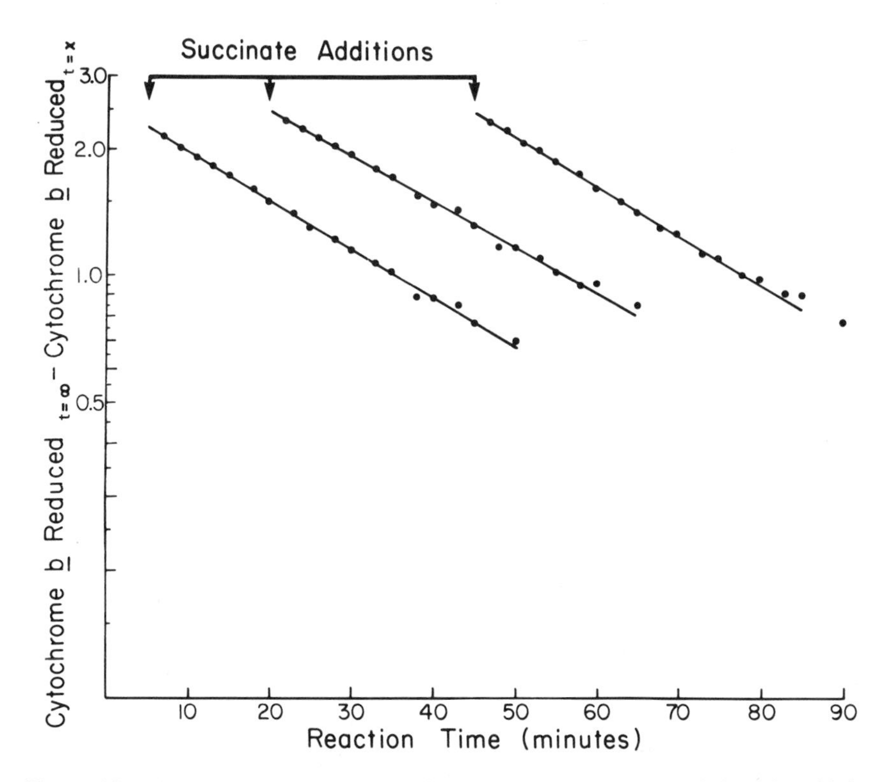

Figure 13 The controlled reduction of cytochrome *b* under conditions in which the start of the reaction by addition of succinate is delayed after the addition of ascorbate and antimycin. The arrows designate the time at which succinate was added to initiate reduction of *b* after preincubation with ascorbate.

result establishes that the system has equilibrated with ascorbate prior to addition of succinate and precludes the possibility that the measured rate is that of a reaction in which ascorbate is rate limiting.

Wikstrom has pointed out that there are two general mechanisms which would explain the oxidant-induced reduction of *b* [46]. As applied to the controlled reduction of *b*, these can be restated as follows. According to one mechanism the controlled reduction of *b* is induced when a high-potential reductant causes a decrease in the midpoint potential of *b*. According to the second mechanism the high-potential reductant increases the apparent potential (E_h) of the redox component which reduces *b* without changing the midpoint potential of *b*.

Wikstrom and Berden [154] have summarized the evidence that a change in midpoint potential is not involved. A change in midpoint potential resulting from energization of the membrane is excluded on the grounds that the oxidant-induced reduction of *b* [154] and the controlled reduction of *b* [153] occur in the presence of uncouplers and both effects are demonstrable in detergent-dispersed preparations such as complex III [147] or reductase complex [18, 150]. A somewhat more interesting point is that a mechanism involving a change in midpoint potential leads to the prediction that anaerobic potentiometric titrations would yield sigmoid shaped curves of E_h versus b_{ox}/b_{red} (see Ref. 46). In other words, the method employed to measure the midpoint potential of the *b* cytochromes would itself alter the midpoint potential. Wikstrom and Berden noted that this effect is generally not observed [154], but the point merits further discussion in light of recent data [63, 157].

Rieske has proposed that oxidation-reduction of a high-potential component ("X") causes a change in the midpoint potential of *b* and that this effect is stabilized by antimycin [147]. Recalling the predicted S-shaped potentiometric titration curve [157], Rieske cited results of Hendler and coworkers [158] in which an anomalous titration curve was obtained with uncoupled rat liver mitochondria and indicated these results were qualitatively what one would predict if X alters the midpoint potential of *b*.

However, Hendler and coworkers ascribed the anomalous potentiometric titration curve to the appearance of a previously masked species of *b* and not to conversion of a low-potential *b* to high-potential *b* [158]. Perhaps more relevant are the technical limitations of the potentiometric technique in which absorbance changes are monitored at a fixed wavelength pair, as was pointed out by these workers [158].

To the evidence cited previously [46] bearing on a change in midpoint potential, we would add the results of the experiment in Figure 11. In the absence of ascorbate and antimycin, succinate rapidly reduces approximately 75% of the *b* in reductase complex (see Fig. 3 above and Ref. 150). In the presence of ascorbate and antimycin the controlled reduction of *b* proceeds until this same percent reduction of *b* is obtained. This is straightforward evidence that the controlled reduction of *b* does not involve a change in the midpoint potential of *b*.

The controlled reduction of *b* may depend on the oxidation-reduction status of a currently unidentified redox component of the bc_1 segment. Although high-potential oxidants and reductants do oxidize and reduce c_1, various investigators have noted that the controlled reduction of *b* is only approximately coincident with the redox status of c_1 [63, 147, 151, 154, 156] and have focused attention on the possible involvement of a component whose redox status is closely related to that of c_1.

To demonstrate the possible involvement of a redox component other than c_1 in the controlled reduction of *b*, we performed a series of experiments in which

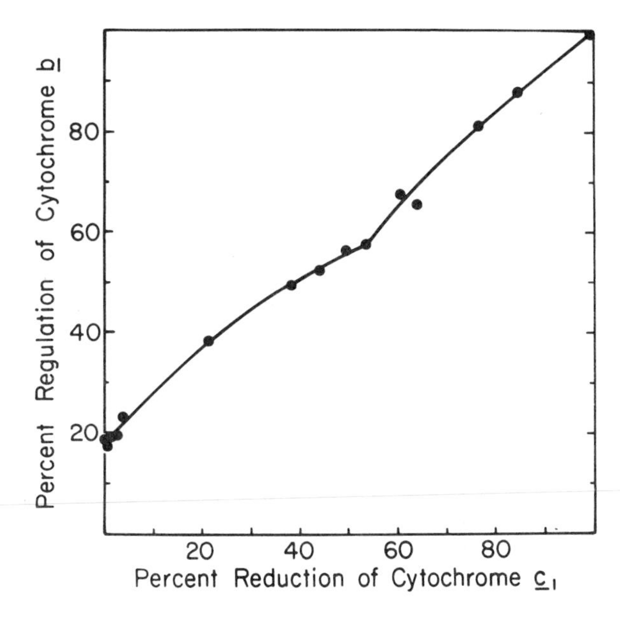

Figure 14 Relationship between the amount of cytochrome *b* which undergoes controlled reduction by succinate and the extent of reduction of cytochrome c_1 by ascorbate. In the presence of excess antimycin, reductase complex was incubated for 10 min with various amounts of ascorbate and the extent of c_1 reduction was then determined by difference spectroscopy. Succinate was then added to initiate the reduction of cytochrome *b* and the extent of *b* reduction was measured after 2 min. In the absence of ascorbate and antimycin 2.6 nmol of *b* was reduced by succinate; the results are expressed as a percent relative to this value.

the proportion of *b* undergoing controlled reduction was compared to the extent of c_1 reduction. The results of such an experiment, as shown in Figure 14, consistently show a discontinuous relationship between the extent of c_1 reduction and the proportion of succinate-reducible *b* which undergoes controlled reduction. This result demonstrates a coincident relationship but does not prove that the reduction of c_1 or the unknown component is prerequisite to the controlled *b* reduction. Nor does this experiment allow estimation of the midpoint potential or relative stoichiometry of the unknown component.

Rieske [63] has proposed that an unknown redox component X controls the midpoint potential of both b_{562} and b_{566} such that when X is oxidized the *b* cytochromes assume a high midpoint potential and are thus more rapidly reduced. This proposal extends the work of Baum and coworkers [144], who predicted the existence of an unknown X which governs the reduction of *b*. Eisenbach and Gutman [151, 152] postulated that a regulatory component "Y" exerts a dynamic control of cytochrome *b* such that when Y is reduced the *b* cytochromes assume a "sluggish" form in which they are slowly reduced by substrate. It was suggested that Y is not identical with the previously postulated X and that Y is not a member of the electron transport chain [152]. It is not clear whether Y is proposed to alter the midpoint potential of *b*, but reference to "sluggish" and "active" forms of *b* does imply a change in the equilibrium properties of *b*. Since the distinction between X and Y is based on midpoint potentials estimated in a steady state [151], and since their operation leads to the same effects on *b*, we doubt that X and Y are different. In any event, for purposes of the following discussion, they can be treated as identical.

The evidence cited above and elsewhere [46] argues against any regulatory mechanism dependent on a change of the midpoint potential of *b*. However, there is extensive circumstantial evidence to suggest that an unidentified redox component controls the oxidation-reduction of *b*. Individual interpretations notwithstanding, the data of Rieske [63, 147], Eisenbach and Gutman [151, 152], Wikstrom and Berden [154], Lee and Slater [153], Lambowitz and Bonner [156], Chance [93], and Trumpower and Katki [150] are consistent with a mechanism in which an unknown component controls the apparent potential of *b*. To emphasize that this regulatory component manifests the operational properties of X and Y, and to later make a point in regard to its identity, we will temporarily refer to this hypothetical component as "XY." In addition, since XY appears to be capable of oxidation-reduction and is present in resolved reductase complex [18, 150, 155] and complex III [147], it seems prudent to first explore mechanisms in which XY is a component of the electron transport chain.

The regulatory component XY should be one of the recognized redox components (see Section I.C) of the bc_1 segment. Anaerobic titrations of the bc_1 segment isolated from pigeon breast muscle mitochondria indicate that the segment accepts 5.2 ± 0.3 electron equivalents per c_1 [134]. If one corrects for the loss of ubiquinone and assumes only one ubiquinone per c_1, this titration accounts for 6 electron equivalents per c_1 in the bc_1 segment. These are accounted for by b_{562} (n = 1), b_{566} (n = 1), c_1 (n = 1), iron-sulfur protein (n = 1), and ubiquinone (n = 2). Oxidation factor would not alter this stoichiometry.

However, one might caution that the stoichiometric titration of redox components was performed with bc_1 segment which was apparently derived from reductase complex of somewhat low activity (see Table 2 and Ref. 134), and loss of a redox component cannot be ruled out. Recalling the above discussion (Section II.A), the stoichiometry of Q/c_1 seems open to question. We would accept the premise, as pointed out by Erecinska and coworkers [134], that all of the redox components in the bc_1 segment have been accounted for. This con-

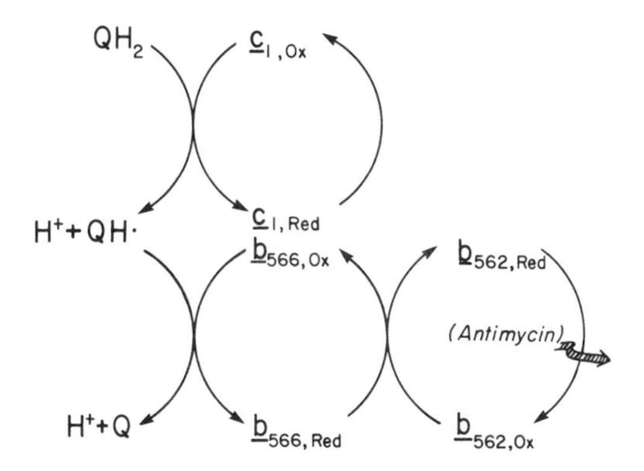

Figure 15 Mechanism of electron transfer for the controlled reduction of cytochrome *b*. The diagram shows the proposed sequence of electron transfer between cytochrome c_1, ubiquinone, and the *b* cytochromes and omits other components of the reductase complex for simplicity.

straint poses no problems to identifying XY and writing a mechanism for the controlled reduction of *b*.

The controlled reduction of *b* can be accounted for by an electron transfer mechanism of the type proposed by Wikstrom and Berden [154], in which the apparent potential of *b* is determined by the two ubiquinone couples, $QH_2/QH\cdot$ and $QH\cdot/Q$, as shown in Figure 15. The proposed electron transfer sequence is one in which c_1 is the acceptor/donor for the high-potential $QH_2/QH\cdot$ couple, and the low-potential $QH\cdot/Q$ couple is the donor for b_{566} (see Ref. 106). As previously pointed out [106], the $QH_2/QH\cdot$ couple would be equivalent to the regulatory component XY. Antimycin would prevent electron transfer between b_{562} and the acceptor which is left undesignated for purposes of current discussion. It is to be noted that the mechanism proposed by Wikstrom and Berden (Fig. 15) subsequently contributed to the formulation of the proton-motive Q cycle [21, 106] and owes certain conceptual features to the earlier work of Baum and coworkers [144].

The oxidant-induced reduction of *b* would result from rapid oxidation of c_1, leading to a transient increased $QH\cdot/Q$ ratio and a resulting increased reduction of both *b* cytochromes. The controlled reduction of *b* would result from conversion of $QH\cdot$ to QH_2, which would raise the potential of the $QH\cdot/Q$ couple to a level ineffective for *b* reduction. Simply stated, formation of $QH\cdot$ is required for reduction of the *b* cytochromes. This mechanism can be expanded to include other components such as oxidation factor [21] and the iron-sulfur protein operating between $QH_2/QH\cdot$ and c_1 or between $QH\cdot/Q$ and b_{566} with little consequence to the essential properties of such a system.

In attempting to identify the regulatory component XY, discussion has centered on the midpoint potential of this redox component. The results of Baum and coworkers [144], Rieske [147], and Wikstrom and Berden [154] are consistent with a E_{m7} value of 100-170 mV for XY. Eisenbach and Gutman have attempted to measure the midpoint potential of XY under steady-state conditions using ascorbate plus TMPD to poise the *c* cytochromes and thus report E_{m7} = +250 mV [151]. The assumption that XY fully equilibrates with the *c* cytochromes under these conditions is critical, since the apparent value of E_m will depend on the relative rate of at least three reactions, including electron transfer between c_1 and XY, TMPD and c_1, and TMPD and XY.

Rather than discuss the limitations of such measurements, we would note that even in the restrictive mechanism described in Figure 15, the $QH_2/QH\cdot$ couple and the c_1 couple would both fulfill the function ascribed to XY. Thus XY would exhibit a continuum of midpoint potentials, whose limits would be $E_m(QH_2/QH\cdot)$ and $E_m(c_{1,red}/c_{1,ox})$, depending on the method of measurement.

Having introduced XY to emphasize that X and Y are probably identical, it is appropriate to note that neither X, Y, nor XY would have a unique identity according to the mechanism in Figure 15, since at least two redox couples, $QH_2/QH\cdot$ and $c_{1,red}/c_{1,ox}$, would regulate the apparent potential of *b*. Hence it might be advisable to discontinue the use of all three terms.

Wikstrom and Berden reported an experiment which supports a mechanism of the type in Figure 15 [154]. The *b* cytochromes were titrated with mixtures of succinate and fumarate in the presence of antimycin and KCN. The high-potential portion of *b* was outside the range of the succinate/fumarate couple (-20 to +60 mV). However, the low-potential portion of *b* titrated with a value of n = 2 in the Nernst equation [154]. In other words, for each electron entering *b* from succinate, a second electron effectively disappeared. This result is explainable by postulating that the transfer of one electron from succinate to *b* is obligatorily linked to the transfer of a second electron to c_1. Such would be the case in the proposed mechanism (Fig. 15).

Rieske performed a somewhat similar experiment and obtained a different result [63, 157]. When detergent-solubilized mitochondria or complex III were titrated with mixtures of reduced and oxidized Q-1 in the presence of antimycin, values of n = 1 in the Nernst equation were obtained [63, 157]. Although further experimentation is required to resolve this discrepancy, we suggest that exogenous QH_2-1 donates electrons to *b* by a different pathway than does succinate. For instance, if QH_2-1 reacts directly with *b* [159] or endogenous Q-10 [159] to generate substantial quantities of $QH\cdot$ [104, 160], the resulting dismutation would effectively transfer electrons to *b* with n = 1. To support this possibility one need only postulate that the semiquinone of endogenous Q-10, as would be generated from succinate, is more protected from dismutation than the semiquinone of exogenous Q-1.

VI. Iron-Sulfur Protein of the bc_1 Segment

The cytochrome bc_1 segment of the electron transport chain contains an iron-sulfur protein whose function in electron transport is poorly understood. This iron-sulfur protein was discovered and purified by Rieske and coworkers [161] and is referred to as the "Rieske" iron-sulfur protein or the "g = 1.90" iron-sulfur center, the latter in reference to the absorption maximum of the EPR signal. In submitochondrial particles, at 100 K, this iron-sulfur center has resonances (first derivative) with g = 2.026, g = 1.887, and g = 1.809 and is distinguishable from the iron-sulfur centers of the primary dehydrogenases by its field position and reactivity to reducing agents [162].

Like all other functional components of the bc_1 segment, the g = 1.90 iron-sulfur center is found in both succinate-cytochrome *c* [164] and NADH-cytochrome *c* reductase [162] complex in addition to complex III [163]. Orme-Johnson and coworkers made the important observation that the iron-sulfur center of the bc_1 segment is demonstrable in whole heart tissue [162], thus virtually excluding the possibility that this signal arises as an artifact during isolation of mitochondria.

There appears to be one iron-sulfur protein, containing a 2Fe 2S cluster, in each bc_1 segment. This stoichiometry is deduced from chemical analysis of nonheme iron and acid-labile sulfide in complex III [163]. In addition, complex III contains a polypeptide whose molecular weight (M_r = 25,000) corresponds to that of the purified Rieske iron-sulfur protein and which is present in amounts comparable to c_1 (see Fig. 5). Since the iron-sulfur protein is located at the surface of the resolved complex and easily disrupted therefrom [163], there remains the question of whether some iron-sulfur protein might be lost during isolation of complex III. However, King and coworkers [165] have reported that resolved succinate-cytochrome *c* reductase complex contains flavin, nonheme iron, and cytochrome c_1 in a ratio of 0.9:12:2.2, which indicates there are 2.1 equivalents of nonheme iron per c_1 in the reductase complex, after correcting for 8 equivalents of nonheme iron per succinate dehydrogenase flavin.

Although the iron-sulfur protein has been purified from the bc_1 segment [161], an assay procedure capable of measuring functional activity of the iron-sulfur protein is not available and thus the most important property of this protein remains undocumented. The purified protein has a nonheme iron content of 60-80 nmol/mg, corresponding to a molecular weight of 25,000-30,000 for a 2Fe 2S type protein. Although the EPR spectrum of the pure protein is generally similar to that found in complex III, there is a broadening and some loss of detail evident in the spectrum of the soluble protein (see figure 2 in Ref. 161). This difference and the nonequivalence of Fe and S in the pure protein [166] suggest that the purification procedure, which involves solubilization of the aggregated protein by succinylation [161, 167], may have modified the iron-sulfur center.

In this regard it might be noted that measurements of midpoint potential and changes in the oxidation-reduction status of the iron-sulfur protein during electron transport are almost exclusively based on measurements of the $g = 1.90$ EPR signal. If the iron-sulfur protein is susceptible to modification in situ, there is the possibility that a damaged iron-sulfur center would contribute part or all of the measured signal and misleading results might be obtained. This possible problem seems especially likely with resolved complexes, in which detergent-induced damage to the cytochromes [27] and loss of functional activity (see Table 2) are known to occur.

We would speculate that the $g = 1.867$ peak reported as a new species of iron-sulfur protein in the bc_1 segment [168] may be derived by damage to the native iron-sulfur center. The $g = 1.867$ peak was found in lyophilized submitochondrial particles [168] and was more readily oxidizable than the Rieske center. Although it was stated that the $g = 1.867$ signal is detectable in phosphorylating ATP-Mg particles from pigeon and beef heart mitochondria, this assurance must be weighed against the fact that such particles are prepared by sonication. In view of the ease with which certain iron-sulfur centers are perturbed [162], it would seem worthwhile demonstrating the $g = 1.867$ signal in intact tissue.

The inherent capability for electron transfer and its occurrence in resolved complexes [162-164] are strongly suggestive that the iron-sulfur protein is an electron transport component in the bc_1 segment. However, Yamashita and Racker [25] found that a reconstituted succinate-cytochrome *c* reductase complex had a surprisingly low content of nonheme iron attributable to the bc_1 segment (0.2 nmol/mg), but yet this reconstituted complex had succinate-cytochrome *c* reductase activity of 2 units/mg [25]. This low content of nonheme iron and relatively high content of succinate-cytochrome *c* reductase activity are not necessarily inconsistent with participation of the iron-sulfur protein in electron transport, since succinate-cytochrome *c* reductase activity is a measure of the total functional activity of the reductase complex and may not be limited by a low content of iron-sulfur protein. For instance, preparations of complex III have a nonheme iron content of 8.2 nmol/mg [62], a ubiquinol-cytochrome *c* reductase activity of 100 units/mg [30], and a low succinate-cytochrome *c* reductase activity, even when the residual succinate dehydrogenase is activated by phospholipid plus ubiquinone (Table 1). Thus it seems possible that the iron-sulfur protein of the bc_1 segment is not rate limiting in the succinate-cytochrome *c* reductase activity of the native or reconstituted complex.

However, if this rationale is correct, an interesting consequence follows. If electron transfer from succinate to cytochrome *c* proceeds at a rate which is not diminished in direct proportion to the loss of iron-sulfur protein, then one molecule of iron-sulfur protein must, in effect, service multiple bc_1 segments. For this to be so, there must be electron transfer from one bc_1 segment to another, either at the level of iron-sulfur protein or at a redox component between the iron-sulfur protein and the rate-limiting step.

Although the iron-sulfur protein is oxidized and reduced during electron transport [166], more supporting evidence is required to establish that the turnover rate of the iron-sulfur center is compatible with the rate of electron transfer through the bc_1 segment. The preliminary work of Orme-Johnson and coworkers [162] suggests that the iron-sulfur center of the bc_1 segment may be reduced with a half-time of approximately 800 msec when NADH is added to NADH-cytochrome *c* reductase complex and at an even faster rate in submitochondrial particles. However, more experiments are required to correlate the reduction kinetics of the iron-sulfur protein with those of cytochromes *b* and c_1. In this regard it is to be noted that reduction kinetics can be misleading if the rate of electron transfer from substrate into the bc_1 segment is rate limiting, since under such conditions the redox components of the bc_1 segment will be reduced in a sequence determined by their midpoint potentials and not necessarily according to their acceptor/donor relationships. Consequently, a substrate such as durohydroquinone, which appears to react rapidly with the bc_1 segment in mitochondria [159], might be suitable for measuring the reduction kinetics of the iron-sulfur protein in intact mitochondria if interference from the semiquinone radical can be circumvented.

It is not known which of the redox components within the bc_1 segment are the acceptor(s) and donor(s) for the iron-sulfur protein. The possibility that the iron-sulfur protein might be a H^+ transporter, whose oxidation-reduction is coincident with H^+ release and uptake, is excluded by the demonstration that the midpoint potential of the g = 1.90 center is pH independent in the range of pH 6.3 to pH 8.3 [169]. Values for the midpoint potential (E_{m7}) of the g = 1.90 center in mammalian mitochondria and resolved complexes vary widely, from 180 to 290 mV (see Ref. 63), but it is generally accepted that this midpoint potential is within range of the cytochrome c_1 couple (E_{m7} = 225-245 mV).

Rieske and coworkers showed that reduction of the g = 1.90 iron-sulfur center by substrate is inhibited by antimycin [166], and Lee and Slater [168] found that antimycin did not inhibit oxidation of the Rieske center. In addition, the iron-sulfur center undergoes oxidation when antimycin is added to pigeon heart mitochondria during steady-state respiration (T. Ohnishi, personal communication). These results support the idea that the iron-sulfur protein, like c_1, is functionally located on the O_2 side of the antimycin block. Also, the oxidation-reduction of the g = 1.90 center by hydrophilic reagents such as ferricyanide and ascorbate is generally coincident with that of c_1 (see figure 4 in Ref. 166). In addition, the dissociation of the iron-sulfur protein from the cytochromes b and c_1 on treatment of complex III with antimycin [63] and evidence that the iron-sulfur protein is digested on treatment of complex III with trypsin [170] indicate that the location of this protein in complex III is structurally equivalent to being at or near the membrane surface [170].

On the basis of these considerations Rieske has concluded that the iron-sulfur protein is located adjacent and peripheral to cytochrome c_1 on the cytosol surface of the inner mitochondrial membrane, and has speculated that if the iron-sulfur protein is an obligatory component of the electron transport chain, it probably functions between cytochromes c_1 and c [63]. The implied electron transfer sequence $c_1 \rightarrow$ iron-sulfur protein $\rightarrow c$ is consistent with the report that the g = 1.90 center is oxidized at a rate comparable to that of cytochrome a in an O_2 pulse experiment [171], but is inconsistent with evidence indicating a direct acceptor/donor relationship between c_1 and c (see Section VII.B).

Although we tend to favor Rieske's view that there is a close structural and functional relationship between the iron-sulfur protein and c_1, the experimental results are only consistent with this view and not conclusive. The sequence of electron transfer components cannot necessarily be predicted from midpoint potentials, and the coincident redox changes of c_1 and the iron-sulfur protein observed on addition of ascorbate or ferricyanide may result from oxidation-reduction of the iron-sulfur protein via c_1. Also, the apparent surface location of the iron-sulfur protein in resolved complex III does not necessarily localize this protein on the cytosol side of the inner membrane. In fact, there appears to be no evidence to exclude the possible location of the g = 1.90 center on the matrix side of the membrane.

To test the presumed location of the iron-sulfur protein on the cytosol surface of the membrane, future experimentation might examine whether the redox poise of the g = 1.90 center responds to the membrane potential in mitochondria in parallel with the redox changes documented for b_{566} (see Section V.D) or c_1 [172]. Likewise, the g = 1.90 center should be destroyed by trypsin treatment of heart mitochondria, whose outer membrane is generally an incomplete permeability barrier to exogenous proteins. And finally, antibodies raised against the purified Rieske protein, when reacted with mitochondria, might be expected to perturb the g = 1.90 signal and would convincingly demonstrate the surface location of this protein and possibly delineate its electron transfer function relative to other redox components in the bc_1 segment.

VII. Cytochrome c_1

A. *Purification and Properties of Cytochrome* c_1

Cytochrome c_1 has been purified from bovine heart [42] and yeast [173] mitochondria and is perhaps the best characterized protein of the bc_1 segment. Cytochrome c_1 appears to be an integral membrane protein [10], and thus differs from peripheral proteins such as cytochrome *c*, in that c_1 cannot be released from the bc_1 segment without destroying the membrane structure. Extraction of c_1 from the membrane is facilitated by treatment of the bc_1 segment with reducing agents such as β-mercaptoethanol [42] or dithionite [10], after which the c_1 is solubilized by cholate in the presence of guanidine or ammonium sulfate. Because of this apparent requirement for reducing agents it is possible that disruption of intra- and/or interprotein disulfide bonds may be required for extraction of c_1. Whether c_1 is involved in the formation of such bonds is not known.

Purified c_1 from heart mitochondria has a heme content of 25 nmol/mg protein and is free of flavin, nonheme iron, and phospholipid [10, 42]. Purified ferrocytochrome c_1 has absorbance indexes $A_{417}/A_{278} = 2.50$ [42] and $A_{417}/A_{370} = 12$ [10]. The isolated c_1 remains soluble after removal of bound detergent but behaves as a heterogeneous, high molecular weight aggregate on gel filtration chromatography [42]. Bovine c_1 contains two polypeptides (see Fig. 5), having $M_r = 30{,}600 \pm 500$ and $M_r = 12{,}000$-$14{,}000$ [10]. The heme of c_1 is covalently bound to the $M_r = 30{,}600$ polypeptide and can be visualized in electrophoresis gels by staining with benzidine-peroxide [10] or by its fluorescence under ultraviolet light [174]. The low molecular weight polypeptide has no established function and may be an impurity.

The heme of c_1, like that of *c*, is linked via thioether linkages to the protein [42]. While the heme of bovine cytochrome *c* is linked through cysteines at positions number 14 and 17 [175], experiments employing controlled digestion with trypsin suggest the heme linkage in c_1 is further removed from the NH_2-terminus [10]. The NH_2-terminal amino acid of c_1, like that of *c*, is blocked from reaction with dansyl chloride [10].

Although the primary amino acid sequence of c_1 is yet to be determined, there is no evidence to suggest homologous amino acid sequences in these two proteins. The tryptic heme peptide of c_1 contains 14 amino acid residues, compared to nine in the heme peptide of *c*, and these include arginine, serine, proline, tyrosine, and aspartic acid which are absent in the tryptic heme peptide of *c* [41]. In contrast to cytochrome *c*, the $M_r = 30{,}600$ heme polypeptide of c_1 is enriched in acidic amino acid residues and purified c_1 has properties consistent with its being an acidic protein [10]. In addition to the acidic amino acid content, the c_1 polypeptide is distinguished by a high content of proline plus glycine residues and by the presence of two cysteine residues in addition to those in the thioether linkage [10]. The occurrence of cysteine residues in c_1 is without precedent in the mammalian *c*-type cytochromes and it will be of interest to see if this property extends to the c_1 proteins from other species.

Cytochrome c_1 purified by Yu and coworkers [42] and in our laboratory [10] contains a low molecular weight polypeptide ($M_r \simeq 14{,}000$) in addition to the heme-containing polypeptide. Our attempts to remove this polypeptide without denaturation of the cytochrome have been uniformly unsuccessful. As a cautionary note, we might point out that with occasional preparations of c_1 this polypeptide was selectively extracted into trichloroacetic acid when the cytochrome was acid precipitated prior to electrophoresis. If undetected, this behavior could be misleading in regard to the polypeptide composition of cytochrome c_1 preparations.

Since the low molecular weight polypeptide does not contain heme and has no established function, it may be an impurity. Evidence in support of this possibility is as follows. Under conditions in which staining is proportional to protein (see Section II.B.3), the M_r 30,600 and $M_r = 14{,}000$ polypeptides are present in a ratio of 2.5:1 in the isolated cytochrome [10]. While the staining of different proteins is not necessarily equivalent, it is to be noted that the heme content of pure c_1 (25 nmol/mg) corresponds to a minimum molecular weight of 40,000. Thus, either the molecular weight of the low molecular weight polypeptide is 9,400 (see Table 6) or this polypeptide is not present in a 1:1 molar ratio with the heme polypeptide in pure c_1. The latter possibility agrees with our results. In addition, cytochrome *f* from spinach chloroplasts has no comparable low molecular weight polypeptide [176], and cytochrome c_1 from yeast can be separated from a $M_r = 18{,}000$ peptide by ion exchange chromatography in the presence of detergent [173].

Whether the $M_r = 14{,}000$ polypeptide is a subunit of c_1 may be a semantic question involving what defines "subunit" when applied to polypeptides derived from a highly organized lipoprotein complex. Obviously, since structure serves function, the subunit character of any polypeptide should be demonstrable in an assay which measures function. In the absence of such information we retain a skeptical view toward the identity of the $M_r = 14{,}000$ polypeptide present in isolated c_1. Taking into account the large quantities of the $M_r = 14{,}000$ polypeptide relative to the c_1 heme polypeptide in all preparations of the bc_1 segment (see Fig. 7), and noting the tenacity with which this polypeptide associates with c_1, it is surprising that more serious consideration has not been given to the possibility that this low molecular weight polypeptide might be apocytochrome *b*.

The oligomeric structure of isolated c_1 includes pentamers and possibly higher polymers [42] and is probably not indicative of the structure of c_1 in the membrane. Detergent-dispersed bc_1 segment has a molecular weight of 177,000-288,000 as determined by various physical methods [63, 177], and this molecular weight is in good agreement with the minimum molecular weight of the bc_1 segment calculated from a heme c_1 content of 4 nmol/mg [63], if one assumes one c_1 per segment. The self association manifested by isolated c_1 is typical of integral membrane proteins from which the dispersing detergent has been removed [178]. In this regard it is possible that the M_r = 14,000 polypeptide and the c_1 heme polypeptide are each inherently insoluble and fulfill a mutually complementary detergent function when isolated from the membrane. Whether these two polypeptides are associated in the bc_1 segment is not known with certainty, but results from chemical crosslinking studies suggest that such is the case [179]. Taking into account the association of two polypeptides in isolated c_1 and their probable proximity in situ [179], and recognizing the lack of a clear functional identity for the low molecular weight polypeptide, it seems preferable to adopt the designation "c_1-associated polypeptide" [179] rather than "subunit" in reference to the M_r = 14,000 polypeptide.

To the extent that it is possible to compare spectra of a cytochrome in solution and in the membrane, the absorption spectrum of pure c_1 is identical to that in the bc_1 segment [42]. The spectrum of c_1 is typically that of a *c*-type cytochrome and includes the 690-nm band indicative of a methionine ligand to the heme [180]. The spectrum of the c_1 hemochromogen is identical to that of cytochrome *c* [42]. The midpoint potential of c_1 is also unchanged during purification. Dutton and coworkers determined E_{m7} = 225 mV for pure c_1 [44], which agrees with the equilibrium constant for electron transfer between soluble c_1 and *c* [181]. In beef heart submitochondrial particles [44] and in succinate-cytochrome *c* reductase complex from pigeon heart mitochondria [149] c_1 has E_{m7} values of 225 and 245 mV, respectively. Interestingly enough, there is less change in the midpoint potential of c_1 during purification than has been observed with cytochrome *c*, whose midpoint potential is reported to drop 60 mV on binding inside of submitochondrial particles [44].

Purified cytochrome c_1 should possess at least three functional activities. These include the ability to transfer electrons to cytochrome *c*, the ability to accept electrons from the redox component which is the electron donor for c_1 in the bc_1 segment, and the ability to reconstitute a lipoprotein complex in which c_1 functions as an integral membrane protein capable of transferring electrons from within the bc_1 segment to membrane-bound cytochrome *c*. A method for measuring these last two activities is not available; thus the full functional activity of purified c_1 has not been established.

Purified cytochrome c_1 appears to be fully active in its ability to transfer electrons to cytochrome *c*. Electron transfer from the soluble ferrocytochrome

c_1 to ferricytochrome c proceeds with a second-order rate constant of 1×10^6 M^{-1} sec^{-1} [181], which is comparable to that measured with succinate-cytochrome c reductase complex (2.8×10^6 M^{-1} sec^{-1}). In addition, purified c_1 forms a stable complex with cytochrome c in which the oligomeric form of c_1 binds a molar equivalent amount of cytochrome c [182, 183]. Formation of the complex appears to be accompanied by conformational changes within the two proteins which weaken the heme crevices [183]. The soluble c_1/c complex is disrupted by greater than 0.07 M ionic strength [183] and electron transfer from soluble c_1 to c is also inhibited by increasing ionic strength.

B. Function of Cytochrome c_1 *in the Reductase Complex*

Cytochrome c_1 is the terminal electron acceptor within the reductase complex and transfers electrons to cytochrome c which is bound to the cytosol surface of the inner mitochondrial membrane. Available evidence is consistent with a model in which c_1 is imbedded in the membrane complex with acidic amino acid residues exposed at the cytosol surface for ionic interaction with cytochrome c. This view is consistent with the acidic and basic character of these two cytochromes and is also supported by evidence that electron transfer between soluble c_1 and c [181] and between membranous c_1 and c [184, 185] is inhibited by high ionic strength. Likewise the rate constant for electron transfer from membranous c_1 to soluble c is lowered by polylysine [185], whose cationic character mimics that of cytochrome c. Experiments with antibodies to the heme-containing polypeptide of c_1 have established that at least some portion of this cytochrome is accessible at the surface of the mitochondrial membrane [186, 187]. However, the lysine- and arginine-containing peptide bonds of c_1, which are accessible and hydrolyzed on treatment of the soluble protein with trypsin [10, 41], are inaccessible in membranous c_1 [10].

Whether cytochrome c_1 moves within the membrane during electron transfer is not known. Purified c_1 does manifest changes in structure associated with changes in oxidation-reduction status. A trypsin-sensitive bond in the NH_2-terminal region of the heme polypeptide is cleaved at a twofold greater rate in ferrocytochrome c_1 [10], and an ellipticity band in the circular dichroism spectrum at 282 nm, which is thought to arise from intramolecular interactions of aromatic amino acids, disappears as c_1 is oxidized [188]. Such changes in protein conformation are not surprising, since one would predict that at least some structural change must accommodate the change in heme ligand bonding which is associated with spin state changes during oxidation-reduction [189, 190].

However, it seems likely that c_1 would either move in the membrane during electron transfer or manifest a change in intramolecular structure more pronounced than necessitated by changes in heme ligand bonding. Cytochrome c_1

appears to be imbedded in the membrane and thus must accept electrons which originate from a donor within the membrane and transfer electrons to cytochrome *c* at the membrane surface. Such electron transfer may involve transfer of an electron through an intraprotein channel formed by aromatic amino acids, axial rotation of c_1 within the membrane, or changes in the structure of c_1 which move the heme group to and from the membrane surface. The purpose of this discussion is not to speculate on which of these possibilities might apply, but to point out that understanding the mechanism of electron transfer to and from cytochrome c_1 poses problems even more challenging than those encountered with cytochrome *c* (see Ref. 191).

It is not known with certainty which redox component within the bc_1 segment is the immediate electron donor for c_1. However, the electron transfer mechanism which was proposed to explain the controlled reduction of cytochrome *b* (Fig. 15) predicts that ubiquinol (QH_2) is the donor for c_1. Chance [93] has described experiments in which he compared the oxidation-reduction kinetics of ubiquinone and cytochromes *b* and c_1, and in our opinion the results of these experiments are consistent with the mechanism shown in Figure 15.

The experimental procedure employed by Chance [93] is one in which pigeon heart mitochondria are equilibrated with substrates in the presence of CO, which inhibits respiration by binding to cytochrome a_3. After adding O_2 to the reaction mixture, CO is photolytically dissociated from a_3 by a laser pulse, thus instantaneously exposing the cytochrome oxidase to O_2 and allowing turnover of the electron transport chain to occur at an initial rate not limited by the O_2 mixing time. The ensuing absorbance changes of b_{562}, b_{566}, and c_1 are followed by a triple dual wavelength spectrophotometer, while the formation of oxidized ubiquinone (Q) is monitored fluorometrically by the membrane probe anthracene palmitic acid, whose fluorescence is quenched by the quinone [93]. Using the regenerative stopped flow apparatus, it is thus possible to simultaneously monitor four redox components with a time scale resolution of 3 msec for the cytochromes and 8 msec for the Q-dependent fluorescence [93].

When mitochondria are preincubated with succinate plus glutamate in the presence of antimycin and CO, the *b* cytochromes and ubiquinone are partially reduced and c_1 is essentially completely reduced [93]. As turnover of the electron transport chain is initiated by the laser pulse, c_1 becomes oxidized in a reaction which is apparently first order with a half-time of 10 msec. During the 10 msec interval in which c_1 oxidation is proceeding, there is no change in the *b* cytochromes (see figure 9 in Ref. 93). After a 1-20 msec lag, by which time oxidation of c_1 is more than 50% complete, b_{566} and b_{562} begin to undergo *reduction* and continue to be reduced for 12 sec, after which time there is a reversal in the redox changes of the *b* cytochromes and they undergo oxidation. The appearance of oxidized ubiquinone begins simultaneous with c_1 oxidation and extends monotonically during the 12-sec interval coincident with the induced *b* reduction.

This experiment illustrates the oxidant-induced reduction of *b* (see Section V.E) and provides information regarding the possible redox donor for c_1. The kinetics measurements demonstrate that oxidized Q is generated at a rate compatible with *b* reduction, and Chance has calculated that 0.4 nmol of Q is generated per 0.4 nmol of *b* reduced [93]. Noting that the fluorescent probe is quenched by Q and thus reports the appearance of Q, but not the status of QH_2 or QH•, these results are consistent with the mechanism in Figure 15.

Chance [93] has interpreted these results to indicate that the induced reduction of *b* results from reducing equivalents stored in reduced Q prior to the laser pulse (i.e., during anaerobiosis). More specifically, it seems likely that reducing equivalents would be stored in both QH_2 and QH• during anaerobiosis (see Fig. 15) and the reductant for *b* would be the QH•/Q couple, whose potential is more negative than that of the QH_2/QH• couple and thus a more likely reductant for the *b* cytochromes (see Section V.E).

As pointed out by Chance, these results exclude the possibility that QH• is the reductant for c_1, since the formation of Q continues for a longer time interval than does oxidation of c_1. When c_1 oxidation has proceeded to half completion, the appearance of Q is only 25% complete (see figure 10 in Ref. 93). In addition, this result shows that c_1 oxidation occurs at a rate compatible with formation of QH• from QH_2, which must precede the formation of Q, and which is undetected by the Q-sensitive fluorescent probe. In other words, the formation of Q lags behind oxidation of c_1 by a time interval in which QH• turnover occurs (see Fig. 15). While these results eliminate QH• as the reductant for c_1 and demonstrate the possible reduction of c_1 by QH_2, they do not exclude the possibility that the iron-sulfur protein or oxidation factor may participate in the reaction between QH_2 and c_1.

VIII. Oxidation Factor

A. Discovery of Oxidation Factor

Oxidation factor is a protein which is required for electron transfer from succinate to cytochrome *c* and which appears to be a currently unrecognized component of the bc_1 segment of the electron transport chain. Oxidation factor was discovered in Racker's laboratory during experiments on reconstitution of the electron transport chain from resolved components [192].

When succinate-cytochrome *c* reductase complex was treated with guanidine plus cholate, it separated into a membranous fraction containing cytochrome *b* and a soluble supernatant containing cytochrome c_1. The membranous *b* fraction and the c_1-containing fraction could be recombined in the presence of succinate dehydrogenase, phospholipid, and ubiquinone to form a reconstituted succinate-cytochrome *c* reductase complex [25]. The reconstituted complex catalyzed electron transfer from succinate to cytochrome *c* with activity equal to the parent reductase complex (2 units/mg), and this activity was fully sensitive to antimycin. The succinate-cytochrome *c* reductase activity of the reconstituted complex was absolutely dependent on the presence of the membranous *b* fraction and the soluble c_1 fraction and formation of a fully active complex occurred in a time-dependent fashion, being essentially complete after 2 hr incubation at 37°C. The heme content of the membranous *b* was 6.6 nmol/mg protein and that of the soluble c_1 was 2.8 nmol/mg protein, and it was recognized that neither of the resolved cytochromes was yet pure [25].

In the course of trying to further purify reconstitutively active cytochrome *b*, it was observed that variable amounts of cytochrome c_1 were retained in the membranous *b* fraction, yet the electron transport activity reconstituted with the membranous *b* exhibited an absolute requirement for the soluble c_1-containing fraction (B. Trumpower and E. Racker, unpublished experiments). While trying to purify reconstitutively active cytochrome c_1, Nishibayashi-Yamashita made the important discovery that the reconstitutive activity originally attributed to c_1 was associated with another protein which was present in the soluble c_1-containing fraction and that this active protein could be partially resolved from c_1 which precipitated upon dialysis [192]. This soluble protein, which restores succinate-cytochrome *c* reductase activity to depleted reductase complex, was named oxidation factor.

B. Assay of Oxidation Factor

Oxidation factor activity is measured by a reconstitution assay in which it is recombined with depleted reductase complex to form a functionally active succinate-cytochrome *c* reductase complex. The reconstitution assay provides a quantitative method for monitoring purification of oxidation factor and has allowed important insight into the possible mechanism of action of this protein.

Succinate-cytochrome *c* reductase complex is depleted of oxidation factor by extraction with guanidine and cholate. If the extraction is performed under carefully controlled conditions, the depleted complex retains both cytochromes *b* and c_1, in a 2:1 ratio, and manifests an absolute requirement for oxidation factor. The activity of the reconstituted succinate-cytochrome *c* reductase complex depends on the quality of the depleted reductase complex which is used for the reconstitution. To preserve reconstitutive activity of the depleted reductase complex, it was found that succinate and glycerol must be included during the treatment with guanidine and cholate [25]. Extending these observations, we found that reconstitutive activity is also preserved by dithionite, in the presence

Table 7 Effect of Succinate and Dithionite on Succinate-Cytochrome *c* Reductase Activity of Depleted Reductase Complex After Reconstitution with Oxidation Factor, Phospholipid, Ubiquinone, and Succinate Dehydrogenase[a]

Reducing agent included during treatment of reductase complex with guanidine and cholate	Activity of depleted reductase complex after reconstitution (units/mg)
None	0.16
Succinate	1.84
Dithionite	2.89

[a]Succinate-cytochrome *c* reductase complex was extracted with 1.30 M guanidine plus 0.5% cholate in the presence of 20% glycerol [25]. Where indicated, 20 mM succinate or 2 mg/ml dithionite was included in the extraction mixture, added before the guanidine. Succinate-cytochrome *c* reductase activity of the depleted reductase complex was measured after reconstitution with phospholipid, ubiquinone, succinate dehydrogenase, and oxidation factor. The activity of the depleted reductase complex was approximately 0.05 units/mg in all three cases if oxidation factor was omitted from the reconstitution mixture.

of glycerol, as shown in Table 7. The succinate-cytochrome *c* reductase activity of the reconstituted complex is approximately 50% greater if dithionite is used in place of succinate to protect the depleted complex during treatment with guanidine and cholate. Inclusion of succinate in addition to dithionite yields depleted reductase complex which can be reconstituted to the same level as obtained with only dithionite.

Rieske and coworkers previously showed that reducing agents protect complex III from cleavage by guanidine [63, 193]. The protective effect of succinate and dithionite described here is similar but not identical to that with complex III. Protection of the depleted reductase complex is measured as a functional activity, whereas Rieske and coworkers have measured the physical dissociation of components such as cytochrome c_1 and iron-sulfur protein from complex III.

Protection of the reconstitutive activity of the depleted reductase complex requires more extensive reduction of the *b* cytochromes than is required for protection of complex III against cleavage. The relative protection afforded by succinate and dithionite in our experiments (Table 7) corresponds to the extent to which they reduce cytochrome *b* (see Fig. 3). By contrast, the cleavage of complex III is 70% inhibited under conditions in which there is no reduction of cytochrome *b* (see table III in Ref. 63). This comparison illustrates that cleavage and dissociation of the bc_1 segment is a multistep process, as pointed out by Rieske [63], and also demonstrates that functional activity is lost under milder dissociating conditions than those required to cleave the bc_1 segment.

The depleted reductase complex which is protected by dithionite and glycerol has a cytochrome *b* and c_1 content of 5.56 and 2.80 nmol/mg protein, respectively. Only 10% of the *b* is reactive to CO, as with the parent complex (see Fig. 3), and 75% of the *b* is rapidly reduced by succinate (see figure 1 in Ref. 21). In the presence of phospholipid, ubiquinone, and succinate dehydrogenase the succinate-cytochrome *c* reductase activity of the depleted complex is absolutely dependent on addition of oxidation factor and is restored to levels equal to that of the fully active, native complex, as shown below.

C. Partial Purification and Properties of Oxidation Factor

Oxidation factor has been partially purified from resolved reductase complex. Although a detailed procedure is more appropriately saved until this protein is completely pure, a brief description is relevant to the following discussion. Resolved reductase complex is extracted with alkali [194] and then extracted with 1.5 M guanidine plus 0.7% cholate in the presence of dithionite. The resulting extract, containing oxidation factor and small amounts of cytochrome c_1, is dialyzed. Most of the c_1 which was extracted precipitates [192] and the oxidation factor which remains in the supernatant is concentrated by ammonium sulfate precipitation. Oxidation factor is then further purified by ammonium sulfate fractionation in the presence of cholate. The fraction containing oxidation factor is passed through a column containing an organic mercurial resin. Trace amounts of c_1 and iron-sulfur protein(s) are absorbed by the mercurial resin and oxidation factor passes through the column, after which the protein is stored at -70°C.

Two problems encountered in the purification of oxidation factor are illustrated in Figure 16. When partially purified extracts of oxidation factor are added in varying amounts to depleted reductase complex, the resulting curve of activity versus oxidation factor added is bell-shaped. On addition of small amounts of oxidation factor, no activity is restored, and after a sigmoidal increase in activity, further addition of oxidation factor results in a progressive loss of activity. This behavior causes difficulties in trying to measure specific activity as required in protein purification and is now known to be due to a removable inhibitor present in the preparation as discussed below.

A second problem, involving low recovery of activity, was solved by the finding that larger amounts of oxidation factor activity are extracted after treating reductase complex with dilute alkali (Fig. 16b versus 16a), and a further increase in activity is realized if the alkali-treated complex is extracted in the presence of dithionite (Fig. 16c). Although these differences may be partially due to differing amounts of endogenous inhibitor, we now routinely extract oxidation factor from alkali-treated reductase complex in the presence of dithionite.

The inhibitor is removed from oxidation factor by fractionation with ammonium sulfate in the presence of cholate as shown in Figure 17. If oxidation

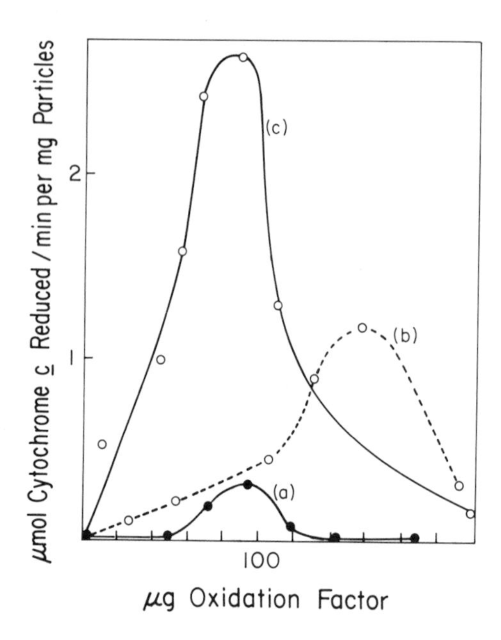

Figure 16 Reconstitution of succinate-cytochrome *c* reductase activity to depleted reductase complex by partially purified oxidation factor showing the variable activity of oxidation factor obtained by different extraction methods and the presence of an endogenous inhibitor. Curve (a) shows the activity reconstituted to depleted reductase complex by oxidation factor extracted from resolved reductase complex. Curve (b) shows the activity reconstituted by oxidation factor extracted from resolved reductase complex after washing the complex with pH 9.5 buffer. Curve (c) shows the activity reconstituted by oxidation factor extracted in the presence of dithionite from washed reductase complex. As discussed in the text, the bell shape of the titration curves is due to an endogenous, removable inhibitor in the oxidation factor.

factor is precipitated as a 25-55% ammonium sulfate fraction, the curve of reconstituted activity versus oxidation factor is bell-shaped (Fig. 17a). If oxidation factor is recovered by precipitation with 25-40% ammonium sulfate, the titration curve is hyperbolic and rises to greater maximum activity (Fig. 17b). The difference between these titration curves (Figs. 17a and 17b) suggests that the more selective ammonium sulfate fractionation has removed or inactivated the inhibitory material.

The inhibitor is recovered by precipitation with 40-55% ammonium sulfate after precipitation of inhibitor-free oxidation factor with 25-40% ammonium sulfate. The material recovered in the 40-55% fraction lowers the activity

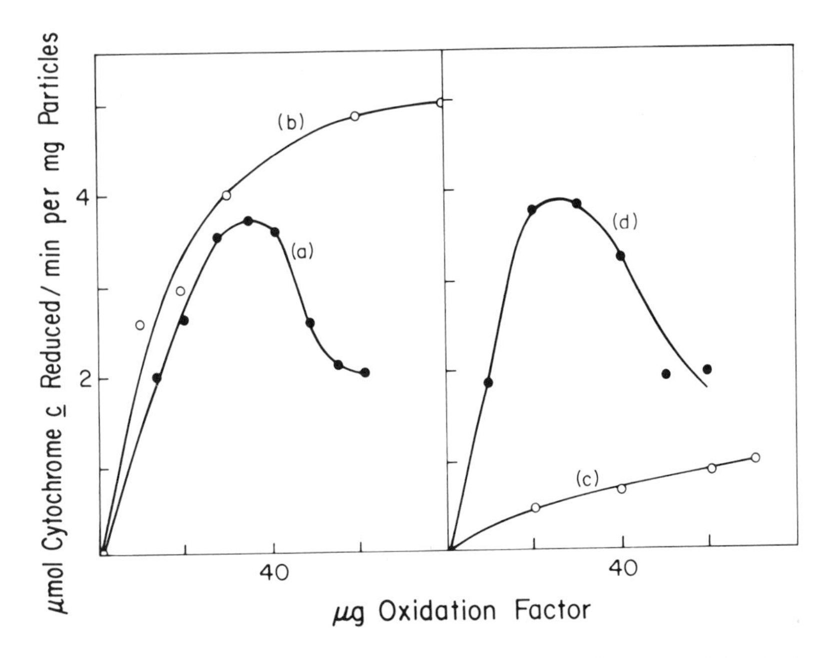

Figure 17 Reconstitution of succinate-cytochrome *c* reductase activity to depleted reductase complex by oxidation factor before and after removal of the inhibitor. Curve (a) was obtained with oxidation factor containing inhibitor. Curve (b) was obtained with oxidation factor from which the inhibitor was removed by fractionation with ammonium sulfate. Curve (c) was obtained by adding the separated inhibitory fraction to the depleted reductase complex, followed by reconstitution of the complex with inhibitor-free oxidation factor. Curve (d) was obtained by mixing the inhibitory fraction and inhibitor-free oxidation factor in the proportion (1:10) in which they were obtained during the fractionation and then using the mixture to reconstitute activity to depleted reductase complex.

of the 25-40% fraction in the reconstitution assay (Fig. 17c). If the active 25-40% fraction and the inhibitory 40-55% fraction are recombined in a fixed proportion identical to that in which they were obtained, the recombined mixture yields a titration curve very similar to that obtained prior to separation of the inhibitor from oxidation factor (Fig. 17d).

The identity of the inhibitory material is unknown. After separation from oxidation factor, the inhibitor becomes inactive on dialysis but is stable to heating for 20 min at 90°C. It is possible that the inhibitor is cholic acid, perhaps bound to a precipitable protein, but this identity is largely speculative. Cholic acid does inhibit oxidation factor activity in the reconstitution assay.

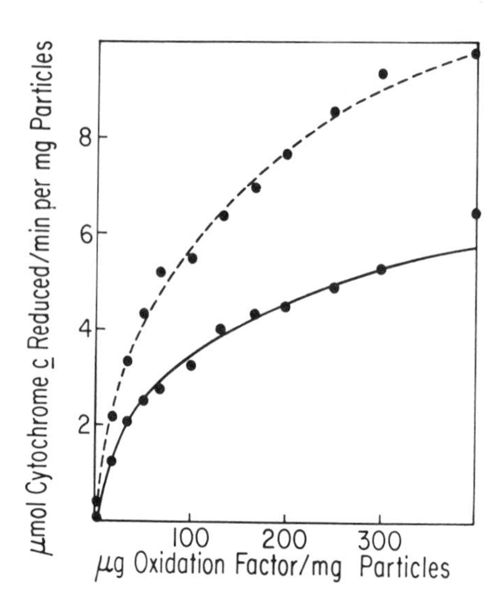

Figure 18 Reconstitution of succinate-cytochrome *c* reductase (solid line) and ubiquinol-cytochrome *c* reductase (dashed line) activities to depleted reductase complex by oxidation factor. The reduced quinone-cytochrome *c* reductase activity was measured with a synthetic analog of ubiquinone-2 [31] and is corrected for an antimycin-insensitive rate of 0.4 units/mg. Both activities were measured at 30°C.

After removal of the inhibitor, oxidation factor restores succinate-cytochrome *c* reductase activity of the depleted reductase complex to 6 units/mg, as shown in Figure 18, which is comparable to the fully active parent reductase complex (see Section II.B.1). The ubiquinol-cytochrome *c* reductase activity is reconstituted to a maximum of 10-12 units/mg, and this activity is completely sensitive to antimycin. The specific activity of oxidation factor can be calculated from the rate of cytochrome *c* reduction in the reconstitution assay, under conditions in which the rate is proportional to the amount of oxidation factor added to the depleted complex. The oxidation factor used for the experiment in Figure 18 had a specific activity of 62 units/mg.

From calculating specific activity and recovery of oxidation factor, an interesting finding arises as illustrated in Table 8. When obtained from reductase complex having an activity of 3.2 units/mg, oxidation factor has activity of 62 units/mg after the mercurial column. If the succinate-cytochrome *c* reductase activity of the parent reductase complex is limited by oxidation factor (see Table 4), this activity should be a close approximation of the specific activity of oxidation factor in the parent reductase complex. Thus it would appear that oxida-

Table 8 Specific Activity and Recovery of Oxidation Factor from Resolved Succinate-Cytochrome *c* Reductase Complex

Sample description	Total protein (mg)	Activity (units/mg)	Total activity (units)
Parent reductase complex	2,790	3.2	8,930
Oxidation factor after ammonium sulfate	40	30.0	1,200
Oxidation factor after mercurial column	18.6	62.5	1,163

tion factor is 20-fold purified with a recovery of 13% after the mercurial column. Assuming there is one molecule of oxidation factor having a molecular weight of 25,000 per each reductase complex of molecular weight 250,000-500,000, it follows that oxidation factor is completely purified when a 20-fold increase in activity is obtained. However, as described below, oxidation factor having a specific activity of 60 units/mg is only 5-10% pure (or less). Although the molecular weights of oxidation factor and the reductase complex are not known, the above assumptions are not likely to be in error by 10- to 20-fold.

The probable answer to this anomaly is that there is latent oxidation factor in the reductase complex, and although this activity is not manifested in the parent complex, it can be extracted and further purified. Two observations support this possibility. Ubiquinone-10 increases the activity of reductase complex whose activity is also increased by oxidation factor, but these increases are not additive (see Table 4). This suggests that Q-10 activates endogenous oxidation factor. In addition, if oxidation factor is extracted from reductase complex whose activity is 6-7 units/mg, the activity of the oxidation factor after the mercurial column is still only 60-70 units/mg. We conclude that the specific activity of oxidation factor in the parent reductase complex is greater than apparent from Table 8, and oxidation factor is correspondingly less purified.

Oxidation factor has the properties of a somewhat labile, soluble protein which is capable of binding to the depleted reductase complex and thus reconstituting electron transfer activity. Oxidation factor is sensitive to trypsin, as shown below, and loses more than 90% of its activity on heating at 80°C for 5 min. On heating at 37°C for 60 min, soluble oxidation factor loses 50% of its activity, but after reconstitution to the depleted reductase complex, it loses no activity during the same treatment. Oxidation factor binds to the depleted reductase complex and the activity remains bound through subsequent washing with 100 mM Na-P_i and centrifugation.

Until oxidation factor has been purified to homogeneity, it is not possible to determine with certainty which of the polypeptides of the bc_1 segment is

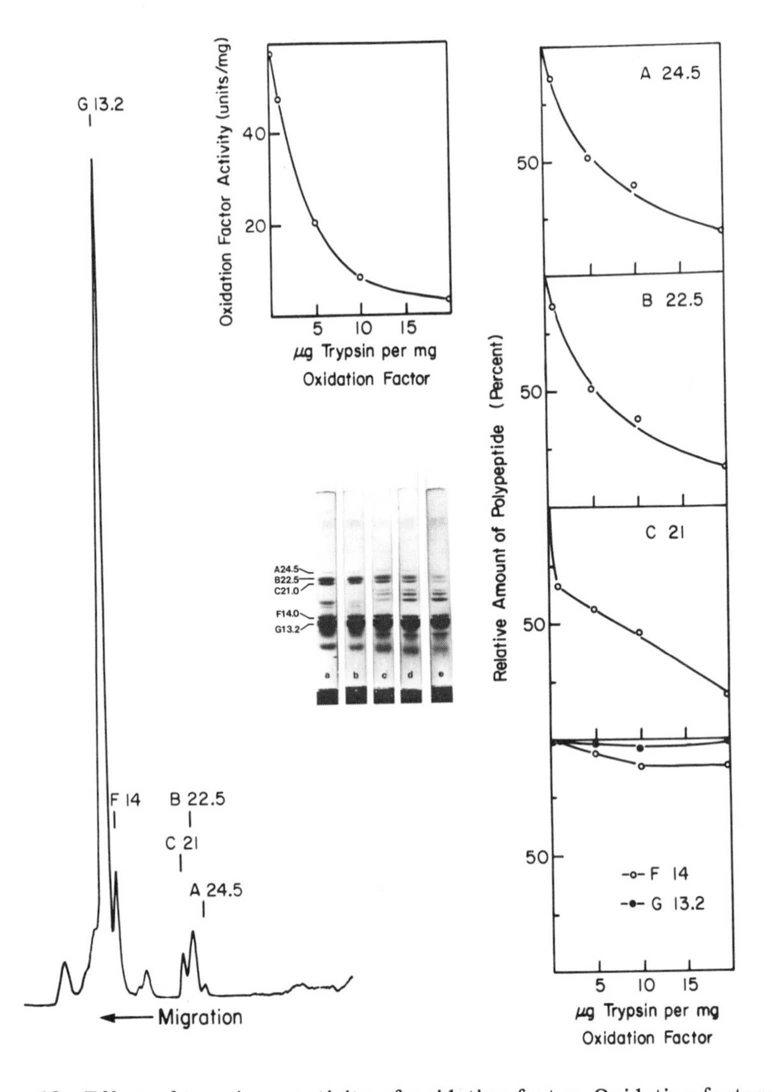

Figure 19 Effect of trypsin on activity of oxidation factor. Oxidation factor was incubated with variable amounts of trypsin for 10 min at 25°C and the reaction stopped by addition of trypsin inhibitor. Samples of oxidation factor were then assayed for activity (shown in the insert, top center) by reconstitution to depleted reductase complex. Comparable samples were analyzed by acrylamide gel electrophoresis in dodecyl sulfate [55]. The electrophoresis profile is of oxidation factor (6.5 μg) not treated with trypsin, showing the migration of five polypeptides (A 24.5, B 22.5, C 21, F 14, and G 13.2) whose intensity of staining was measured from gel scans after trypsin treatment. The amounts of these polypeptides, based on staining intensity, are shown in the inserts on the right. The intensity of staining of polypeptides A 24.5, B 22.5, and C 21 was determined from electrophoresis gels containing twofold more protein than was applied to the gel whose profile is shown. The inserted photograph shows the electrophoresis gel pattern of the oxidation factor samples after the treatment with variable amounts of trypsin.

oxidation factor. However, an experiment relevant to this question is shown in Figure 19. After mercurial column chromatography oxidation factor contains nine polypeptides (Fig. 19). One of these polypeptides, G 13.2, accounts for approximately 80% of the protein, and another, A 24.5, accounts for less than 1%. When oxidation factor is submitted to controlled digestion with trypsin, activity is lost in proportion to the amount of trypsin employed (Fig. 19). Of the nine polypeptides present before trypsin treatment, three are digested at a rate which coincides with loss of activity within the limits of experimental error. These three polypeptides are A 24.5, B 22.5, and C 21. The major polypeptide, G 13.2, is unaffected by trypsin under conditions in which 90% of the activity is lost.

This experiment indicates that oxidation factor is not polypeptide G 13.2 or F 14 and that A 24.5, B 22.5, or C 21 may be oxidation factor. This result is consistent with two other unpublished experiments. Depleted reductase complex lacks polypeptides B 22.5 and C 21, and when activity is reconstituted with oxidation factor, these polypeptides bind to the depleted complex and are retained through washing and centrifugation. By comparison, polypeptides F 14 and G 13.2 are present in the depleted reductase complex. If oxidation factor is applied to gel filtration chromatography, the oxidation factor activity coincides with fractions containing polypeptides B 22.5, C 21, and small amounts of A 24.5, although there is extensive loss of activity. Polypeptide G 13.2 is recovered as a homogeneous species and has no oxidation factor activity.

We have been aware of the possibility that oxidation factor may be the iron-sulfur protein of the bc_1 segment. This cannot be determined unequivocally until oxidation factor is purified to homogeneity, but relevant information is as follows. Oxidation factor is extracted from reductase complex by guanidine and cholate under reducing conditions which retard cleavage of the bc_1 segment but allow, and possibly enhance (see Ref. 63), extraction of the iron-sulfur protein. Soluble oxidation factor is digested by trypsin and is somewhat unstable to prolonged dialysis and to gel filtration chromatography. Oxidation factor has the properties of a functional component required for electron transfer between cytochromes b and c_1. Activity of oxidation factor appears to be associated with one or more polypeptides of molecular weight 21,000-24,500 (see Fig. 19), which is close to that of the iron-sulfur protein (25,000).

However, at the current degree of purity, oxidation factor is not enriched in nonheme iron or acid-labile sulfide. When purified, the iron-sulfur protein of the bc_1 segment should contain 80 nmol of nonheme iron and an equal amount of acid-labile sulfide. After ammonium sulfate fractionation, oxidation factor contains 4-7 nmol of nonheme iron per mg of protein and equivalent amounts of acid-labile sulfide. After further purification by passage through the mercurial resin (Table 8), the protein has approximately 0.8 nmol of nonheme iron and no detectable acid-labile sulfide. Since serum albumin gives a nonheme iron content of 0.8 nmol/mg in the assay for nonheme iron, we consider this low value insignificant. It does seem significant that recovery of activity from the mercurial column is essentially quantitative (Table 8) under conditions in which the column adsorbs iron-sulfur protein and cytochrome c_1. In addition, treatment of oxidation factor with mersalyl, 100 nmol /mg protein, followed by ammonium sulfate precipitation to remove the excess mersalyl, causes no significant loss of activity. On the basis of these considerations, it seems unlikely to us that oxidation factor is the iron-sulfur protein of the bc_1 segment, but we remain open to this possibility until the protein is pure.

D. Function of Oxidation Factor in the Reductase Complex

Oxidation factor is required for electron transfer within the bc_1 segment to cytochrome c_1 and, under certain conditions, is also required for reduction of both b cytochromes [21]. Thus the function of oxidation factor appears to preclude a linear sequence of electron transfer such as succinate $\rightarrow b \rightarrow c_1$. In addition, oxidation factor appears to have no prosthetic group such as heme, flavin, or iron-sulfur center capable of oxidation-reduction. The probable explanation for these observations is that oxidation factor may participate in transfer of reducing equivalents from reduced quinone to cytochrome c_1 and to b_{566} [21]. Thus the prosthetic group of oxidation factor would be ubiquinone or ubiquinol, or stated alternatively, oxidation factor may be a ubiquinol-binding protein.

Oxidation factor is required for reduction of cytochrome c_1 by durohydroquinone[192]. When durohydroquinone is added to depleted reductase complex, cytochrome b is reduced but c_1 is not. Addition of oxidation factor allows reduction of c_1 [192]. Likewise, when succinate is added to depleted reductase complex, there is rapid reduction of 75% of the cytochrome b and no reduction of c_1 [21]. Addition of oxidation factor, either before or after succinate, allows rapid reduction of c_1. The extent of b reduction is unchanged by addition of oxidation factor and is identical to that in the parent reductase complex (see Fig. 3 and Ref. 21). A point we would emphasize is that in the absence of oxidation factor c_1 is not reduced for periods as long as 30 min after addition of succinate. This suggests that in the absence of oxidation factor, there is no bypass whereby reducing equivalents accumulated at low reduction potential might cross over from the b cytochromes in one reductase complex to reduce the high-potential c_1 in another reductase complex.

The above results could be explained by the electron transfer sequence succinate $\rightarrow$ Q $\rightarrow b \rightarrow$ oxidation factor $\rightarrow c_1$. However, the following observation is not so readily explained. When succinate is added to depleted reductase complex in the presence of antimycin, there is no reduction of either cytochrome b or c_1

[21]. If succinate is added to depleted reductase complex in the presence of antimycin, and the complex has been reconstituted with oxidation factor either before or after addition of succinate, both cytochromes b and c_1 are reduced and the extent of their reduction increases with increasing amounts of oxidation factor [21]. In other words, in the presence of antimycin oxidation factor is required for reduction of both cytochromes b and c_1, but in the absence of antimycin oxidation factor is only required for reduction of c_1.

One possible explanation for the function of oxidation factor is that it might participate in a sequence of electron transfer of the type succinate $\rightarrow$ Q $\rightarrow$ $b \rightarrow$ oxidation factor $\rightarrow c_1$, and in addition it may regulate the midpoint potential of cytochrome b. Thus oxidation factor would be required for electron transfer between b and c_1, and in its absence cytochrome b would acquire a low midpoint potential and could not be reduced by succinate.

This type of mechanism, in which oxidation factor modulates the midpoint potential of b, seems unlikely for several reasons. First, it would require the midpoint potential of b to be altered by oxidation factor only in the presence of antimycin, since in the absence of antimycin oxidation factor is not required for reduction of b. It would also require oxidation factor to change the midpoint potential of both b_{562} and b_{566} in the presence of antimycin and allow for evidence which indicates that antimycin binds to only b_{562} [110]. And finally, a change in midpoint potential would have to occur only when the b cytochromes were oxidized, in the presence of antimycin and absence of oxidation factor. This restriction arises from the fact that if antimycin is added to depleted reductase complex in which the b cytochromes have been reduced by succinate, there is no oxidation of b following addition of antimycin as would occur if b were to assume a low midpoint potential.

Another explanation for the function of oxidation factor is that it oxidizes ubiquinol by a two-step process in which an electron is transferred from QH_2 to c_1 and $QH\cdot$ is generated as the reductant for cytochrome b [21]. This proposal assigns to oxidation factor a uniquely important role in which it diverts two electron equivalents from QH_2 to two different acceptors and at the same time participates directly in the electron transfer reaction(s) whereby H^+ are released at the outer surface of the inner mitochondrial membrane. Thus, as explained below, the proposed function of oxidation factor is consistent with a mechanism of electron transfer through the bc_1 segment known as the protonmotive Q cycle.

IX. The Protonmotive Q Cycle

A. Origins and Evolution of the Q Cycle

A central premise of the chemiosmotic mechanism of energy conservation is that the electron transport chain includes redox components which are hydrogen carriers and that these are arranged sequentially and topographically such that electron transfer is accompanied by outward translocation of H^+ [81]. Thus each coupling site consists of a "redox loop" within which a component of the electron transport chain is reduced with uptake of H^+ from the matrix phase and oxidized with liberation of H^+ to the cytosol phase of the inner mitochondrial membrane. By this now classic redox loop system, one H^+ would be transported outward across the inner membrane for each electron passing through a coupling site. Identifying the redox components which function as hydrogen carriers in loops 2 and 3 (corresponding to coupling sites 2 and 3) has remained an important unsolved problem of the chemiosmotic mechanism [195, 196]. As a possible solution to this problem, Mitchell has formulated the protonmotive Q cycle [106, 197, 198].

The protonmotive Q cycle is a mechanism whereby electron transfer through the succinate-cytochrome *c* reductase complex causes active transport of four H^+ outward across the mitochondrial membrane as two electrons are transferred from succinate to cytochrome *c*. This stoichiometry of $4H^+/2e^-$ is sufficient to account for H^+ translocation associated with two coupling sites in terms of a chemiosmotic mechanism. The net stoichiometry of two H^+ per coupling site required for a chemiosmotic mechanism would be obtained in that electron transfer from succinate to cytochrome *c* would transport the four H^+ required for the second and third coupling sites. Thus the Q cycle may solve the problem of the missing hydrogen carrier required for the third coupling site [106].

The Q cycle mechanism is presently in a formulative stage. Although the essential features of the mechanism are consistent with available experimental results, several unanswered questions can be recognized and it is likely that the Q cycle will evolve to accommodate new information relevant to these questions. It is thus useful to examine the development of the Q cycle first in terms of its H^+ translocating function and then in terms of a more complete mechanism of electron transfer in the reductase complex.

To describe the H^+ translocating properties of the Q cycle, we have outlined in Figure 20a the oxidation-reduction reactions required of ubiquinone and omitted the other redox components of the reductase complex. The essential feature of the Q cycle is that as one electron is transferred from substrate to cytochrome *c*, two H^+ are actively transported outward across the inner mitochondrial membrane. This electron enters the Q cycle on the matrix side of the membrane and leaves the Q cycle on the cytoplasm side. On the matrix side of the membrane the reduction of Q to QH_2 is accompanied by uptake of two H^+. One of the two electrons required for reduction of Q to QH_2 is that which enters the Q cycle from the substrate; the second electron comes from an unspecified donor. On the cytoplasm side of the membrane QH_2 is oxidized to Q, resulting in release of two H^+. As QH_2 is oxidized to Q, one electron is diverted to cytochrome *c*

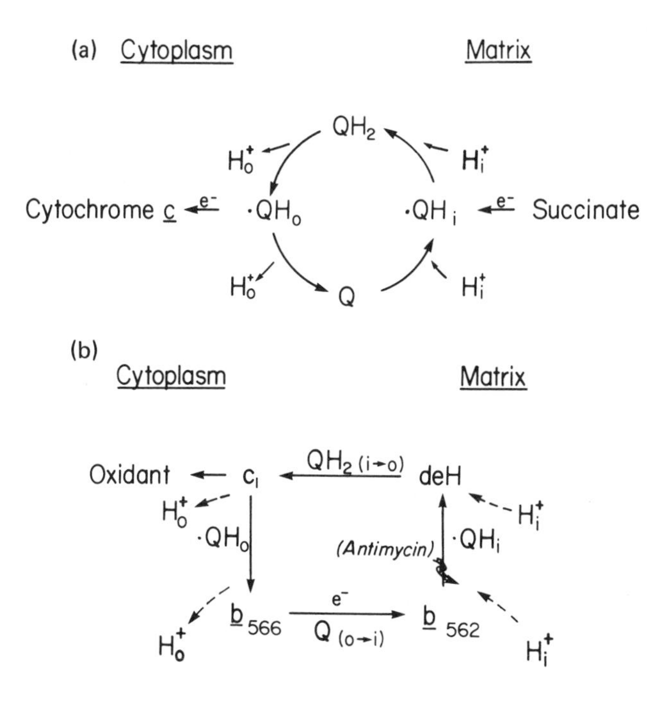

Figure 20 (a) General formulation of the protonmotive Q cycle. The essential feature of the Q cycle is that as one electron is transferred from succinate dehydrogenase to cytochrome *c*, $2H^+$ are actively transported from the matrix to the cytoplasm region of the inner mitochondrial membrane. To illustrate the activities which must be manifested by ubiquinone, the general formulation omits other components of the reductase complex which may function as acceptor/donor groups in the redox reactions of ubiquinone. (b) A formulation of the protonmotive Q cycle showing the proposed sequential arrangement of cytochromes *b* and c_1 and location of the antimycin inhibition site. It is proposed that antimycin prevents oxidation-reduction of a *b* cytochrome by the Q/QH· couple [197]. Note that the protonmotive activity of ubiquinone as diagrammed in (a) is preserved in this formulation. As discussed in the text, this formulation only accounts for electron transfer under steady-state conditions.

and O_2. The second electron is diverted inward to the matrix side of the membrane, thus replacing the electron which was contributed by the unspecified donor referred to above.

As two electrons from succinate enter the Q cycle the two H^+ which are formed along with fumarate in the matrix are, in effect, subtracted from the four H^+ which are transported outward by the two electrons passing through the Q cycle. Thus the succinate-cytochrome *c* reductase reaction proceeds with a *net* $H^+/2e^- = 2$. If the succinate-cytochrome *c* reductase reaction is combined with the cytochrome *c* oxidase reaction ($\frac{1}{2}O_2 + 2H^+ \rightarrow H_2O$) in which two H^+ are consumed in the matrix, the resulting succinate oxidase reaction proceeds with a *net* $H^+/2e^- = 4$. In other words, if cytochrome oxidase consumes two H^+ which are liberated from succinate in the matrix, the transfer of two electrons through the potential span of the second and third coupling sites actively transports four H^+ outward across the mitochondrial membrane. Thus ubiquinone would function as the hydrogen carrier for the second and third coupling sites as suggested by Mitchell [106, 141].

Before considering the function of other redox components in the Q cycle it is useful to examine the requirements placed upon the redox behavior of ubiquinone. There must be two segregated sites, asymmetrically located toward the cytoplasm and matrix regions of the membrane, for localized redox reactions of ubiquinone. Mitchell has referred to the Q reaction sites at the cytoplasm and matrix regions of the membrane as center *o* and center *i*, respectively, and has described their thermodynamic behavior in a recent review [106]. A key property of these redox centers is that they must each function as acceptor/donor sites for both the $QH_2/QH\cdot$ and $QH\cdot/Q$ couples and must do so in a manner such that the semiquinone intermediates cannot communicate directly between centers *o* and *i*. In terms of the general formulation (Fig. 20a) reaction between $\cdot QH_o$ and $\cdot QH_i$ must be prohibited, otherwise the ensuing dismutation reaction would shortcircuit the Q cycle. Alternatively stated, the inner membrane must act as a permeability barrier between two semiquinone binding sites.

While the transmembrane exchange of semiquinone is prohibited, the Q cycle requires transmembrane movement of QH_2 and Q or the operative equivalent of such movement. Depending on the number of ubiquinone molecules in each bc_1 segment, the equivalent of transmembrane movement could result from hydrogen transfer through the membrane between two ubiquinone-binding sites. If each bc_1 segment contains only one ubiquinone molecule, actual transmembrane movement is required. In this regard it is interesting to recall the dimensions of the ubiquinone molecule (Section IV.D and Fig. 10). It is conceivable that the isoprenoid sidechain of ubiquinone may be bound in the median region of the membrane, possibly to a hydrophobic site on a protein, and that the quinone functional group undergoes directed excursions between two reaction centers juxtaposed across the membrane.

The general formulation (Fig. 20a) leaves unspecified which redox components might function as acceptors and donors for the $QH_2/QH\cdot$ and $QH\cdot/Q$ couples on the matrix and cytoplasm sides of the membrane. For instance, the entry of an electron at center *i* might reduce Q to $\cdot QH_i$, or $\cdot QH_i$ to QH_2. Likewise, at center *o* the electron which is directed to O_2 via cytochrome *c* may come from oxidation of QH_2 to $\cdot QH_o$ or $\cdot QH_o$ to Q. To specify the redox components of the Q reaction centers leads to a more complete mechanism of electron transfer involving all of the functional components of the reductase complex.

A more detailed form of the Q cycle which includes the cytochromes *b* and c_1 and succinate dehydrogenase is shown in Figure 20b. This mechanism is based on that first proposed by Mitchell [197] and retains the H^+ translocating properties of the general formulation (see Fig. 20a and Ref. 198]. Although this mechanism is incomplete, it illustrates how the Q cycle may account for several important functional properties of the reductase complex.

Electrons enter the Q cycle from succinate dehydrogenase and exit via cytochrome c_1. In the aerobic steady state an electron is transferred from iron-sulfur center S-3 of the dehydrogenase to reduce $\cdot QH_i$ to QH_2. The QH_2 formed on the matrix side of the membrane is then oxidized on the cytoplasm side by transfer of an electron to c_1, thus generating $\cdot QH_o$ as the reductant for cytochrome b_{566}. Electron transfer from $\cdot QH_o$ to b_{566} would form Q, which would then move to the matrix side of the membrane. The electron from b_{566} would be transferred through the membrane to b_{562}. To complete one turn of the cycle, b_{562} would reduce Q to regenerate $\cdot QH_i$.

This form of the Q cycle incorporates the mechanism proposed for the succinate-ubiquinone reductase reaction (see Fig. 9). As discussed earlier (Section III.B), TTB would inhibit the oxidation-reduction of iron-sulfur center S-3 by the $QH\cdot/QH_2$ couple. Likewise the Q cycle incorporates the mechanism proposed to explain the controlled reduction of *b* (see Fig. 15). Antimycin would inhibit the oxidation-reduction of b_{562} by the $Q/QH\cdot$ couple (Fig. 20b). In the presence of antimycin, which would prevent reduction of the *b* cytochromes by $\cdot QH_i$ in a reversal of the cycle, reduction of c_1 by ascorbate would prevent the formation of $\cdot QH_o$ from QH_2. Thus the controlled reduction of *b* would include both *b* cytochromes as observed (Section V.E).

In the aerobic steady state, addition of antimycin leads to a "crossover" phenomenon, in which the *b* cytochromes become more reduced and c_1 more oxidized. The Q cycle accounts for this behavior in that while electrons are removed from c_1 by the cytochrome *c* oxidase reaction, the continuous oxidation of c_1 generates reducing equivalents for the *b* cytochromes which are trapped therein by antimycin. Interestingly enough, the site of antimycin inhibition and the location of the crossover point are not identical. If this mechanism is correct, it may be somewhat misleading to say that antimycin inhibits electron transfer between cytochromes *b* and c_1.

The driving force for the Q cycle comes from the transfer of an electron across the complete potential span of the reductase complex, from the dehydrogenase center S-3 to cytochrome c_1 (see Fig. 20b). The reduction of c_1 by the high-potential $QH_2/QH\cdot$ couple drives the reduction of the *b* cytochromes by the low-potential QH•/Q couple. Thus the redox components at center *o* are aptly described as manifesting a "see-saw" relationship [106]. Viewed alternatively, the reduction of cytochrome *b* driven by an obligatory reduction of c_1 is simply the coupling of an endergonic reaction to an exergonic one, a phenomenon whose precedent is well established in other reactions of intermediary metabolism.

B. Function of Oxidation Factor in the Q Cycle

A comprehensive description of the Q cycle including the proposed function of oxidation factor is shown in Figure 21. This Q-cycle mechanism is identical to that previously proposed to explain the function of oxidation factor [21], except it is here extended to include the iron-sulfur protein of the bc_1 segment. For purposes of later discussion, we have also added to the Q cycle a proposed mechanism whereby membranous succinate dehydrogenase might initiate the Q cycle.

It is suggested that oxidation factor transfers an electron from QH_2 to the iron-sulfur protein en route to cytochrome c_1 in a reaction which generates $\cdot QH_o$ as the reductant for b_{566}. It follows that oxidation factor may also participate in electron transfer from $\cdot QH_o$ to b_{566}. In the terminology used to describe the general formulation of the Q cycle, oxidation factor may be a uniquely important component of center *o* which participates in electron transfer from the high-potential $QH_2/QH\cdot$ couple to the high-potential acceptors, iron-sulfur protein and c_1, and from the low-potential QH•/Q couple to the low-potential acceptor, b_{562}.* Thus this novel protein may liberate from ubiquinol the protons required for chemiosmotic coupling at the second and third coupling sites.

When succinate is added to depleted reductase complex, reduction of c_1 is precluded by lack of the oxidation factor-dependent reaction, $QH_2 \rightarrow \cdot QH_o$, while the *b* cytochromes are reduced by $\cdot QH_i$, whose possible formation by succinate dehydrogenase is discussed below. If antimycin inhibits oxidation-reduction of b_{562} by the Q/QH• couple as proposed by Mitchell [197], one would expect that antimycin would inhibit reduction of both cytochromes *b* and c_1 in the absence of oxidation factor. This prediction was confirmed [21].

When antimycin is added to mitochondria in the aerobic steady state, cytochrome c_1 becomes oxidized while the *b* cytochromes and ubiquinone become reduced [199], and in the presence of antimycin the redox behavior of the iron-sulfur protein appears to follow that of cytochrome c_1, as if both of these components are on the "O_2 side" of the antimycin block (see Section VI). These

*We wish to acknowledge that Dr. Peter Mitchell first brought to our attention the importance of the relationship between oxidation factor and the *b* cytochrome at center *o*, and likewise the possibility that oxidation factor may bind and protect ubisemiquinone from dismutation.

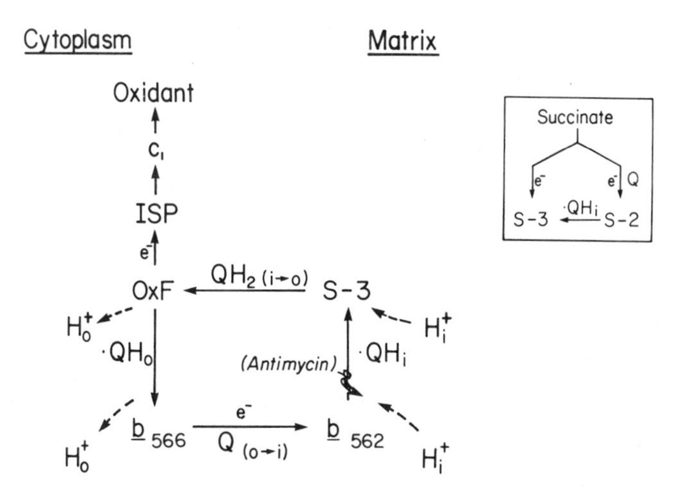

Figure 21 A comprehensive formulation of the protonmotive Q cycle showing the proposed function of oxidation factor (OxF). This formulation is an extension of that in Figure 20b and accounts for the participation of all currently established functional components of the bc_1 segment, including the iron-sulfur protein (ISP). The inset shows a proposed mechanism for the interaction of succinate dehydrogenase with the bc_1 segment in which iron-sulfur center S-2 may catalyze the priming reaction ($Q \rightarrow QH\cdot$) to initiate electron transfer through the reductase complex.

results would be predicted from the proposed mechanism (Fig. 21) since oxidation of c_1 and iron-sulfur protein generates reducing equivalents for the b cytochromes, whose reoxidation by Q is blocked by antimycin. The suggested site of antimycin inhibition within the Q cycle [197] also accounts for the important finding that antimycin inhibits electron transfer from reduced cytochrome b to ubiquinone [200].

To explain how antimycin blocks electron flow from substrates to O_2 under conditions in which QH_2 accumulates, one must explain why oxidation factor does not continue to oxidize QH_2, in which case electron transfer to c_1 would sustain O_2 consumption while the b cytochromes would remain reduced. The proposed explanation is that as antimycin traps reducing equivalents in the b cytochromes, stoichiometric quantities of $\cdot QH_o$ accumulate and remain bound to oxidation factor, thus preventing its further reaction with QH_2. If this explanation is correct, it leads to the prediction that oxidation factor also participates in electron transfer from $\cdot QH_o$ to b_{566}.

If oxidation factor participates in both high-potential and low-potential redox reactions as suggested above, it must bind ubisemiquinone in a manner which stabilizes the semiquinone and protects it from dismutation. During respiration the apparent potentials of cytochromes c_1 and b_{566} are typically +300 mV and -100 mV, respectively (see Section V.D and references therein). For the high-potential $QH_2/\cdot QH_o$ and low-potential $\cdot QH_o/Q$ couples to participate in the redox reactions at center *o* as proposed, these two couples must exist in sufficient concentrations and at the appropriate potentials to be effective donors. Thus the two ubisemiquinone couples must be segregated by approximately 400 mV.* From equations relating redox potentials and stability constants of semiquinones one can calculate that ubisemiquinone must manifest a stability constant of approximately 10^{-7} at center *o*, while the estimated value for ubisemiquinone in a lipophilic environment is approximately 10^{-10} [106]. Thus the function of oxidation factor requires that it must stabilize ubisemiquinone by a factor of 10^3. It is our view, as discussed in Section IV.D, that the ubiquinone molecule is uniquely suited for such stabilization.

To account for the efficacy of antimycin under conditions in which QH_2 accumulates, the oxidation factor-independent reduction of c_1 and/or the iron-sulfur protein by QH_2 must be a relatively slow reaction. Respiration proceeds in the presence of antimycin at approximately 10% of the control rate in mitochondria [201]. It is conceivable that this antimycin-insensitive respiration results from the noncatalyzed reaction of QH_2 with c_1 or the iron-sulfur protein. However, this explanation is difficult to reconcile with our finding that when succinate is added to reductase complex depleted of oxidation factor, there is no reduction of c_1 for periods as long as 30 min [21]. In view of the function proposed for oxidation factor, it is also conceivable that the antimycin-insensitive respiration results from a dismutation reaction whereby QH_2 might be generated from $\cdot QH_o$, both being bound to oxidation factor. This also seems unlikely since inhibition of succinate-cytochrome *c* reductase activity by antimycin in the resolved reductase complex is virtually 100%. To explain these observations we suggest that the mobility of QH_2 in the reductase complex is sufficiently limited to preclude its reaction with c_1 or the iron-sulfur protein in the absence of oxidation factor.

To extend the previous explanation (Section IX.A) of how the Q cycle accounts for the controlled reduction of cytochrome *b* it is useful to recall the discussion (Section V.E) regarding the identity of the component whose redox status governs the reducibility of *b*. It was suggested that the formation of $\cdot QH_o$, which is the required reductant for the *b* cytochromes in the presence of antimycin, is prevented by reduction of a high-potential redox component which thus appears to control the reduction of *b*. If the Q cycle is extended to include

*For a more rigorous discussion of the thermodynamic properties of the electron transfer reactions at center *o* see Ref. 106.

the iron-sulfur protein and oxidation factor (Fig. 21), it is obvious that reduction of the iron-sulfur protein, cytochrome c_1, or $\cdot QH_o$ bound to oxidation factor would prevent reduction of *b* and the hypothetical regulatory component XY would have no unique identity.

An interesting property of the Q cycle is that when succinate is added to resolved reductase complex, cytochrome c_1 should undergo reduction, even in the presence of antimycin, if succinate dehydrogenase can initiate the Q cycle by catalyzing the reactions $Q \rightarrow \cdot QH_i \rightarrow QH_2$. Although we have not carried out measurements on a millisecond time scale (see Ref. 21), we have found that when succinate is added to resolved reductase complex, cytochrome c_1 is rapidly reduced in the presence of antimycin. Although this same observation has been previously attributed to a leak through the antimycin block [19], it is our view that it provides important evidence in support of the Q cycle.

The redox behavior of the cytochromes during a single turnover of the reductase complex also provides evidence in support of the proposed location of the iron-sulfur protein (Fig. 21). For purposes of discussion, consider the response of the cytochromes which would be predicted from a Q cycle of the type shown in Figure 21 if succinate is added to the reductase complex in the presence of antimycin and if there were no iron-sulfur protein in the bc_1 segment. Under such circumstances one would predict that oxidation factor would reduce cytochrome c_1 by transfer of an electron from QH_2 and generate one equivalent of $\cdot QH_o$. The electron equivalent from $\cdot QH_o$ would in turn equilibrate between the two *b* cytochromes according to their midpoint potentials.

A second possibility to consider is that the bc_1 segment contains an iron-sulfur protein whose function is as described in Figure 21. In such a case, with succinate as substrate and in the presence of antimycin, oxidation factor would successively transfer two electrons from two molecules of QH_2, thus reducing c_1 and the iron-sulfur protein and generating two equivalents of $\cdot QH_o$. The two equivalents of $\cdot QH_o$ would equilibrate with the two *b* cytochromes according to their midpoint potentials. As a third alternative, consider the result which would be obtained if the iron-sulfur protein is located between oxidation factor and b_{566} rather than as in Figure 21. In the presence of antimycin, addition of succinate would reduce c_1 and the one equivalent of $\cdot QH_o$ generated would equilibrate between the iron-sulfur protein and the two *b* cytochromes. Because of the high midpoint potential of the iron-sulfur protein [63], the reduction of c_1 would be accompanied by little or no reduction of cytochrome *b*.

We have tested these three alternatives with resolved reductase complex. In the absence of antimycin, addition of succinate caused rapid reduction of 74-78% of the dithionite-reducible *b*, in agreement with previous findings [150]. In the presence of antimycin, addition of succinate caused rapid reduction of only 62-68% of the cytochrome *b*. Both in the presence and absence of antimycin the extent of *b* reduction remained constant for at least 10 min after succinate addition, and antimycin did not change the amount of dithionite-reducible *b*. This slight inhibition of *b* reduction caused by antimycin during a single turnover of the reductase complex is also evident in earlier experiments on the controlled reduction of *b* (see figure 2 in Ref. 150 and Fig. 14). We interpret these results as evidence that a redox component in addition to cytochrome c_1 is located on the high-potential O_2 side of oxidation factor in a Q-cycle arrangement (Fig. 21). It is reasonable to postulate that this redox component is the iron-sulfur protein, since this location is equivalent to placing cytochrome c_1 and the iron-sulfur protein on the O_2 side of the antimycin block [63].

C. Problems and Prospects for the Q Cycle

One particularly attractive feature of the Q cycle is that it allows one to design experiments to test the proposed electron transfer mechanism. With this in mind it is appropriate to consider several unanswered questions which the proposed Q cycle raises and to draw attention to several predictions which might be tested by experimentation.

One question to consider is, How does the Q cycle start and stop? Formulations of the Q cycle such as those in Figure 20 apply only to a steady-state system of electron transfer, in which a pre-existing quantity of ubisemiquinone acts as a catalyst which is regenerated in each turn of the cycle. Obviously the Q cycle must be modified to account for changes of the redox poise of ubiquinone. Thus the Q cycle must describe the events which initiate electron transfer from succinate and which generate ubisemiquinone prerequisite to the steady state. In other words, the Q cycle must accommodate a priming reaction in which reducing equivalents from succinate bring about the reactions $Q \rightarrow QH\cdot \rightarrow QH_2$.

We suggest that iron-sulfur center S-2 of succinate dehydrogenase may initiate the Q cycle by catalyzing the reaction $Q \rightarrow \cdot QH_i$, while the reaction $\cdot QH_i \rightarrow QH_2$ is catalyzed by center S-3 as shown in Figure 21. This mechanism was proposed and discussed in Section III.B for the succinate-ubiquinone reductase reaction. When the proposed succinate-ubiquinone reductase reaction is incorporated into the Q cycle, one consequence is that $\cdot QH_i$ may be generated simultaneously by two redox donors, iron-sulfur center S-2 and cytochrome b_{562}. If such is the case, center S-3 must be capable of turnover independent of S-2. Alternatively, during steady-state electron transfer the concentration of $\cdot QH_i$ would be stoichiometric with respect to the dehydrogenase. If center S-3 receives $\cdot QH_i$ from both center S-2 and b_{562}, the structural association of these components in center *i* assumes a unique importance as pointed out by Mitchell [106]. Insight into the nature of this interaction at center *i* may come from developing a succinate-ubiquinone-cytochrome *b* reductase assay of the type discussed in Section V.C.

A second question to consider is, How does the Q cycle account for ubiquinol-cytochrome *c* reductase activity associated with the bc_1 segment. As shown here the Q cycle includes iron-sulfur center S-3 of the dehydrogenase as a required component (Fig. 21), but preparations of complex III which are extensively depleted of dehydrogenase catalyze the antimycin-sensitive ubiquinol-cytochrome *c* reductase reaction with high activity. We would suggest two possible mechanisms whereby complex III might catalyze electron transfer from reduced quinone to cytochrome *c*. The first possibility includes the proposal that center S-3 of the dehydrogenase is required for ubiquinol-cytochrome *c* reductase activity. The second possibility is that the noncatalyzed dismutation reaction ($2 \cdot QH_i \rightarrow Q + QH_2$) may be a functional replacement for center S-3.

The ubiquinol-cytochrome *c* reductase reaction may involve a rather unique function for center S-3 and to illustrate this point we have summarized in Figure 22 a sequence of reactions by which electron transfer from reduced quinone to cytochrome *c* might occur by a Q-cycle mechanism. For simplicity we have omitted the iron-suflur protein and oxidation factor and have considered the H^+ liberation reactions as topographical events as they might occur when complex III is inserted into liposomes with cytochrome c_1 exposed to the outside medium and when exogenous durohydroquinone is used as substrate [30].

Electron transfer is initiated by the oxidation factor-catalyzed reaction in which an electron is transferred from QH_2 to c_1 (Equation 1) and a second electron is transferred from $\cdot QH_o$ to b_{566} (Equation 2). From durohydroquinone oxidation factor thus introduces two electrons into the bc_1 complex and liberates two H^+ directly to the external medium [30, 202, 203]. The electron from b_{566} then proceeds through part of the Q cycle (Equations 3-5) to form $\cdot QH_i$, which may be oxidized by center S-3 (Equation 6). This series of reactions transfers one electron from durohydroquinone to c_1 and leads to net reduction of center S-3 (Equation 7). If catalytic activity is to continue, center S-3 must be oxidized, and this may occur by a second turnover of the cycle as follows. As c_1 is oxidized, a second molecule of durohydroquinone may react with oxidation factor and by the same series of reaction as above may form $\cdot QH_i$. The $\cdot QH_i$ may then oxidize center S-3, thus regenerating S-3 as a catalytic component (Equation 12). Thus, what is proposed is that center S-3 alternately oxidizes and reduces $\cdot QH_i$. This proposal does not require center S-3 to carry out the low-potential reduction reaction, $Q \rightarrow QH\cdot$, as would be prohibited by the mechanism for the succinate-ubiquinone reductase reaction. This sequence of reactions proceeds through the antimycin-sensitive oxidation of b_{562} and results in net active transport of two H^+ per molecule of durohydroquinone. The extra two H^+ which are liberated to the external medium during oxidation of durohydroquinone [30, 202] are accounted for.

In preparations of complex III which contain residual succinate dehydro-

$$
\begin{array}{lll}
(1) & QH_2 + \underline{c}_1 & \rightarrow \cdot QH_o + \cdot \underline{c}_1 + H_o^+ \\
(2) & \cdot QH_o + \underline{b}_{566} & \rightarrow Q_o + \cdot \underline{b}_{566} + H_o^+ \\
(3) & \cdot \underline{b}_{566} + \underline{b}_{562} & \rightarrow \underline{b}_{566} + \cdot \underline{b}_{562} \\
(4) & Q_o & \rightarrow Q_i \\
(5) & Q_i + \cdot \underline{b}_{562} + H_i^+ & \rightarrow \underline{b}_{562} + \cdot QH_i \\
(6) & \cdot QH_i + deH & \rightarrow Q_i + \cdot deH + H_i^+ \\
\hline
(7) & QH_2 + \underline{c}_1 + deH & \rightarrow \cdot \underline{c}_1 + \cdot deH + Q_i + 2H_o^+
\end{array}
$$

$$
\begin{array}{lll}
(8) & QH_2 + \underline{c}_1 & \rightarrow \cdot QH_o + \cdot \underline{c}_1 + H_o^+ \\
(9) & \cdot QH_o + \underline{b}_{566} & \rightarrow Q_o + \cdot \underline{b}_{566} + H_o^+ \\
(10) & \cdot \underline{b}_{566} + \underline{b}_{562} & \rightarrow \underline{b}_{566} + \cdot \underline{b}_{562} \\
(11) & Q_i + \cdot \underline{b}_{562} + H_i^+ & \rightarrow \underline{b}_{562} + \cdot QH_i \\
(12) & H_i^+ + \cdot QH_i + \cdot deH & \rightarrow QH_2 + deH \\
\hline
(13) & Q_i + \cdot deH + \underline{c}_1 + 2H_i^+ & \rightarrow \cdot \underline{c}_1 + 2H_o^+ + Q_o + deH
\end{array}
$$

$$
\begin{array}{lll}
(7) & QH_2 + \underline{c}_1 + deH & \rightarrow \cdot \underline{c}_1 + \cdot deH + Q_i + 2H_o^+ \\
(13) & Q_i + \cdot deH + \underline{c}_1 + 2H_i^+ & \rightarrow \cdot \underline{c}_1 + 2H_o^+ + Q_o + deH \\
\hline
(14) & QH_2 + 2\underline{c}_1 + 2H_i^+ & \rightarrow \cdot 2\underline{c}_1 + Q_o + 4H_o^+
\end{array}
$$

Figure 22 A sequence of electron transfer reactions by which the bc_1 segment may catalyze antimycin-sensitive reduced quinone-cytochrome *c* reductase activity. The reaction sequence refers to the diagram shown in Figure 21 but the iron-sulfur protein and oxidation factor have been omitted for simplicity. deH, succinate dehydrogenase.

genase such a mechanism is obviously feasible. In preparations in which succinate dehydrogenase activity is latent it seems possible that center S-3 may be capable of turnover (see Section III.B) in the oxidation of reduced quinone. This suggestion is amenable to testing. In addition, in preparations of the bc_1 segment which are demonstrably free of the FP subunit of the dehydrogenase, the IP subunit containing center S-3 may remain as a functionally active contaminant. Since the IP subunit can be detected by electrophoresis, this suggestion can also be tested.

The midpoint potential of center S-3 and the apparent potential of the $\cdot QH_i/Q$ couple would determine the extent to which $\cdot QH_i$ accumulates or is oxidized by the dehydrogenase (Equation 6). If $\cdot QH_i$ is not oxidized by the dehydrogenase, the ensuing dismutation reaction would fulfill the functional requirements of the catalytic center. If it can be established that TTB inhibits electron transfer by blocking the interaction of center S-3 and ubisemiquinone, this would provide a basis to test the extent to which $\cdot QH_i$ is catalytically oxidized and reduced or undergoes a noncatalyzed dismutation.

It was originally suggested that oxidation factor catalyzes electron transfer from ubiquinol to cytochrome c_1 [21]. We have extended this suggestion to include the intermediate participation of the iron-sulfur protein between oxidation factor and c_1. If this proposal is correct, the availability of purified c_1 [10, 42], synthetic quinone substrate [31], and oxidation factor may provide an assay which could be used to purify the iron-sulfur protein. An obvious extension of this rationale is that the structural and functional relationships among the components at center *o* may form the basis on which to develop a complete reconstitution of the bc_1 segment from purified components.

While this may seem a highly ambitious undertaking, we would note that the proposed function of oxidation factor requires that it is likely structurally associated with a *b* cytochrome, the iron-sulfur protein, and possibly cytochrome c_1. Although circumstantial evidence may be misleading, it has not escaped notice that the low molecular weight polypeptide G 13.2, which we have implied might be apocytochrome *b*, is present as a persistent contaminant in the most purified preparations of cytochrome c_1 and our preparations of oxidation factor.

To conclude on an even more speculative note, we wish to suggest a mechanism whereby oxidation factor may regulate electron flow through the bc_1 segment in response to the energization state of the mitochondrial membrane. It was previously noted [21] that the proposed function of oxidation factor would require that it be located on or near the cytoplasmic surface of the inner mitochondrial membrane. If oxidation factor were capable of controlling the apparent potential or effective concentration of the two semiquinone couples at center *o*, it could regulate electron flow to cytochromes c_1 and b_{566}. Such a mechanism would be feasible if oxidation factor responded to the electrochemical potential across the membrane in a manner such that it varied the stability constant of bound ubisemiquinone.

X. Summary

1. We have reviewed what is currently understood and noted several unanswered questions in regard to the mechanism of electron transfer and H^+ translocation in the succinate-cytochrome *c* reductase segment of the mitochondrial electron transport chain. We have proposed that succinate-cytochrome *c* reductase complex is a discrete lipoprotein complex and thus a fundamental structural unit of the inner mitochondrial membrane.

2. As a possible explanation of how reducing equivalents are transferred from succinate dehydrogenase to the bc_1 segment we have proposed that iron-sulfur centers S-2 and S-3 of the dehydrogenase sequentially reduce the two ubi-semiquinone couples in a reaction which involves an allotopic interaction of the iron-sulfur groups with the bc_1 segment.

3. In considering the properties of ubiquinone we have suggested that it is not freely mobile in the inner membrane, and that any lateral or transverse mobility of this molecule very likely involves processes more complex than free diffusion. In addition, we have suggested that ubiquinone is uniquely suited to the formation of stable semiquinone intermediates if a chroman-type conformation is stabilized by binding to a protein.

4. A newly discovered protein named oxidation factor is required for electron transfer from succinate and reduced quinone to cytochrome *c*. We have proposed that oxidation factor transfers one electron from ubiquinol to the iron-sulfur protein of the bc_1 segment and a second electron from ubisemiquinone to cytochrome b_{566}, thus liberating protons at the cytoplasmic surface of the inner mitochondrial membrane.

XI. Recent Developments

A somewhat clearer picture is now emerging in regard to how succinate dehydrogenase interacts with the cytochrome bc_1 segment and how electrons are transferred from succinate to cytochrome *c*. By two-dimensional electrophoresis it has been shown that the most highly resolved preparations of complex II contain two low molecular weight polypeptides (M_r = 7000 and 13,500) in addition to the FP and IP subunits of succinate dehydrogenase [206]. It seems likely that these polypeptides are required for reconstitution of succinate dehydrogenase with the bc_1 segment. Likewise one or both of these polypeptides may account for the fact that preparations of complex II have TTB-sensitive succinate-ubiquinone reductase activity, whereas succinate dehydrogenase does not (see Section III.B).

One aspect of this work not mentioned is that these preparations of complex

II contain cytochrome *b* (3-4 nmol/mg) in amounts almost stoichiometric with the dehydrogenase flavin (4-5 nmol/mg) and cytochrome c_1 in small amounts (0.3-1.5 nmol/mg). This distribution of cytochromes is consistent with the possibility noted above (Section I.D), i.e., that complex II may be a cleavage product derived from an otherwise structurally intact reductase complex and that the association of cytochrome *b* with the dehydrogenase may reflect a functionally important structural relationship in the parent complex. In any event, it would be useful to know which polypeptide(s) in such preparations of complex II represent(s) the *b* cytochrome which is present in relatively large amounts.

If these succinate dehydrogenase-associated polypeptides [206] are required for interaction of succinate dehydrogenase with the bc_1 segment, it should be possible to use the two-dimensional electrophoresis procedure to show that they are missing from preparations of complex III, which do not reconstitute with the dehydrogenase (see Sections III.B and V.C) and are present in preparations of the bc_1 complex [11] which do. In this regard, it appears that the M_r = 13,500 polypeptide associated with succinate dehydrogenase is not identical to the low molecular weight cytochrome c_1-associated polypeptide (see Refs. 10 and 42 and Table 6 above). If the two succinate dehydrogenase-associated polypeptides are not accounted for by the nine polypeptides of complex III (see Section II.B.3), the reductase complex consists of a minimum of 13 polypeptides, including the subunits of the dehydrogenase.

It seems very likely that one of the two succinate dehydrogenase-associated polypeptides (M_r = 13,500) is identical to a polypeptide isolated by Yu and coworkers [207]. These coworkers recognized that the apparent functional difference between complex III and their preparation of bc_1 complex might be attributable to the presence of a binding protein in their preparation [11, 207]. Subsequently they extracted their bc_1 complex with Triton in the presence of urea and purified from this extract a low molecular weight protein which they refer to as a "ubiquinone-binding protein" [207-209].

This Q-binding protein (QP_S) was recombined with succinate dehydrogenase to reconstitute a complex which had succinate-ubiquinone reductase activity [207]. The reconstituted activity was inhibited by TTB [208], and the rate was comparable to that obtained with preparations of complex II [208]. Coincident with this reconstitution, the high-affinity ferricyanide site of the dehydrogenase (see Section II.B) became masked, and the reconstitutive activity of the dehydrogenase was apparently stabilized. In addition, the reconstituted succinate-ubiquinone reductase activity did not require addition of phospholipid, although the endogenous content of phospholipid in the reconstituted complex, possibly originating as a contaminant in the QP_s preparation, was not reported.

In a recent communication Yu and coworkers examined the time course of reduction of cytochrome *b* and c_1 and the formation of ubisemiquinone free radical (g = 2.00) in a reconstituted system consisting of succinate dehydrogenase, QP_S and cytochrome bc_1 complex were depleted of the former two components [209]. To interpret the results of these experiments one must understand that the time course of reduction of the cytochromes and formation of semiquinone are very likely determined by the midpoint potentials of the redox components and not by the turnover number of the individual reactions.

Demonstration of the semiquinone free radical apparently required the presence of the bc_1 complex in addition to QP_S and the dehydrogenase [209]. This suggests that the presumed QP_S did not stabilize QH• to a sufficient extent to allow detection of the semiquinone and that the bc_1 complex either stabilized QH• or acted as a QH• regenerating system. The latter explanation is fully consistent with the protonmotive Q cycle (see Section IX.A and Figs. 9, 15, 20, 21).

Another interesting feature of these experiments is that the g = 2.00 signal disappeared after addition of antimycin [209]. This effect of antimycin is not what one would predict from linear electron transfer sequences of the type succinate → Q → *b* → c_1. On the other hand, one would expect antimycin to cause the disappearance of a semiquinone whose regeneration was otherwise catalyzed by a cyclic protonmotive Q mechanism as shown above. It is especially intriguing that the time course of cytochrome *b* reduction in these experiments showed an unusual cyclic response [209]. During the initial time interval cytochrome *b* appeared to undergo partial reduction, followed by reoxidation prior to the appearance of the semiquinone signal. A second, more extensive reduction of *b* then proceeded approximately coincident with formation of the semiquinone. This cyclic response of *b* raises the question of whether oxidation of *b* is prerequisite to formation of ubisemiquinone, or whether ubisemiquinone is the required oxidant for *b* (see Figs. 20, 21).

It should be pointed out that there is no direct evidence that QP_S is in fact a ubiquinone-binding protein. Nor has it been reported whether the isolated protein contains endogenous ubiquinone. Further insight into how QP_S confers succinate-ubiquinone reductase activity on succinate dehydrogenase may come from establishing what changes, if any, are caused in the iron-sulfur centers. It is also of interest to know if QP_S specifically binds TTB and carboxins [210], either as isolated or after reconstitution with the dehydrogenase. Likewise, although it is explicable that the bc_1 complex is required to catalytically regenerate free semiquinone (g = 2.00), it would be informative to know whether in the absence of the bc_1 complex, QP_S plus dehydrogenase does form stoichiometric amounts of the stable semiquinone pair which is spin-coupled to center S-3 [76-78]. This latter point is especially interesting in view of evidence that the S-3 associated semiquinone pair is oriented perpendicular to the plane of the membrane in a transmembranous orientation [79, 211].

There is now further evidence that TTB inhibits succinate-ubiquinone reductase activity by inhibiting reduction of ubisemiquinone to ubiquinol and that the latter reaction may be catalyzed by iron-sulfur center S-3 of the dehydrogenase (see Fig. 9 above). Ackrell and coworkers used the rapid freeze technique to examine the oxidation-reduction kinetics of succinate dehydrogenase in preparations of complex II [212]. Using a synthetic analog of ubiquinone-1, whose reduction was previously shown to be inhibited by TTB and carboxins [213], they showed that this quinone oxidized the previously reduced EPR detectable components of the dehydrogenase within the turnover time of the enzyme [212].

It was found that carboxins, which were previously shown to inhibit succinate-ubiquinone reductase at the same site as TTB [213], inhibited reoxidation of iron-sulfur centers S-1 and S-3 while there was no change in the rate of reduction of center S-1 and only a slight decrease in the rate of reduction of center S-3 [213].

Two additional findings which relate to the mechanism by which succinate dehydrogenase reduces ubiquinone were the diminished efficacy of TTB in the presence of antimycin [214] and the production of superoxide anion by addition of TTB if electron transfer through the reductase complex is extensively inhibited by antimycin [215].

To explain these and earlier observations, it has been proposed that TTB (and presumably carboxins) inhibits the reaction $QH\cdot \rightarrow QH_2$ by intercalating between the iron-sulfur cluster of center S-3 and the spin-coupled $QH\cdot$ pair [215]. It has been further proposed that the amount of TTB required to inhibit respiration is inversely related to the amount of $QH\cdot$ associated with center S-3 [214]. Thus it would be expected that the efficacy of TTB should be inversely related to the ubiquinone content of particles, as observed by Nelson and coworkers [107], and likewise inversely dependent on the succinate dehydrogenase content, as reported by Baginsky and Hafeti [20]. If TTB and similar compounds do in fact inhibit by such a mechanism, their apparent K_i values would not necessarily correlate with their iron chelation constants [213] but would instead reflect a more complex structural relationship between iron-sulfur center, spin-coupled semiquinone, and inhibitor.

There have been several additional reports [216-218] regarding the polypeptide composition of resolved bc_1 complex (see Table 6). Marres and Slater [216] tested the validity of the electrophoretic method for measuring apparent molecular weights of the polypeptides of the bc_1 complex by measuring their retardation coefficients and free electrophoretic mobilities. They found that the core protein polypeptides and a third polypeptide thought to be apocytochrome *b* showed anomalous migration rates, indicating that the commonly used electrophoretic method [204] does not reliably relate relative mobility to apparent molecular weight.

Yeast complex III has now been purified in active form for the first time

[218], and a cytochrome *b* polypeptide has been purified from yeast mitochondria [219]. Likewise, Von Jagow and coworkers [220] have now apparently purified a *b* cytochrome from their highly resolved bc_1 complex (see Section V.B). The yeast cytochrome *b* has an apparent molecular weight of 28,000 by the dodecyl sulfate-acrylamide gel electrophoresis method and 28,800 by sucrose gradient centrifugation [219]. The apparent molecular weight of the beef heart *b* is 31,000 by the electrophoretic method (see above and Ref. 216), 33,000 by heme content, and 62,000 by equilibrium centrifugation [220]. The *b* cytochrome from both of these preparations [219, 220] is reactive to carbon monoxide. In addition, a method for measuring the biological activity of isolated cytochrome *b* has not yet been developed (see Section V.C).

Finally, there still remains the problem that the apocytochrome *b* polypeptides have not been convincingly identified with one of the polypeptides of the resolved bc_1 complex from which they presumably originate. Several workers have noted that this may reflect inherent limitations of the dodecyl sulfate-acrylamide gel electrophoresis method (see Section II.3 and Ref. 216). In view of the fact that both the yeast [219] and mammalian [220] *b* polypeptides can be detected by electrophoresis of the isolated proteins, it may be informative to mix the purified proteins [219, 220] with their respective bc_1 complexes [218, 12] and to compare the resulting electrophoretic profile with the *b* apoprotein and the bc_1 complex.

The last development to report is that a student in this laboratory has now purified oxidation factor. We have identified it as a reconstitutively active form of the iron-sulfur protein of the bc_1 segment (B. Trumpower and C. Edwards, submitted for publication). This is the first time this protein has been purified in biologically active form, and this finding provides good evidence that this iron-sulfur protein participates in some obligatory fashion, most likely as an electron carrier, in the mitochondrial respiratory chain. Earlier work on the iron-sulfur protein was discussed above (Section VI) and in a review by Rieske [63], who pointed out that oxidation factor and the iron-sulfur protein may be identical.

The progressive purification of the iron-sulfur protein was documented by a reconstitution assay, as described above (Section VIII.B), in which the amount of succinate-cytochrome *c* reductase activity restored to depleted reductase complex was first-order with respect to this protein. A major problem in purifying the iron-sulfur protein was that partially purified preparations were extensively contaminated by a low molecular weight polypeptide (polypeptide G 13.2 in Fig. 19). The reconstitutively active iron-sulfur protein was tenaciously associated with cytochrome c_1.

Two findings were thus critical to the eventual purification. The first was that a more selective extraction of oxidation factor was achieved when antimycin was added to the alkaline-washed reductase complex prior to extraction in the presence of dithionite (see Fig. 16 above). The second was that oxidation factor activity was recovered from chromatography in pure form when the partially purified extract was passed through a mercurial resin and then chromatographed on hydroxyapatite (B. Trumpower and C. Edwards, submitted for publication).

The interesting aspect of this chromatography was that sequential treatment with both of these resins appeared to be necessary for purification. In view of the fact that cytochrome c_1 contained two sulfhydryl groups (see Section VII.A), whose modification by mercurials or N-ethyl maleimide caused the cytochrome to become more auto-oxidizable (B. Trumpower, A. Katki, and C. Edwards, unpublished observations), it seemed possible that the iron-sulfur protein was dissociated from c_1 upon denaturation of the cytochrome by the mercurial resin. This may have interesting consequences relating to the structural-functional relationship of these two proteins and thus merits further investigation.

During the purification the specific activity of oxidation factor, as measured in the reconstitution assay, increased to 700 units/mg (for a comparison, see Table 8). We have identified oxidation factor as a reconstitutively active form of the iron-sulfur protein of the bc_1 segment on the basis of the following findings.

The purified protein, based on activity, is a homogenous polypeptide on acrylamide gel electrophoresis and has an apparent molecular weight of 24,500 (B. Trumpower and C. Edwards, submitted for publication). When mixed with complex III the purified protein has the same electrophoretic mobility as the polypeptide which several laboratories have identified as the iron-sulfur protein of the bc_1 complex (see Table 6), whose migration position can be recognized as being immediately ahead of the heme-containing polypeptide of c_1 (see Fig. 5). In addition, the purified protein contains 55-60 nmol of nonheme iron/mg and 0.62-0.65 equivalents of acid-labile sulfide. Comparison of this nonheme iron content to that of earlier preparations (see Section VIII.C) demonstrates the extent to which this protein has been further purified.

The purified, reconstitutively active protein has the characteristic EPR spectrum of the "g = 1.90" protein (B. Trumpower, C. Edwards, and T. Ohnishi, in preparation). We have found that the g = 1.90 signal is not detectable in the depleted reductase complex upon addition of succinate, ascorbate, or dithionite. After cytochrome *c* reductase activity is restored to the depleted complex by reconstitution with oxidation factor, addition of succinate to the reconstituted complex causes the rapid appearance of the g = 1.90 signal (B. Trumpower, C. Edwards, and T. Ohnishi, in preparation). In these reconstitution experiments we found that the M_r = 24,500 polypeptide was not detectable by electrophoresis of the depleted complex, and this polypeptide then reappeared coincident with reconstitution of the EPR signal and cytochrome *c* reductase activities.

Although the nonheme iron and acid-labile sulfide content are less than expected (see Section VIII.C), suggesting that a portion of the purified protein has lost its iron-sulfur cluster and may thus be partially inactive, the data on hand are strong evidence that oxidation factor is a purified, reconstitutively active form of the iron-sulfur protein first discovered by Rieske and coworkers [161].

Acknowledgments

We are indebted to Drs. Donald Schneider and Mårten Wikström, who contributed many suggestions which have helped to clarify our writing. We are especially grateful for those unrestrained criticisms which have challenged our views. We also wish to thank Professor Efraim Racker, who introduced us to oxidation factor in his laboratory and who has continued to encourage us by his interest and enthusiasm.

The research from this laboratory has been supported by a National Institutes of Health Research Grant, GM20379, and by a Biomedical Research Support Grant, RR-05392, from the National Institutes of Health.

References

1. M. Erecinska, R. L. Veech, and D. F. Wilson, *Arch. Biochem. Biophys. 160:* 412 (1974).
2. D. F. Wilson, P. L. Dutton, M. Erecinska, and J. G. Lindsay, in *Mechanisms in Bioenergetics* (G. F. Azzone, L. Ernster, S. Papa, E. Quagliariello, and N. Siliprandi, eds.), Academic, New York, 1973, pp. 527-533.
3. Y. Hatefi, in *Comprehensive Biochemistry* Vol. 14 (M. Florkin and E. H. Stotz, eds.), Elsevier, Amsterdam, 1966, pp. 199-231.
4. A. Kroger and M. Klingenberg, in *Current Topics in Bioenergetics* Vol. 2 (D. R. Sanadi, ed.), Academic, New York, 1967, pp. 151-193.
5. M. Klingenberg, in *Biological Oxidations* (T. P. Singer, ed.), Interscience, New York, 1968, pp. 3-54.
6. C. I. Ragan, *Biochim. Biophys. Acta 456:* 249 (1976).
7. J. F. Hare and F. L. Crane, *Subcell. Biochem. 3:* 1 (1974).
8. R. A. Capaldi, R. L. Bell, and T. Branchek, *Biochem. Biophys. Res. Commun. 74:* 425 (1977).
9. P. Gellerfors and B. D. Nelson, *Eur. J. Biochem. 52:* 433 (1975).
10. B. L. Trumpower and A. Katki, *Biochemistry 14:* 3635 (1975).
11. C. A. Yu, L. Yu, and T. E. King, *J. Biol. Chem. 249:* 4905 (1974).
12. P. Riccio, H. Schagger, W. D. Engel, and G. von Jagow, *Biochim. Biophys. Acta 459:* 250 (1977).
13. K. A. Davis and Y. Hatefi, *Biochem. Biophys. Res. Commun. 44:* 1338 (1971).
14. C. A. Yu, L. Yu, and T. E. King, *Biochim. Biophys. Acta 267:* 300 (1972).

15. K. A. Davis and Y. Hatefi, *Biochemistry 10,* 2509 (1971).
16. T. Ohnishi, J. C. Salerno, D. B. Winter, J. Lim, C. A. Yu, L. Yu, and T. E. King, *J. Biol. Chem. 251:* 2094 (1976).
17. H. Beinert, B. A. C. Ackrell, E. B. Kearney, and T. P. Singer, *Biochem. Biophys. Res. Commun. 58:* 564 (1974).
18. K. A. Davis, Y. Hatefi, K. L. Poff, and W. L. Butler, *Biochim. Biophys. Acta 325:* 341 (1973).
19. Y. Hatefi, in *The Enzymes of Biological Membranes* Vol. 4 (A. Martonosi, ed.), Plenum, New York, 1976, pp. 3-41.
20. M. L. Baginsky and Y. Hatefi, *J. Biol. Chem. 244:* 5313 (1969).
21. B. L. Trumpower, *Biochem. Biophys. Res. Commun. 70:* 73 (1976).
22. N. R. Orme-Johnson, R. E. Hansen, and H. Beinert, *Biochem. Biophys. Res. Commun. 45:* 871 (1971).
23. H. D. Tisdale, D. C. Wharton, and D. E. Green, *Arch. Biochem. Biophys. 102:* 114 (1963).
24. H. D. Tisdale, in *Methods in Enzymology* Vol. 10 (R. W. Estabrook and M. E. Pullman, eds.), Academic, New York, 1967, pp. 213-215.
25. S. Yamashita and E. Racker, *J. Biol. Chem. 244:* 1220 (1969).
26. L. Yu, C. A. Yu, and T. E. King, *Biochemistry 12:* 540 (1973).
27. M. Erecinska, R. Oshino, H. Oshino, and B. Chance, *Arch. Biochem. Biophys. 157:* 431 (1973).
28. D. Kleiner, *Arch. Biochem. Biophys. 165:* 121 (1974).
29. J. Rieske, in *Methods in Enzymology* Vol 10 (R. W. Estabrook and M. E. Pullman, eds.), Academic, New York, 1967, pp. 239-245.
30. K. H. Leung and P. C. Hinkle, *J. Biol. Chem. 250:* 8467 (1975).
31. Y. P. Wan, R. H. Williams, K. Folkers, K. H. Leung, and E. Racker, *Biochem. Biophys. Res. Commun. 63:* 11 (1975).
32. J. F. Hare and F. L. Crane, *Bioenergetics 2:* 317 (1971).
33. D. Ziegler and J. S. Rieske, in *Methods in Enzymology* Vol 10 (R. W. Estabrook and M. E. Pullman, eds.), Academic, New York, 1967, pp. 231-235.
34. M. L. Baginsky and Y. Hatefi, *Biochem. Biophys. Res. Commun. 32:* 945 (1968).
35. W. W. Wainio, *The Mammalian Mitochondrial Respiratory Chain,* Academic, New York, 1970.
36. T. P. Singer, in *Methods of Biochemical Analysis* Vol. 22 (D. Glick, ed.), Wiley, New York, 1974, pp. 123-175.
37. T. E. King, in *Methods in Enzymology* Vol. 10 (R. W. Estabrook and M. E. Pullman, eds.), Academic, New York, 1967, pp. 322-331.
38. A. D. Vinogradov, E. V. Gavrikova, and V. G. Goloveshkina, *Biochem.* (Moscow) *41:* 1155 (1976); pp. 943-954 in translation.
39. I. Sekuzu, Y. Oru, and K. Okunuki, *J. Biochem.* (Tokyo) *48:* 214 (1960).
40. R. Bomstein, R. Goldberger, and H. D. Tisdale, *Biochim. Biophys. Acta 50:* 527 (1961).
41. L. Yu, Y. L. Chiang, C. A. Yu, and T. E. King, *Biochim. Biophys. Acta 379:* 33 (1975).

42. C. A. Yu, L. Yu, and T. E. King, *J. Biol. Chem. 247:* 1012 (1972).
43. D. E. Green, J. Järnefelt, and H. D. Tisdale, *Biochim. Biophys. Acta 31:* 34 (1960).
44. P. L. Dutton, D. F. Wilson, and C. P. Lee, *Biochemistry 9:* 5077 (1970).
45. E. E. Jacobs and R. Sanadi, *J. Biol. Chem. 235:* 531 (1960).
46. M. K. F. Wikstrom, *Biochim. Biophys. Acta 301:* 155 (1973).
47. M. K. F. Wikstrom and A. M. Lambowitz, *FEBS Lett. 40:* 149 (1974).
48. G. Hubscher, M. Kiese, and R. Nicolas, *Biochem. Z. 325:* 223 (1954).
49. R. E. Basford, H. D. Tisdale, J. L. Glenn, and D. E. Green, *Biochim. Biophys. Acta 24:* 107 (1957).
50. J. S. Rieske, in *Methods in Enzymology* Vol. 10 (R. W. Estabrook and M. E. Pullman, eds.), Academic, New York, 1967, pp. 488-493.
51. R. Goldberger, A. L. Smith, H. Tisdale, and R. Bomstein, *J. Biol. Chem. 236:* 2788 (1961).
52. R. Goldberger, A. Pumphrey, and A. L. Smith, *Biochim. Biophys. Acta 58:* 307 (1962).
53. J. A. Berden and E. C. Slater, *Biochim. Biophys. Acta 216:* 237 (1970).
54. B. D. Nelson and P. Gellerfors, *Biochim. Biophys. Acta 357:* 358 (1974).
55. U. K. Laemmli, *Nature 227:* 680 (1970).
56. R. L. Bell, Ph.D. Thesis, University of Oregon, 1977.
57. P. Gellerfors, M. Lunden, and B. D. Nelson, *Eur. J. Biochem. 67:* 463 (1976).
58. W. W. Fish, J. A. Reynolds, and C. Tanford, *J. Biol. Chem. 245:* 5166 (1970).
59. J. G. Williams and W. B. Gratzer, *J. Chromatogr. 57:* 121 (1971).
60. R. T. Swank and K. D. Munkres, *Anal. Biochem. 39:* 462 (1971).
61. C. A. Yu, L. Yu, and T. E. King, *Biochem. Biophys. Res. Commun. 66:* 1194 (1975).
62. U. Das Gupta and J. S. Rieske, *Biochem. Biophys. Res. Commun. 54:* 1247 (1973).
63. J. S. Rieske, *Biochim. Biophys. Acta 456:* 195 (1976).
64. W. H. Walker and T. P. Singer, *J. Biol. Chem. 245:* 4224 (1970).
65. W. G. Hanstein, K. A. Davis, M. A. Ghalambar, and Y. Hatefi, *Biochemistry 10:* 2517 (1971).
66. T. Ohnishi, J. Lim, D. B. Winter, and T. E. King, *J. Biol. Chem. 251:* 2105 (1976).
67. E. Racker, *Fed. Proc. 26:* 1335 (1967).
68. E. Racker, *A New Look at Mechanisms in Bioenergetics,* Academic, New York, 1976.
69. A. D. Vinogradov, E. V. Gavrikova, and V. G. Goloveshkina, *Biochem. Biophys. Res. Commun. 65:* 1264 (1975).
70. L. C. McPhail and C. C. Cunningham, *Biochemistry 14:* 1122 (1975).
71. H. Beinert, B. A. C. Ackrell, E. B. Kearney, and T. P. Singer, *Eur. J. Biochem. 54:* 185 (1975).
72. A. Bruni and E. Racker, *J. Biol. Chem. 243:* 962 (1968).

73. C. A. Yu, L. Yu, and T. E. King, *Fed. Proc. Abstr. 36:* 815 (1977).
74. P. G. Mowery, D. J. Steenkamp, B. A. C. Ackrell, T. P. Singer, and G. A. White, *Arch. Biochem. Biophys. 178:* 495 (1977).
75. E. R. Redfearn, P. A. Whittaker, and J. Burgos, in *Oxidases and Related Redox Systems* (T. E. King, H. S. Mason, and M. Morrison, eds.), Wiley, New York, 1965, pp. 943-959.
76. F. J. Ruzicka, H. Beinert, K. L. Schepler, W. R. Dunham, and R. H. Sands, *Proc. Nat. Acad. Sci. U. S. 72:* 2886 (1975).
77. W. J. Ingledew and T. Ohnishi, *FEBS Lett. 54:* 617 (1975).
78. W. J. Ingledew, J. C. Salerno, and T. Ohnishi, *Arch. Biochem. Biophys. 177:* 176 (1976).
79. W. J. Ingledew and T. Ohnishi, *Biochem. J. 164:* 167 (1977).
80. F. L. Crane, Y. Hatefi, R. L. Lester, and C. Widmer, *Biochim. Biophys. Acta 25:* 220 (1957).
81. P. Mitchell, *Chemiosmotic Coupling in Oxidative and Photosynthetic Phosphorylation,* Glynn Research Ltd., Bodmin, Cornwall, England, 1966.
82. A. T. Diplock and G. A. D. Haslewood, *Biochem. J. 104:* 1004 (1967).
83. F. L. Crane, in *Biochemistry of Quinones* (R. A. Morton, ed.), Academic, New York, 1965, pp. 183-206.
84. G. L. Sottocasa and G. Sandri, *Ital. J. Biochem. 17:* 17 (1968).
85. F. L. Crane, C. Widmer, R. L. Lester, and Y. Hatefi, *Biochim. Biophys. Acta 31:* 476 (1959).
86. R. L. Lester and S. Fleischer, *Biochim. Biophys. Acta 47:* 358 (1961).
87. L. Szarkowska, *Arch. Biochem. Biophys. 113:* 519 (1966).
88. G. B. Cox, A. M. Snoswell, and F. Gibson, *Biochim. Biophys. Acta 153:* 1 (1968).
89. M. Klingenberg and A. Kroger, in *Biochemistry of Mitochondria* (E. C. Slater, Z. Kaniuga, and L. Wojtczak, eds.), Academic, New York, 1967, pp. 11-27.
90. H. S. Penefsky, *Biochim. Biophys. Acta 58:* 619 (1962).
91. J. C. Catlin, R. S. Pardini, G. D. Daves, J. C. Heidker, and K. Folkers, *J. Amer. Chem. Soc. 90:* 3572 (1968).
92. C. Hall, M. Wu, F. L. Crane, H. Takahashi, S. Tamura, and K. Folkers, *Biochem. Biophys. Res. Commun. 25:* 373 (1966).
93. B. Chance, in *Dynamics of Energy Transducing Membranes* (L. Ernster, R. W. Estabrook, and E. C. Slater, eds.), Elsevier, Amsterdam, 1974, pp. 553-578.
94. D. E. Green, *Comp. Biochem. Physiol. 4:* 81 (1962).
95. D. E. Green and D. C. Wharton, *Biochem. Z. 338:* 335 (1963).
96. D. E. Green and I. Silman, *Ann. Rev. Plant Physiol. 18:* 147 (1967).
97. L. Ernster, I. Y. Lee, B. Norling, and B. Persson, *Eur. J. Biochem. 9:* 299 (1969).
98. A. Kroger and M. Klingenberg, *Eur. J. Biochem. 34:* 358 (1973).
99. A. Kroger and M. Klingenberg, *Eur. J. Biochem. 39:* 313 (1973).
100. D. Backstrom, B. Norling, A. Ehrenberg, and L. Ernster, *Biochim. Biophys. Acta 197:* 108 (1970).

101. A. Kroger, *FEBS Lett. 65:* 278 (1976).
102. R. Bensasson and E. J. Land, *Biochim. Biophys. Acta 325:* 175 (1973).
103. N. K. Bridge and G. Porter, *Proc. Roy. Soc., Ser. A 244:* 276 (1958).
104. A. Boveris, E. Cadenas, and A. O. M. Stoppani, *Biochem. J. 156:* 435 (1976).
105. E. Cadenas, A. Boveris, C. I. Ragan, and A. O. M. Stoppani, *Arch. Biochem. Biophys. 180:* 248 (1977).
106. P. Mitchell, *J. Theor. Biol. 62:* 327 (1976).
107. B. D. Nelson, B. Norling, B. Persson, and L. Ernster, *Biochim. Biophys. Acta 267:* 205 (1972).
108. T. Ohnishi, W. J. Ingledew, and S. Shiraishi, *Biochem. J. 153:* 39 (1976).
109. H. G. Lawford and P. B. Garland, in *Dynamics of Energy Transducing Membranes* (L. Ernster, R. W. Estabrook, and E. C. Slater, eds.), Elsevier, Amsterdam, 1974, pp. 159-167.
110. E. C. Slater, *Biochim. Biophys. Acta 301:* 129 (1973).
111. R. E. Olson and G. H. Dialameh, *Biochem. Biophys. Res. Commun. 2:* 198 (1960).
112. C. F. Fox, in *Biochemistry of Cell Walls and Membranes* (C. F. Fox, ed.), University Park Press, Baltimore, Md., 1975, pp. 279-306.
113. M. Edidin, *Ann. Rev. Biophys. Bioeng. 3:* 179 (1974).
114. R. J. Cherry, in *Biological Membranes* Vol. 3 (D. Chapman and D. F. H. Wallach, eds.), Academic, New York, 1976, pp. 47-102.
115. L. F. Fieser and M. Fieser, *Advanced Organic Chemistry,* Rheinhold, New York, 1961.
116. A. Langemann and O. Isler, in *Biochemistry of Quinones* (R. A. Morton, ed.), Academic, New York, 1965, pp. 89-147.
117. D. L. Breen, *J. Theor. Biol. 53:* 101 (1975).
118. H. Morimoto and I. Imada, *Biochim. Biophys. Acta 275:* 10 (1972).
119. P. E. Brumby and V. Massey, in *Methods in Enzymology* Vol. 10 (R. W. Estabrook and M. E. Pullman, eds.), Academic, New York, 1967, pp. 463-474.
120. D. Keilin, *Proc. Roy. Soc., Ser. B 98:* 312 (1925).
121. M. B. Katan, N. van Harten-Loosebroeck, and G. S. P. Groot, *Eur. J. Biochem. 70:* 409 (1976).
122. K. Ohnishi, *J. Biochem. 59:* 1 (1966).
123. H. Weiss and B. Ziganke, *Eur. J. Biochem. 41:* 63 (1974).
124. H. Weiss and B. Ziganke, in *Electron Transfer Chains and Oxidative Phosphorylation* (E. Quagliariello, S. Papa, F. Palmieri, E. C. Slater, and N. Siliprandi, eds.), North Holland, Amsterdam, 1973, pp. 15-22.
125. Y. C. Awasthi, R. Berezney, F. J. Ruzicka, and F. L. Crane, *Biochim. Biophys. Acta 189:* 457 (1969).
126. N. C. Robinson and R. A. Capaldi, *Biochemistry 16:* 375 (1977).
127. B. Chance, *J. Biol. Chem. 233:* 1223 (1958).
128. B. Chance and B. Schoener, *J. Biol. Chem. 241:* 4567 (1966).
129. B. Chance, C. P. Lee, and B. Schoener, *J. Biol. Chem. 241:* 4574 (1966).

130. B. Chance and B. Schoener, *J. Biol. Chem. 241:* 4577 (1966).
131. E. C. Slater, J. A. Berden, H. J. Wegdam, *Nature 226:* 1248 (1970).
132. M. K. F. Wikstrom, in *Energy Transduction in Respiration and Photosynthesis* (E. Quagliariello, S. Papa, and C. S. Rossi, eds.), Adriatica Editrice, Bari, 1977, pp. 693-709.
133. T. Higuti, S. Mizune, and S. Muraoka, *Biochim. Biophys. Acta 396:* 36 (1975).
134. M. Erecinska, D. F. Wilson, and Y. Miyata, *Arch. Biochem. Biophys. 177:* 133 (1976).
135. P. L. Dutton and D. F. Wilson, *Biochim. Biophys. Acta 346:* 165 (1974).
136. J. G. Lindsay, P. L. Dutton, and D. F. Wilson, *Biochemistry 11:* 1937 (1972).
137. D. F. Wilson and M. Erecinska, *Arch. Biochem. Biophys. 167:* 116 (1975).
138. N. Sato, D. F. Wilson, and B. Chance, *Biochim. Biophys. Acta 253:* 88 (1971).
139. P. L. Dutton, J. G. Lindsay, and D. F. Wilson, in *Biochemistry and Bio-Physics of Mitochondrial Membranes* (G. F. Azzone, E. Carafoli, A. L. Lehninger, E. Quagliariello, and N. Siliprandi, eds.), Academic, New York, 1972, pp. 165-176.
140. P. Mitchell and J. Moyle, *Eur. J. Biochem. 4:* 530 (1968).
141. P. Mitchell, *FEBS Symp. 28:* 353 (1972).
142. B. Chance, *Abstr. 2nd Int. Cong. Biochem. (Paris),* 1952, p. 32.
143. A. M. Pumphrey, *J. Biol. Chem. 237:* 2384 (1962).
144. H. Baum, J. S. Rieske, H. I. Silman, and S. H. Lipton, *Proc. Nat. Acad. Sci. U. S. 57:* 798 (1967).
145. M. Erecinska, B. Chance, D. F. Wilson, and P. L. Dutton, *Proc. Nat. Acad. Sic. U. S. 69:* 50 (1972).
146. H. Baum and J. S. Rieske, *Biochem. Biophys. Res. Commun., 24:* 1 (1966).
147. J. S. Rieske, *Arch. Biochem. Biophys. 145:* 179 (1971).
148. M. Erecinska, *Fed. Proc. Abstr. 31:* 415 (1972).
149. D. F. Wilson, M. Erecinska, J. S. Leigh, and M. Koppelman, *Arch. Biochem. Biophys. 151:* 112 (1972).
150. B. L. Trumpower and A. G. Katki, *Biochem. Biophys. Res. Commun. 65:* 16 (1975).
151. M. Eisenbach and M. Gutman, *FEBS Lett. 46:* 368 (1974).
152. M. Eisenbach and M. Gutman, *Eur. J. Biochem. 52:* 107 (1975).
153. I. Y. Lee and E. C. Slater, *Biochim. Biophys. Acta 283:* 395 (1972).
154. M. K. F. Wikstrom and J. Berden, *Biochim. Biophys. Acta, 283:* 403 (1972).
155. M. Eisenbach and M. Gutman, *FEBS Lett. 61:* 247 (1976).
156. A. M. Lambowitz and W. D. Bonner, Jr., *Biochem. Biophys. Res. Commun., 52:* 703 (1973).
157. J. S. Rieske, J. O. Alben, and H. T. Liao, in *Electron Transfer Chains and*

Oxidative Phosphorylation (E. Quagliariello, S. Papa, F. Palmieri, E. C. Slater, and M. Siliprandi, eds.), North Holland, Amsterdam, 1975, pp. 119-126.

158. R. W. Hendler, D. W. Towne, and R. I. Shrager, *Biochim. Biophys. Acta 376:* 42 (1975).
159. A. Boveris, R. Oshino, M. Erecinska, and B. Chance, *Biochim. Biophys. Acta 245:* 1 (1971).
160. M. Erecinska and D. F. Wilson, *Arch. Biochem. Biophys. 174:* 143 (1976).
161. J. S. Rieske, D. H. MacLennan, and R. Coleman, *Biochem. Biophys. Res. Commun. 15:* 338 (1964).
162. N. R. Orme-Johnson, R. E. Hansen, and H. Beinert, *J. Biol. Chem. 249:* 1928 (1974).
163. J. S. Rieske, W. S. Zaugg, and R. E. Hansen, *J. Biol. Chem. 239:* 3023 (1964).
164. J. S. Leigh and M. Erecinska, *Biochim. Biophys. Acta 387:* 95 (1975).
165. T. E. King, C. A. Yu, L. Yu, and Y. L. Chiang, in *Electron Transfer Chains and Oxidative Phosphorylation* (E. Quagliariello, S. Papa, F. Palmieri, E. C. Slater, and N. Siliprandi, eds.), North Holland, Amsterdam, 1975, pp. 105-118.
166. J. S. Rieske, R. E. Hansen, and W. S. Zaugg, *J. Biol. Chem. 239:* 3017 (1964).
167. J. S. Rieske, in *Methods in Enzymology* Vol. 10 (R. W. Estabrook and M. E. Pullman, eds.), Academic, New York, 1967, pp. 357-362.
168. I. Y. Lee and E. C. Slater, *Biochim. Biophys. Acta 347:* 14 (1974).
169. R. C. Prince, J. G. Lindsay, and P. L. Dutton, *FEBS Lett. 51:* 108 (1975).
170. H. Baum, H. I. Silman, J. S. Rieske, and S. H. Lipton, *J. Biol. Chem. 242:* 4876 (1967).
171. J. S. Leigh and B. Chance, *Fed. Proc. Abstr. 33:* 1289 (1974).
172. P. L. Dutton and J. G. Lindsay, in *Mechanisms in Bioenergetics* (G. F. Azzone, L. Ernster, S. Papa, E. Quagliariello, and N. Siliprandi, eds.), Academic, New York, 1973, pp. 535-544.
173. E. Ross and G. Schatz, *J. Biol. Chem. 251:* 1991 (1976).
174. M. B. Katan and G. S. P. Groot, in *Electron Transfer Chains and Oxidative Phosphorylation* (E. Quagliariello, S. Papa, F. Palmieri, E. C. Slater, and N. Siliprandi, eds.), North Holland, Amsterdam, 1975, pp. 127-132.
175. R. V. Eck and M. O. Dayhoff, *Atlas of Protein Sequence and Structure,* National Biomedical Research Foundation, Silver Springs, Md., 1966.
176. N. Nelson and E. Racker, *J. Biol. Chem. 247:* 3848 (1972).
177. W. J. Vail and J. S. Rieske, *FEBS Lett. 58:* 33 (1975).
178. S. J. Singer, in *Structure and Function of Biological Membranes* (L. I. Rothfield, eds.), Academic, New York, 1971, pp. 145-222.
179. R. J. Smith and R. A. Capaldi, *Biochemistry 16:* 2629 (1977).
180. H. A. Harbury, J. R. Cronin, M. W. Fanger, T. P. Hettinger, A. J. Murphy, J. P. Myer, and S. N. Vinogradov, *Proc. Nat. Acad. Sci. U. S. 54:* 1658 (1963).

181. C. A. Yu, L. Yu, and T. E. King, *J. Biol. Chem. 248:* 528 (1973).
182. L. S. Kaminsky, Y. L. Chiang, C. A. Yu, and T. E. King, *Biochem. Biophys. Res. Commun., 59:* 688 (1974).
183. Y. L. Chiang, L. S. Kaminsky, and T. E. King, *J. Biol. Chem. 251:* 29 (1976).
184. L. Smith and K. Minnart, *Biochim. Biophys. Acta 105:* 1 (1965).
185. L. Smith, H. C. Davies, and M. Nava, *J. Biol. Chem. 249:* 2904 (1974).
186. D. L. Schneider and E. Racker, in *Oxidases and Related Redox Systems* (T. E. King, H. S. Mason, and M. Morrison, eds.), Wiley, New York, 1973, pp. 799-821.
187. D. L. Schneider, in *Progress in Surface and Membrane Science* Vol. 8 (D. A. Cadenhead, J. F. Danielli, and M. D. Rosenberg, eds.), Academic, New York, 1974, pp. 209-243.
188. C. A. Yu, F. C. Yong, and T. E. King, *Biochem. Biophys. Res. Commun. 45:* 508 (1971).
189. J. L. Hoard, M. J. Hamor, T. A. Hamor, and W. S. Caughey, *J. Amer. Chem. Soc. 87:* 2312 (1965).
190. J. L. Hoard, *Science 174:* 1295 (1971).
191. R. E. Dickerson, T. Takano, D. Eisenberg, O. B. Kaffai, L. Samson, A. Cooper, and E. Margoliash, *J. Biol. Chem. 246:* 1511 (1971).
192. H. Nishibayashi-Yamashita, C. Cunningham, and E. Racker, *J. Biol. Chem. 247:* 698 (1972).
193. J. S. Rieske, H. Baum, C. D. Stoner, and S. H. Lipton, *J. Biol. Chem. 242:* 4854 (1967).
194. T. E. King, in *Methods in Enzymology* Vol. 10 (R. W. Estabrook and M. E. Pullman, eds.), Academic, New York, 1967, pp. 216-225.
195. E. C. Slater, *Quart. Rev. Biophys. 4:* 35 (1971).
196. G. D. Greville, in *Current Topics in Bioenergetics* Vol 3 (D. R. Sanadi, ed.), Academic, New york, 1969, pp. 1-78.
197. P. Mitchell, *FEBS Lett. 56:* 1 (1975).
198. P. Mitchell, *FEBS Lett. 59:* 137 (1975).
199. B. Chance and G. R. Williams, *Adv. Enzymol. 17:* 65 (1956).
200. G. von Jagow and C. Bohrer, *Biochim. Biophys. Acta 387:* 409 (1975).
201. E. C. Slater, in *Methods in Enzymology* Vol. 10 (R. W. Estabrook and M. E. Pullman, eds.), Academic, New York, 1967, pp. 48-57.
202. P. C. Hinkle and K. H. Leung, in *Membrane Proteins in Transport and Phosphorylation* (G. F. Azzone, M. E. Klingenberg, E. Quagliariello, and N. Siliprandi, eds.), American Elsevier, New York, 1974, pp. 73-78.
203. F. Guevieri and B. D. Nelson, *FEBS Lett. 54:* 339 (1975).
204. K. Weber and M. Osborn, *J. Biol. Chem. 244:* 4406 (1969).
205. A. H. Lichtman and J. L. Howland, *FEBS Lett. 34:* 256 (1973).
206. R. A. Capaldi, J. Sweetland, and A. Merli, *Biochemistry 16:* 5707 (1977).
207. C. A. Yu, L. Yu, and T. E. King, *Biochem. Biophys. Res. Commun. 78:* 259 (1977).
208. C. A. Yu, L. Yu, and T. E. King, *Biochem. Biophys. Res. Commun. 79:* 939 (1977).

209. C. A. Yu, S. Nagaoka, L. Yu, and T. E. King, *Biochem. Biophys. Res. Commun. 82:* 1070 (1978).
210. P. C. Mowery, B. A. C. Ackrell, T. P. Singer, G. A. White, and G. D. Thorn, *Biophys. Res. Commun. 71:* 354 (1976).
211. T. Ohnishi, J. C. Salerno, H. Blum, J. S. Leigh, and W. J. Ingledew, in *Bioenergetics of Membranes* (L. Packer, G. C. Papageorgiou, and A. Trebst, (eds.), Elsevier, Amsterdam, 1977, pp. 209-216.
212. B. A. C. Ackrell, E. B. Kearney, C. J. Coles, T. P. Singer, H. Beinert, Y.-P. Wan, and K. Folkers, *Arch. Biochem. Biophys. 182:* 107 (1977).
213. P. C. Mowery, D. J. Steenkamp, B. A. C. Ackrell, T. P. Singer, and G. A. White, *Arch. Biochem. Biophys. 178:* 495 (1977).
214. B. L. Trumpower and Z. Simmons, *Biochem. Biophys. Res. Commun. 82:* 289 (1978).
215. B. L. Trumpower, in *Frontiers of Biological Energetics: From Electrons to Tissues* (L. Dutton, J. L. Leigh, and A. Scarpa, eds.), Academic, New York 1978, pp. 965-973.
216. C. A. M. Marres and E. C. Slater, *Biochim. Biophys. Acta 462:* 531 (1977).
217. L. Yu, C. A. Yu, and T. E. King, *Biochim. Biophys. Acta 495:* 232 (1977).
218. J. N. Siedow, S. Power, F. F. De La Rosa, and G. Palmer, *J. Biol. Chem. 253:* 2392 (1978).
219. L. F. H. Lin and D. S. Beattie, *J. Biol. Chem. 253:* 2412 (1978).
220. G. Von Jagow, H. Schagger, W. D. Engel, W. Machleidt, and I. Machleidt, *FEBS Lett. 91:* 121 (1978).

3

Structure of Cytochrome *c* Oxidase

Roderick A. Capaldi

University of Oregon
Eugene, Oregon

I. Introduction

Cytochrome *c* oxidase or cytochrome *c*:oxygen oxidoreductase (CH 1.9.3.1) is the terminal oxidase in respiratory metabolism of all aerobic organisms and is responsible for catalyzing the reduction of dioxygen to water in the reaction

$$4H^+ + 4e^- + O_2 \rightleftharpoons 2H_2O$$

The electrons for this reaction are provided by reduced cytochrome *c*. The free energy developed in the oxidation process is used to promote oxidative phosphorylation and in consequence becomes available to the cell as ATP.

II. Isolation Procedures

Purification schemes for cytochrome *c* oxidase from yeast, *Neurospora,* and *Locusta migratoria,* as well as from mammalian sources, take advantage of the fact that the enzyme is less readily solubilized by detergents than are most other components of the mitochondrial membrane. Several detergents including Triton X-114 [1], Triton X-100 [2-4], cholate [5-7], and deoxycholate [8, 9] have been used to obtain an initial separation of cytochrome *c* oxidase from other cytochromes, ATP-synthesizing components, and assorted transport proteins. Further purification usually involves detergent solubilization with cholate [5, 6, 8, 9] or Triton X-100 followed by several ammonium sulfate precipitation steps [1-4, 7]. The properties of several different preparations of cytochrome *c* oxidase are listed in Table 1.

One of the simplest approaches to purification of the enzyme uses deoxycholate and cholate as solubilizing agents [9]. This method takes only 6-8 hr to complete and yields an enzyme with a heme content as high as 11.7 nmol/mg protein. A preparation procedure which yields cytochrome *c* oxidase with a heme *a* content of 14 nmol/mg protein has been reported by Hartzell and Beinert [7] but this takes considerably longer to complete. As a simple approach to increasing the purity of crude enzyme preparations we use proteolytic digestion. Samples with high electron transfer activity and with a heme *a* content of 14 nmol/mg protein have been obtained by digestion with trypsin. Preparations of cytochrome *c* oxidase which contain more than 14-15 nmol heme per mg protein have not been reported.

III. Cytochrome *c* Oxidase as a Hemoprotein

As isolated, cytochrome *c* oxidase contains heme *a* and copper in approximately a 1:1 molar ratio. The structure of heme *a* as determined by Caughey and associ-

Table 1 Composition and Spectral Characteristics of Cytochrome *c* Oxidase Preparations

Authors	Species	Heme *a* content	Phospholipid (%)	Spectral maxima		α Band	References
				ox Soret[a]	red Soret[a]		
Yonetani	Beef	7.2	10	422.3	443.5	603	5
Sun et al.	Beef	8.2	22	–	–	–	1
Fowler et al.	Beef	8.4–8.7	–	423	444	605	8
Mason et al.	Yeast	9.4	2	422	443	603	3
Kuboyama et al.	Beef						6
Capaldi and Hayashi	Beef	9.0–11.7	10–16	421	442	605	9
Hartzell and Beinert	Beef	14.0	4	–	–	–	7
Rubin and Tzagoloff	Yeast	15.0	3.8	428	445	603	4

[a]Maxima of Soret bands in the oxidized (ox) and reduced (red) forms.

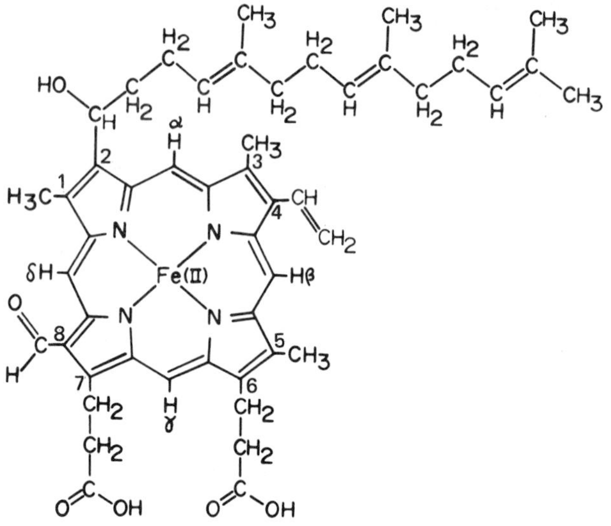

Figure 1 The structure of heme *a*.

ates [10], is shown in Figure 1. Attached to the protein, the heme behaves as two spectrally distinct entities present in equimolar amounts. One fraction, called heme a_3, reacts with CO and CN; the second fraction, heme *a*, does not [11, 12]. Copper atoms also behave as two spectrally distinct components and these are present in equimolar amounts in the cytochrome *c* oxidase complex. One-half of the copper is identifiable by a characteristic electron paramagnetic resonance (EPR) signal while the other half is not seen in EPR and thus is called the EPR-invisible copper [12]. The redox properties of cytochrome *c* oxidase have been investigated both by anaerobic reductive titrations and potentiometric titrations [13-22]. Results are generally interpreted in terms of low-potential (heme *a*) and high-potential (heme a_3) centers. In a recent study Mackey et al. generated oxidant and reductant coulometrically and followed the potential as a function of the number of electron equivalents added [22]. From a computer fitting of the resultant potential-composition curves they obtained midpoint potentials for the different redox centers as follows: heme *a*, 215 ± 10 mV; heme a_3, 340 ± 10 mV; coppers, 215 ± 10 and 350 ± 10 mV.

Based on magnetic circular dichroism measurements, Palmer et al. [23] have proposed that one heme (heme *a*) and one copper are magnetically isolated (low) spin S = 1/2 centers, while the second heme (heme a_3) and second copper form a (high) spin S = 2 center. They propose that in the high-spin center heme a_3 and copper are bridged by an imidazole group. This model would allow for two-electron reduction of molecular oxygen to a peroxide. The presence of a peroxide intermediate has been detected recently in the low-temperature studies of electron transfer conducted by Chance et al. [24].

IV. Spectral Properties

Cytochrome *c* oxidase in the mitochondrial inner membrane is distinguished by a Soret band at 422 nm in the oxidized form which shifts to 442 nm upon reduction and by an α band at 605 nm in the reduced form of the enzyme [25]. There is also a band in the infrared at 830 nm which has been attributed to the coppers. These same spectral characteristics are shown by most preparations of purified cytochrome *c* oxidase as listed in Table 2. It is generally agreed that the two hemes do not contribute equally to all spectral bands. According to Wickstrom et al. [25], the α band of reduced cytochrome oxidase is due to heme *a* with no more than 20% contribution from heme a_3. The Soret band is thought to have about an equal contribution from both hemes.

V. Electron Transfer by the Enzyme

The electron transfer activity of membrane-bound and isolated cytochrome *c* oxidase can be assayed polarographically by following the uptake of oxygen or spectrophotometrically by following the rate of oxidation of reduced cytochrome *c*. Using either assay the turnover rate of cytochrome *c* oxidase in intact mitochondria is low. There are several reasons for this. In coupled mitochondria, the overall rate of electron transfer is controlled by the rate of ATP synthesis. This so-called respiratory control is lost in the presence of uncouplers. However, in uncoupled mitochondria the rate of electron transfer through cytochrome *c* oxidase (with NADH or succinate as substrates) is still limited by slow electron transfer step in the dehydrogenase segments of the respiratory chain. It is possible to bypass NADH-cytochrome *c* reductase or succinate-cytochrome *c* reductase by adding electrons through reduced cytochrome *c* or ascorbate plus cytochrome *c*. However, to do this, mitochondria must be disrupted with detergents to allow for optimal interaction with substrate [26, 27]. When mitochondria or submitochondrial particles are incubated in 1% deoxycholate before assay, cytochrome *c* oxidase activities as high as 500 μmol cytochrome *c* oxidized/sec per heme a_3 (at 25°C, pH 7.0) can be obtained [26, 27].

The activities reported for isolated cytochrome *c* oxidase are generally much lower than those obtained in detergent-disrupted mitochondria. Vanneste et al. [26] showed that the loss of activity during isolation of the enzyme increased

Table 2 Range of Molecular Weight Estimates for Cytochrome *c* Oxidase

Molecular weight	Method	Active or inactive	Detergent	References
67,000	Sedimentation analysis	Inactive	SDS	37
67,000	Irradiation with high-energy electrons calculated by target theory	In mitochondria		38
72,000	Sedimentation analysis	Inactive	SDS	39
100,000	Light scattering	n.d.[a]	4-6 M urea + 0.33% DOC[a]	40
200,000	Sedimentation analysis	Active	pH 7.4, 0.1% Emasol 4130	41
228,000	Sedimentation analysis	Active	2% DOC, pH 8.0	42
230,000	Light scattering	Inactive	0.033% DOC	40
340,000	Sedimentation equilibrium	Active	1.03 mM Triton X-100	43
530,000	Sedimentation analysis	Active	0.25% Emasol 1130, pH 7.4	44

[a]n.d., not determined; DOC, deoxycholate.

with each ammonium sulfate precipitation step. They argued that the protein must be being denatured during purification. More likely, however, the low activities reported result from not using optimal conditions of assay. We have found that cytochrome *c* oxidase must be both monodispersed (as a dimer) and must be surrounded by the appropriate lipophilic environment for highest activity. The detergents, Tween 80 or Emasol 1130, which are generally used to assay cytochrome *c* oxidase, do not provide these conditions [28, 29] (See Section XII). Our assay procedure involves diluting the enzyme in Triton X-100 to disperse the protein and then assaying in 1 oleoyllysolecithin which we find to be the best activating detergent for the protein [30]. Using such assay conditions, we routinely obtain cytochrome *c* oxidase activities which are 70-80% of those reported for unfractionated but detergent-solubilized mitochondria.

Ferguson-Miller et al. [31] have recently used tetramethyl-p-phenylenediamine (TMPD) in the cytochrome *c* oxidase assay as a mediator of electron transfer between ascorbate and cytochrome *c*. This reagent effects a very rapid re-reduction of oxidized cytochrome *c* bound to the cytochrome oxidase complex. Eadie-Hofstee Scatchard plots of cytochrome *c* oxidase activities, obtained at low cytochrome *c* concentration and in the presence of TMPD and ascorbate, show a biphasic kinetic plot with a high (10^{-8} M sec^{-1}) and low (10^{-6} M sec^{-1}) affinity phases. Ferguson-Miller et al. attribute these two phases to two different cytochrome *c* binding sites on the cytochrome *c* oxidase molecule [31]. Both phases are seen in membrane-bound and isolated enzyme.

Wickstrom [32] has recently shown that in the mitochondrion the oxidation of ferrocytochrome *c* by oxygen is coupled to the activity of an electrogenic proton pump, i.e., there is a net increase of protons on one side of the mitochondrial inner membrane (outside) and a net decrease of protons on the inner or matrix side. He finds a H^+/e^- stoichiometry of 1, meaning that transfer of two electrons to 1/2 O_2 leads to separation of four charges across the mitochondrial inner membrane. While the quantitative aspects of proton movements with isolated enzyme have not been studied, several suthors have reported that purified cytochrome *c* oxidase can be reconstituted with lipids to form vesicles in which the rate of electron transfer is limited by the formation of both a proton gradient and an electrical potential across the membrane [33-36]. Enzyme prepared with cholate or deoxycholate but not with Triton X-100 (which is itself an ionophore) is effective in the reconstitution studies. Reagents such as uncouplers or the combination of valinomycin and nigericin increase the rate of cytochrome *c* oxidase electron transfer activity by dissipating these gradients [33-36].

Table 3 Estimates of the Molecular Weight of the Subunits of Beef Heart Cytochrome *c* Oxidase

Method	I	II	III	IV	V	VI	VII
Gel electrophoresis							
Weber-Osborn [61]	38,000	19,000	25,000	13,800	6,000	8,600	4,300
Swank-Munkres [58]	35,300	24,200	21,000	16,200	12,100	6,700	3,400
Fairbanks et al. [65]	33,000	21,300	19,000	14,000	12,500	8,500	4,900
Column chromatography							
In 6 M guanidine-HCl	–	–	–	17,000	12,500	9,700	5,300
In SDS	35,000	23,000	–	–	–	–	–

Sources: Downer et al. [53], Capaldi et al. [54], and Briggs et al. [62].

VI. Size and Shape of the Cytochrome *c* Oxidase Complex

Based on a maximal heme content of 14 nmol/mg protein, it can be calculated that the minimum molecular weight of the two heme/two copper complex is 140,000. As listed in Table 3, there have been many attempts to measure the minimum molecular weight of cytochrome *c* oxidase with extremely variable results. A major difficulty is that in dealing with a membrane-bound enzyme it is necessary to correct for bound lipid and bound detergent in molecular weight determinations, and most studies have not done this. We have performed sedimentation equilibrium studies on cytochrome *c* oxidase dispersed in either Triton X-100 or deoxycholate, using samples of known detergent and lipid composition [43]. In Triton X-100, the enzyme was found to be monodisperse, with a molecular weight of 345,000, i.e., in this nonionic detergent, cytochrome *c* oxidase was a four-heme aggregate or dimer of the complex (Fig. 2). In deoxycholate, the enzyme was a mixture of monomers and dimers.

It has been known for some years that cytochrome *c* oxidase forms two-dimensional crystals under some conditions of isolation [46-49]. We have been using electron microscopy and image reconstitution methods to study two such arrays, one obtained with enzyme isolated in Triton detergents (TX crystals) [50] and a second obtained with deoxycholate as the fractionating detergent (DOC crystals) (Fuller, Capaldi, and Henderson, unpublished results). The TX crystals show $P_12_12_12$ space group while the DOC crystals show a $P2_1$ space group. The arrangement of proteins in the two arrays are shown schematically in Figure 3. In the TX crystals, cytochrome *c* oxidase molecules are arranged as dimers spanning the lipid bilayer; the preparation is vesicular, and the vesicles are flattened so that the two membranes are apposed. A major portion of the protein is exposed in the inside of these vesicles. In the DOC lattice the cytochrome *c* oxidase molecules are monomeric (two-heme complex) and arranged in a sheet with no continuous lipid bilayer. Alternate rows of the DOC lattice are of opposite orientation.

Our preliminary results show clearly that the cytochrome *c* oxidase molecule is highly asymmetric, approximately 110-120 Å × 50-60 Å. The long axis of the molecule extends through the membrane bilayer. There is some indication that the protein may be Y-shaped with the arms of the Y extending through the bilayer.

VII. Subunit Structure

The subunit structure of cytochrome *c* oxidase has proved very difficult to determine. It has been claimed that there are anywhere from one [51] to ten [52]

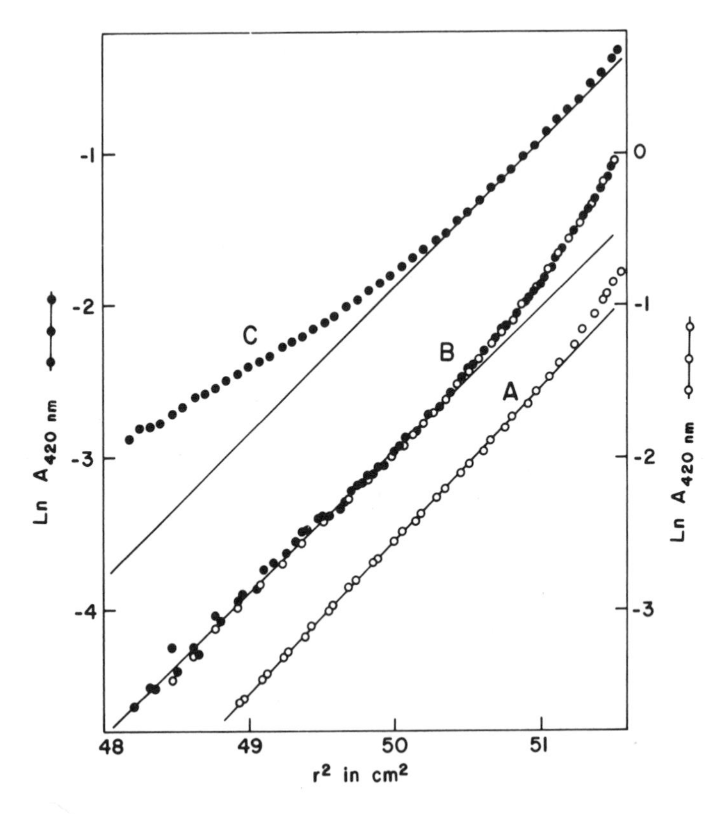

Figure 2 Sedimentation equilibrium of cytochrome *c* oxidase / detergent complexes. A, cytochrome *c* oxidase / Triton X-100 complex; B, cytochrome *c* oxidase / deoxycholate complex (this is for a sample eluting at the excluded volume of a Sephadex G-100 column); C, cytochrome *c* oxidase / deoxycholate complex (this is for a complex eluting at the peak and on the trailing edge through a Sepharose 4B column, $K_d = 0.50$). The molecular weight for the enzyme in a is 345,000 after correcting for bound lipid and bound detergent. Estimations for protein in B and C are 200,000 or below at the meniscus and 360,000 at the bottom of the cell. Reproduced with permission from Robinson and Capaldi [43].

different polypeptides in the enzyme complex. Our studies indicate that there are at least seven different polypeptides in all preparations of the beef heart enzyme [53, 54], and the same number of components have been seen in yeast [4, 55], *Neurospora* [56], and *L. migratoria* [57] cytochrome *c* oxidase. We have used many different procedures of SDS polyacrylamide gel electrophoresis

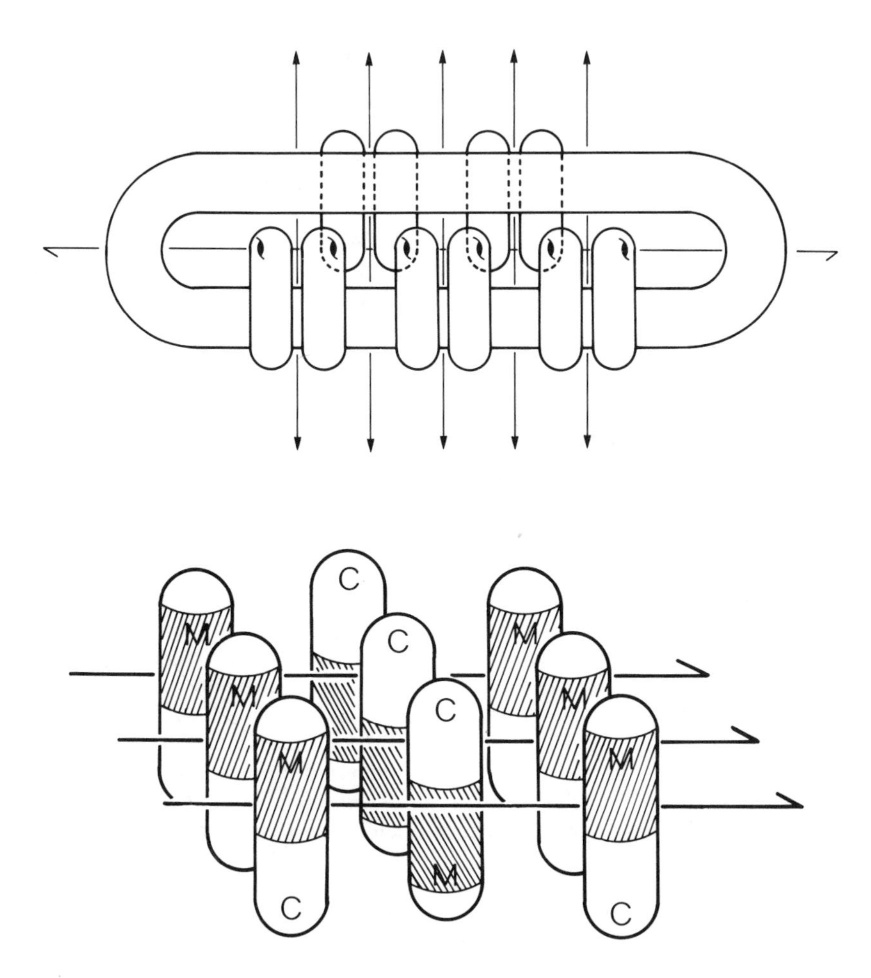

Figure 3 A schematic diagram of the packing of cytochrome *c* oxidase molecules in TX and DOC crystals. Reproduced with permission from Henderson et al. [50]; and Fuller, Capaldi, and Henderson, R., manuscript in preparation.

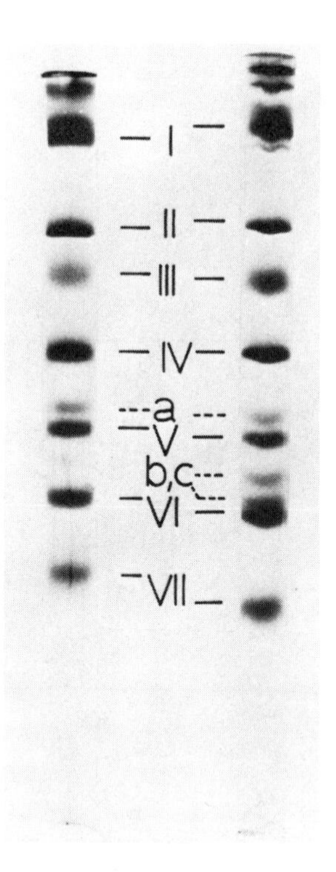

Figure 4 The polypeptide composition of beef heart cytochrome *c* oxidase. The gel on the right is enzyme-isolated according to Capaldi and Hayashi [8]. The gel on the left is of trypsin-treated cytochrome *c* oxidase. The enzyme was incubated with trypsin (1/50 wt/wt) for 2 hr at room temperature. The reaction was stopped with trypsin inhibitor and the mixture was eluted through a Sepharose 4B column in Triton X-100 to remove trypsin, trypsin inhibitor, and any small, cleaved fragments. The left-hand gel is a standard of unreacted enzyme.

in studying the beef heart enzyme [53, 54]. The best resolution was obtained when samples of enzyme dissolved at 37°C in sodium dodecyl sulfate (SDS) and β-mercaptoethanol were electrophoresed in the Swank-Munkres buffer system [58] on 12.5% gels which contained high levels of crosslinker (bisacrylamide). In our scheme, polypeptides (or bands on gels) are numbered I-VII in order of descending molecular weight (Fig. 4). Recent experiments in our laboratory indicate that band VII is heterogeneous and contains one major and two or three

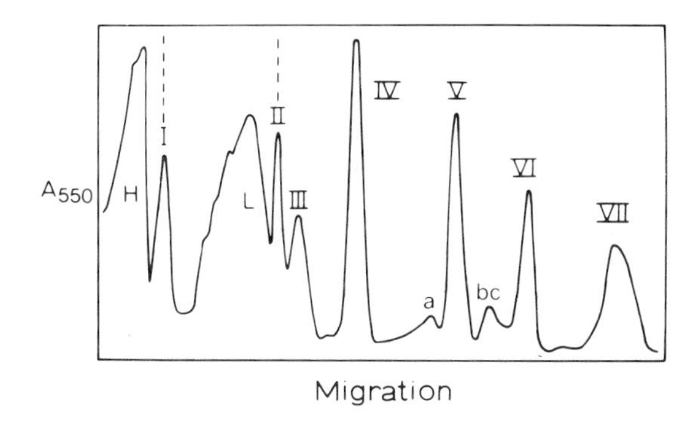

Figure 5 Polypeptide composition of cytochrome *c* oxidase immunoprecipitated from submitochondrial particles. The gel was 15% acrylamide and the buffer conditions for electrophoresis were those of Swank and Murkres [58]. H, L: heavy and light subunit of IgG.

minor components. Other minor components of preparations include those labeled a, b, and c in Figure 4. These are present in variable and always substoichiometric amount in different preparations, and this is one reason for thinking that they are impurities of the enzyme. The minor components are further diminished in amount when the enzyme is treated with trypsin, a treatment which does not affect activity. Additionally, they are very much reduced in cytochrome *c* oxidase immunoprecipitated from solution with antibody prepared against the holoenzyme (Fig. 5).

One important fact to emerge from a comparison of different gel systems is that the polypeptide profile obtained for cytochrome *c* oxidase preparations depends very much on the electrophoretic conditions used [54]. This is clearly demonstrated by the two-dimensional gel in Figure 6 in which a sample of enzyme has been run in the Weber-Osborn gel system [61] in the horizontal dimension and in the Swank-Munkres system in the vertical direction. It can be seen that polypeptide III is only poorly resolved in the Weber-Osborn gels and would not be seen at all if normal amounts of protein (60 μg or less) were loaded on the gel. Polypeptides IV, V, VI, and VII along with the small molecular weight impurities are not well separated on Weber-Osborn gels. Further, the polypeptides of pairs V and VI and II and III changed places so that the order of migration in this gel system is VI before V and III before II. Several other factors affect the polypeptide profile obtained on gels. The mode of preparing the sample for electrophoresis is important. If lipid-depleted enzyme is heated to 100°C in SDS for 3 min before running gels, polypeptides I and III are missing from the profile. The length of the gel used can also affect the profile seen, as can changes in the

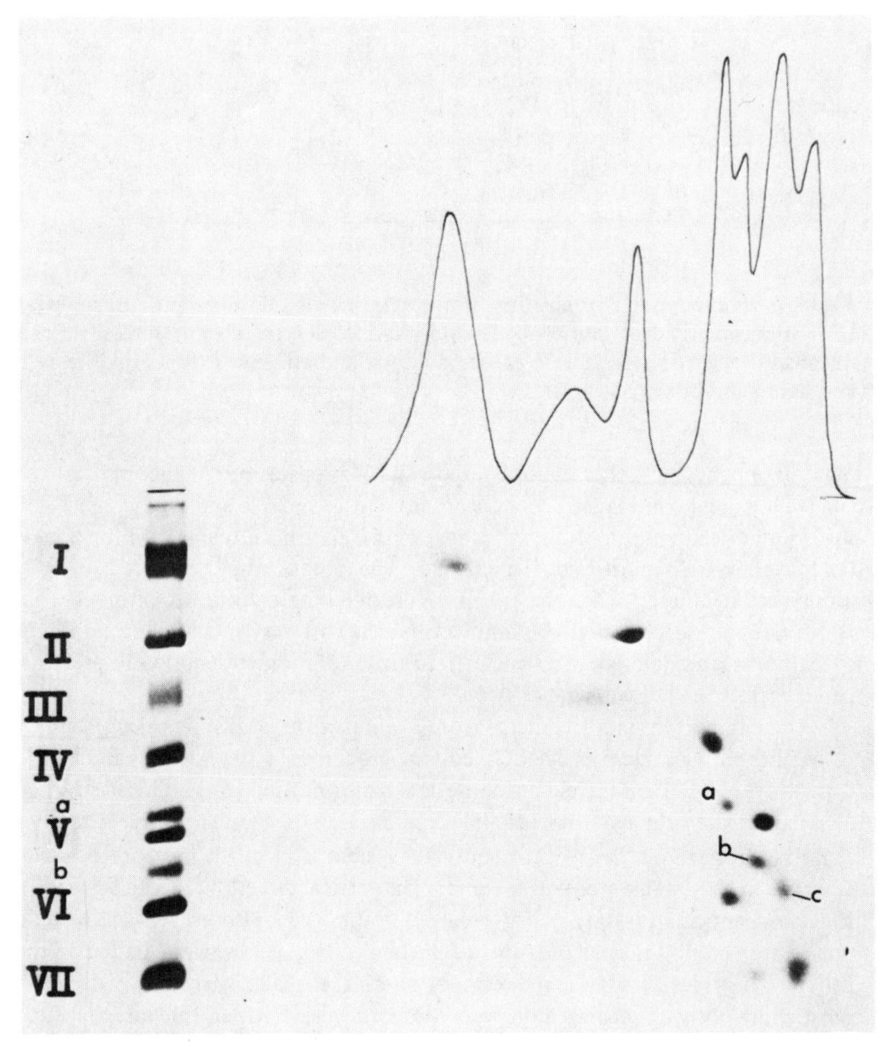

Figure 6 Two-dimensional SDS polyacrylamide gel electrophoresis of cytochrome *c* oxidase showing the relationship of bands observed in Weber-Osborn gels with those seen in the Swank-Munkres gel system. Reproduced with permission from Capaldi et al. [54].

concentration of acrylamide in the gel. When cytochrome *c* oxidase is run on a 9% gel (using 1:10 bisacrylamide to acrylamide) in the Swank-Munkres buffer, polypeptide III runs close to IV while on 12.5% gels it runs closer to II.

The variable migration of polypeptides on gels means that different apparent molecular weights are obtained in different gel procedures. Table 3 gives the values obtained for the seven polypeptides of beef heart cytochrome *c* oxidase under different gel conditions and from gel filtration studies in SDS or 6 M guanidine-HCl. The aggregate molecular weight of these seven polypeptides is 125,000 based on the average value obtained for each component. This indicates that there can only be one copy of most of the polypeptides in the unit (two-heme) complex. We have attempted to monitor the stoichiometry of the polypeptides by several methods including using the affinity of each subunit for coomassie blue and by reacting the SDS dissociated and denatured enzyme with a large excess of [^{35}S] DABS (diazobenzenesulfonate) and comparing the number of counts in each subunit divided by the number of DABS reactive groups (obtained from amino acid composition) (Ludwig and Capaldi, unpublished results). Such studies indicate that there is one copy each of the six largest polypeptides.

One question that is often asked is whether the small molecular weight polypeptides represent proteolytic cleavage products of larger subunits which were generated during the isolation procedure. To test this we have prepared mitochondria from fresh beef hearts in the presence of inhibitors of proteolysis. Cytochrome *c* oxidase was then isolated from these mitochondria in the presence of the same protease inhibitors by immunoprecipitation with a cytochrome *c* oxidase specific antibody. All of the subunits, including VII(s), were present in the same relative proportions as in conventionally prepared samples.

The molecular weights of the seven components of cytochrome *c* oxidase isolated from yeast, *Neurospora,* and *L. migratoria* are listed in Table 4. For the yeast enzyme there is evidence from isoelectric focusing studies that each of the seven subunits is homogeneous [55]. For the *Neurospora* enzyme, in contrast, there is a clear indication that VII is heterogeneous [56].

All of the polypeptides of beef heart cytochrome *c* oxidase (including VII as a mixture) [53, 60, 62] and several of the components in yeast [55] and *Neurospora* [56] have been purified and their amino acid compositions determined. The complex as a whole is one of the more hydrophobic membrane proteins isolated thus far (Table 5). In all species, the three larger subunits are more hydrophobic than subunits IV-VII(s) (Tables 6 and 7). SubunitI, for example, has a polarity of 36% and in all species examined it is characterized by a high mole percent of leucine and low mole percent of glutamic acid [55, 56, 60, 62]. The amino acid sequences of subunits in both the beef heart and yeast enzyme are being examined in several laboratories. So far the full sequence of polypeptides V and VII and the partial equence of II and VI from beef heart cytochrome *c* oxidase have been published [63, 64]. These are shown in Figure 7.

Table 4 Estimates of the Molecular Weights of the Subunits of Cytochrome *c* Oxidase in Various Species

	S. cerevisiae		*N. crassa*	*L. migratoria*
Subunit	Ref. 55	Ref. 4	Ref. 56	Ref. 57
I	40,000	40,000	41,000	38,000
II	33,000	27,300	28,500	24,000
III	22,000	25,000	21,000	19,000
IV	14,500	13,800	16,000	14,500
V	12,700	13,000	14,000	12,500
VI	12,700	10,200	11,500	10,000
VII	4,600	9,500	10,000	8,000

Table 5 Amino Acid Composition of Cytochrome *c* Oxidase Complex

	Number of residues		
Amino acid	Ref. 6	Ref. 45	Ref. 81
Lys	28	39	33
His	20	30	23
Arg	21	31	25
Asp	52	60	59
Thr	51	53	46
Ser	53	54	42
Glu	52	60	64
Pro	48	46	38
Gly	53	59	60
Ala	55	62	60
Cys	7	7	7
Val	45	51	46
Met	13	35	26
Ile	40	43	40
Leu	79	87	73
Tyr	29	33	28
Phe	43	47	46

Table 6 Amino Acid Compositions of Subunits of Beef Heart Cytochrome *c* Oxidase

Amino acid	Polypeptides						
	I	II	III	IV	V	VI	VII
Lys	3.3	3.8	4.0	9.3	6.9	7.5	8.5
His	2.8	2.8	3.1	2.9	2.5	3.0	2.7
Arg	2.1	3.0	2.7	3.3	5.7	5.7	3.7
Asp	7.8	7.5	7.2	9.3	9.9	8.7	7.7
Thr	7.5	7.6	8.4	4.4	5.7	4.5	5.5
Ser	7.6	10.8	8.1	7.0	5.7	8.4	8.5
Glu	4.9	10.7	9.7	11.7	11.4	10.8	8.2
Pro	5.5	5.8	5.7	6.1	7.4	3.7	6.5
Gly	11.2	7.6	9.8	8.2	8.5	12.2	10.5
Ala	8.3	5.9	8.6	8.2	9.2	10.5	9.3
Val	6.4	4.9	4.9	6.3	5.6	4.8	5.4
Met	4.1	4.2	2.9	3.0	1.1	1.8	2.6
Ile	6.2	4.6	4.8	3.8	4.7	3.3	2.6
Leu	11.4	12.8	9.8	8.0	9.5	7.2	9.7
Tyr	3.7	4.4	3.3	4.3	2.9	2.3	3.3
Phe	7.2	3.6	7.0	4.2	3.4	5.6	5.2

VIII. Are There Preparations of Cytochrome *c* Oxidase with Less than Seven Polypeptides Which Retain Electron Transfer Activity?

There are now several reports in the literature which suggest that only a subfraction of the subunits in the cytochrome *c* oxidase complex are necessary for electron transfer activity [66-69]. In two separate studies, repeated detergent solubilization and ammonium sulfate precipitation steps were used, and this treatment apparently removed the large subunits I, II, and III [66, 67]. However, the heme content of these preparations did not exceed 14 nmol/mg protein, indicating very little extraction of "total protein." The most likely explanation for the gel pattern seen is that I, II, and III become aggregated, are not dissociated by SDS, and therefore do not enter the gels. Usually undissociated material in samples can be detected at the top of gels, but this aggregate behaves differently and must remain in the reservoir buffer of the electrophoresis machine. Evidence for this explanation comes from our gel filtration studies (Capaldi, unpublished results). As much as half of the protein on gels may run in the void volume when

Table 7 Amino Acid Composition of the Polypeptides of Yeast and *Neurospora* Cytochrome *c* Oxidase

	I		II		III	IV	VI
Amino acid	Yeast[a]	*Neurospora*[b]	Yeast[a]	*Neurospora*[b]	*Neurospora*[b]	Yeast[a]	Yeast[a]
Lys	4.4	1.5	7.2	2.5	1.2	8.0	6.7
His	2.1	2.6	2.1	2.7	4.6	2.2	1.4
Arg	2.3	2.8	2.5	3.1	2.6	3.1	4.2
Asp	7.8	6.5	8.2	8.6	7.2	11.7	12.0
Thr	5.1	4.8	4.9	4.2	6.5	6.6	3.8
Ser	7.2	10.1	8.1	9.4	7.6	5.3	5.6
Glu	5.9	4.3	9.2	9.8	6.1	11.3	15.5
Pro	4.3	6.8	4.3	7.3	4.7	6.4	4.3
Gly	12.0	10.2	14.8	6.8	9.6	8.4	4.6
Ala	7.8	7.5	7.0	4.9	7.5	6.8	9.5
Val	6.3	7.1	5.4	7.7	6.4	7.4	7.0
Met	3.2	1.3	1.7	2.1	1.3	2.0	0.8
Ile	7.2	8.7	6.5	9.2	9.4	4.7	3.3
Leu	11.5	13.4	8.5	11.7	12.6	8.7	9.3
Tyr	3.5	4.2	2.4	4.9	3.7	2.5	3.5
Phe	6.3	8.3	3.8	5.4	9.2	2.2	5.0

[a]Values from Poyton and Schatz [55].
[b]Values from Sebald et al. [56].

Subunit II (Ref. 64)

F-Met-Ala-Tyr-Pro-Met-Gln-Leu-Gly-Phe-Gln-Asp-Ala-Thr-Ser-Pro-Ile-Met-Glu-Glu-Leu
Leu-His-Phe-
40
Arg-Ile-Leu-Tyr-Met-Met-Asp-Glu-Ile-Asn-Asn-Pro-Ser-Leu-Thr-Val-Lys-Thr-Met-Gly-
His-Gln-Trp-Tyr-Trp-Ser-Tyr-Glu-Tyr-Thr-Asp-Tyr-Glu-Asp-Leu-Ser-Phe-Asp-Ser-Tyr-
Met-Ile-Pro-Thr-Ser-Glu-Leu-Lys-Pro-Gly-Glu-Leu-Arg-Leu-Leu-Glu-Val-Asp-Asn-Arg-
Val-Val-Leu-Pro-Met-Glu-Met-Thr-Ile-Arg-Met-Leu-Val-Ser-Ser-Glu-Asp-Val-Leu-His-
Ser-Trp-Ala-Val-Pro-Ser-Leu-Gly-Leu-Lys-Thr-Asp-Ala-Ile-Pro-Gly-Arg-Leu-Asn-Gln-
Thr-Thr-Leu-Met-Ser-Ser-Arg-Pro-Gly-Leu-Tyr-Tyr-Gly-Gln-Cys-Ser-Glu-Ile-Cys-Gly-
Ser-Asn-His-Ser-Phe-Met-Pro-Ile-Val-Leu-Glu-Leu-Val-Pro-Leu-Lys-Tyr-Phe-Glu-Lys-
Trp-Ala-Ser-Ser-Met-Leu

Subunit V (Ref. 63)

Ser-His-Gly-Ser-His-Glu-Thr-Asp-Glu-Glu-Phe-Asp-Ala-Arg-Trp-Val-Thr-Tyr-Phe-Asn-
Lys-Pro-Asp-Ile-Asp-Ala-Trp-Glu-Leu-Arg-Lys-Gly-Met-Asn-Thr-Leu-Val-Gly-Try-Asp-
Leu-Val-Pro-Glu-Pro-Lys-Ile-Ile-Asp-Ala-Ala-Leu-Arg-Ala-Cys-Arg-Arg-Leu-Asn-Asp-
Phe-Ala-Ser-Ala-Val-Arg-Ile-Leu-Glu-Val-Val-Lys-Asp-Lys-Ala-Gly-Pro-His-Lys-Glu-
Ile-Tyr-Pro-Tyr-Val-Ile-Gln-Glu-Leu-Arg-Pro-Thr-Leu-Asn-Glu-Leu-Gly-Ile-Ser-Thr-
Pro-Glu-Glu-Leu-Gly-Leu-Asp-Lys-Val

Subunit VII(a) (Ref. 64)

Ser-His-Try-Glu-Glu-Gly-Pro-Gly-Lys-Asn-Ile-Pro-Phe-Ser-Val-Glu-Asn-Lys-Trp-Arg-
Leu-Leu-Ala-Met-Met-Thr-Leu-Phe-Phe-Gly-Ser-Gly-Phe-Ala-Ala-Pro-Phe-Phe-Ile-Val-
Arg-His-Gln-Leu-Leu-Lys-Lys

Figure 7 Sequences of cytochrome *c* oxidase subunits.

some samples of cytochrome *c* oxidase (showing only small molecular weight components on gels) are eluted through Sephadex G-200 in 0.1% SDS. Gels of the void volume material show no bands whatsoever on gels but have the hydrophobic amino acid composition expected of a mixture of polypeptides I, II, and III.

Another study has claimed to remove subunits I, II, III, V, and VII without loss of electron transfer activity by trypsin or chymotrypsin cleavage of cytochrome *c* oxidase [68]. In our hands proteolytic digestion removes impurities but does not cleave any subunits of the complex (see Fig. 4).

Recently, Phan and Mahler [69] claim to have removed the large molecular weight subunits I, II, and III from cytochrome *c* oxidase by hydrophobic chromatography, again without loss of electron transfer activity. To date, no confirmation of these results has appeared, and the question of how many subunits are required for cytochrome *c* oxidase electron transfer activity remains undecided.

More certain at least for the yeast enzyme is the fact that the seven polypeptides are all needed for integration of the complex into the mitochondrial membrane. Mutants are available which are missing one or other of most of the polypeptides in cytochrome *c* oxidase. In all cases the complex is not organized and inserted in the membrane (for review see Ref. 70).

IX. Site of Translocation of Subunits

It is now clear that three of the subunits of yeast and *Neurospora* cytochrome *c* oxidase (I, II, and III) are made on mitochondrial ribosomes and coded for by mitochondrial DNA while the four smaller are coded for on cytoplasmic DNA and made in the cytoplasm [70]. Just how the different components come together with heme and copper atoms to form the structural unit of cytochrome *c* oxidase is not clear. The available data on the synthesis of components of the mitochondrial inner membrane including cytochrome *c* oxidase has been reviewed recently [70] and will not be considered further here.

X. Where are the Hemes and Coppers in Cytochrome *c* Oxidase?

The location of heme and copper atoms among the different subunits of cytochrome *c* oxidase has proved difficult to determine because conditions needed to separate the subunits of the enzyme are ones which for the most part release the noncovalently-bound prothetic groups. Gutteridge et al. [71] have reported that the subunits of cytochrome *c* oxidase can be separated by gel electrophoresis in low concentrations of SDS with the prosthetic groups still bound to protein. They assign subunits I and III as hemoproteins and subunit VII as the copper-binding protein. Yu et al. [72] have shown that 50% aqueous pyridine splits the cytochrome *c* oxidase complex with the heme moieties remaining bound to subunits I and V. These workers also claim that subunit II is the copper-binding protein based on fractionation studies [73]. The conclusion that subunit II is a copper-binding subunit has also been reached by Buse and Steffens based on sequence data [59, 64]. They find that this polypeptide has a remarkable sequence homology with other copper-binding proteins such as azurin and plastocyanin. Buse et al. [74] have also used the argument of sequence homology to identify subunit VI as a heme-bearing subunit.

XI. Identification of the Binding Site

The binding site(s) of cytochrome *c* on the cytochrome *c* oxidase complex has been studied recently in several different laboratories. Bisson et al. [75] have prepared arylazido cytochrome *c* modified at Lys 13, a residue near the heme edge which has been shown to be involved in the binding of cytochrome *c* to cytochrome *c* oxidase [76, 77]. This modified cytochrome *c* bound to both beef heart and yeast cytochrome *c* oxidase at subunit II. These workers have also prepared arylazido cytochrome *c* modified at Lys 22, a residue at the periphery of the presumed cytochrome *c* oxidase binding site. This modified cytochrome *c* did not covalently bind to the cytochrome *c* oxidase.

Erecinska [78] has also prepared azido cytochrome *c* derivatives but has not characterized the position of these modifications. Her azido cytochrome *c* covalently linked to cytochrome *c* oxidase at subunits V and VII (in our terminology) without affecting the cytochrome *c* oxidase activity of the preparation. Birchmeier et al. [79] have modified yeast cytochrome *c* at Cys 107 with 5′,5-dithiobis(2-nitrobenzoate). (This residue is on the back side of the cytochrome *c* molecule from the heme edge; see Ref. 76.) This modified cytochrome cross-linked to cytochrome *c* oxidase at subunit III.

We have taken a somewhat different approach to finding the cytochrome *c* binding site on cytochrome *c* oxidase [80, 81]. Firstly, we made a cytochrome *c*/cytochrome *c* oxidase complex with unmodified cytochrome *c* (by gel filtration in buffer of low ionic strength) and then crosslinked this preformed complex with the cleavable bifunctional reagent, disuccinimidylpropionate. In our studies, cytochrome *c* was found to bind to cytochrome *c* oxidase at subunit II in agreement with the results of Bisson et al. [75].

Table 8 **Phospholipid Composition of Different Preparations of Beef Heart Cytochrome *c* Oxidase**

Detergents used in preparation	Total phospholipid (P/mg protein)	Total phospholipid P (%)					References
		PE[a]	PC[a]	DPG[a]	PI[a]	Others	
Triton X-114 and X-100, method of Jacobs et al. [82]	11.8	13.4	30.1	50.4	–	6.0	83
As above using different levels of detergent	1.59	–	–	73.0	–	27.0	83
DOC[a] and cholate as in method of Fowler et al. [8]	15.7	30.6	32.0	30.0	–	7.0	83
DOC and cholate	9.8	21	27	31	11	9	89
After acetone extraction of above	2.3	13	13	47	13	13	89
DOC and cholate as in method of Capaldi and Hayashi [9]	4.8	36	23	48	–	3	43
As above after TX exchange	2.3	22	10	56	–	12	43

[a]PE, phosphatidylethanolamine; PC, phosphatidylcholine; DPG, cardiolipin; PI, phosphatidylinositol; DOC, deosycholate.

XII. Binding of Lipids to Cytochrome *c* Oxidase

As isolated, cytochrome *c* oxidase contains bound lipid in an amount which depends on the isolation procedure used. This associated lipid is mainly cardiolipin (Table 8). Most but not all of the bound lipid can be exchanged for nondenaturing detergents such as cholate, Triton X-100, Tween 80, or deoxycholate [43], or can be removed by phospholipase digestion [83]. However, a small amount of cardiolipin (from one to two molecules) is tightly bound to the enzyme and is not removed by any of these procedures. The tightly bound cardiolipin is not extracted by chloroform-methanol or when the complex is dissociated with denaturing detergents such as SDS [43, 83].

Spin-labeling studies with doxylstearic acid, cholestane, and spin-labeled phosphalidylcholine have revealed that cytochrome *c* oxidase binds and immobilizes approximately 54 molecules of lipid, (assuming two chain lipids rather than the four chain cardiolipin) [84-87]. Recently, interaction of lipids with cytochrome *c* oxidase has been measured by nuclear magnetic resonance using 1,2(12,12′-difluorostearoyl)phosphalidylcholine and 1(7′,7′-difluorpalmitoyl)-2-palmitoleoylphosphatidylcholine as probes. Vesicles were reconstituted from cytochrome *c* oxidase and fluorinated phospholipid in the ratio of 1 mg protein to 5 mg lipid. These studies have shown that the influence of protein on lipid motion goes considerably beyond the boundary layer [88].

XIII. Lipid Requirements for Electron Transfer Activity

Cytochrome *c* oxidase loses activity when lipid is removed by phospholipase digestion or by detergent treatment, and this activity is only regenerated when lipid or other activating amphiphiles (which contain fatty acids) are added back [30, 43, 83]. The importance of various lipid head groups for cytochrome *c* oxidase activity has been examined in several laboratories but with variable results [30, 83, 89, 90]. This may be due to the fact that different delipidation procedures have been used in different studies, and these may denature the enzyme to different extents. Further, all of the methods used left the enzyme in an insoluble form requiring dispersal for optimal reaction with lipids. As discussed already, it is difficult to disperse cytochrome *c* oxidase once it is aggregated in a lipid-free and detergent-free form. Thus differences in the way in which the protein was dispersed could lead to different results.

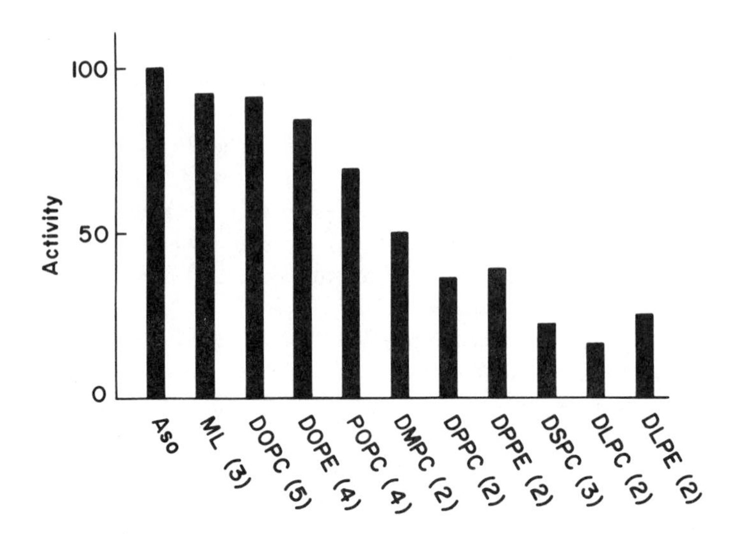

Figure 8 Lipid requirements for cytochrome *c* oxidase electron transfer activities. Enzyme was mixed with synthetic lipids (1 mg/5 mg lipid) and vesicles were generated by dilution of the mixture. Activities measured with an oxygen electrode are expressed as a percentage of the activity obtained with asolectin. The numbers in parentheses indicate the number of experiments averaged to give the data reported. Reproduced with permission from Vik and Capaldi [80].

Our approach has been to delipidate the enzyme while keeping it monodisperse, and this can be achieved by exchanging endogenous lipid for detergents such as Triton X-100 [30, 43]. The delipidated enzyme is inactive in Triton X-100, but when this detergent is replaced by lipids or certain other detergents such as Tween 80, activity is regenerated. In our study, lipid-depleted enzyme (containing only tightly bound cardiolipin) was reconstituted with synthetic lipids of defined head group and defined fatty acid composition [30]. As shown in Figure 8, lipids of different head group but the same fatty acid composition all regenerated at the same level of activity, indicating that the enzyme does not select between any of the lipid head groups found naturally in the mitochondrial inner membrane.

With respect to fatty acid requirements, optimal cytochrome *c* oxidase activity was obtained with unsaturated fatty acids [30]. This was true whether diacyl lipids, lysolipids, or detergents containing different fatty acids were used in the reconstitution studies. Of all the amphiphiles tested, 1-oleyllysolecithin gave the highest activity. This lipid provides an appropriate head group and fatty acid and also acts as a detergent, maintaining the enzyme in a disperse state for efficient interaction with substrate cytochrome *c*. We now use this detergent in preference to Tween 80 or Emasols in cytochrome *c* oxidase assays.

XIV. Arrangement of Polypeptides in the Cytochrome *c* Oxidase Complex

In order to define the topology of the cytochrome *c* oxidase, it is important to know the following: (1) Whether any of the subunits of the complex are buried in the interior of the protein or are shielded from the aqueous medium by lipid; (2) which subunits extend to the cytoplasmic surface of the mitochondrial inner membrane and which to the matrix side; (3) what the juxtaposition of subunits is in the complex; (4) where the hemes and copper are located and how the hemes are oriented in the membrane. There is recent data on the first three of these questions. The relative exposure of subunits with yeast and beef heart cytochrome *c* oxidase has been studied by Eytan and Schatz [91]. The authors have labeled the yeast enzyme, both in the detergent solubilized form and in membranes reconstituted with added lipid, using several different water-soluble lipid-insoluble protein modifying reagents. They concluded from this work that subunits I and II were predominantly buried in the complex with the other subunits comprising the exposed portion of the protein. Eytan et al. [92] have also examined the arrangement of subunits in beef heart cytochrome *c* oxidase, this time using [^{35}S] DABS as the probe for exposed areas of the complex. They found that the polypeptide I plus one of the smaller subunits was only poorly labeled by this reagent and were thus presumably buried within the complex and/or shielded from solvent by lipids. Their study was done using Weber-Osborn gels, which as discussed earlier do not adequately resolve smaller molecular weight subunits, and it is not clear from their report which of the small molecular weight subunits is shielded from solvent. We have also used [^{35}S] DABS to label detergent solubilized and membrane-bound cytochrome *c* oxidase [93]. In agreement with Eytan et al. [92] we found that subunit I was poorly labeled. By using Swank-Munkres type gels we were able to show that subunit VI is also shielded from labeling.

Eytan et al. [92] have studied the arrangement of cytochrome *c* oxidase in the mitochondrial inner membrane. Intact mitochondria (with the cytoplasmic side of the membrane outermost) and submitochondrial particles (with the matrix side outermost) were each reacted with [^{35}S] DABS, cytochrome *c* oxidase was isolated by immunoprecipitation, and the relative labeling of subunits was then examined. They found that subunits II and/or III (which comigrated on Weber-Osborn gels) were exposed on the cytoplasmic side and there was label in the lower molecular weight region of the gel containing subunits V-VII plus impurities. Subunit IV was the only component labeled from the matrix side,

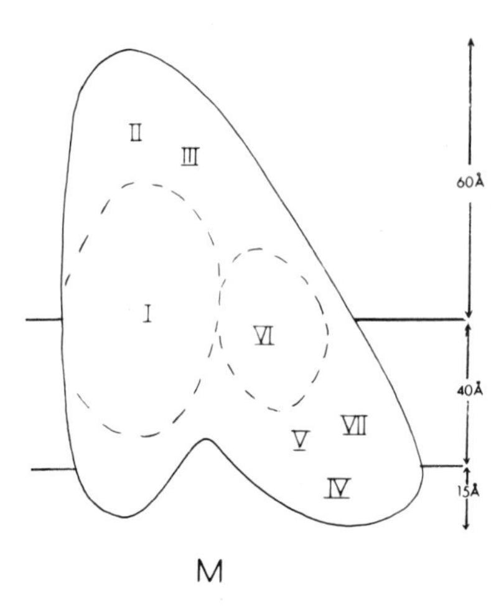

Figure 9 Schematic structure for beef heart cytochrome *c* oxidase.

while subunit I was unreactive from either surface of the membrane. We have recently completed similar studies and confirm that subunits II and III are exposed on the cytoplasmic side of the membrane; we also find that IV is labeled from the matrix side along with II, III, V, and VII [93]. Subunits I and VI were either poorly labeled or not labeled at all from either surface of the membrane.

Two cleavable bifunctional reagents, dimethyl-3,3′dithiobiopropionimidate (DTBP) and dithiobissuccinimidylpropionate (DSP) have been employed recently to identify near neighbors in cytochrome *c* oxidase [81, 94]. These crosslinkers react predominantly with lysine residues and serve to covalently link subunits which are within 11 Å of each other. Crosslinked products can be resolved from monomeric subunits in SDS polyacrylamide gel electrophoresis, and because the reagents can be cleaved (by sulfydryl reagents) the components in these products can be identified unambiguously. We have found that cytochrome *c* oxidase is crosslinked in several discrete steps by both reagents. Reaction with low levels of crosslinker leads to formation of subunit dimers including II + V, III + V, V + VII, I + V, IV + VI, and IV + VII. Reaction with high levels of reagent generates a major product with a molecular weight of 140,000. This is the cytochrome *c* oxidase monomer containing all seven different subunits. Small amounts of dimer (280,000 daltons) were also seen in preparations of enzyme (dissolved in Triton X-100) reacted with high levels of DSP. No higher aggregates, i.e., trimers or tetramers, were seen.

XV. Summary

Several excellent reviews have appeared on the spectral properties of cytochrome *c* oxidase [25], its reaction with various ligands [25, 95], and the biosynthesis of this important enzyme [70]. For this reason, I have concentrated in this chapter on recent advances in an understanding of the structure of the enzyme. The available data now suggest a picture of beef heart cytochrome *c* oxidase shown schematically in Figure 9. Although not explicitly shown in this figure, cytochrome *c* oxidase must be considered a lipoprotein because of its tightly bound cardiolipin. Whether this lipid is bound to a specific subunit is not known. As discussed earlier, the stable species of cytochrome *c* oxidase in nonionic detergents and in bilayer membranes as judged from TX crystals (See Section VI) is the dimer or four-heme complex. It is not clear at present whether the two-heme or four-heme complex is the physiologically functioning unit. There is evidence that the two-heme and two-copper atoms in the monomer can accept four electrons and reduce dioxygen to water [22], but this requires strong reducing conditions which may not be reached in mitochondria. It is possible instead that the low potential cytochrome *a* and its associated copper act to funnel electrons into the high potential cytochrome a_3 (the CN and CO binding site and presumably the O_2 binding site). Two $a_3 \cdot$ Cu pairs as envisaged by Palmer et al. [23] (one in each monomer of the dimer) could then reduce molecular oxygen to form two water molecules.

In relation to the mechanism of electron transfer, very little is known at the moment about the conformational rearrangements that occur during oxido-reduction. The structural work described here was all conducted on oxidized enzyme. The reduced form of cytochrome *c* oxidase now deserves experimental attention.

References

1. F. F. Sun, K. S. Prezbindowski, F. L. Crane, and E. E. Jacobs, *Biochim. Biophys. Acta 153:* 804 (1968).
2. F. F. Sun and E. E. Jacobs, *Biochim. Biophys. Acta 143:* 639 (1967).
3. T. L. Mason, R. O. Poyton, D. C. Wharton, and G. Schatz, *J. Biol. Chem. 248:* 1346 (1973).
4. M. S. Rubin and A. Tzagoloff, *J. Biol. Chem. 248:* 4269 (1973).
5. T. Yonetani, *J. Biol. Chem. 236:* 1680 (1961).

6. M. Kuboyama, F. C. Yong, and T. E. King, *J. Biol. Chem. 247:* 6375 (1972).
7. C. R. Hartzell and H. Beinert, *Biochim. Biophys. Acta 368:* 318 (1974).
8. L. R. Fowler, S. W. Richardson, and Y. Hatefi, *Biochim. Biophys. Acta 64:* 170 (1962).
9. R. A. Capaldi and H. Hayashi, *FEBS Lett. 26:* 261 (1972).
10. W. S. Caughey, G. A. Smythe, D. H. O'Keefe, J. Maskasky, and M. L. Smith, *J. Biol. Chem. 250:* 7602 (1975).
11. D. Keilin and E. F. Hartree, *Nature 141:* 870 (1938).
12. H. Beinert, R. E. Hansen, and C. R. Hartzell, *Biochim. Biophys. Acta 423:* 339 (1976).
13. D. C. Wharton and M. A. Cusanovich, *Biochem. Biophys. Res. Commun. 37:* 111 (1969).
14. D. F. Wilson and P. L. Dutton, *Arch. Biochem. Biophys. 136:* 583 (1970).
15. B. F. Van Gelder, *Biochim. Biophys. Acta 118:* 36 (1966).
16. W. R. Heineman, T. Kuwana, and C. R. Hartzell, *Biochem. Biophys. Res. Commun. 49:* 1 (1972).
17. A. O. Muijsers, R. H. Tusjema, R. W. Henderson, and B. F. Van Gelder, *Biochim. Biophys. Acta 267:* 216 (1972).
18. J. S. Leigh, D. F. Wilson, C. S. Owen, and T. E. King, *Arch. Biochem. Biophys. 160:* 476 (1974).
19. D. F. Wilson, J. G. Lindsay, and E. S. Brocklehurst, *Biochim. Biophys. Acta 256:* 277 (1972).
20. T. Tsudzuki and D. F. Wilson, *Arch. Biochem. Biophys. 145:* 149 (1971).
21. Q. H. Gibson and C. Greenwood, *J. Biol. Chem. 240:* 2694 (1965).
22. L. N. Mackey, T. Kuwana, and C. R. Hartzell, *FEBS Lett. 36:* 326 (1973).
23. G. Palmer, G. T. Babcock, and L. E. Vickery, *Proc. Nat. Acad. Sci. U. S. 73:* 2006 (1976).
24. B. Chance, C. Saronio, and J. S. Leigh, *J. Biol. Chem. 250:* 9226 (1975).
25. M. K. F. Wickstrom, H. J. Harmon, W. J. Ingledew, and B. Chance, *FEBS Lett. 65:* 259 (1976).
26. W. H. Vanneste, M. Ysebaeri-Vanneste, and H. Mason, *J. Biol. Chem. 249:* 7390 (1974).
27. L. Smith and P. W. Camerino, *Biochemistry 2:* 1432 (1963).
28. C. A. Yu, L. Yu, and T. E. King, *J. Biol. Chem. 250:* 1383 (1975).
29. B. Karlsson, B. Lanne, B. G. Malmstrom, G. Berg, and R. Ekholm, *FEBS Lett. 84:* 291 (1977).
30. S. Vik and R. A. Capaldi, *Biochemistry 16:* 5755 (1977).
31. S. Ferguson-Miller, D. L. Brautigan, and E. Margoliash, *J. Biol. Chem. 251:* 1104 (1976).
32. M. K. F. Wickstrom, *Nature, 266:* 271 (1977).
33. P. C. Hinkle, J. J. Kim, and E. Racker, *J. Biol. Chem. 247:* 1138 (1972).
32. M. K. F. Wickstrom, *Nature 266:* 271 (1977).
33. P. C. Hinkle, J. J. Kim, and E. Racker, *J. Biol. Chem. 247:* 1138 (1972).
34. J. M. Wrigglesworth and P. Nicholls, *Biochem. Soc. Trans. 3:* 168 (1975).
35. D. R. Hunter and R. A. Capaldi, *Biochem. Biophys. Res. Commun. 56:* 623 (1973).

36. E. Racker, *J. Membr. Biol. 10:* 221 (1972).
37. Y. Orii and K. Okunuki, *J. Biochem.* (Tokyo) *61:* 388 (1967).
38. Y. Kagawa, *Biochim. Biophys. Acta 131:* 586 (1967).
39. R. S. Criddle and R. M. Bock, *Biochem. Biophys. Res. Commun. 1:* 138 (1959).
40. A. Tzagoloff, P. C. Yang, D. C. Wharton, and J. S. Rieske, *Biochim. Biophys. Acta 961:* 1 (1965).
41. B. Love, S. H. P. Chan, and E. Stotz, *J. Biol. Chem. 245:* 6664 (1970).
42. W. W. Wainio, T. Laskowska-Klita, J. Rosman, and D. Grebner, *Bioenergetics 4:* 453 (1973).
43. N. C. Robinson and R. A. Capaldi, *Biochemistry 16:* 375 (1977).
44. S. Takemori, I. Sekuzu, and K. Okunuki, *Biochim. Biophys. Acta 51:* 464 (1961).
45. H. Matsubara, Y. Orii, and K. Okunuki, *Biochim. Biophys. Acta 97:* 61 (1965).
46. T. Wakabayashi, A. E. Senior, O. Hatase, H. Hayashi, and D. E. Green, *Bioenergetics 3:* 339 (1972).
47. G. Vanderkooi, A. E. Senior, R. A. Capaldi, and H. Hayashi, *Biochim. Biophys. Acta 274:* 38 (1972).
48. H. Hayashi, G. Vanderkooi, and R. A. Capaldi, *Biochem. Biophys. Res. Commun. 49:* 92 (1972).
49. S. Seki, H. Hayashi, and T. Oda, *Arch. Biochem. Biophys. 138:* 110 (1970).
50. R. Henderson, R. A. Capaldi, and J. S. Leigh, *J. Mol. Biol. 112:* 631 (1977).
51. Y. Orri, M. Manabe, and M. Yoneda, *J. Biochem. 81:* 505 (1977).
52. J. R. Bucher and R. Penniall, *FEBS Lett. 60:* 180 (1975).
53. N. W. Downer, N. C. Robinson, and R. A. Capaldi, *Biochemistry 15:* 2930 (1976).
54. R. A. Capaldi, R. L. Bell, and T. Branchek, *Biochem. Biophys. Res. Commun. 74'* 425 (1977).
55. R. O. Poyton and G. Schatz, *J. Biol. Chem. 250:* 752 (1975).
56. W. Sebald, W. Machleidt, and J. Otto, *Eur. J. Biochem. 38:* 311 (1973).
57. H. Weiss, B. Lorenz, and W. Kleinow, *FEBS Lett. 25:* 49 (1972).
58. R. T. Swank and K. D. Munkres, *Anal. Biochem. 39:* 462 (1971).
59. G. Buse and G. Steffens, in *The Genetics and Biogenesis of Chloroplasts and Mitochondria* (T. H. Bucher et al., eds.), North-Holland, Amsterdam, p. 189.
60. G. Steffens and G. Buse, *Hoppe-Seyler's Z. Physiol. Chem. 357:* 1125 (1976).
61. K. Weber and M. Osborn, *J. Biol. Chem. 244:* 4406 (1969).
62. M. M. Briggs, P. F. Kamp, N. C. Robinson, and R. A. Capaldi, *Biochemistry 14:* 5123 (1975).
63. M. Tanaka, M. Hanio, K. T. Yasunobu, C. A. Yu, L. Yu, Y. H. Wei, and T. E. King, *Biochem. Biophys. Res. Commun. 76:* 1014 (1977).

64. G. Buse and G. J. Steffens, *Hoppe-Seyler's Z. Physiol. Chem. 359:* 1005 (1978).
65. G. Fairbanks, T. L. Steck, and D. F. H. Wallach, *Biochemistry 10:* 2606 (1961).
66. H. Komai and R. A. Capaldi, *FEBS Lett. 30:* 272 (1973).
67. J. F. Hare and F. L. Crane, *Subcell. Biochem. 3:* 1 (1974).
68. T. Yamamoto and Y. Orii, *J. Biochem. 75:* 1081 (1974).
69. S.-H. Phan and H. Mahler, *J. Biol. Chem. 251:* 257 (1976).
70. G. Schatz and T. L. Mason, *Ann. Rev. Biochem. 43:* 51 (1974).
71. S. Gutteridge, D. B. Winter, W. J. Bruyninckx, and H. Mason, *Biochem Biophys. Res. Commun. 78:* 945 (1977).
72. C. A. Yu, L. Yu, and T. K. King, *Biochem. Biophys. Res. Commun. 74:* 670 (1977).
73. M. Tanaka, M. Haniu, K. T. Yasunobu, C. A. Yu, L. Yu, and T. K. King, *Biochem. Biophys. REs. Commun. 66:* 357 (1976).
74. G. Buse, G. J. Steffens, and G. C. M. Steffens, *Hoppe-Seyler's Z. Physiol. Chem. 359:* 1011 (1978).
75. R. Bisson, A. Azzi, H. Gutweniger, R. Colonia, C. Montecucco, and A. Zanotti, *J. Biol. Chem. 253:* 1874 (1978).
76. R. E. Dickerson and R. Timkovich, in *The Enzymes,* 3rd ed., vol 11 (P. D. Boyer, ed.), Academic, New York, 1975, p. 397.
77. E. Margoliash, S. Ferfuson-Miller, D. L. Brautigan, and A. N. Chaviano, in *Structure-Function Relationships of Proteins* (R. Markahm and R. W. Horne, eds.), Elsevier, Amsterdam, 1976, p. 145.
78. M. Erecinska, *Biochem. Biophys. Res. Commun. 76:* 495 (1977).
79. W. Birchmeier, C. E. Kohler, and G. Schatz, *Proc. Nat. Acad. Sci. U. S. 73:* 4334 (1976).
80. M. M. Briggs and R. A. Capaldi, *Biochem. Biophys. Res. Commun. 80:* 553 (1978).
81. M. M. Briggs, Ph.D. Thesis, University of Oregon, 1977.
82. E. E. Jacobs, E. C. Andrews, W. P. Cunningham, and F. L. Crane, *Biochem. Biophys. Res. Commun. 25:* 87 (1966).
83. Y. C. Awasthi, T. F. Chuang, T. W. Keenan, and F. L. Crane, *Biochim. Biophys. Acta 226:* 42 (1971).
84. P. C. Jost, O. H. Griffith, R. A. Capaldi, and G. Vanderkooi, *Proc. Natl. Acad. Sci. U. S. 70:* 480 (1973).
85. P. C. Jost, R. A. Capaldi, G. Vanderkooi, and O. H. Griffith, *J. Supramol. Structure, 1:* 269 (1973).
86. P. C. Jost, O. H. Griffith, R. A. Capaldi, and G. Vanderkooi, *Biochim. Biophys. Acta 311:* 141 (1973).
87. P. C. Jost, K. K. Nadakavukaren, and O. H. Griffith, *Biochemistry 16:* 3110 (1977).
88. K. Longmuir, R. A. Capaldi, and F. W. Dahlquist, *Biochemistry 16:* 5746 (1977).

89. G. P. Brierley and A. J. Merola, *Biochim. Biophys. Acta 64:* 205 (1962).
90. C. A. Yu, L. Yu, and T. E. King, *J. Biol. Chem. 250:* 1383 (1975).
91. G. D. Eytan and G. Schatz, *J. Biol. Chem. 250:* 767 (1975).
92. G. D. Eytan, R. C. Carroll, G. Schatz, and E. Racker, *J. Biol. Chem. 250:* 8598 (1975).
93. B. Ludwig, N. W. Downer, and R. A. Capaldi, *Biochemistry,* in press (1979).
94. M. M. Briggs and R. A. Capaldi, *Biochemistry 16:* 73 (1977).
95. M. Erecinska and D. F. Wilson, *Arch. Biochem. Biophys. 188:* 1 (1978).

4

The Mitochondrial ATPase

A. E. Senior

University of Rochester Medical Center
Rochester, New York

I. Introduction

I wrote a review in 1973 entitled "The structure of Mitochondrial ATPase" [1] in which I drew together all the information then available on the structure of this remarkably complicated membrane enzyme. In that article I alluded also to other features which I thought were of topical interest. Among these were the similarity of bacterial and chloroplast ATPase to mitochondrial ATPase, the interesting observations of conformational changes in the enzyme which seemed to be linked with energy transduction, the coming use of reconstituted liposomes to study the properties of the enzyme, and the somewhat unusual nucleotide binding and kinetic characteristics of the enzyme. Since 1973 there has been published a lot of experimentation and much of it has dealt with these very points.

It has become obvious that mitochondrial, bacterial, and chloroplast ATPase are very similar indeed, although not identical. In mitochondria the enzyme functions to synthesize ATP as the terminal event in oxidative phosphorylation. In chloroplasts the enzyme catalyses ATP synthesis as the terminal step in photophosphorylation. It is now generally agreed that in both oxidative and photophosphorylation the primary event is the establishment of an electrochemical gradient across the membrane, achieved by proton pumping, and that ATP synthesis is driven in some (as yet unknown) fashion by proton movement down this gradient through the ATPase. While no single piece of experimental evidence directly and independently proves this, the great bulk of supporting evidence provides compelling reason to accept it.

In bacteria the enzyme catalyzes oxidative phosphorylation or photophosphorylation and it also plays another major physiological role, that of driving uptake of nutrients (e.g., cations, amino acids). In this role the enzyme hydrolyzes ATP and sets up an electrochemical gradient across the membrane, down which nutrients move in coupled transport systems. That this is a major and evolutionarily very old role of the enzyme is shown by the fact that the enzyme is found in anaerobic *Clostridium* [2, 3] and *Streptococcus* [4] species.

In fact then, this ATPase is a universal and fundamental feature of living cells, and the perfection of it was clearly a giant step in the evolution of living things which allowed large-scale, efficient generation and utilization of energy.

Here I shall first describe the structure of the mitochondrial ATPase, pointing out the similarities and differences between the mitochondrial, chloroplast, and bacterial enzymes. Then I shall review the new data on functional aspects of the enzyme, and to this end the second part of this chapter will deal extensively with the nucleotide binding and kinetic properties of the enzyme, the properties of inhibitors and activators, the chemistry of the active site, and the conformational changes which have been investigated recently. Current ideas on the mechanism of ATP synthesis will be criticized. Other reviews in the field which have appeared since 1973 include those of Abrams and Smith [4] (on the bacterial ATPase), Simoni and Postma [5] (on the energetics of bacterial transport), Pedersen [6] (on mitochondrial ATPase), Nelson [7] (on the chloroplast ATPase), and Panet and Sanadi [8] (a comparison of the three ATPases).

II. Structure of Mitochondrial ATPase

A. Introduction

By 1973 it was clear that the mitochondrial ATPase was a large membrane complex as shown and described in the Figure 1. Part of it resides in the inner membrane of the mitochondrion and is called the membrane sector. There is a stalk sector, visible in electron micrographs, which binds the F_1 sector to the membrane sector. Evidence for this structure was summarized in my previous review [1]. The F_1 sector is relatively easily detached from the membrane and may be purified in soluble form. It is the sector which carries the site(s) of ATP hydrolysis and synthesis (it is not yet known how many sites there are) and the purified soluble F_1 is an active ATPase. The soluble F_1 may be readily bound back on the membrane sector and when re-bound it regains the ability to catalyze ATP synthesis. The ATPase activity of the whole ATPase complex shows somewhat different characteristics to that of the purified soluble F_1, for instance the specificity for nucleoside triphosphates is altered [1] and the soluble F_1 is not inhibited by agents such as oligomycin and N,N′-dicyclohexylcarbodiimide (DCCD) which bind to the membrane sector of the complex.

Another component which can be purified in soluble form is OSCP, (oligomycin-sensitivity conferring protein), which is involved in binding F_1 to the membrane [1]. Other components can also be purified in soluble form, as discussed in detail later. As in the past, recent work has leaned heavily on the ability to dissociate (resolve) the complex into components and to reassociate (reconstitute) them with regain of both normal structural and functional properties (see Ref. 9 for a recent discussion). Section II.B is a summary of the recent work on the structure of the complex.

B. Newer Preparative Methods

1. Preparation of F_1

Common features of all methods for preparation of F_1 are excessive consumption of time and mitochondria. Beechey et al. [10] recently introduced a technique which overcomes these limitations somewhat. Submitochondrial particles are shaken with chloroform. Soluble F_1 is extracted into the aqueous phase in

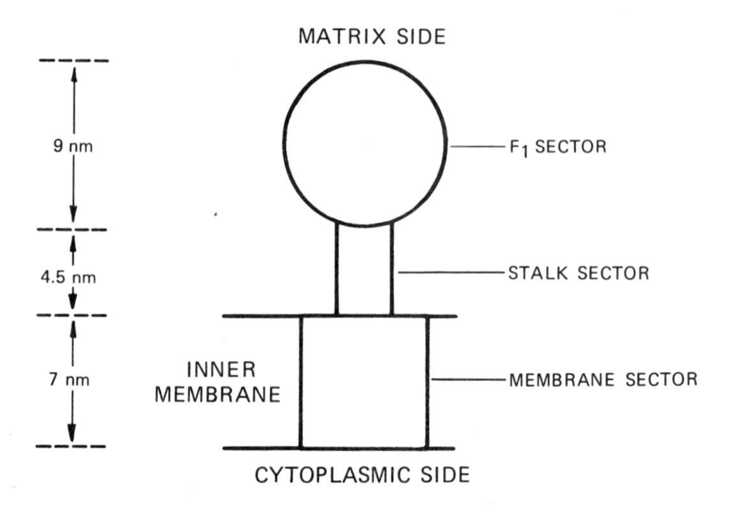

Figure 1 Diagrammatic representation of the mitochondrial ATPase complex, showing the three sectors. (1) The F_1 sector: This may be detached and purified in soluble form. It contains no lipid, nor any known prosthetic groups. It carries the site of ATP synthesis and hydrolysis, and may be re-bound to the membrane after solubilization and purification. The soluble form catalyzes ATP hydrolysis. The membrane-bound form, before detachment from the membrane or after reconstitution with the membrane, catalyzes ATP hydrolysis, ATP synthesis, and all the isotope-exchange reactions related to oxidative phosphorylation. This sector has been well characterized as described further in Figure 3. (2) The stalk sector: First let it be made clear that the very existence of a stalk sector is a matter of controversy. Some workers deny its existence, although the majority appear to favor it. Its constitution is a further area of discussion (see Fig. 3). (3) The membrane sector: This forms part of the inner membrane and contains at least four protein subunits plus phospholipid. The proteins are "hydrophobic." Removal of F_1 and/or the stalk renders the membrane leaky to protons. Such leaks are blocked by restoration of F_1 and stalks, or by addition of inhibitors such as DCCD or oligomycin or trialkytins, all of which bind with high affinity to the membrane sector. Hence the membrane sector is thought to act as a proton-conducting pathway or channel.

high yield. The F_1 may then be purified further from the aqueous layer by conventional techniques [11, 12]. This enzyme has high activity and contains all the five subunits of F_1 seen in the older preparations [12]. Other methods for preparation of F_1 from mitochondria have also been described recently. One involves "hydrophobic chromatography" on substituted agarose, derived from the ability of F_1 to bind alkylguanidines [13]. Other procedures involve use of the French press [14] to detach the F_1 from the membrane (instead of sonication as in the older procedures) and use of polyethylene glycols [15] to precipitate F_1 from solution. Workers preparing soluble bacterial F_1 have introduced use of the polyethylene glycols [16], and affinity chromatography on ADP-agarose [17] and on 6-[(3-carboxy-4-nitrophenyl)thio]-inosine triphosphate [18]. While these developments may replace the older methods in time, many workers continue to use the methods of Knowles and Penefsky [19], Senior and Brooks [20], or Horstman and Racker [21] for soluble F_1 from beef heart; or the methods of Lambeth and Lardy [22] or Catterall and Pedersen [23] for soluble F_1 from rat liver; or that of Tzagoloff [24] for soluble F_1 from yeast.

2. Preparation of the Whole ATPase Complex

Preparations of the whole ATPase complex from mitochondria and from bacteria are summarized in Table 1. These preparations are also referred to in the literature by the names DCCD-sensitive ATPase, oligomycin-sensitive ATPase, rutamycin-sensitive ATPase, or proton-translocating ATPase. All are solubilized with detergent and remain soluble only in the presence of detergent. The whole ATPase complex has not yet been purified from chloroplasts.

These detergent-solubilized preparations of the whole complex contain a good deal of lipid and many protein components, as discussed below. In the soluble form they do not catalyze ATP synthesis or any of the isotope-exchange reactions (e.g., ATP-P_i exchange) characteristic of energy-transducing systems. However, they do retain many features of the in situ enzyme; the ATPase activity is sensitive to oligomycin, DCCD, mercurials, and trialkyltins, and the specificity for nucleotide triphosphates is similar. Moreover, when incorporated into vesicles ("reconstituted liposomes") these preparations do catalyze ATP-P_i exchange and proton transport [38, 40, 43].

The purest preparations are the ones from the thermophilic bacterium PS3 [40], yeast [35, 38], and *Xenopus* oocytes [39]. These preparations contain about eight to ten protein components as shown by sodium dodecyl sulfate (SDS) gel electrophoresis, which of course gives only a separation based on molecular size. The beef heart preparations are less pure. When small amounts (up to 30 μg) are examined on SDS gels about 10-14 bands are seen (Ref. 44 and Senior, unpublished results). When larger amounts are analyzed many more bands are seen.

C. Structure of the F_1 Sector

1. Molecular Weight and Tightly Bound Subunits

Our original observation that F_1 from beef heart and rat liver contains five tightly bound subunits [45, 46] has been fully confirmed [19, 22, 47]. The same five-

Table 1 Preparations of the Whole ATPase Complex

Authors	Source	Method	References
Early Mitochondrial Preparations			
Kagawa and Racker	Beef heart	Extraction with cholate, $(NH_4)_2SO_4$ fractionation	25, 26
Kopaczyk et al.	Beef heart	As above	27, 28
Tzagoloff et al. MacLennan and Tzagoloff	Beef heart	Extraction with deoxycholate and cholate, $(NH_4)_2SO_4$	29, 30 31
Swanljung and Ernster Swanljung	Beef heart	Extraction with deoxycholate, gel filtration through Sepharose 6B	32 33
Swanljung and Frigeri	Beef heart	Extraction with Triton X-100, affinity chromatography on Sepharose 4B containing bound F_1 inhibitor protein	34
Tzagoloff	Yeast	Extraction with deoxycholate and cholate, $(NH_4)_2SO_4$ fractionation	24
Tzagoloff and Meagher	Yeast	Extraction with Triton X-100, density gradient centrifugation in Triton X-100 or Tween 80	35

Recent Mitochondrial Preparations			
Hatefi et al.	Beef heart	Extraction with deoxycholate and cholate, $(NH_4)_2SO_4$	36
Serrano et al.	Beef heart	Extraction with cholate, $(NH_4)_2SO_4$ fractionation, density gradient centrifugation in lysolecithin and deoxycholate	37
Ryrie	Yeast	As Ref. 35 with additional gel filtration through Sepharose 4B	38
Koch	*Xenopus*	As Ref. 35 with additional ion-exchange chromatography	39
Bacterial Preparations			
Sone et al.	PS3 (thermophile)	Extraction with Triton X-100, ion-exchange chromatography in Triton X-100, gel filtration in Sepharose 4B	40
Lee et al.	*M. phlei.*	Extraction with Triton X-100, density gradient centrifugation, affinity chromatography on ADP-Sepharose	41
Bragg and Hou	*E. coli*	Extraction with cholate, $(NH_4)_2SO_4$ fractionation, gel filtration on Sepharose 6B and 4B, extraction with Triton X-100, density gradient centrifugation	42

subunit pattern has been noted in F_1 from yeast [35], chloroplasts [7], and many species of bacteria [4]. Often where workers have noted only the three larger subunits, this was due to insufficient amounts being analyzed on gels. At least 50 μg must be analyzed since the two smaller subunits constitute only a small proportion of the mass.

In all cases the molecular weight of F_1 is around 360,000 [48]. There is, however, quite a variation in this number as reported from various laboratories; values range from about 325,000 to 385,000. Reported values for the molecular weights of the individual subunits also vary. Table 2 lists "best values" for the subunits of beef heart F_1. There is evidence that these values are identical in rat liver [22, 45]. Yeast, chloroplast, and bacterial F_1 show some variation from these values [1, 4, 7]. It seems likely from surveying the literature that the size of the subunits of bacterial F_1 may vary among species. The five tightly bound subunits have been denoted F_1-1 through F_1-5, A through E, and α through ϵ by various workers. I prefer the first terminology but will defer to the chloroplast and bacterial students by using also the α-ϵ notation here.

2. The Specific ATPase Inhibitor Protein (F_1 Inhibitor)

A small heat-stable protein which specifically inhibits the ATPase activity of soluble or membrane-bound F_1 has been isolated from beef heart [21, 49, 50], rat liver [51], and yeast mitochondria [52, 53]. The inhibitor has been shown to be a loosely bound sixth subunit of F_1 in beef heart [54]. Its molecular weight is intermediate between that of subunits F_1-4 (δ) and F_1-5 (ϵ) as shown by SDS gel electrophoresis experiments (coelectrophoresis with F_1 [54]), and reported values are 7,500-10,500 [50, 51, 52, 54]. The amino acid composition of the F_1 inhibitor from beef heart is quite singular, lacking methionine, proline and threonine [50, 54], and cysteine [55]. The yeast protein lacks methionine and cysteine [52]. In beef heart and rat liver the inhibitor protein is a very basic protein [51, 56].

Samples of soluble beef heart F_1 prepared by methods which use low pH (~7) throughout and avoid gel filtration contain six subunits–the five tightly bound subunits plus the F_1 inhibitor [14, 54]. The inhibitor is readily dissociated from F_1 by heat, high pH, or gel filtration [21, 57], and some soluble F_1 preparations are completely inhibitor-free [54]. In situ the inhibitor protein associates tightly with F_1 when the ATP/ADP ratio is high (and Mg^{2+} is present) [21] and tends to dissociate when electron transfer is operative, the pH rises, or the ATP/ADP ratio is low [21, 58]. It is thought that the inhibitor protein is an important regulatory subunit of the enzyme which prevents ATP hydrolysis ("backflow of ATP") but not ATP synthesis [49, 58, 59]. The inhibition of ATP hydrolysis is noncompetitive (beef heart [58]) and the K_d for binding to membrane-bound F_1 was around 10^{-7} M (yeast [52]).

Table 2 Molecular Weights of Components[a] of the Beef Heart Mitochondrial ATPase Complex

Component	Molecular weight	References
F_1-1 (α)	53,000	1, 45, 56
F_1-2 (β)	50,000	1, 45, 56
F_1-3 (γ)	33,000	1, 56
F_1-4 (δ)	17,000	1, 56
F_1-5 (ϵ)	~7,500	1, 45, 56
F_1 inhibitor	10,000	1, 50, 54, see also 96
OSCP	18,000	31, 91, 92
F_6 (Fc_2)	8,000	96
DCCD binding protein	~10,000-14,000	99, 100
Factor B	12,000	119
	29,000	116,117
	44,000	118

[a]Here are listed molecular weights of those component proteins which have been purified and characterized.

The tightness of binding may well vary depending upon the conformational state of the F_1. A scheme involving a tight (inhibitory) binding configuration and a looser (noninhibitory) binding configuration of the inhibitor subunit has been proposed [58] and seems attractive. Soluble F_1 preparations which contain the inhibitor protein nevertheless have high ATPase activity, thus the inhibitor must be in the loosely bound configuration in these cases. It is worth pointing out in passing that an increase in ATPase activity is seen with inhibitor-free soluble F_1 [60, 61] after heating or trypsin treatment. This type of evidence does not therefore constitute proof of the presence of an inhibitor protein. Several reports could be cited which have erroneously assumed this to be so.

It is not known how many molecules of inhibitor protein normally bind to F_1, nor where on F_1 the inhibitor protein binds. Some "cross-reaction" is seen, for instance the yeast inhibitor protein is active against beef heart F_1 [52], as is the rat liver inhibitor protein [51]. The beef heart inhibitor is active against both rat liver and yeast F_1 [51, 52].

Chloroplast F_1 and bacterial F_1 also contain an inhibitor subunit [50, 62, 63] which is the ϵ or F_1-5 subunit. The situation here is somewhat different because the inhibitory subunit is normally tightly bound to the F_1 and remains bound through all purification steps. However, soluble F_1 samples containing the inhibitor may still have high ATPase activity, again suggesting that "inhibitory" and "noninhibitory" configurations may exist. It has been tacitly assumed that the inhibitor subunit of chloroplast or bacterial F_1 plays the same regulatory role as that of the mammalian inhibitor protein. This is not yet fully established experimentally, although of course an unregulated ATPase activity would probably be debilitating if not lethal. The chloroplast inhibitor protein is not active against beef heart F_1, neither is the beef heart inhibitor active against chloroplast F_1 [50].

It was recently reported that component TN-I, the inhibitory subunit of troponin, inhibits ATPase activity of both chloroplast and beef heart soluble F_1 [64-66]. The inhibitor is noncompetitive with K_i around 10^{-6} M. This is a very interesting observation. There could be a simple explanation, which is that TN-I is a basic protein like the F_1 inhibitor of beef heart and it has within it a segment of basic residues which may form the "actin-binding site" [67]. As has recently been shown, polycations may nonspecifically inhibit the ATPase activity of F_1 [63]. On the other hand the inhibition seen with TN-I was more complete and was achieved at quite low levels of added TN-I.

3. Stoichiometry and Juxtaposition of Subunits in the F_1 Sector

The question is, How many of each subunit are there per F_1? The answer at the moment is that we do not know. It may be surprising that at this time when many protein sequences and even three-dimensional structures are known, there is still difficulty in assessing the subunit stoichiometry of a soluble enzyme. But the fact is that different experimental approaches have given completely different answers.

When Jack Brooks and I first ran beef heart F_1 and rat liver F_1 on 10% SDS gels, we attempted to quantitate the Coomassie blue stain in each band by measuring the area under the curves [45]. This suggested an approximate subunit stoichiometry of 3:3:1:1:1 for subunits F_1-1 through F_1-5, which agreed well with the total molecular weight of 360,000 which we had found [48]. However, as we noted at the time, large variation was seen [45]. Later, Catterall et al. [47] refined this approach somewhat by estimating specific staining capacity for each subunit, and agreed with 3:3:1:1:1 for rat liver F_1. Yoshida et al. [68] using a similar approach report 3:3:1:1:1 for PS3 (bacterial) F_1. I then attempted to prove that this was correct by labeling the cysteine groups of beef heart F_1 with N-ethylmaleimide (NEM) [70]. Beef heart F_1, like chloroplast F_1 [69], contains eight cysteine and two cystine [60, 70]. I first cleaved the disulfide bonds, reacted the reduced enzyme with radioactive NEM, then assessed the ratio of binding among subunits. Subunits F_1-1, F_1-3, and F_1-5 bind NEM; subunits F_1-2 and F_1-4 do not [70] (see also Ref. 71). By then analyzing the cysteine content of each purified subunit after disulfide bond cleavage, I could calculate the stoichiometry of the subunits. The answer, unexpectedly, was 2:?:2:?:2. Such a stoichiometry gives a total molecular weight lower than

360,000 (around 320,000 if one assumes that two chains of F_1-2 and two chains of F_1-4 are also present).

In the chloroplast and bacterial systems a different approach is feasible. The organism may be grown in the presence of radioactive substrate, which becomes incorporated into the F_1, and the distribution of label in the subunits of purified F_1 may be assessed after running SDS gels. Using this approach Nelson [7] suggests a subunit stoichiometry of 2:2:1:1:2 for chloroplast F_1; Bragg and Hou [72] suggest ratios of 3:3:1:1:1 for both *Escherichia coli* and *Salmonella typhimurium* F_1; Kagawa et al. [73] suggest 3:3:1:1:1 for the PS3 F_1.

Another experimental approach was reported by Vogel and Steinhart [16]. By freezing and thawing they were able to cleave *E. coli* F_1 into two large fragments (I_A and I_B) each of molecular weight aroung 100,000, and a fragment II of molecular weight around 50,000. Reassociation of the whole F_1 occurred when the three fragments were incubated together under carefully defined conditions [16]. Titration showed that fragments I_A and II recombined in a 1:1 ratio. Fragment I_A contained subunits α (F_1-1), γ (F_1-3), and ϵ (F_1-5). Fragment contained β (F_1-2) only. Fragment I_B contained α, γ, δ (F_1-4), and ϵ. The data strongly suggest a stoichiometry in the reconstituted F_1 of 2:2:2:1 or 2:2 because if there were three α chains and three β chains the minimal molecular weight is around 460,000. Summarizing the problem of the stoichiometry, it seems agreed that the two large subunits F_1-1 (α) and F_1-2 (β) are present in equal amounts. There agreement ends, and either a 3:3:1:1:1 or a 2:2:2:2:2 or similar pattern seems likely. There is no reason to think that there is a difference between mitochondrial, chloroplast, and bacterial F_1, as the two patterns emerged for both mitochondrial and bacterial F_1. The differences in results could be due however to dissociation of subunits from the F_1 under various conditions. Kagawa et al. make this point [73] and stress that the PS3 F_1 is very stable in this regard.

Juxtaposition of subunits may be explored by crosslinking. Bragg and Hou crosslinked *E. coli* F_1 subunits [72, 74] and reported that crosslinking gave a component (called "Y" in Ref. 72) which had a molecular weight around 130,000 and yielded about equal amounts of α and β on cleavage. They concluded that Y was an α-β adduct and denied the possibility that α-α or β-β dimers were present. From reading the paper it seems that Y could contain these dimers in equal amounts and so no firm conclusions regarding the neighborliness of α and β can be drawn. They also have noted β-γ, α-δ, β-ϵ, and γ-ϵ adducts [74].

Wingfield and Boxer [75] crosslinked beef heart F_1 and saw α-α (F_1-1 $\rightarrow$ F_1-1) crosslinks, but no α-β. Satre et al. [71] crosslinked beef heart F_1, using as additional tools the facts that radioactive 7-chloro-4-nitrobenzo-2-oxa-1,3-diazole chloride (NBD-Cl) binds to the F_1-2 but not to the F_1-1 subunit and that radioactive NEM binds to the F_1-1 but not the F_1-2 subunit. The enzyme was crosslinked (ATPase activity was inhibited by this procedure), the adducts were labeled, and the labeling patterns were studied before and after cleavage of the crosslinks. The results suggested that F_1-1 → F_1-1 (α-α) and F_1-1 → F_1-2 (α-β) crosslinks formed, but that no F_1-2 → F_1-2 (β-β) crosslinks formed.

Baird and Hammes [76] have studied the crosslinks formed between chloroplast F_1 subunits and have seen multiple crosslinked species. They found that the heat-activated and latent enzymes were very similar in their behavior in crosslinking experiments. Among the multiple species seen were species containing β-β links and they concluded that the two β subunits are close together in chloroplast F_1.

From all this data on crosslinking it is not easy to draw firm conclusions. In many cases validity of interpretation rests quite heavily on the correctness of the measured molecular weights of the adducts and the individual subunits. It should be borne in mind that in any asymmetric arrangement, F_1-1 (α) of *E. coli* F_1 could be equivalent to (i.e., play the functional role of and occupy the same physical polition of) F_1-2 (β) in chloroplast F_1 or mitochondrial F_1 and vice versa; the molecular weights of F_1-1 and F_1-2 are very similar. However, I think we can firmly conclude that a planar hexagonal arrangement of the two larger subunits of the enzyme as shown in Figure 2A seems unlikely since α-α or β-β adducts have been noted. Several of the workers cited above have proposed their own models [7, 16, 72, 73, 76] which serve as good stimulants for future work and, incidentally, are all different.

4. Functional Roles of Subunits of the F_1 Sector

The inhibitor subunit has already been described. Work on the roles of the other subunits has also progressed recently. Digestion of chloroplast F_1 or bacterial F_1 with trypsin yields an active ATPase preparation which contains only α (F_1-1) and β (F_1-2) subunits [77, 78]. A butanol-extracted F_1 preparation from *Micrococcus lysodeikticus* contains only α and β subunits and has full ATPase activity [79]. NBD-Cl, an inhibitor of ATPase activity, reacts with one tyrosine residue in the β (F_1-2) subunit in mitochondrial F_1 [80-82]. Thus the β (F_1-2) subunit is clearly implicated in ATP hydrolysis.

Several studies suggest that the δ (F_1-4) subunit is concerned with binding F_1 to the membrane. Bragg et al. [83] and later Nelson et al. [78] and Futai et al. [84] prepared *E. coli* F_1 which lacked only this subunit. Full ATPase activity was preserved but this enzyme did not rebind to membranes. Addition of pure δ (F_1-4) subunit restored the binding [62]. The work of Vogel and Steinhart [16] also fully supports the idea that δ is required for rebinding in *E. coli* since a reassociated enzyme containing α, β, γ, and ε had ATPase but no coupling factor activity. Addition of δ restored coupling factor activity. In *Streptococcus faecalis* it has been known for some time that rebinding of F_1 to membrane re-

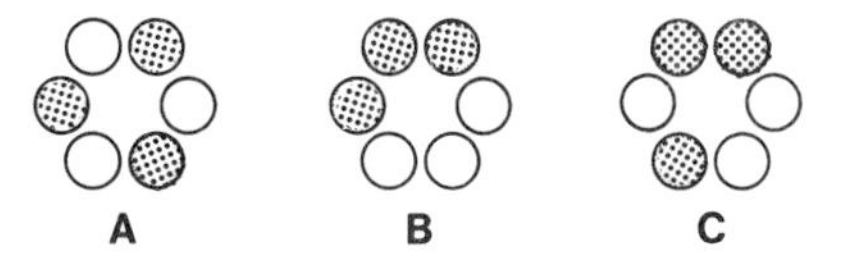

Figure 2. Possible arrangements of the two large F_1 subunits, F_1-1 and F_1-2 (α and β). Shaded circles represent α (F_1-1) subunits and open circles represent β (F_1-2) subunits. Many authors have cited the possibility that the two larger subunits are arranged as a planar hexagon. Crosslinking studies now tend to refute this. In mitochondrial F_1, only α-α and α-β crosslinks have been seen. This eliminates model A and throws doubt on models B and C where some β-β should be seen as well. In chloroplast F_1, β-β crosslinks were seen, again eliminating model A. In any case, as argued in the text, there may only be a total of four of the larger subunits (the stoichiometry is not yet established).

quires Mg^{2+} and a protein "nectin" [4]. Nectin has recently been shown to be the δ subunit [85]. Further, terminal portions of the α chains of *S. faecalis* F_1 also seem to be necessary for binding [85], but not for ATPase activity. In *M. lysodeikticus*, trypsin cleavage of short terminal portions of the α subunits from F_1 did not reduce ATPase activity [86].

Work with chloroplast F_1 suggests that, here too, δ is involved in binding. For instance a digitonin-treated F_1, containing α, β, and γ subunits only [50], was an active ATPase but did not rebind to membranes [77]. Recently Younis et al. [87] showed that pure δ subunit conferred membrane binding capability on chloroplast F_1 depleted of this subunit. A role for the γ subunit is suggested by the fact that in *E. coli* and chloroplast F_1 the ϵ subunit (inhibitor subunit) binds only to γ-containing preparations and not to the α and β (trypsin-treated) preparations [62, 77].

I have purposefully discussed mostly the *E. coli* and chloroplast F_1 work first in this section because very little is known of the roles of mitochondrial F_1 subunits. Kozlov et al. [88] have described a beef heart F_1 preparation which contains only α, β, and ϵ subunits. ATPase activity (which was low to begin with) was retained but the enzyme could not rebind to membranes [89]. Leimgruber and I [61] found that trypsin treatment does not clearly deplete beef heart F_1 of specific subunits in contrast to the results with chloroplast and bacterial F_1.

Recently Yoshida et al. [68] reported reconstitution of full ATPase activity and structure after dissociation of F_1 from PS3 into its individual subunits under denaturing conditions. Here then is an experimental system which opens up many new approaches. The dissociation-reassociation accomplished by Vogel and Steinhart [16] may also be very applicable to chloroplast and mitochondrial F_1.

D. *Structure of Stalk and Membrane Sectors*

The exact structure of the stalk sector is a continuing point of contention. Indeed some authors question the existence of the stalk. It is seen, however, in electron micrographs of negatively-stained material, and MacLennan and Asai [90] presented both biochemical and electron microscopic evidence that a protein "OSCP" (oligomycin-sensitivity conferring protein) is the stalk in beef heart ATPase. OSCP has been obtained in pure soluble form from beef heart [91, 92], rat liver [93], and yeast [94]. It is a basic protein, molecular weight ~18,000, whose amino acid composition differs substantially from that of any of the F_1 subunits (compare Refs. 91 and 46). It binds to soluble F_1 [95] and to the membrane sector [94]. No work has been reported on the chemical nature of these binding sites.

A protein F_6 has recently been purified to homogeneity from beef heart mitochondria [96]. Its molecular weight is around 8,000, it is extremely heat-stable, and it seems to be the same as the protein formerly referred to as Fc_2. It is involved in binding to the membrane [96, 97] and may be part of the stalk or part of the membrane sector (one cannot be more definite as yet). A similar protein has been noted in yeast [98] and in rat liver [93]. It should be noted that while F_6 alone (in the absence of OSCP) can bind F_1 to the membrane, this effect seems to be nonphysiological because the bound F_1 is not oligomycin-sensitive and is removed easily on washing [97]. For the F_1 to be bound tightly in an oligomycin- or DCCD-sensitive fashion both F_6 and OSCP must be present. Apart from their binding function, OSCP and F_6 may also play a role in proton conduction. There are at present no reports on the validity of this hypothesis.

OSCP and F_6 have not yet been noted in bacterial or chloroplast systems. If they were absent this would of course represent a point of marked structural difference from the mitochondrial enzyme. In the PS3 ATPase complex, which had eight components (see below), the component of molecular weight ~19,000 might correspond to OSCP and that of ~13,500 might correspond to F_6, but this is purely speculation at the moment. There does appear to be a stalk in this organism [73].

Purification and characterization of the individual subunits of the membrane sector has recently begun. As noted above, F_6 may originate from the membrane sector. The DCCD-binding protein has been purified from mitochondria [99, 100] and *E. coli* [101]. It has a large number of nonpolar amino acids and behaves like a "proteolipid." Its molecular weight is reported as 10,000-14,000 in beef heart [99, 100] and 8,400 in *E. coli* [101]. Factor B may also be a membrane sector protein (v.i). A fraction corresponding to the band seen in SDS gels of molecular weight around 29,000 has been purified recently [102].

The total number of protein subunits in the membrane sector of mitochondria is not known with certainty since only the SDS gel technique has been used to identify components. Capaldi [103] concluded that there were four components in beef heart with molecular weights 55,000, 29,000, 20,000, and 10,000. Tzagoloff and coworkers found four components in yeast with molecular weights 29,000, 22,000, 12,000, and 7,800 [35, 104]. By comparison the thermophilic bacterium PS3 has apparently three components with molecular weights 19,000, 13,500, and 5,400 [40, 68]. From radioactive labeling experiments a stoichiometry of 1:2:5 has been calculated for these components [73]. DCCD inhibits in PS3 [40] but its site of binding is not yet established. In *E. coli* and chloroplasts the membrane sector has not been characterized yet. The DCCD-binding protein is present in both, and there is evidence for a component of molecular weight 54,000 in *E. coli* membrane sector [105]. Oligomycin does not inhibit bacterial or chloroplast ATPase and this is a point of difference between the membrane sector of the mitochondrial enzyme (to which oligomycin is known to bind) and that of bacteria and chloroplasts. Another inhibitor which binds to the membrane sector of mitochondria is triphenyltin, and this inhibitor does bind to one of the membrane sector proteins in chloroplasts [106].

In mitochondria the relationship between the adenine nucleotide translocating protein and the ATPase is of importance. Vignais et al. [107] found that ADP transported into mitochondria did not mix freely with the matrix ADP pool, but was preferentially phosphorylated. This suggested some ordered "delivery system." The adenine nucleotide translocase has a molecular weight around 30,000 in SDS gels [108, 109], thus it could be one of the components of the membrane sector of the ATPase. However, the recent evidence of Serrano et al. [37] suggests that it is not. Vignais et al. discuss possible spatial relationships between the translocase and the ATPase complex [107].

The membrane sector contains lipid whereas the stalk and F_1 sectors do not. The role of this lipid has recently been studied in more detail. Bertoli et al. [110] have concluded that the lipid which remains bound to the membrane sector during purification of the yeast ATPase complex is somewhat more viscous than the lipid in the membrane as a whole, and it probably corresponds to "boundary lipid" [240]. Arrhenius plits showed a change in activation energy of ATPase activity at around 15-19°C, probably due to a liquid crystalline → gel phase transition in this lipid [110]. The effects of length and degree of saturation of acyl chains of the lipids of the membrane sector on activation of mitochondrial whole ATPase complex were studied by Bruni et al. [111] and by Cunningham and George [112]. Added lipids activated the ATPase and induced expression of oligomycin sensitivity when a stable fluid lamellar phase was established. The work of Swanljung et al. [113] documented the effects of various lipids as activators and also found that oligomycin inhibition was markedly reduced below the temperature-induced phase change.

Several inhibitors of the whole ATPase bind to the membrane sector. These include DCCD, oligomycin, the trialkytins, and several other antibiotics such as venturicidin. The involvement of lipid in the inhibitory effects of these compounds is demonstrated by experiments aimed at perturbing the lipids. For instance Triton X-100 relieves inhibition due to DCCD [114] although DCCD remains bound. Organic solvents such as diethyl ether, chloroform, carbon tetrachloride, and ethyl acetate relieve inhibition caused by all the above compounds [115]. Apparently in these cases the sites of inhibitor binding remain, and the inhibitor is not dislodged, but in some way perturbation or removal of lipid has disrupted communication between the membrane sector and the F_1 sector.

E. Coupling Factors

Coupling factors are soluble components extracted from the mitochondrial (or chloroplast or bacterial) membranes which, when added back to the extracted membranes, restore energy coupling. Purified F_1, OSCP, and F_6 all act as coupling factors. I concluded previously [1] that coupling factors may be assumed to be components of the ATPase complex. This is certainly true of F_1, OSCP, and F_6. The other coupling factor which has been obtained in pure form is factor B. Factor B was first found by Lam et al. and has now been purified and characterized [116, 117]. Other factors very similar to factor B in behavior have recently been found [118, 119] and it is possible that factor B occurs in several polymeric forms with a monomer size around 12,000 daltons [119].

Factor B is known to contain an essential sulfhydryl group(s) [116]. The whole ATPase complex also contains an essential sulfhydryl group(s) which is not on F_1 [60] or OSCP [91]. Chen et al. [120] recently labeled a sulfhydryl group(s) in situ in the membrane and localized the label in a factor B-like protein. Factor B, which has no intrinsic enzyme activity, may act much like oligomycin, i.e., stimulating coupling by repairing proton leaks in the membrane. Addition of factor B to reconstituted liposomes containing purified whole ATPase complex stimulated energy coupling [121]. On balance, therefore, it seems that factor B is probably a membrane sector component.

Many other coupling factor preparations have been reported over the years, and after careful analysis have been found to contain either OSCP, F_6, or B as active principle. Thus they may now be dismissed. Among them are Fc_1, Fc_2, F_2, F_3, F_4, and F_5 (see Refs. 1 and 241). Factors C and D are as yet uncharacterized and factor A is a latent form of F_1 (see Ref. 1). The term "F_0" has recently come to be used in the same sense as the term "membrane sector" in the literature. As originally described, however, F_0 was "a suspension of small fragments containing the entire respiratory chain and phospholipids" [122]. The "hydrophobic protein" fraction [123] is a crude fraction which has many components [102], high ATPase activity, and high NADH and succinate dehydrogenase activity [37].

F. Morphology of the ATPase Complex

Figure 3 and its legend, showing the ATPase complex with likely localization of some of the subunits, is meant as a summary statement to this section on the structure of the ATPase. This structure has been demonstrated recently in fixed, stained, sectioned mitochondrial and chloroplast membranes [124, 125], and for soluble yeast and bacterial ATPase complex [35, 73]. The whole complex is thought to have a molecular weight of around 460,000 [35, 73].

It is of interest to consider forces that hold the complex together, and the functional sequellae of removal of components. Vadineanu et al. [126] have made some quantitative estimates of F_1, OSCP, and F_6 binding in submitochondrial particles depleted of these components. The total amount of F_1 that could bind to particles was around 0.55 nmol/mg protein, about stoichiometric with cytochrome c_1, as has also been calculated by aurovertin titration [127] and NBD-Cl reaction [82]. OSCP and F_1 appear to be about stoichiometric with each other in 1:1 ratio [126]. F_1 bound with a K_d of around 0.1 μM in the presence of OSCP (+ EDTA). The K_d for OSCP binding to the membrane was around 20-40 nM and the K_d for F_6 binding to the membrane appeared to be higher (less tight) than this [126]. In the presence of Mg^{2+}, however, F_1 binding was very much tighter (K_d around 2-4 nM) [126]. The role of Mg^{2+} in binding the F_1 seems paramount. Uncoupled submitochondrial particles made in the presence of EDTA (e.g., "A" particles, "ETP") are almost certainly uncoupled because they have lost some F_1 and/or OSCP by dissociation and are leaky to protons. The "structural" role of F_1 previously described [128] seems to be a leak-plugging effect. Vadineanu et al. [126] showed that the oligomycin titer efficacious in restoring respiratory control in uncoupled particles was exactly equal to the number of F_1 molecules removed during preparation. This lends weight to the view that when oligomycin is added to submitochondrial particles which are uncoupled because they are partly depleted of OSCP and F_1, it binds preferentially to those ATPase complexes which are partly depleted [129], thus first plugging leaks and restoring coupling at low concentration before inhibiting at higher levels. The total oligomycin binding capacity in whole mitochondria is known to be about the same as the aurovertin binding site concentration [127], i.e., stoichiometric with F_1.

Montecucco and Azzi [130] have studied NCCD, a spin-label analog of DCCD. It bound to submitochondrial particles with a stoichiometry of 0.5

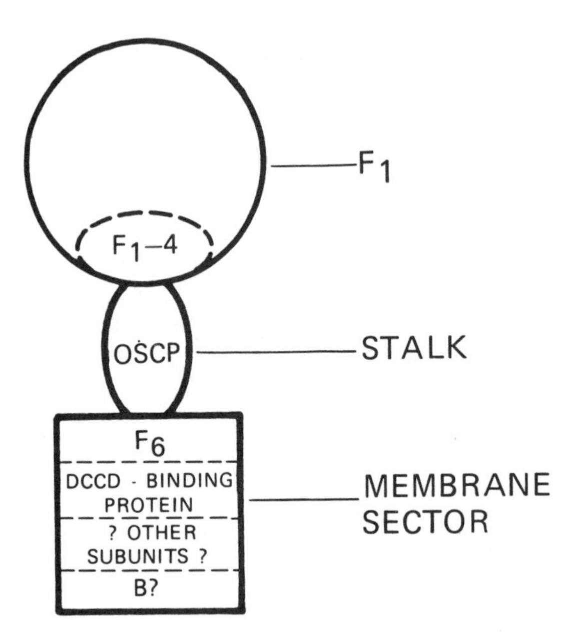

Figure 3 Possible arrangement of subunits in the mitochondrial ATPase complex. (1) F_1 sector: Mitochondrial F_1 contains five types of tightly bound subunit along with the loosely bound F_1 inhibitor protein. Each of the tightly bound subunits and the F_1 inhibitor have been purified and the amino acid composition determined. The stoichiometry of these subunits is undecided at present. A previously suggested stoichiometry of 3:3:1:1:1 for the five tightly bound subunits has been confirmed by some work and denied by other work, which indicated 2:2:2:2:2 or some similar arrangement. The inhibitor protein is a regulatory subunit which prevents ATP "backflow," i.e., it prevents hydrolysis but allows synthesis. Some functional roles may be ascribed to other individual subunits. The F_1-4 (δ) subunit is probably concerned with binding F_1 to the membrane. The active site of ATP hydrolysis is present on enzyme containing only the F_1-1, F_1-2, and F_1-5 (α, β, and ϵ) subunits, and the inhibitory reagent NBD-Cl reacts with the F_1-2 (β) subunit. In *E. coli* and chloroplast F_1, the inhibitor subunit is the F_1-5 or ϵ subunit and the F_1-3 or γ subunit is concerned with binding the inhibitor subunit. This identity of the inhibitor subunit seems to represent a difference between mitochondrial F_1 on the one hand and bacterial and chloroplast F_1 on the other. (2) Stalk sector: The protein OSCP has been purified from mitochondria and its amino acid composition established. It is required for physiological binding of F_1 to the membrane. It has not yet been noted in chloroplast and bacteria. It is probably all or part of the stalk sector of the mitochondrial enzyme; the protein F_6 may also be in the stalk. (3) The membrane sector: F_6 is a small heat-stable protein which is required for binding F_1 to thoroughly depleted membrane preparations. I have placed it in

nmol/mg (same as F_1, aurovertin, NBD-Cl, oligomycin, and cytochrome c_1) and inhibited the ATPase activity. The spectrum of the bound species suggested that it was extremely immobilized, although pH sensitive. Sequential removal of the F_1 inhibitor protein and F_1 had no effects but removal of OSCP changed the spectrum considerably and at the same time the NCCD became exposed to the hydrophilic environment as shown by increased reducibility by ascorbic acid. This work suggests the importance of protection of the membrane sector subunits from the aqueous environment, and also gives further support to the structural scheme shown in Figure 3. On leaving this section on structure one may well ask, Why does this enzyme have this strange shape?

III. Biosynthesis of the ATPase Complex

This has been an extremely active research area. Here I wish only to summarize the main points. Two reviews [131, 132] and a recent book [133] are good sources of background information.

Using the yeast system, Tzagoloff and his coworkers have made substantial progress on this problem. Some of the subunits of mitochondrial ATPase are coded for by nuclear DNA and are made on cytoribosomes. These are OSCP and all of the F_1 subunits. The membrane sector subunits are coded for by mitochondrial DNA and synthesized in mitochondria. One of these, the so-called "subunit 9" is a very hydrophobic protein, a "proteolipid" in fact (see Ref. 131).

In systems other than yeast the situation is somewhat similar although not identical. For instance in *Neurospora* ATPase only two of the membrane sector subunits are synthesized by mitochondria [134]. Evidence is accruing that membrane sector subunits in other organisms are mitochondrially synthesized, e.g., in *Xenopus laevis* [39] and *Dictyostelium discoideum* [135].

There are several reasons why this is an exciting field. First it is one way of studying genetic regulation in eukaryotes. The mitochondrial and nuclear systems are clearly integrated, and regulation of the mitochondrial system with its

---

Figure 3 (continued)
the membrane sector, interacting with OSCP; it might have been placed in the stalk with equal justification. The DCCD binding protein has been purified and is extremely hydrophobic, probably deep-seated in the lipid. Factor B has been purified. Its location is very tentatively assigned as the membrane sector. There are at least two other mitochondrial membrane sector proteins of molecular weight ~29,000 and ~22,00 which have not yet been characterized, and a high molecular weight subunit (~55,000) may also be present.

smaller DNA and its specific products may be more amenable to study. Secondly, the mechanism of assembly of the membrane is an important problem. In this regard one may well ask why the eukaryotes have retained *any* mitochondrial protein-synthesizing machinery. Is it in some way a necessary adjunct for construction of the membrane? Thirdly, there is the application of mutants. In yeast, both Tzagoloff [136] and Linnane [137] and their coworkers have been able, by use of mutants, to map the mitochondrial DNA. Several ATPase complex mutants occur and may be related to specific changes in protein composition of the membrane sector (e.g., Refs. 138, 139). This approach then opens up the possibility of studying the functional roles of individual subunits and their relationship to the whole complex without the difficult and sometimes ambiguous tasks of purification, dissociation, reconstitution, etc.

This approach to the problem has already been fruitful in *E. coli.* The *unc A* mutants of *E. coli* are mutants which have some lesion in the F_1 sector [5, 140, 141]. *Unc B* mutants have a lesion in the membrane sector. There are various examples of each type of mutant. One can anticipate that careful scrutiny of the structural and functional characteristics of the ATPase complex from these mutants, combined with resolution and reconstitution of individual subunits using methods already worked out for *E. coli* (e.g., Refs. 16, 62, 78), will have a large impact on our grasp of the mechanism of action and construction of this enzyme. (See also Chapter 6.)

IV. Functions of the ATPase Complex

A. *Introduction*

The whole ATPase complex is a coupled reversible proton pump. ATP hydrolysis is coupled to proton transport and the protons are extruded on the side of the membrane opposite to the F_1. Conversely, passage of protons down an electrochemical gradient through the ATPase toward the F_1 drives ATP synthesis. This of course is Mitchell's chemiosmotic hypothesis, which has constituted one of the more entertaining biochemical battlefields since around 1960. Much of the evidence has been accrued in intact membrane systems of mitochondria, chloroplasts, and bacteria from measurement of pH changes, use of uncouplers, and use of the inhibitors DCCD and oligomycin which have been found to block proton transport (see Refs. 5, 142, and 143 for reviews). Other persuasive evidence has come from "pH-jump" or "acid-base transition" experiments which show that artificially imposed pH gradients can drive ATP synthesis [144, 145] and from fast reaction experiments which show that the acid-base transitions are kinetically competent to drive ATP synthesis [146, 231]. In reconstituted systems ATP synthesis may be coupled to a proton gradient generated by the purple membranes [147, 148] and ATP hydrolysis generates a transmembrane emf [43, 149]. Impure preparations of ATPase complex from beef heart acted as a proton pump after reconstitution into liposomes [150, 151]. Moreover, recently Sone et al. showed that the purified PS3 ATPase complex acted as a proton pump after reconstitution [43]. The beef heart ATPase preparation of Serrano et al. [37], the purity of which is as yet uncertain, also acts as a proton pump.

The problem therefore is, How does the ATPase complex couple transport of protons down electrochemical gradients to the synthesis of ATP and vice versa? No clear-cut answers are obvious yet but a good deal of evidence has been accumulated. I will review this evidence step by step before summarizing with a brief critique of current models.

B. *Chemistry of the Active Site(s)*

Early work suggested that ATP hydrolysis on mitochondrial F_1 did not involve sulfhydryl groups, but did involve tyrosine groups [60]. The reagent NBD-Cl was then applied [77, 80, 81] and it has been found that reaction of NBD-Cl with one tyrosine residue on the β (F_1-2) subunit of F_1 completely inhibits ATPase activity. This is true for both soluble and membrane-bound F_1 [82]. The situation is not simple however. Whereas ATPase activity is inhibited, the binding of ATP, as measured by ATP-induced quenching of aurovertin fluorescence, is not inhibited [152]. It was therefore suggested that the tyrosine to which NBD-Cl binds is essential in a step subsequent to a conformational change produced initially by ATP binding. The authors invoke the transient occurrence of an intermediate in which ADP is bound on the enzyme [152].

Other work suggests that there is at least one arginine group on soluble mitochondrial F_1 and membrane-bound F_1 which is essential for ATPase activity. This was deduced from chemical modification experiments utilizing phenylglyoxal and butanedione [153]. In this respect F_1 resembles other nucleoside triphosphate-binding enzymes. The light-activated ATP analog, arylazidopropionyl ATP seems to be a potentially useful labeling reagent. This compound was a substrate for F_1 and on exposure to light bound in the ratio 0.8 mol to 1 mol of F_1 in a form that was not released after SDS denaturation [154].

In contrast to the situation in Ca^{2+} ATPase and Na^+-K^+ ATPase, no phosphorylated enzyme intermediate has been found in F_1 from any source.

C. *Binding Sites for Nucleotides; Kinetics*

1. Equilibrium Binding Sites

In soluble beef heart F_1, Hilborn and Hammes [155] have found that ADP binds at one site [or perhaps two sites (see Ref. 156, p. 2974)] with dissociation constants (K_d) of 280 nM (+Mg^{2+}) or 11 μM (+EDTA). Other nucleoside diphosphates such as IDP or 6-mercaptopurineriboside diphosphate did not bind at this site(s), but ε-ADP (the fluorescent analog, 1,N^6-ethenoadenosine diphosphate) did bind. ADP also bound at a site with K_d = 30-47 μM (±Mg^{2+}) which was thought at that time to be the catalytic ATPase site because the K_d was similar to the K_i of ADP. No high-affinity equilibrium binding of ATP, ITP, UTP (+EDTA), or AMP (±Mg^{2+}) was observed. Garret and Penefsky [157] have noted that at 50 μM ADP, two sites on soluble beef heart F_1 were saturated. Pedersen [158] has noted a binding site for ADP on soluble rat liver F_1, K_d = 1-2 μM, whereas the K_i of ADP in this enzyme was 310 μM [159]. AMP-PNP (adenyl-5′-imidodiphosphate) bound reversibly to two sites on soluble beef heart F_1 with K_d around 1.3 μM [157]. However, even high concentrations of AMP-PNP were not able to displace all the ADP from the enzyme in either beef heart [157] or rat liver F_1 [160].

In soluble chloroplast F_1, ADP, ε-ADP, and AMP-PNP bound to two sites with $K_d < 10$ μM on the latent and heat-activated enzyme [156]. The heat-activated chloroplast F_1 had in addition a site which bound ADP with K_d around 100 μM and a site which bound AMP-PNP with K_d around 4-8 μM. The facts that NBD-Cl eliminated this site and that the K_i for AMP-PNP inhibition of ATPase is similar to the K_d of AMP-PNP binding suggested that this was the catalytic site. On the heat-activated enzyme there were two additional ADP binding sites with affinities in the range of 76 to 100 μM. Roy and Moudrianakis [161] and Girault et al. [162] have also noted two ADP binding sites on chloroplast F_1 with binding affinities in the range of 2 to 35 μM (former) and 1 to 2.5 μM (latter authors). Livne and Racker [163] noted two ATP binding sites on chloroplast F_1. Adolfsen and Moudrianakis [164] have noted competition between ADP and AMP-PNP for binding to the soluble F_1 from *Alcaligenes faecalis.* Dissociation constants measured were around 15 μM for ADP and 3 μM for AMP-PNP, and the possibility that there was more than one site cannot be excluded from the data. No binding of P_i or AMP was seen. These workers noted the slowness of binding of ADP, a fact previously noted by Tondre and Hammes for binding of ε-ADP to soluble beef heart F_1 [165]. Both groups attribute this slowness to the requirement for an induced conformational change.

No studies have het been reported of equilibrium binding of nucleotides to the whole ATPase complex.

2. Tightly Bound Nucleotide

Soluble F_1 and membrane-bound F_1 contain ATP and/or ADP which does not readily exchange with nucleotide in the medium and is not readily released from the enzyme. Thus it remains bound during purification and during such treatments as repeated ammonium sulfate precipitation, charcoal treatment, gel filtration, or ion-exchange chromatography (soluble enzyme), or repeated washing by homogenization and centrifugation (membrane-bound enzyme). There were early hints that such nucleotide was present. For instance Warshaw et al. [57] had found ADP in purified factor A, and Roy and Moudrianakis found ATP and ADP in soluble chloroplast F_1 [161]. Subsequently, however, much impetus was given to study of these nucleotides by the publication of speculative schemes of oxidative phosphorylation which demanded their involvement. Slater and co-workers found that soluble beef heart F_1 contained approximately 3 mol ATP and 2 mol ATP per mol [166]. Soluble chloroplast F_1 (trypsin-activated) contained approximately 0.8 mol ATP and 0.7 mol ADP per mol [167]. Both ATP and ADP were also found tightly bound to submitochondrial particles, chloroplast membranes, and bacterial membranes, and the membrane-bound F_1 was shown to be the site of ATP and ADP binding by release of the F_1, and by examination of an *E. coli* mutant which was F_1 deficient (see Refs. 167-169). Bachofen et al. found tightly bound ADP and ATP in chloroplast and chromatophore membranes [170]. Garret and Penefsky [157] found 1 mol ATP and 2 mol ADP per mol tightly bound to soluble beef heart F_1. Leimgruber and Senior [61] found zero ATP and 2 mol ADP per mol of soluble beef heart F_1, and confirmed the presence of both ATP and ADP in submitochondrial particles [171].

Leimgruber [172] has also found tightly bound ATP and ADP in the whole ATPase complex from beef heart made by the procedure of Tzagoloff et al. [29] and in sarcoplasmic reticulum (Leimgruber, Scott, Senior, and Shamoo, unpublished results). The tightly bound ATP and ADP has recently been reported to be present in rat liver soluble F_1 [158] and also in soluble F_1 of *S. faecalis* [173]. It should be noted that in both cases, although the nucleotide remained bound through purification, a considerable portion of it was removed by repeated ammonium sulfate precipitation.

It is clear, however, that this tightly bound nucleotide is common to energy-transducing membranes. In the beef heart system, which is the best characterized so far, soluble F_1 has 2 mol ADP/mol and a variable amount of ATP. In the membrane-bound F_1, ATP is definitely present in amounts equal to or greater than the ADP. As stated above, under "non-energized" conditions the rate of exchange of the tightly bound nucleotide with the medium is very slow. However, on induction of "energization," for instance by shining light transiently on chloroplast membranes [167, 174, 175] or by adding ATP to submitochondrial particles [169], at least some of the nucleotide exchanges rapidly with the medium. In effect the binding affinity of the sites is reduced. Magnusson and McCarty [175] calculated a K_d of around 5-7 μM for the ADP and ATP sites on chloroplast F_1 under these energized conditions, whereas the K_d in the nonenergized conditions must be considerably lower (Harris et al. [166] quote an estimate of $K_d < 0.1$ μM for ATP exchange on soluble F_1). The nucleotide which becomes incorporated into the chloroplast or mitochondrial F_1 under the energized conditions is effectively sequestered and cannot be removed by repeated washing [169, 176] although a further burst of energization will cause it to become exchangeable again [176, 177].

All the measurements of bound ATP and ADP described so far in this section were done by denaturing the enzyme to release the nucleotide. Two groups have found conditions which release the nucleotide without concomitant denaturation. Garret and Penefsky [157] found that gel filtration of soluble beef heart F_1 in glycerol-containing buffer yielded a preparation ("glycerol-treated F_1") which contained no tightly bound ATP or ADP. The enzyme retained both ATPase and coupling activity [157, 178] but was unstable in aqueous buffer. It could readily "reload" nucleotide, some of it in a form which was "tightly bound." A total of ~5 mol ADP and AMP-PNP could be taken up per mol of F_1 and 2-3 mols remained "tightly bound" after gel filtration in aqueous buffer. The soluble enzyme which contained ~2 mol AMP-PNP tightly bound had inhibited ATPase activity, showing that the nucleotide did not readily exchange from the enzyme in this form and implicating these sites in control of ATPase activity. On the other hand the nucleotide-depleted F_1 which was fully saturated with AMP-PNP by reloading and contained ~5 mol AMP-PNP per mol F_1 was still active as a coupling factor when re-bound to particles [157, 178]. This result may not be as strange as it seems, however, if it is assumed that energization during the assay (oxidative phosphorylation) spurs rapid exchange of the tightly bound nucleotide with ATP or ADP in the medium.

Leimgruber and Senior [61] completely removed the tightly bound nucleotide from beef heart F_1 either by treatment with very small amounts of trypsin or by gel filtration in buffer containing K_2SO_4. The nucleotide-depleted F_1 preparations retained full ATPase activity and were stable in aqueous buffers, but in contrast to the glycerol-treated enzyme were devoid of coupling activity. This lack of coupling activity was not due to an inability to rebind to the membrane. Other properties of the F_1 which were unperturbed by the removal of the tightly bound nucleotides were the inhibition by the F_1-inhibitor protein and the inhibition by ADP. These soluble nucleotide-depleted F_1 preparations were unable to reload tightly bound nucleotides.

Further work showed that tightly bound ATP and ADP could also be removed from membrane-bound F_1 without reduction of non-energy-linked ATPase activity [171], but complete loss of energy coupling occurred. On the basis of these and the foregoing results, we have proposed that a series or "cluster" of tightly bound nucleotides is in some way necessary for ATP synthesis and energy-linked ATP hydrolysis [61].

3. Kinetics

Earlier work on kinetics was discussed in my previous review [1] and the results may be briefly restated. The K_m for ATP for soluble or membrane-bound beef heart mitochondrial F_1 is in the range of 0.2 to 1.25 mM depending on the assay and conditions. Several other nucleoside triphosphates including ITP, GTP, UTP, CTP, SHTP (6-mercaptopurineriboside triphosphate), ϵ-ATP (1,N^6-ethenoadenosine triphosphate), and the arylazidoaminopropionyl ATP (see Ref. 154) serve as substrates for hydrolysis. ADP is a competitive inhibitor with K_i in the range of 30 to 100 μM against soluble beef heart F_1, again depending on assay and conditions.

Lardy and coworkers have recently attacked this problem and have studied both beef heart and rat liver soluble F_1 [179-183]. They have postulated that both a catalytic site (or sites) and a regulatory site (or sites) are present. They do not specify the number of each of these sites. ATP binds to both catalytic and regulatory sties and inhibits its own hydrylysis. ADP was said to bind at the regulatory site, as do Cr-ATP and Cr-ADP (the chromium complexes), AMP-PNP, and IMP-PNP [182, 183]. There was some evidence that ITP could also bind at this regulatory site [182]. At the catalytic site, ATP, ITP, GTP, and GMP-PNP bind.

One is impressed by the complexity of the system. The effects of the imidodiphosphate analogs are particularly difficult to understand. AMP-PNP inhibits ATPase activity of soluble and membrane-bound F_1 competitively and potently [184-186]. It also inhibits ATP-P_i exchange and ATP-driven energized processes, but does not inhibit ATP synthesis [184]. This has led some authors to suggest a dual site hypothesis, that is to say that ATP synthesis and ATP hydrolysis may occur at different sites [184, 188]. In support of this the K_m for ADP phosphorylation is known to be rather low ($\sim$3.8 μM, see Ref. 187), suggesting initial binding of ADP is to a relatively high-affinity site. ADP binding to high-affinity sites on soluble F_1 does occur (see above) and is not reversed by AMP-PNP. Garret and Penefsky [157] noted that AMP-PNP could not completely displace ADP from soluble beef heart F_1 and Pedersen [188] noted a similar effect in rat liver F_1. This might rationalize the inability of AMP-PNP to inhibit oxidative phosphorylation.

On the other hand GMP-PNP was thought to bind at the catalytic site on soluble F_1 [180] since it competitively inhibits ITP and GTP hydrolysis, and it does inhibit oxidative phosphorylation in submitochondrial particles [183]. This can be taken as evidence for a single site of both ATP synthesis and hydrolysis. (The fact that NBD-Cl reacts at a single tyrosine and may inhibit both ATP hydrolysis and synthesis at the same rates might also be taken as evidence for a single site [189].) To add to the complicated story, IMP-PNP was an inhibitor of ATP and ITP hydrolysis, but did not inhibit ATP synthesis, or ATP-P_i or ITP-P_i exchange [182].

Additionally, as Ebel and Lardy [179] and Lardy and coworkers [183] have shown, and as Pedersen also reported [190], anions affect the inhibition patterns caused by the various nucleotides and may themselves activate or inhibit ATPase. The anions appear to diminish the effect of ATP binding at the regulatory site, decreasing the negative cooperativity and making the kinetic plots of ATP hydrolysis linear [179] or monophasic [190].

Pedersen [190] working on the rat liver enzyme has reported somewhat similar findings to Lardy's group both in respect to ITP, GTP, and ATP hydrolysis and the effects of anions. ITP and GTP appeared to bind only at the catalytic site, whereas ATP bound to at least two sites which were kinetically distinct and negative cooperativity was seen. In bicarbonate-containing buffer the ATPase reactions of soluble F_1 and membrane-bound F_1 were very similar in kinetic behavior. In chloride-containing buffer they differed considerably.

Kinetic analyses of the soluble F_1 from chloroplasts have been summarized recently by Nelson [7] and there the situation is also complicated. The K_m for ATP hydrolysis is around 0.1-1.1 mM; it is proposed that there are cooperative ATP binding sites and two sites for ADP binding. ADP inhibits and there is cooperativity between the ADP sites.

4. Comments on the Nucleotide Binding and Kinetic Studies

We are confronted with a complicated situation. First, some minimal statements.

1. There are multiple binding sites for nucleotides on F_1. They appear to fall broadly into three classes. First, there are very high-affinity binding sites for the so-called "tightly bound" nucleotides. These are normally occupied by ADP and ATP, but AMP-PNP may also occupy them. Secondly, there are high-affinity binding sites of K_d in the range of 1 to 50 μM at which ADP and adenine nucleotide analogs (e.g., ϵ-ADP, dADP, AMP-PNP) may bind. Thirdly, there is at least one somewhat looser site ($K_d > 100\ \mu$M) which is the catalytic site for ATP hydrolysis.

2. The number of each type of site is not known with certainty. Different authors report different results depending on enzyme source and conditions. Penefsky et al. [191] suggest that a total of five adenine nucleotide binding sites are present on F_1. This is definitely a minimum estimate—there may be additional ATP-specific sites.

3. There is a nucleotide binding site or sites which modulate or "regulate" ATP hydrolysis in a negative fashion. ATP itself can bind at this site.

4. It is not clear whether ATP hydrolysis and ATP synthesis occur at a single site, or at distinct sites. A third possibility, that ATP synthesis and ATP-

driven energy functions require the sequential participation of two or more sites in a "cluster" is also compatible with the data.

Now I shall add some general comments to these minimal statements. The nucleotide binding studies have been performed for the most part of soluble F_1. The membrane-bound F_1 may behave differently. Work must be directed toward the whole complex. Running through the discussion of many of the papers cited above is the suggestion that the situation is dynamic, i.e., that conformational changes on F_1 may occur to alter kinetic properties or to expose or modify binding sites, and that such conformational changes may be quite subtle, continuous, and sensitive to the ionic composition of the medium. Moyle and Mitchell [192] have put this on a formal basis, proposing and providing evidence for "active" to "inactive state" transitions in F_1, dependent on the concentration of Mg^{2+} and several anions. Adolfsen and Coworkers [14, 63] have noted that the soluble F_1 from mitochondria, chloroplasts, and bacteria shows electrophoretic heterogeneity. This has been correlated with activity changes and the authors have coined the term "functional polymorphism" for this phenomenon. While it is too early to estimate the significance of these conformational rearrangements (Do they occur in the membrane-bound enzyme?) it is clearly important to control conditions carefully so that comparisons can be made between different reports. In our laboratory we are currently concentrating on the tightyl bound nucleotides. Their ubiquity in energy-transducing systems is impressive and the evidence that we have gathered so far suggests that they must be present for energy transduction to occur.

D. Inhibitor and Activators

1. Acting at the F_1 Sector

Inhibitors and activators of mitochondrial ATPase may act either on the F_1 sector or on the membrane sector. The inhibition by the specific inhibitor protein is at the level of F_1 and has already been discussed in detail (see Section II.C.2). Activation and inhibition by anions also occurs by binding to F_1 and has been introduced above (Section IV.C.3). Anions which activate are listed by Ebel and Lardy and include selenite, sulfite, and bicarbonate [179]. Inhibitory anions are azide, cyanate, and thiocyanate [179]. These anions are effective on soluble beef heart F_1 and rat liver F_1 and on the membrane-bound enzyme [179, 182, 190]. The physiological role of these anions is not clear at the moment. It seems likely that F_1 is usually surrounded by a relatively high bicarbonate concentration in active respiring mitochondria since the matrix is the major site of body CO_2 production. Plasma $[HCO_3^-]$ is around 28 mM, and Ebel and Lardy [179] found 5.8-fold activation of rat liver F_1 by bicarbonate with K_a = 5.8 mM. It would seem that bicarbonate should normally be included in assays and binding measurements. It is notable that an ATPase enzyme in gastric mucosa and pancreas which may be related to secretion of H^+ and HCO_3^-, respectively, in the two organs shows a remarkable similar pattern of responses to the same anions which stimulate or inhibit F_1 [193, 194].

It has long been known that 2,4-dinitrophenol activates soluble F_1 directly, and Ebel and Lardy included it in their list of activating anions. There are some reasons, however, to suspect that it may be more specific than this and may in fact mimic a yet unrecognized physiological activator as suggested by Cantley and Hammes [195]. Moyle and Mitchell also discuss this point [192]. The main reason for suggesting this is that an uncharged structural analog, 4-fluoro-3-nitrophenylazide (FNPA), was also found to activate soluble F_1 [196] and competed with 2,4-dinitrophenol for a binding site ($K_d \simeq 100\ \mu M$) on F_1. This analog was transformed by light to a nitrene derivative and inhibited the ATPase activity of F_1 potently after photoactivation. This inhibition was partly prevented by 2,4-dinitrophenol. FNPA was not an uncoupler, so that its activation of F_1-ATPase activity is not related to uncoupling. Cantley and Hammes [195] documented the actions of 2,4-dinitrophenol on F_1, showing that it activated both soluble and membrane-bound forms by lowering the K_m of ATP and that it did not affect ADP binding. It has recently been suggested that mitochondria from hepatoma [197, 198], ascites tumor [198], and rat mammary tumor [199] do not evince the activation of ATPase by 2,4-dinitrophenol. This might reflect some structural change in these tumor mitochondria as discussed in Ref. 199.

Quercetin is an inhibitor of F_1. It is one of a group of flavonoid compounds. It inhibits both soluble and membrane-bound mitochondrial F_1 but has no effect on oxidative phosphorylation [200] and therefore resembles behaviorally the F_1 inhibitor protein. Binding of quercetin to chloroplast F_1 was documented by Cantley and Hammes [201]. There were two high-affinity quercetin binding sites ($K_d = 20\text{-}35\ \mu M$), which were distinct from nucleotide or NBD-Cl binding sites. These sites appear to be on the two largest subunits, α and β, as shown by the fact that quercetin was inhibitory to both intact and trypsin-treated chloroplast F_1 [77]. It is worth noting in passing that quercetin has been useful in experimental studies of tumor metabolism [202, 203].

Aurovertin is a specific inhibitor of mitochondrial ATPase, which has been useful mainly in measuring conformational changes of F_1. These conformational studies are discussed in Section IV.E. Aurovertin binds to beef heart soluble F_1 with a stoichiometry of 2:1 with ATP present or 1:1 in the presence of ADP or Mg^{2+}. Evidence suggested that these sites were present also on membrane-bound F_1 [204, 205]. They are high-affinity sites, with K_d values of 13 nM with Mg^{2+}, 70 nM with ADP, and 520 nM with ATP (both sites). Only one site is thought to be involved in the inhibition of ATPase [204]. The inhibition was said to be uncompetitive [204] although Ebel and Lardy [206] point out that the data suggested noncompetitive inhibition. The inhibition patterns in rat liver F_1 were variable. There was definite interaction between aurovertin and stimulating anions (bicarbonate) and even stimulation by aurovertin was seen under some conditions [206]. Bertina et al. [127] also studied binding of aurovertin to soluble beef heart F_1 ($+Mg^{2+}$) and found one site per F_1 with K_d = 60 nM. The aurovertin binding sites have not been further characterized as yet but one interesting feature is that the binding of F_1 inhibitor protein in some way blocks aurovertin binding at one site in membrane-bound and soluble F_1 [205].

Bathophenanthroline inhibits soluble F_1 from mitochondria [207] but not from chloroplasts or *E. coli* [208, 209]. The inhibition is interesting in that it is potent (1-10 μM concentrations inhibit by 80-90%) and is reversed by uncouplers [207]. The reversal by uncouplers is probably due, however, to chemical interaction between uncouplers and the bathophenanthroline [210]. This inhibitor is a well-known chelating reagent, especially for Fe, and the observations raised the possibility that F_1 might contain Fe. So far this has not been confirmed [210]. There have been suggestions that F_1 contains Mg^{2+} [166, 211] but this cannot explain the bathophenanthroline inhibition. At the moment therefore this is an interesting effect without an explanation.

Other inhibitors that act at the level of F_1 are triphenylsulfonium [11] and alkylguanidines [13]. Pedersen has reported that *p*-chloromercuribenzoate (pCMP) prevents anion activation by reacting at the anion-binding site(s) [212]. Effects of sulfhydryl reagents on kinetic behavior of F_1 have not yet been studied, but since anions may bind at a regulatory site [183] it might be possible to characterize the regulatory site by labeling with sulfhydryl reagents.

Obviously, F_1 itself is a very complicated molecule indeed. As well as the various and numerous nucleotide binding sites discussed above, it has binding sites for OSCP and the F_1 inhibitor protein, and possibly F_6 as well. Further it binds many other small molecules, including anions, aurovertin, quercetin, an iron chelator, substituted phenols, alkylguanidines, and Mg^{2+}.

2. Acting at the Membrane Sector

The main thrust of recent work on membrane sector inhibitors has been to try to pin down the actual sites of action. Bastos [213] reported that diazidoethidium bromide, a photoactivated analog of the inhibitor ethidium bromide, covalently binds to subunit 9 of yeast ATPase complex. It competed with DCCD but not with oligomycin for this site. Trialkyltins bind to the membrane sector and have been further studied recently by Dawson and Selwyn [129]. These authors favor the view that trialkyltins and oligomycin act at different sites, since oligomycin does not affect trialkyltin binding. This is also supported by the work of Linnett et al. [114] in which trialkyltin inhibition but not oligomycin inhibition was relieved by Triton X-100, and by the work of Lancashire and Griffiths [214] who isolated an oligomycin-sensitive, trialkyltin-insensitive yeast mutant.

Many antibiotics inhibit by binding to the membrane sector. The best known and most used is oligomycin. Enns and Criddle [215] have recently shown that oligomycin interacts with the smallest subunit (subunit 9) of the yeast ATPase complex by reducing the oligomycin-inhibited complex with tritium-labeled sodium borohydride and characterizing the adduct formed. The sites of action of the other antibiotics (e.g., leucinostatin, venturicidin) relative to that of oligomycin have been discussed by Lardy and coworkers [183, 216].

The characteristics of DCCD inhibition have already been discussed (Section II.D); its site of action was relatively easier to characterize because it does form a covalent bond with a membrane sector protein. The nature of the chemical reaction between DCCD and the DCCD-binding protein in the membrane sector has become apparent from Fillingame's recent work [101]. The group modified appears to be a carboxyl group and it is therefore possible that this carboxyl group plays a role in proton transfer across the membrane sector. The relationship between the DCCD binding site and that of the other inhibitors named above is not yet decided. Finally one other inhibitor which binds at the membrane sector is tetradifon, a commercially available acaricide [217].

E. Conformational Changes in the ATPase Complex

Rearrangements of conformation occur on F_1 during energy transduction, and they may be induced either by electron transfer chain activity or by direct imposition of proton gradients.

In mitochondria the evidence is of three kinds. First, aurovertin binds to F_1 (Section IV.D.1) and when it does so it forms a highly fluorescent complex. The fluorescence of this complex may be altered by other ligands, e.g., ATP quenches the fluorescence somewhat in soluble F_1 [204, 205] whereas ADP enhances it. In mitochondria or submitochondrial particles previously complexed with aurovertin, addition of succinate increases the fluorescence of the complex and this increase is abolished by uncouplers, anaerobiosis, and oligomycin [127, 204]. Chang and Penefsky confirmed that this change in fluorescence could be taken as an indication of conformational change in the membrane-bound F_1 [205]. They also concluded that the conformational change involved dissociation of ATP and Mg^{2+} from the enzyme. One notable feature of this paper [205] was that the conformational changes were quite rapid—the rapid phase of the (two-phased) energized fluorescence change had a half-time of around 40 msec. In their kinetic experiments in which respiration-induced ATP synthesis was measured over the first 100 msec, Thayer and Hinkle [146] reported a lag in ATP production when NADH was used to initiate the reaction, and it would seem that this was compatible in time with the aurovertin-reported conformational change. Secondly, there is evidence that on induction of respiration the forces which bind the F_1 inhibitor protein to membrane-bound F_1 become considerably weakened, and that partial dissociation of the inhibitor protein from the F_1 may occur [58]. This change in binding affinity was counteracted by uncouplers and is presumably a reflection of conformational change on F_1. A third kind of evidence for conformational change on F_1 in mitochondria comes from the increased exchangeability of the tightly-bound nucleotide under energized conditions (Section IV. C.2; see also Ref. 169).

There is excellent evidence that membrane-bound chloroplast F_1 undergoes conformational change which is linked to phosphorylation and may be evoked by either light or "pH jump." Direct chemical evidence was provided first by Ryrie and Jagendorf [218] who showed that either light, acid-base transition, or ATP hydrolysis caused membrane-bound F_1 to transiently incorporate ^3H from ^3H_2O. The ^3H (50-90 atoms/mol F_1) was effectively sequestered. This ^3H incorporation was largely abolished by uncouplers and inhibitors of photophosphorylation.

The same authors also studied inhibition of photophosphorylation by sulfate [219]. This is ADP dependent, and it was concluded that the sulfate inhibition was due to a reaction between sulfate and groups on F_1 which could only occur during coupled electron transport. A similar and more definitive chemical approach involved the use of NEM, which inhibits photophosphorylation in a light-dependent fashion [220]. Incubation of chloroplast membranes in the dark with NEM produces no effect, but as soon as the light is turned on, NEM inhibits and reacts rapidly with a group(s) in the γ subunit of the chloroplast F_1 [221]. Uncouplers abolished the inhibition and largely prevented the incorporation of NEM into the chloroplast F_1. The conclusion was that the light generates electron flow which causes a conformational change in F_1, transiently exposing a group(s), most probably sulfhydryl, which reacts with NEM [221]. ADP and P_i gave considerable protection against incorporation. Other direct chemical evidence for the causation of a conformational change in membrane-bound chloroplast F_1 by energization comes from modification by $KMnO_4$ [222], internal oxidation of –SH to S–S catalyzed by *o*-iodosobenzoate [223], and modification of amino groups by trinitrobenzenesulfonate [224].

In Section IV.C.2 I described how the tightly bound nucleotides on chloroplast F_1 become more exchangeable after induction of electron flow by light or after a pH jump. This constitutes evidence that the conformation of the F_1 is changing. The exchange of ATP and ADP was found to be uncoupler-sensitive, and several factors which affect photophosphorylation (e.g., pH, Mg^{2+} dependence, light intensity, sensitivity to adenosine-5′-phosphosulfate) had parallel effects on the nucleotide exchange [175-177]. One interesting fact that emerged from Magnusson and McCarty's work was that the capability to exchange nucleotide persisted for some time in the dark (up to several minutes). The light-induced pH gradient would have decayed long before this so that the conformational change which is rapid in onset (half of the nucleotide exchange occurred in less than 100 msec) seems slow to relax.

F. The Mechanism of ATP Synthesis: Current Views

1. The Chemiosmotic Mechanism

Mitchell, in what one might call the second part of the chemiosmotic hypothesis (the first part being that H^+ gradients are generated by the respiratory chain and drive ATP synthesis) has proposed that protons are carried through a proton-conducting pathway (channel) in the membrane sector to the active site on F_1. The active site sits in a field of high protonic potential. Mitchell himself has listed the evidence that such a proton channel exists [143]. The evidence is derived from direct measurement of proton permeability in intact and F_1-depleted membranes; from the use of oligomycin and DCCD which act a low concentration to plug proton leaks in uncoupled particles and, at higher concentration, to block proton permeability altogether; and from the leak-plugging effects of enzymatically-inactive F_1. However, there is no knowledge yet as to the location of the active site(s) of F_1 in relation to the membrane sector.

In 1974 Mitchell proposed specific chemical events to account for ATP hydrolysis and synthesis at the active site, coupled to H^+ translocation across the site [225]. Subsequently Boyer [226] disparaged this proposal using the arguments that (1) the ionic species (of P_i and ADP) that were proposed by Mitchell as reactants were unlikely to occur around pH 7, (2) the proposal required the whole electrochemical potential to be in the form of a proton potential, (3) the reaction pathway was unorthodox and unlikely in the light of other established pathways known for phosphate ester hydrolysis, and (4) the scheme appeared to utilize different pathways of synthesis and hydrolysis.

This limited warfare then escalated as Mitchell countered [227] with the arguments that his proposal did not rigidly specify unlikely anionic species, rather the relative charge on ADP + P_i as compared to ATP was the important point—there had to be the ability for net translocation of the O^{2-} anion across the site in the opposite direction to the H^+ movement. Consequently the net charge on the reacting ionic species of ADP + P_i must be more negative by two units than the net charge on the ATP. The same pathway could be used for hydrolysis and synthesis. Mitchell has expanded his proposals recently [143] and points out that the proposed reaction pathway need not be unorthodox if certain neutralizing effects of Mg^{2+} or charged protein groups on the F_1 itself are postulated.

Whether or not Mitchell's scheme turns out to be correct, it has certainly achieved a great deal already by focusing attention on the central question.

2. The Conformational Model

Terminology is important here. Some time ago Boyer [228] proposed a scheme in which redox reactions in the respiratory chain were coupled to ATP synthesis by protein conformational changes. Other workers subscribed to this idea for a while but it has now been generally abandoned. The model discussed here requires that the proton gradient triggers a conformational change in the F_1, which leads to release of preformed, tightly bound ATP. The model is depicted and described in Figure 4. It has been much canvassed recently by Boyer, and also by Slater's group who have been concerned mostly with the role of the tightly bound nucleotides. In essence this model proposes an indirect mode of interaction between the proton gradient and the chemical events of ATP synthesis, in contrast to Mitchell's proposal. The chemical events at the active site are unspecified. More detailed and speculative versions of this scheme may be found in Refs. 167, 169, and 229. Evidence which is compatible with this model is as follows.

1. Conformational changes are induced in F_1 in response to pH gradients as discussed above. From the limited evidence so far available they seem quite rapid in onset, approximately as fast as onset of phosphorylation.

2. ADP and ATP do occur in a tightly bound form on F_1 in all energy-transducing membranes. Their presence appears to be required for energy transduction; their binding affinity changes on energization, and factors which modulate their exchange under energized conditions are found to have parallel effects on phosphorylation. However, the scheme requires that the binding affinity of ATP and ADP should vary in a reciprocal fashion during one conformational cycle, and this has not yet been shown. Also the nucleotide exchange is puzzlingly insensitive to Dio-9 and venturicidin [169], and only a fraction of the tightly bound nucleotide is usually seen to exchange.

3. Tightly bound ADP can act as initial phosphate acceptor in chloroplasts [167, 229] and submitochondrial particles [229, 230]. In the presence of added ADP the rate is comparable to normal phosphorylation rates. However, in the absence of added ADP the rate of phosphorylation is slow. This might simply be a reflection of loss of ADP from the tight-binding sites or failure of ATP to be released (in absence of replacement ADP from the medium) during the conformational cycle [230, 231].

4. Experiments by Boyer's group [232, 233] have confirmed earlier findings [234] that small amounts of bound ATP are formed in mitochondria in an uncoupler-insensitive fashion. Studies of the isotope-exchange reactions that

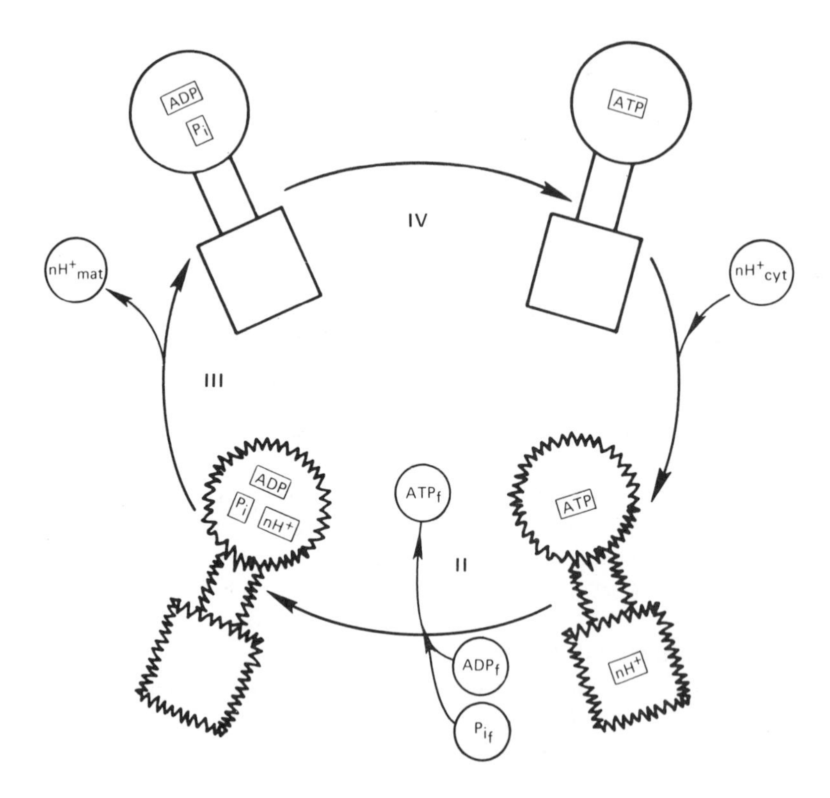

Figure 4 The conformational model. This scheme incorporates recent ideas of Boyer, Slater, and others. Step I: The ATPase complex has ATP tightly bound (ATP) and there is an electrochemical gradient across the inner membrane. Free protons (nH^+_{cyt}) approach from the cytoplasmic side, triggering a conformational change on the ATPase complex by binding to specific groups on the membrane sector. Step II: The conformational change on F_1 alters the binding affinity of ATP and the binding affinity of ADP at the sites which normally bind ATP and ADP very tightly. Incoming ADP and P_i from the medium (ADP_f and P_{if}) become tightly bound as tightly bound ATP becomes released to the medium (ATP_f). The bound protons (nH^+) migrate along a proton-conducting pathway to F_1. Step III: The protons are released on the matrix side (nH^+_{mat}). The ATPase complex relaxes, with ADP and P_i tightly bound. Step IV: Condensation forms ATP, tightly bound. *Comments.* (1) Energy input: Step IV is energy independent. (2) Sequence of events: Clearly steps I, II, and III may be concerted or ordered in a manner different to that described.

occur on membrane-bound F_1 suggested that the reaction ADP + $P_i \rightleftharpoons H_2O$ + ATP occurs rapidly in an energy-independent, uncoupler-insensitive fashion. The energy-requiring steps are primarily the release of bound ATP from the enzyme and also the initial binding of P_i [229, 232, 233, 235]. A partial analogy to this model which is often quoted is the molecule hemoglobin, which senses $[H^+]$ over a narrow range and responds by changing its O_2 binding affinity, with conformational change being the mediator.

Although all this evidence may seem to provide persuasive support for the conformational model, much of it is also compatible with Mitchell's model. Conformational changes are usual in proteins, and it cannot be ruled out yet that the tightly bound nucleotides play a predominantly structural or regulatory role, or react in a Mitchell-type mechanism.

3. The Chemical Pathway

Several groups have attempted to deduce likely pathways from known features of the reaction in mitochondria, such as pH dependence and associated isotope-exchange rates, or by analogy with known reactions of phosphates in organic chemistry. No phosphorylated intermediate has ever been found; ADP is required for the $H_2{}^{18}O \rightleftharpoons P_i$ exchange [236]; and labeled P_i appears very rapidly in ATP [231]. These facts point to a concerted reaction mechanism in which ADP and P_i condense. Korman and McClick [237] favor a mechanism involving pseudorotation of a pentacovalent intermediate; this they feel best explains the oxygen exchange data. In this scheme oxygen exchange occurs by addition of H_2O to enzyme-bound ATP or P_i. Boyer [226, 233] feels that the isotope exchange data are adequately explained by reversible cleavage of the β-γ P–O bond of tightly bound ATP at the active site.

Some years ago Roy and Moudrianakis [161] proposed that the following reaction occurred on soluble chloroplast F_1:

$$2ADP \rightleftharpoons AMP + ATP$$

Figure 4 (continued)
(3) Details: The number of P_i, ADP, ATP, and proton binding sites is not defined. Protons need not migrate into F_1—they could be released on the matrix side of the membrane sector. At any one time the F_1 could contain both ADP and ATP. It is required, however, that at step II, ATP sites become looser while ADP sites remain tighter. In hydrolysis these sites are required to reverse their roles (either by the same site binding ATP instead of ADP, or by the same site becoming looser instead of tighter).

The ATP and AMP remained tightly bound and could be recovered only by denaturation of the enzyme. When chloroplasts were incubated in the light with an electron acceptor in the presence of labeled AMP or P_i, and the F_1 was then solubilized, labeled ADP was found bound to the F_1. This suggested that AMP was the acceptor of P_i in photophosphorylation. Subsequently a scheme was formulated as follows [238]:

$$F_1 + AMP + P_i \xrightarrow{\text{light, electron transport}} F_1{-}ADP \quad (1)$$

$$ADP + F_1{-}ADP \longrightarrow F_1\begin{matrix} \diagup ADP \\ \diagdown ADP \end{matrix} \quad (2)$$

$$F_1\begin{matrix} \diagup ADP \\ \diagdown ADP \end{matrix} \longrightarrow F_1\begin{matrix} \diagup AMP \\ \diagdown ATP \end{matrix} \quad (3)$$

$$F_1\begin{matrix} \diagup AMP \\ \diagdown ATP \end{matrix} \longrightarrow F_1 + ATP + AMP \quad (4)$$

This overall scheme then accounted for photophosphorylation. Reaction 3 occurs on soluble chloroplast F_1; the other reactions occur only on the membrane-bound F_1. Soluble chloroplast F_1 did not bind added AMP. This model has been examined recently (after an initial period of unjustified rejection) and has been criticized on two grounds. First, the rate of production of labeled ADP from $[^{32}P]P_i$ in chloroplasts has been found to be too slow to account for rates of photophosphorylation [231]. The evidence strongly supports ADP as the primary P_i acceptor. Production of labeled ADP did occur, however [231], and was considered to arise by reversal of Reaction 3, since added medium AMP was not incorporated into ADP, neither did it affect the rate of formation of labeled ADP. Similarly Harris and Slater [167] saw incorporation of $[^{32}P]P_i$ into the β position of bound ADP and ATP during energized exchange of bound nucleotide. However, this occurred in the absence of added AMP, so again reversal of reaction 3 seems the most straightforward explanation. Formation of labeled ADP from $[^{32}P]P_i$ did not occur in submitochondrial particles [231].

Secondly, Reaction 3 and its reversal have been thought to arise from adenylate kinase contamination of the soluble F_1 preparations. In a recent report [239] Moudrianakis and Tiefert strongly reassert that this transphosphorylation is intrinsic to chloroplast F_1. Several lines of evidence were cited including activity measurements, effects of Mg^{2+}, and use of differential inhibitors of adenylate kinase and chloroplast F_1. Catterall and Pedersen [187] saw no evidence for the transphosphorylation reaction in soluble rat liver F_1. However, Penefsky et al. [191] working with soluble beef heart F_1 found that both glycerol-treated (nucleotide-depleted) and native F_1 formed AMP after incubation with either ATP or ADP ($+Mg^{2+}$). This AMP was not, however, enzyme bound. The transphosphorylation also occurred in the reverse direction, giving bound $[^3H]$ADP derived from $[^3H]$AMP. Penefsky et al. tentatively concluded that this adenylate kinase-like reaction was most probably due to a contaminant [191].

It should finally be noted that when chloroplasts are allowed to incorporate $[^{14}C]$ADP in the light by exchange of tightly bound nucleotide on F_1 [177] the ^{14}C label on F_1 is recovered in AMP, ADP, and ATP in the approximate ratio 1:2:1. These nucleotides are released to the medium on energization by light. This confirms that Reaction 4 of Roy and Moudrianakis' scheme can occur under energized conditions.

To summarize this section, therefore, the evidence points to the fact that ADP is the primary acceptor of P_i in phosphorylation in chloroplasts and mitochondria. A concerted mechanism for direct union of ADP and P_i seems very likely. The transphosphorylation reaction seen on pure soluble chloroplast F_1 appears to be real, but its role on the membrane-bound F_1 is currently obscure since the rate of the reaction cannot account for normal rates of ATP formation.

G. Postscript

I have felt for some time that one route to understanding ATP synthesis lies through careful characterization of the ATPase complex, contemporaneous with establishment of structure-function correlations as they become feasible. Now that the ubiquity of the enzyme has become apparent, persons with greatly varied interests and backgrounds, for example geneticists and researchers interested primarily in transport phenomena, have begun to examine the enzyme, and we may expect the combination of different talents to accelerate progress.

As I have expounded to students and seminar groups in recent years, here we have found a machine which has facilitated the bulk of energy transduction on this planet. It is highly efficient at ambient temperature, nonpolluting, nonexplosive, and we know that it works. We may be able to apply the principles, when we understand them, to our own devices.

Acknowledgments

The author gratefully acknowledges financial support from NSF (Grant Number PCM76-04991). Part of the work described here and done in our laboratory in

Rochester was also supported by research grant AM-16366 from NIAMDD, USPHS, and by research grant GB-38350 from NSF.

References

1. A. E. Senior, *Biochim. Biophys. Acta 301:* 249 (1973).
2. V. Riebeling and K. Jungermann, *Biochim. Biophys. Acta 430:* 434 (1976).
3. D. J. Clarke and J. G. Morris, *Biochem. J. 154:* 725 (1976).
4. A. Abrams and J. B. Smith, in *The Enzymes* Vol. 10 (P. D. Boyer, ed.), 1974, pp. 395-429.
5. R. D. Simoni and P. W. Postma, *Ann. Rev. Biochem. 44:* 523 (1975).
6. P. L. Pedersen, *J. Bioenergetics 6:* 243 (1975).
7. N. Nelson, *Biochim. Biophys. Acta 456:* 314 (1976).
8. R. Panet and D. R. Sanadi, *Current Topics in Membranes and Transport* Vol. 8, (D. R. Sanadi, ed.), Academic, New York, 1976, pp. 99-150.
9. A. E. Senior, in *The Molecular Biology of Membranes* (S. Fleischer, Y. Hatefi, D. H. MacLennan, and A. Tzagoloff, eds.), Plenum, New York, 1977.
10. R. B. Beechey, S. A. Hubbard, P. E. Linnett, A. D. Mitchell, and E. A. Munn, *Biochem. J. 148:* 533 (1975).
11. R. H. Barrett and M. J. Selwyn, *Biochem. J. 156:* 315 (1976).
12. V. Spitsberg and J. E. Blair, *Biochim. Biophys. Acta 492:* 237 (1977).
13. M. T. de Gomez-Puyou, A. Gomez-Puyou, and M. Beigel, *Arch. Biochem. Biophys. 173:* 326 (1976).
14. R. Adolfsen, J. A. McClung, and E. N. Moudrianakis, *Biochemistry 14:* 1727 (1975).
15. B. Cannon and G. Vogel, *FEBS Lett. 76:* 284 (1977).
16. G. Vogel and R. Steinhart, *Biochemistry 15:* 208 (1976).
17. T. Higashi, V. K. Kalra, S. H. Lee, E. Bogin, and A. F. Brodie, *J. Biol. Chem. 250:* 6541 (1975).
18. F. W. Hulla, M. Hockel, S. Risi, and K. Dose, *Eur. J. Biochem. 67:* 469 (1976).
19. A. F. Knowles and H. S. Penefsky, *J. Biol. Chem. 247:* 6617 (1972).
20. A. E. Senior and J. C. Brooks, *Arch. Biochem. Biophys. 140:* 257 (1970).
21. L. L. Horstman and E. Racker, *J. Biol. Chem. 245:* 1336 (1970).
22. D. O. Lambeth and H. A. Lardy, *Eur. J. Biochem. 22:* 355 (1971).
23. W. A. Catterall and P. L. Pedersen, *J. Biol. Chem. 246:* 4987 (1971).
24. A. Tzagoloff, *J. Biol. Chem. 244:* 5020 (1969).
25. Y. Kagawa and E. Racker, *J. Biol. Chem. 241:* 2467 (1966).
26. Y. Kagawa and E. Racker, *J. Biol. Chem. 241:* 2475 (1966).
27. K. Kopaczyk, J. Asai, D. W. Allmann, T. Oda, and D. E. Green, *Arch. Biochem. Biophys. 123:* 602 (1968).
28. K. Kopaczyk, J. Asai, and D. E. Green, *Arch. Biochem. Biophys. 126:* 358 (1968).

29. A. Tzagoloff, K. H. Byington, and D. H. MacLennan, *J. Biol. Chem. 243:* 2405 (1968).
30. A. Tzagoloff, D. H. MacLennan, and K. H. Byington, *Biochemistry 7:* 1596 (1968).
31. D. H. MacLennan and A. Tzagoloff, *Biochemistry 7:* 1603 (1968).
32. P. Swanljung and L. Ernster, in *Energy Transduction in Respiration and Photosynthesis* (E. Quagliariello, S. Papa, and C. S. Rossi, eds.), Adriatica Editrice, Bari, 1971, pp. 839-850).
33. P. Swanljung, *Anal. Biochem. 43:* 382 (1971).
34. P. Swanljung and L. Frigeri, *Biochim. Biophys. Acta 283:* 391 (1972).
35. A. Tzagoloff and P. Meagher, *J. Biol. Chem. 246:* 7328 (1969).
36. Y. Hatefi, D. L. Stiggall, Y. Galante, and W. G. Hanstein, *Biochem. Biophys. Res. Commun. 61:* 313 (1974).
37. R. Serrano, B. I. Kanner, and E. Racker, *J. Biol. Chem. 251:* 2453 (1976).
38. I. J. Ryrie, *Arch. Biochem. Biophys. 168:* 712 (1975).
39. G. Koch, *J. Biol. Chem. 251:* 6097 (1976).
40. N. Sone, M. Yoshida, H. Hirata, and Y. Kagawa, *J. Biol. Chem. 250:* 7917 (1975).
41. S. H. Lee, N. S. Cohen, and A. F. Brodie, *Proc. Nat. Acad. Sci. U. S. 73:* 3050 (1976).
42. P. D. Bragg and C. Hou, *Arch. Biochem. Biophys. 174:* 553 (1976).
43. N. Sone, M. Yoshida, H. Hirata, H. Okamoto, and Y. Kagawa, *J. Membr. Biol. 30:* 121 (1976).
44. F. S. Stekhoven, *Biochem. Biophys. Res. Commun. 47:* 7 (1972).
45. A. E. Senior and J. C. Brooks, *FEBS Lett. 17:* 327 (1971).
46. J. C. Brooks and A. E. Senior, *Biochemistry 11:* 4675 (1972).
47. W. A. Catterall, W. A. Coty, and P. L. Pedersen, *J. Biol. Chem. 248:* 7427 (1973).
48. D. O. Lambeth, H. A. Lardy, A. E. Senior, and J. C. Brooks, *FEBS Lett. 17:* 330 (1971).
49. M. E. Pullman and G. C. Monroy, *J. Biol. Chem. 238:* 3762 (1963).
50. N. Nelson, H. Nelson, and E. Racker, *J. Biol. Chem. 247:* 7657 (1972).
51. S. H. P. Chan and R. L. Barbour, *Biochim. Biophys. Acta 430:* 426 (1976).
52. M. Satre, M. B. de Jerphanian, J. Huet, and P. V. Vignais, *Biochim. Biophys. Acta 387:* 241 (1975).
53. Y. Landry and A. Goffeau, *Biochim. Biophys. Acta 376:* 470 (1975).
54. J. C. Brooks and A. E. Senior, *Arch. Biochem. Biophys. 147:* 467 (1976).
55. D. A. Harris, J. Rosing, and E. C. Slater, *FEBS Lett. 47:* 236 (1974).
56. A. F. Knowles and H. S. Penefsky, *J. Biol. Chem. 247:* 6624 (1972).
57. J. B. Warshaw, K. W. Lam, B. Nagy, and D. R. Sanadi, *Arch. Biochem. Biophys. 123:* 385 (1968).
58. R. J. VandeStadt, B. L. DeBoer, and K. VanDam, *Biochim. Biophys. Acta 292:* 338 (1973).
59. K. Asami, K. Juntti, and L. Ernster, *Biochim. Biophys. Acta 205:* 307 (1970).

60. A. E. Senior, *Biochemistry 12:* 3622 (1973).
61. R. M. Leimgruber and A. E. Senior, *J. Biol. Chem. 251:* 7103 (1976).
62. J. B. Smith and P. C. Sternweiss, *Biochemistry 16:* 306 (1977).
63. R. Adolfsen and E. N. Moudrianakis, *Biochemistry 15:* 4163 (1976).
64. H. Takisawa, S. Yamazaki, Y. Tamaura, S. Hirose, and Y. Inada, *Arch. Biochem. Biophys. 170:* 743 (1975).
65. S. Yamazaki, H. Takisawa, Y. Tamaura, S. Hirose, and Y. Inada, *FEBS Lett. 56:* 248 (1975).
66. S. Yamazaki, H. Takisawa, Y. Tamaura, S. Hirose, and Y. Inada, *FEBS Lett. 66:* 23 (1976).
67. J. M. Wilkinson and R. J. A. Grand, *Proceedings of the 9th FEBS Meeting 31:* 137 (1975).
68. M. Yoshida, N. Sone, H. Hirata, and Y. Kagawa, *J. Biol. Chem. 250:* 7910 (1975).
69. F. Farron and E. Racker, *Biochemistry 9:* 3829 (1970).
70. A. E. Senior, *Biochemistry 14:* 660 (1975).
71. M. Satre, G. Klein, and P. B. Vignais, *Biochim. Biophys. Acta 453:* 111 (1976).
72. P. D. Bragg and C. Hou, *Arch. Biochem. Biophys. 167:* 311 (1975).
73. Y. Kagawa, H. Sone, M. Yoshida, H. Hirata, and H. Okamoto, *J. Biochem. 80:* 141 (1976).
74. P. D. Bragg and C. Hou, *Biochem. Biophys. Res. Commun. 72:* 1042 (1976).
75. P. T. Wingfield and D. H. Boxer, *Biochem. Soc. Trans. 3:* 763 (1975).
76. B. A. Baird and G. G. Hammes, *Biochemistry 251:* 6953 (1976).
77. D. W. Deters, E. Racker, N. Nelson, and H. Nelson, *J. Biol. Chem. 250:* 1041 (1975).
78. N. Nelson, B. I. Kanner, and D. L. Gutnick, *Proc. Nat. Acad. Sci. U. S. 71:* 2720 (1974).
79. M. R. J. Salton and M. T. Schorr, *Biochem. Biophys. Res. Commun. 49:* 350 (1972).
80. S. J. Ferguson, W. J. Lloyd, M. H. Lyons, and G. K. Radda, *Eur. J. Biochem. 54:* 117 (1975).
81. S. J. Ferguson, W. J. Lloyd, and G. K. Radda, *Eur. J. Biochem. 54:* 127 (1975).
82. S. J. Ferguson, W. J. Lloyd, and G. K. Radda, *Biochem. J. 159:* 347 (1976).
83. P. D. Bragg, P. L. Davies, and C. Hou, *Arch. Biochem. Biophys. 159:* 664 (1973).
84. M. Futai, P. C. Sternweiss, and L. A. Heppel, *Proc. Nat. Acad. Aci. U. S. 71:* 2725 (1974).
85. A. Abrams, D. Morris, and C. Jensen, *Biochemistry 15:* 5560 (1976).
86. M. Hockel, F. W. Hulla, S. Risi, and K. Dose, *Biochim. Biophys. Acta 429:* 1020 (1976).
87. H. M. Younis, G. D. Winget, and E. Racker, *J. Biol. Chem. 252:* 1814 (1977).

88. I. A. Kozlov and H. N. Mikelsaar, *FEBS Lett. 43:* 212 (1974).
89. I. A. Kozlov, A. A. Kondrashin, V. A. Konarenko, and S. T. Metelsky, *J. Bioenergetics 8:* 1 (1976).
90. D. H. MacLennan and J. Asai, *Biochem. Biophys. Res. Commun. 33:* 441 (1968).
91. A. E. Senior, *J. Bioenergetics 2:* 141 (1971).
92. A. E. Senior, in *Methods in Enzymology* Vol. 32D (S. Fleischer and L. Packer, eds.), Academic, New York, 1977.
93. L. K. Russell, S. A. Kirkley, T. R. Kleyman, and S. H. P. Chan, *Biochem. Biophys. Res. Commun. 73:* 434 (1976).
94. A. Tzagoloff, *J. Biol. Chem. 245:* 1545 (1970).
95. R. J. VandeStadt, R. J. Kraaipoel, and K. VanDam, *Biochim. Biophys. Acta 267:* 25 (1972).
96. B. I. Kanner, R. Serrano, M. A. Kandrach, and E. Racker, *Biochem. Biophys. Res. Commun. 69:* 1050 (1976).
97. A. F. Knowles, R. J. Guillory, and E. Racker, *J. Biol. Chem. 246:* 2672 (1971).
98. J. M. Fessenden-Raden and A. M. Hack, *Biochemistry 11:* 4609 (1972).
99. K. J. Cattell, C. R. Lindop, I. G. Knight, and R. B. Beechey, *Biochem. J. 125:* 169 (1971).
100. F. S. Stekhoven, R. F. Waitkus, and H. Th. B. VanMoerkerk, *Biochemistry 11:* 1144 (1972).
101. R. H. Fillingame, *J. Biol. Chem. 251:* 6630 (1976).
102. R. A. Capaldi, *Biochem. Biophys. Res. Commun. 55:* 655 (1973).
103. R. A. Capaldi, *Biochem. Biophys. Res. Commun. 53:* 1331 (1973).
104. A. Tzagoloff and A. Akai, *J. Biol. Chem. 247:* 6517 (1972).
105. R. D. Simoni and A. Shandell, *J. Biol. Chem. 250:* 9421 (1975).
106. J. M. Gould, *Eur. J. Biochem. 62:* 567 (1976).
107. P. V. Vignais, P. M. Vignais, and J. Doussiere, *Biochim. Biophys. Acta 376:* 219 (1975).
108. P. Riccio, H. Aquila, and M. Klingenberg, *FEBS Lett. 56:* 133 (1975).
109. G. Lauguin, G. Brandolin, and P. Vignais, *FEBS Lett. 67:* 306 (1976).
110. E. Bertoli, J. B. Finean, and D. E. Griffiths, *FEBS Lett. 61:* 163 (1976).
111. A. Bruni, P. W. VanDijck, and J. DeGier, *Biochim. Biophys. Acta 406:* 315 (1975).
112. C. C. Cunningham and D. T. George, *J. Biol. Chem. 250:* 2036 (1975).
113. P. Swanljung, L. Frigeri, K. Ohlson, and L. Ernster, *Biochim. Biophys. Acta 305:* 519 (1973).
114. P. E. Linnett, A. D. Mitchell, and R. B. Beechey, *FEBS Lett. 53:* 180 (1975).
115. M. Partis, A. D. Mitchell, J. Whitehead, J. F. Donnellan, P. E. Linnett, and R. B. Beechey, *Biochem. Soc. Trans. 2:* 205 (1974).
116. K. W. Lam, D. Swann, and M. Elzinga, *Arch. Biochem. Biophys. 130:* 175 (1969).
117. K. W. Lam and S. S. Yang, *Arch. Biochem. Biophys. 133:* 366 (1969).

118. T. Higashiyama, R. C. Steinmeier, B. C. Serrianne, S. L. Knoll, and J. H. Wang, *Biochemistry 14:* 4117 (1975).
119. K. S. Sou and Y. Hatefi, *Biochim. Biophys. Acta 423:* 398 (1976).
120. C. Chen, M. Yang, H. D. Durst, D. R. Saunders, and J. H. Wang, *Biochemistry 14:* 4122 (1975).
121. S. Joshi, F. Shaikh, and D. R. Sanadi, *Biochem. Biophys. Res. Commun. 65:* 1371 (1975).
122. Y. Kagawa and E. Racker, *J. Biol. Chem. 241:* 2461 (1966).
123. Y. Kagawa and E. Racker, *J. Biol. Chem. 246:* 5477 (1971).
124. J. N. Telford and E. Racker, *J. Cell Biol. 57:* 580 (1973).
125. S. Oleszko and E. N. Moudrianakis, *J. Cell Biol. 63:* 936 (1974).
126. A. Vadineanu, J. A. Berden, and E. C. Slater, *Biochim. Biophys. Acta 449:* 468 (1976).
127. R. M. Bertina, P. I. Schrier, and E. C. Slater, *Biochim. Biophys. Acta 305:* 503 (1973).
128. H. S. Penefsky, *J. Biol. Chem. 242:* 5789 (1967).
129. A. P. Dawson and M. J. Selwyn, *Biochem. J. 152:* 333 (1975).
130. C. Montecucco and A. Azzi, *J. Biol. Chem. 250:* 5020 (1975).
131. A. Tzagoloff, M. R. Rubin, and M. F. Sierra, *Biochim. Biophys. Acta 301:* 71 (1973).
132. G. Schatz and T. L. Mason, *Ann. Rev. Biochem. 43:* 51 (1974).
133. A. Tzagoloff (ed.), *Membrane Biogenesis,* Plenum, New York, 1975.
134. G. Jackl and W. Sebald, *Eur. J. Biochem. 54:* 97 (1975).
135. R. N. Stuchell, B. I. Weinstein, and D. S. Beattie, *J. Biol. Chem. 250:* 570 (1975).
136. A. Tzagoloff, in *The Molecular Biology of Membranes* (S. Fleischer, Y. Hatefi, D. H. MacLennan, and A. Tzagoloff, eds.), Plenum, New York, 1977.
137. A. W. Linnane, in *The Molecular Biology of Membranes* (S. Fleischer, Y. Hatefi, D. H. MacLennan, and A. Tzagoloff, eds.), Plenum, New York, 1977.
138. A. Tzagoloff, A. Akai, and R. B. Needleman, *J. Biol. Chem. 250:* 8236 (1975).
139. A. Tzagoloff, A. Akai, and F. Foury, *FEBS Lett. 65:* 391 (1976).
140. G. B. Cox and F. Gibson, *Biochim. Biophys. Acta 346:* 1 (1974).
141. H. U. Schairer, P. Friedl, B. L. Schmid, and G. Vogel, *Eur. J. Biochem. 65:* 257 (1976).
142. F. M. Harold, *Bacteriol. Rev. 36:* 172 (1972).
143. P. Mitchell, *Biochem. Soc. Trans. 4:* 399 (1976).
144. A. T. Jagendorf, *Fed. Proc. 26:* 1361 (1967).
145. W. S. Thayer and P. C. Hinkle, *J. Biol. Chem. 250:* 5330 (1975).
146. W. S. Thayer and P. C. Hinkle, *J. Biol. Chem. 250:* 5336 (1975).
147. E. Racker and W. Stoeckenius, *J. Biol. Chem. 249:* 662 (1974).
148. M. Yoshida, N. Sone, N. Hirata, Y. Kagawa, Y. Takeuchi, and K. Ohno, *Biochem. Biophys. Res. Commun. 67:* 1295 (1975).
149. L. A. Drachev, A. A. Jasaitis, H. Mikelsaar, I. B. Nemecek, A. Y. Semenov, E. G. Semenova, I. I. Severina, and V. P. Skulachev, *J. Biol. Chem. 251:* 7077 (1976).

150. Y. Kagawa, in *The Enzymes of Biological Membranes* Vol. 4 (A. Martonosi, ed.), Plenum, New York, 1976, pp. 125-142.
151. Y. Kagawa, *Biochim. Biophys. Acta 265:* 297 (1972).
152. S. J. Ferguson, W. J. Lloyd, G. K. Radda, and E. C. Slater, *Biochem. Biophys. Acta 430:* 189 (1976).
153. F. Marcus, S. M. Schuster, and H. A. Lardy, *J. Biol. Chem. 251:* 1775 (1976).
154. J. Russell, S. J. Jeng, and R. J. Guillory, *Biochem. Biophys. Res. Commun. 70:* 1225 (1976).
155. D. A. Hilborn and G. G. Hammes, *Biochemistry 12:* 983 (1973).
156. L. C. Cantley and G. G. Hammes, *Biochemistry 14:* 2968 (1975).
157. N. E. Garret and H. S. Penefsky, *J. Biol. Chem. 250:* 6640 (1975).
158. P. L. Pedersen, *J. Supramol. Structure 3:* 222 (1975).
159. W. A. Catterall and P. L. Pedersen, *Biochem. Soc. Spec. Publ. 4:* 63 (1974).
160. P. L. Pedersen, *Biochem. Biophys. Res. Commun. 64:* 610 (1975).
161. H. Roy and E. N. Moudrianakis, *Proc. Nat. Acad. Sci. U. S. 68:* 464 (1971).
162. G. Girault, J. M. Galmiche, M. Michel-Villaz, and J. Thierv, *Eur. J. Biochem. 38:* 473 (1973).
163. A. Livne and E. Racker, *J. Biol. Chem. 244:* 1332 (1969).
164. R. Adolfsen and E. N. Moudrianakis, *Arch. Biochem. Biophys. 172:* 425 (1976).
165. C. Tondre and G. G. Hammes, *Biochim. Biophys. Acta 314:* 245 (1973).
166. D. A. Harris, J. Rosing, R. J. VandeStadt, and E. C. Slater, *Biochim. Biophys. Acta 314:* 149 (1973).
167. D. A. Harris and E. C. Slater, *Biochim. Biophys. Acta 387:* 335 (1975).
168. E. C. Slater, J. Rosing, D. A. Harris, R. J. VandeStadt, and A. J. Kemp, in *Membrane Proteins in Transport and Phosphorylation* (G. F. Azzone, M. E. Klingenberg, E. Quagliariello, and N. Siliprandi, eds.), North Holland, Amsterdam, 1974, pp. 137-147.
169. D. A. Harris and E. C. Slater, in *Electron Transfer Chains and Oxidative Phosphorylation* (E. Quagliariello, S. Papa, F. Palmieri, E. C. Slater, and N. Siliprandi, eds.), North Holland, Amsterdam, 1975, pp. 379-384.
170. R. Bachofen, W. Beyeler, and Ch. Pflugshaupt, in *Electron Transfer Chains and Oxidative Phosphorylation* (E. Quagliariello, S. Papa, F. Palmieri, E. C. Slater, and N. Siliprandi, eds.), North Holland, Amsterdam, 1975, pp. 167-172.
171. R. M. Leimgruber and A. E. Senior, *J. Biol. Chem. 251:* 7110 (1976).
172. R. M. Leimgruber, Ph.D. Thesis, University of Rochester, 1977.
173. A. Abrams, E. A. Nolan, C. Jensen, and J. B. Smith, *Biochem. Biophys. Res. Commun. 55:* 22 (1973).
174. R. P. Magnusson and R. E. McCarty, *Biochem. Biophys. Res. Commun. 70:* 1283 (1976).

175. R. P. Magnusson and R. E. McCarty, *J. Biol. Chem. 251:* 7417 (1976).
176. H. Strotmann, S. Bickel, and B. Huchzermeyer, *FEBS Lett. 61:* 194 (1976).
177. S. Bickel-Sandkotter and H. Strotmann, *FEBS Stee. 65:* 102 (1976).
178. N. E. Garret and H. S. Penefsky, *J. Supramol. Structure 3:* 469 (1975).
179. R. E. Ebel and H. A. Lardy, *J. Biol. Chem. 250:* 191 (1975).
180. S. M. Shuster, R. E. Ebel, and H. A. Lardy, *J. Biol. Chem. 250:* 7848 (1975).
181. S. M. Shuster, R. E. Ebel, and H. A. Lardy, *Arch. Biochem. Biophys. 171:* 656 (1975).
182. S. M. Shuster, R. J. Gertschen, and H. A. Lardy, *J. Biol. Chem. 251:* 6705 (1976).
183. L. A. Lardy, S. M. Shuster, and R. E. Ebel, *J. Supramol. Structure 3:* 214 (1975).
184. H. S. Penefsky, *J. Biol. Chem. 249:* 3579 (1974).
185. R. D. Philo and M. J. Selwyn, *Biochem. J. 143:* 745 (1974).
186. R. L. Melnick, J. T. DeSousa, J. Maguire, and L. Packer, *Arch. Biochem. Biophys. 166:* 139 (1975).
187. W. A. Catterall and P. L. Pedersen, *J. Biol. Chem. 247:* 7969 (1972).
188. P. L. Pedersen, *Biochem. Biophys. Res. Commun. 64:* 610 (1975).
189. S. J. Ferguson, P. John, W. J. Lloyd, G. K. Radda, and F. R. Whatley, *Biochim. Biophys. Acta 357:* 457 (1974).
190. P. L. Pedersen, *J. Biol. Chem. 251:* 934 (1976).
191. H. S. Penefsky, A. Schwab, and N. E. Garrett, in *Electron Transfer Chains and Oxidative Phosphorylation* (E. Quagliariello, S. Papa, F. Palmieri, E. C. Slater, and N. Siliprandi, eds.), North Holland, Amsterdam, 1975, pp. 135-147.
192. J. Moyle and P. Mitchell, *FEBS Lett. 56:* 55 (1975).
193. A. E. Blum, G. Shah, T. St. Pierre, H. F. Helender, C. P. Sung, V. D. Wiebelhaus, and G. Sachs, *Biochim. Biophys. Acta 249:* 101 (1971).
194. B. Simon, R. Kinne, and G. Sachs, *Biochim. Biophys. Acta 282:* 293 (1972).
195. L. C. Cantley and G. G. Hammes, *Biochemistry 12:* 4900 (1973).
196. A. E. Senior and A. M. Tometsko, in *Electron Transfer Chains and Oxidative Phosphorylation* (E. Quagliariello, S. Papa, F. Palmieri, E. C. Slater, and N. Siliprandi, eds.), North Holland, Amsterdam, 1975, pp. 155-160.
197. P. L. Pedersen, T. Eska, H. P. Morris, and W. A. Catterall, *Proc. Nat. Acad. Sci. U. S. 68:* 1079 (1971).
198. P. L. Pedersen and H. P. Morris, *J. Biol. Chem. 249:* 3327 (1974).
199. A. E. Senior, S. E. McGowan, and R. Hilf, *Cancer Res. 35:* 2061 (1975).
200. D. R. Lang and E. Racker, *Biochim. Biophys. Acta 333:* 180 (1974).
201. L. C. Cantley and G. G. Hammes, *Biochemistry 15:* 1 (1976).
202. E. M. Suolinna, R. N. Buchsbaum, and E. Racker, *Cancer Res. 35:* 1865 (1975).

203. E. M. Suolinna, D. R. Lang, and E. Racker, *J. Nat. Cancer Inst. 53:* 1515 (1974).
204. T. M. Chang and H. S. Penefsky, *J. Biol. Chem. 248:* 2746 (1973).
205. T. M. Chang and H. S. Penefsky, *J. Biol. Chem. 249:* 1090 (1974).
206. R. E. Ebel and H. A. Lardy, *J. Biol. Chem. 250:* 4992 (1975).
207. D. C. Phelps, K. Nordenbrand, B. D. Nelson, and L. Ernster, *Biochem. Biophys. Res. Commun. 63:* 1005 (1975).
208. C. L. Bering, R. A. Dilley, and F. L. Crane, *Biochem. Biophys. Res. Commun. 63:* 736 (1975).
209. I. L. Sun, D. C. Phelps, and F. L. Crane, *FEBS Lett. 54:* 253 (1975).
210. D. C. Phelps, K. Nordenbrand, T. Hundal, C. Carlsson, B. D. Nelson, and L. Ernster, in *Electron Transfer Chains and Oxidative Phosphorylation* (E. Quagliariello, S. Papa, F. Palmieri, E. C. Slater, and N. Siliprandi, eds.), North Holland, Amsterdam, 1975, pp. 385-400.
211. A. Abrams, C. Jensen, and D. H. Morris, *Biochem. Biophys. Res. Commun. 69:* 804 (1976).
212. P. L. Pedersen, *Biochem. Biophys. Res. Commun. 71:* 1182 (1976).
213. R. D. N. Bastos, *J. Biol. Chem. 250:* 7739 (1975).
214. W. E. Lancashire and D. E. Griffiths, *Eur. J. Biochem. 51:* 377 (1975).
215. R. K. Enns and R. S. Criddle, *Fed. Proc. 35:* 1586 Abstr. (1976).
216. H. A. Lardy, P. Reed, and C. H. C. Lin, *Fed. Proc. 34:* 1707 (1975).
217. E. Bustamente and P. L. Pedersen, *Biochem. Biophys. Res. Commun. 51:* 292 (1973).
218. I. J. Ryrie and A. T. Jagendorf, *J. Biol. Chem. 247:* 4453 (1972).
219. I. J. Ryrie and A. T. Jagendorf, *J. Biol. Chem. 246:* 582 (1971).
220. R. E. McCarty, P. R. Pittman, and Y. Tsuchiya, *J. Biol. Chem. 247:* 3048 (1972).
221. R. E. McCarty and J. Fagan, *Biochemistry 12:* 1503 (1973).
222. I. J. Ryrie and A. T. Jagendorf, *J. Biol. Chem. 249:* 4404 (1974).
223. R. H. Vallejos and C. S. Andrea, *FEBS Lett. 61:* 95 (1976).
224. D. Oliver and A. T. Jagendorf, *J. Biol. Chem. 251:* 7168 (1976).
225. P. Mitchell, *FEBS Lett. 43:* 189 (1974).
226. P. D. Boyer, *FEBS Lett. 50:* 91 (1975).
227. P. Mitchell, *FEBS Lett. 50:* 95 (1975).
228. P. D. Boyer, in *Oxidases and Related Redox Systems* Vol. 2 (T. E. King, H. S. Mason, and M. Morrison, eds.), Wiley, New York, 1965, pp. 994-1017.
229. P. D. Boyer, D. J. Smith, J. Rosing, and C. Kayalar, in *Electron Transfer Chains and Oxidative Phosphorylation* (E. Quagliariello, S. Papa, F. Palmieri, E. C. Slater, and N. Siliprandi, eds.), North Holland, Amsterdam, 1975, pp. 361-372.
230. J. Rosing, D. J. Smith, C. K. Kayalar, and P. D. Boyer, *Biochem. Biophys. Res. Commun., 72:* 1 (1976).
231. D. J. Smith, B. O. Stokes, and P. D. Boyer, *J. Biol. Chem. 251:* 4165 (1976).

232. R. L. Cross and P. D. Boyer, *Biochemistry 14:* 392 (1975).
233. P. D. Boyer, R. L. Cross, and W. Momsen, *Proc. Nat. Acad. Sci. U. S. 70:* 2837 (1973).
234. R. H. Eisenhardt and O. Rosenthal, *Biochemistry 7:* 1327 (1968).
235. P. D. Boyer, B. O. Stokes, R. G. Wolcott, and C. Degani, *Fed. Proc. 34:* 1711 (1975).
236. D. H. Jones and P. D. Boyer, *J. Biol. Chem. 244:* 5767 (1969).
237. E. F. Korman and J. McClick, *J. Bioenergetics 3:* 147 (1972).
238. H. Roy and E. N. Moudrianakis, *Proc. Nat. Acad. Sci. U. S. 68:* 2720 (1971).
239. E. N. Moudrianakis and M. A. Tiefert, *J. Biol. Chem. 251:* 7796 (1976).
240. P. C. Jost, O. H. Griffith, R. A. Capaldi, and G. Vanderkooi, *Proc. Nat. Acad. Sci. U. S. 70:* 480 (1973).
241. R. B. Beechey and K. J. Cattel, in *Current Topics in Bioenergetics* Vol. 5 (D. R. Sanadi, ed.), Academic, New York 1973, pp. 305-357.

5

Photosynthetic Reaction Centers

John M. Olson
Brookhaven National Laboratory
Upton, New York

J. Philip Thornber
University of California
Los Angeles, California

*This chapter was written under the auspices of the U.S. Energy Research and Development Administration. By acceptance of this paper, the publisher and/or recipient acknowledges the U.S. Government's right to retain a nonexclusive, royalty-free license in and to any copyright concerning this paper.

I. Introduction*

Photosynthetic reaction centers are the places where primary photochemistry occurs in photosynthetic membranes [1]. Every reaction center consists of a primary electron donor (usually a chlorophyll dimer) and a primary electron acceptor associated with a specific protein. The primary donor in the excited state transfers an electron to the primary acceptor, and electronic excitation energy is thereby converted to chemical free energy. In some bacterial reaction centers an intermediate electron carrier (probably a bacteriopheophytin molecule) is known to be involved in the electron transfer from primary donor to primary acceptor [3-12]. In photosynthetic membranes reaction centers are combined with light-harvesting pigments which absorb most of the light and funnel the excitation energy to the reaction centers for photochemical conversion. Most of the chlorophyll in a membrane collects light and only a small fraction (1-10%) functions in reaction centers [1].

Reaction centers use either bacteriochlorophyll (Bchl) *a* or *b* or chlorophyll (Chl) *a* as primary donor [1]. All other types of chlorophyll and the bulk of Bchl *a* oe *b* or Chl *a* serve only as light-harvesting pigments. Reaction centers of purple bacteria contain either Bchl *a* or *b*, whereas those of green bacteria contain Bchl *a* only (Fig. 1). Cyanobacteria (blue-green algae) and chloroplasts of eukaryotes contain two types of reaction centers corresponding to photosystems I and II, as shown in Figure 2; both types contain Chl *a* as a primary donor.

When a chlorophyll dimer in a reaction center loses an electron to the primary acceptor (or the intermediate carrier), the red or far-red absorption band (Q_y transition) largely disappears [1]. Upon reduction the band reappears. The most sensitive method of detecting oxidation-reduction reactions of reaction center chlorophylls is optical absorption spectrophotometry. The names of the various reaction-center chlorophylls are therefore based on the wavelengths in nanometers of their Q_y bands: P680 or P690 (photosystem II), P700 (photosystem I), P840 (green bacteria), P870 and P890 (purple bacteria, Bchl *a*), and P960 (purple bacteria, Bchl *b*).

Besides being linked to light-harvesting pigments, reaction centers are also components of electron transfer chains (Figs. 1 and 2). The electron transfer components are assembled in the membrane along with the reaction centers and the light-harvesting pigments. In cyclic electron transfer chains, electron flow is coupled to phosphorylation and ATP is the net product of photochemistry.

*Three excellent reviews are recommended as an introduction to this chapter: "The Primary Photochemical Reaction of Bacterial Photosynthesis" by Parson and Cogdell [13], "Primary Events and the Trapping of Energy" by Sauer [1], and "Primary Electron Acceptors in Green Plant Photosystem I and Photosynthetic Bacteria" by Ke [12].

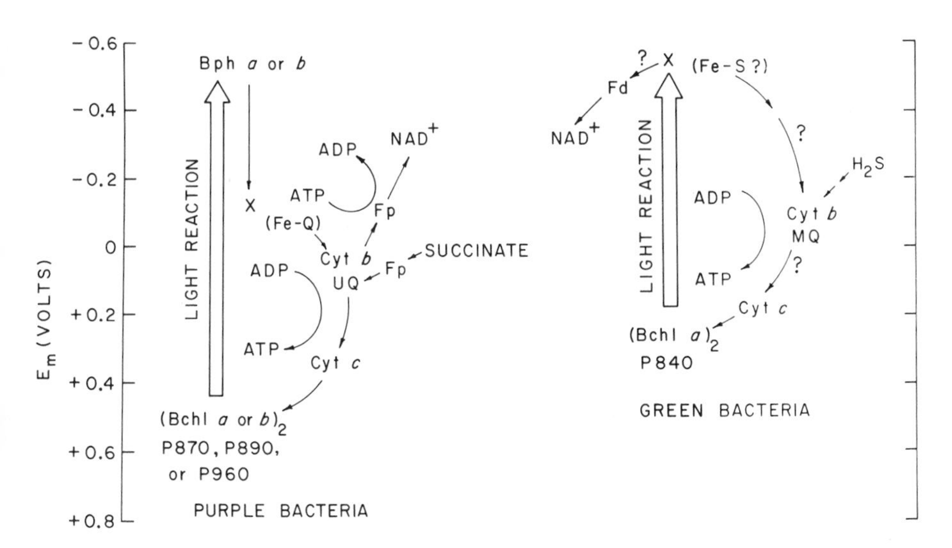

Figure 1 Electron transport diagrams for bacterial photosynthesis. Abbreviations used are Cyt, cytochrome; Fp, flavoprotein; Fd, ferredoxin; X, primary electron acceptor; UQ, ubiquinone; MQ, menaquinone.

In noncyclic electron flow, electrons (and protons) from an exogenous source are "pumped" up to the level of NADH (or NADPH). The NAD(P)H and ATP are consumed in CO_2 fixation and other energy-requiring biochemical reactions. In order to understand the primary event better and to obtain a description of the three-dimensional structure of the components involved, photosynthesis research has been aimed at isolation of photochemical reaction centers free of contaminating antenna chlorophylls and secondary electron donors.

Partial separation of the primary reactants from the antenna (light-harvesting) chlorophylls was first obtained by Kok [14] who found that P700 could be enriched in respect to bulk chlorophyll by careful extraction of photosynthetic membranes with organic solvents. However, it was studies on some purple bacteria [15-17] that first made it possible to describe tentatively the absorption spectrum of a reaction center and to attempt to fractionate the reaction center from the associated antenna. These studies showed that the antenna pigments could be destroyed photochemically or by oxidation while the reaction center remained functional and therefore presumably intact. (It is still not clear why the antenna pigments are more susceptible to destruction than those in the reaction centers.) In 1968 removal of the antenna pigments from the reaction center, rather than their destruction, was achieved by the use of a nonionic detergent, Triton X-100 [18, 19], and a new era in studies on the primary photochemical event began.

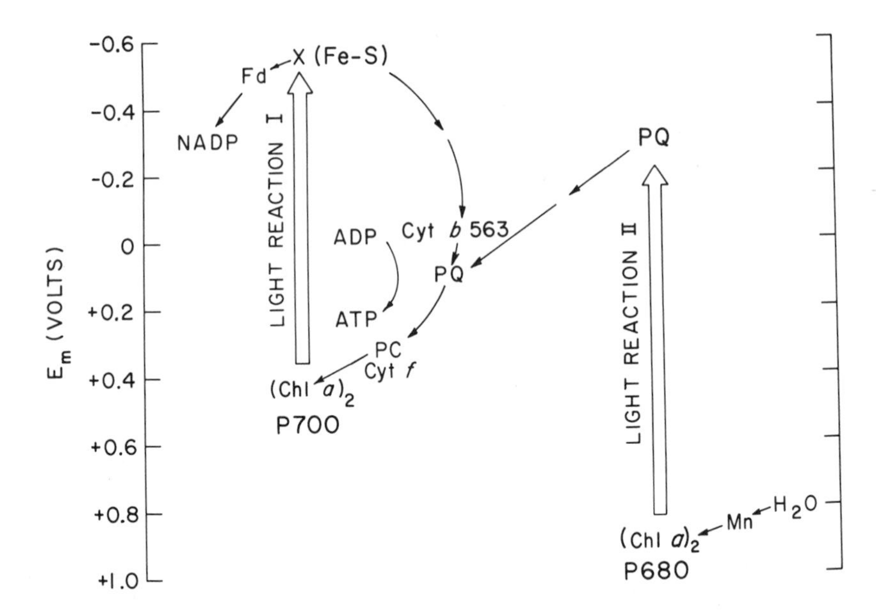

Figure 2 Electron transport diagram for oxygen-evolving photosynthesis. Abbreviations are the same as for Figure 1 and in addition PC, plastocyanin; and PQ, plastoquinone.

At present reaction centers from some purple bacteria can be isolated by the use of detergents and purified to a high degree. The reaction center from *Rhodopseudomonas sphaeroides* R-26 is the best characterized entity to date. The reaction-center particles from other purple bacteria still contain cytochromes and other bits and pieces of the electron transport chain. The best reaction-center preparations from cyanobacteria (blue-green algae) and plant chloroplasts still are contaminated with light-harvesting chlorophyll, and reaction centers from green bacteria have not even been separated from their unit membranes.

II. Purple Bacteria

The photochemical reaction center free of all antenna pigments has been obtained only from some purple sulfur and nonsulfur bacteria. A photochemical reaction center was first isolated from two mutants lacking colored carotenoids [18, 19]. It is not certain why the absence of colored carotenoids facilitates dissociation of the reaction-center entity from the antenna pigments by detergents, but it appears that the carotenoids might act as a "glue" between the different chlorophyll-containing entities. The absence of carotenoids was later shown *not* to be a prerequisite for successful isolation of the reaction center since reaction centers were obtained also from wild-type strains by fractionation of their detergent solubilized membranes with certain modifications in technique. However, highly purified photochemical reaction centers are obtained more readily from carotenoidless mutants than from their wild-type counterparts.

The fractionation techniques used to isolate the required component from solubilized membranes, and the purple bacteria that have yielded reaction-center preparations, are described below. Further details are available elsewhere [20].

A. *Rhodopseudomonas sphaeroides*

1. Isolation Procedures

Isolation of a component exhibiting near-infrared absorption peaks at 760, 800, and 870 nm (the 870-nm peak underwent reversible photobleaching), which previous experiments [14, 15, 17] had suggested were due to reaction-center pigments, was first carried out using a carotenoidless mutant, *Rps. sphaeroides* R-26. Reed and Clayton [19] used Triton X-100 as the membrane-solubilizing agent to obtain the first reaction-center preparation. Later, its replacement by the weakly zwitterionic detergent dodecyl dimethylamine oxide (LDAO) permitted subsequent fractionation techniques to remove all extraneous components from the final product [21, 22]. Typically, the reaction center is isolated by high-speed centrifugation of LDAO-solubilized photosynthetic membranes (chromatophores) followed by ammonium sulfate precipitation from the supernatant. The desired component can be further purified by mixing the precipitate with diatomaceous earth in a column and then eluting with an ammonium sulfate gradient of decreasing concentration [23]. The reaction center elutes at an ammonium sulfate concentration of 1 M. After removal of salt by dialysis, passage through a DEAE-cellulose column yields a homogeneous preparation of the reaction center.

Three different detergents have been used to solubilize the chromatophores of the wild-type organism. Fractionation of the resulting solution has yielded a reaction-center preparation as pure as that obtained from the R-26 mutant. The first detergent used was cetyltrimethylammonium bromide (CTAB) [24-26]; the next was sodium dodecyl sulfate (SDS), used alone [27] or in conjunction with alkaline-urea-Triton treatment [27-29]; more recently, LADO has been found to be preferable for strain Y [30]. Solubilization of the chromatophores is best achieved by incubation at 26°C in 100 mM sodium phosphate plus 0.25% LADO at pH 7.5. The pure component is isolated by ammonium sulfate fractionation, or by high-speed centrifugation followed by chromatography of the suprenatant on Sepharose 6B.

Reaction centers obtained with CTAB or LDAO can be further purified by

Table 1 Composition of Reaction Centers from *Rps. sphaeroides* R-26

Component	Molecular weight × 10^{-3}
1 Protein subunit L (light)	~ 21
1 Protein subunit M (medium)	~ 24
1 Protein subunit H (heavy)[a]	~ 28
4 Bchl *a*	3.6
2 Bph *a*	1.8
2 UQ_{10}	1.7
1 Fe^{2+}[a]	0.06
Total	≃ 80

[a]Not essential for photochemistry.

Sources: Feher and Okamura [31], Okamura et al. [23], Reed and Peters [173], Straley et al [36].

removal of 90% of the phospholipids present [169]. Incubation with sodium deoxycholate (11 mg/g protein) is used to solubilize the phospholipids, which are then separated on Sephadex G-75.

2. R-26 Reaction Center

a. *Composition.* The reaction center of mutant R-26 [23, 31] is composed of Bchl *a*, bacteriopheophytin (Bph) *a*, UQ_{10} (ubiquinone with 10 isoprenoid units in the side chain), iron, and protein in the ratios shown in Table 1. Since mutant R-26 lacks carotenoids, the isolated reaction center lacks the sphaeroidene present in the wild-type reaction center [32, 169]. Carbohydrate, labile sulfide, pteridine, and phospholipid are also missing from R-26 reaction centers.

Upon complete dissociation of the reaction center, three polypeptides (L, light; M, medium; H, heavy) are observed on SDS polyacrylamide gel electrophoresis (SDS-PAGE). The amino acid compositions of the individual subunits (Table 2) and of the reaction center as a whole (Table 3) show that the protein has a high content of nonpolar residues (71% for L*, 70% for M, and 62% for H). This is consistent with the reaction center being insoluble in water in the absence of detergent and with its location inside a membrane in vivo. Molecular weight estimates for the isolated reaction center range from 70 to 95 kilodaltons (kD).

If the H subunit is removed from the reaction center, the LM unit can still function as a photochemical entity, as shown in Figure 3. Almost the only

*The NH_2-terminal amino acid sequence of the L subunit is H_2N-Ala-Leu-X-Phe-Glu-Arg-Lys-Tyr-Arg-Val-Pro-Gly-Gly-Thr-Leu-Val-Gly-Gly-Asn-Leu-Phe-Asp-Phe- [33].

Table 2 Amino Acid Composition (mol %) of Reaction Center Subunits from *Rps. sphaeroides* R-26

	Subunit		
Amino acid	L	M	H
Gly[a]	12.0	11.7	10.3
Leu[a]	11.2	11.9	9.9
Ala[a]	9.7	9.9	10.5
Phe[a]	7.5	7.7	4.2
Asp	6.3	6.4	8.2
Ile[a]	6.2	5.1	5.4
Val[a]	6.0	5.5	7.8
Pro[a]	5.9	4.7	7.8
Trp[a]	5.7	7.3	0.8
Thr	5.2	4.4	4.7
Glu	4.8	6.5	8.2
Ser	4.4	5.6	5.0
Tyr[a]	4.3	3.3	2.5
Arg	3.3	4.0	4.4
Lys	2.4	0.9	5.1
His	2.3	2.1	2.3
Met[a]	1.6	3.1	2.0
½Cys[a]	1.3	0.0	0.9

[a]Nonpolar residues.
Source: Steiner et al. [156].

spectral difference between the LM unit and the LMH unit is, as expected, the height of the protein band at ~280 nm. The LM unit (~50 kD) is the smallest entity so far obtained which can be called a reaction center.

b. Spectral Properties. Absorption spectra of Reed's [34] 650-kD reaction-center preparation are shown in Figure 4. The absorption bands at 365 and 385 nm reflect contributions from both Bchl *a* and Bph *a* whereas the bands at 597, 803, and 865 nm belong to Bchl *a* alone, and the bands at 532 and 757 nm belong to Bph *a* alone. The bands at 277 and 681 nm belong to protein and irreversibly oxidized Bchl *a*, respectively. Upon illumination of the reaction center, the 803-nm band shifts to 790 nm, the 865-nm band disappears, and the 1245-nm band appears because of the formation of oxidized Bchl *a*. Associated with the oxidized Bchl *a* is an electron spin resonance (ESR) signal at $g = 2.0025 \pm$

Table 3 Amino Acid Composition (mol %) of Some Reaction-center-containing Preparations

Amino acid	Reaction center *Rps. sphaeroides* R-26[a]	P700-Chl *a*-Protein[b] *Phormidium luridum*	*Beta vulgaris*
Gly[c]	11.0	10.0	11.0
Leu[c]	10.8	11.2	11.4
Ala[c]	10.4	10.0	9.8
Asp	6.9	8.8	8.4
Glu	6.7	6.5	7.8
Phe[c]	6.7	6.2	6.8
Val[c]	6.5	6.2	6.3
Pro[c]	6.0	4.7	4.2
Ile[c]	5.5	6.2	6.5
Ser	4.9	5.9	5.5
Thr	4.8	5.9	5.2
Trp[c]	4.3	1.8	1.1
Arg	3.8	3.4	3.6
Tyr[c]	3.1	3.1	2.7
Lys	2.7	3.4	2.9
Met	2.6	1.9	1.4
His	2.4	4.0	5.9
½Cys[c]	0.7	0.9	0.3

[a]Data from Steiner et al. [156].
[b]Data from Thornber [103, 127].
[c]Nonpolar residues

0.0002 with a bandwidth of 9.6 ± 0.2 gauss [35]. The narrow bandwidth suggests that the odd electron of the oxidized state is shared between two Bchl *a* molecules. Electron nuclear double resonance (ENDOR) experiments have proved that this is so [31].

The 802- and 865-nm bands may be used to determine the reaction-center (4 Bchl *a* + 2 Bph *a*) concentration in a reduced preparation, since $\epsilon_{802} = 288 \pm 14$ mM^{-1} cm^{-1} and $\epsilon_{865} = 128 \pm 6$ mM^{-1} cm^{-1}. Alternatively, reversible bleaching of the 865-nm band [$\Delta\epsilon = 112 \pm 6$ mM^{-1} cm^{-1}] can be used [36].

c. *Function* [3, 5, 31, 37]. When a photon is absorbed in a photosynthetic unit (reaction center and its associated light-harvesting pigments), an excited singlet state (neutral exciton) is created which finds its way to the reaction center where the excited state becomes localized on a Bchl *a* dimer, the primary electron donor. In less than 6 psec an electron moves from $(\text{Bchl } a)_2{}^*$, the ex-

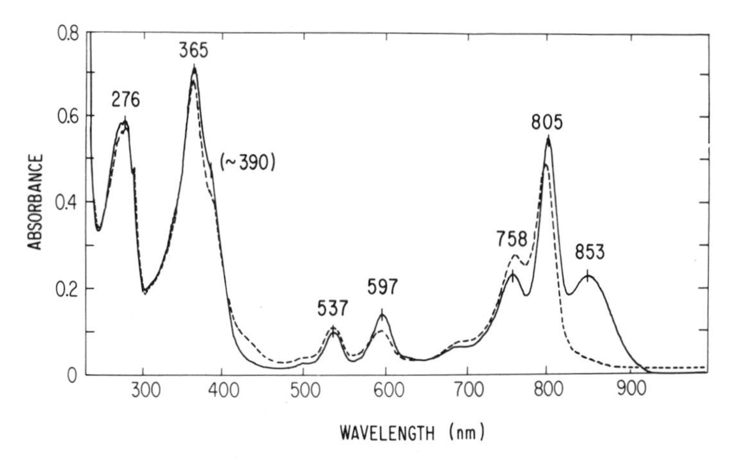

Figure 3 Absorption spectra of the LM reaction-center from *Rps. sphaeroides* R-26 [23]. The solid curve with labeled maxima is for reaction centers in the dark. The dashed curve is for reaction centers illuminated by white actinic light. Reproduced with permission from Feher and Okamura [31].

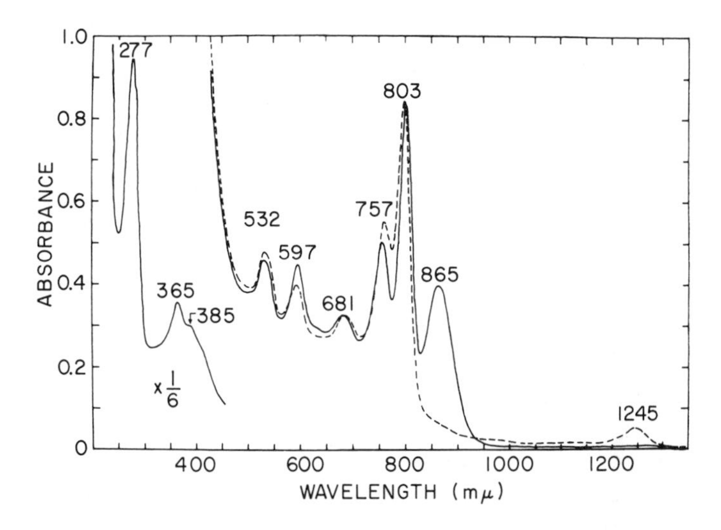

Figure 4 Absorption spectra of the 650-kD reaction-center complex from *Rps. sphaeroides* R-26 in 10 mM Tris, pH 7.5. The solid curve with absorption maxima is for the dark. The dashed curve with a maximum at 1245 nm is for illumination with 800-nm actinic light. Reproduced with permission from Reed [34].

cited dimer, to a single Bph *a* molecule (I), an intermediate electron carrier. Approximately 200 psec later the electron moves on to a single UQ_{10} molecule, the primary electron acceptor (X), to form ubisemiquinone. The UQ_{10} is magnetically coupled to an Fe^{2+}, which may be removed or replaced by Mn without complete loss of function.

The ESR signal associated with the reduced primary acceptor ($UQ_{10}{}^{\cdot -}$ Fe) is very broad at g = 1.82 [38] whereas the reduced intermediate carrier (Bph $a^{\cdot -}$) [31] has a much narrower signal ($\Delta H = 12.9 \pm 0.3$ gauss) at $g = 2.0036 \pm 0.0002$. These ESR characteristics together with the results of ENDOR experiments show that the reduced intermediate carrier is a monomeric tetrapyrrole anion [39].

The energy of the excited state, (Bchl $a)_2{}^*$, is 1.4 eV. When the electron moves from (Bchl $a)_2{}^*$ to Bph *a* to form, about 30% of the energy is lost as heat, and the remaining 1.0 eV is stored as charge separation [4]. (See Table 5, which appears at the end of Section II.F, for the redox potentials of various electron carriers.) As the electron moves to the primary acceptor X, a further 0.4-0.5 eV is lost, so that the amount converted to chemical free energy in the isolated reaction center is ~0.6 eV (overall efficiency $\simeq$ 40%). (For a discussion of the "lost" 0.4-0.5 eV, see Dutton et al. [3] and Fajer et al. [5].) This sequence of electron transfers and energy conversion steps is summarized in Figure 1 (*purple bacteria*).

d. Localization. The H subunit, the least hydrophobic subunit of the reaction center, is at least partially exposed to the outer (cytoplasmic) surface of the chromatophore membrane, whereas the LM subunit appears to span the entire membrane [31]. According to the model of Dutton and coworkers [3], the primary acceptor UQFe is close to the outer surface, as shown in Figure 5, and donates its electron to the second acceptor (also UQ), which then takes up a proton from the outer aqueous phase. The primary donor (Bchl $a)_2$ is within the membrane, and cytochrome c_2 is at the interface of the membrane with the inner aqueous phase.

3. Wild-type (Strain Y) Reaction Center

a. Composition [30, 169]. The reaction centers made with LDAO or CTAB contain 4 Bchl *a* and 2 Bph *a*, as does the R-26 reaction center. The delipidated wild-type reaction center also has on the average ~0.7 molecule of sphaeroidene, which may or may not be an integral part of it [169] (cf. Ref. 21). The conformation of the sphaeroidene bound to the wild-type reaction center probably differs from that of the bulk carotenoid of the chromatophore [32]. The wild-type reaction center also contains three protein subunits (L = 22 kD, M = 24 kD, and H = 27 kD). Before delipidation the wild-type reaction center contains 13%

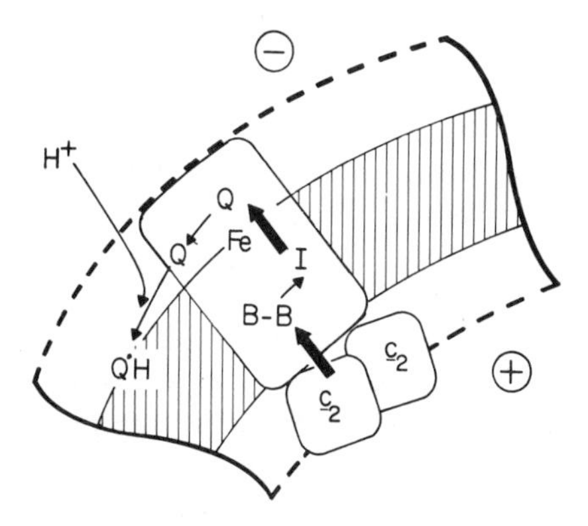

Figure 5 Schematic representation of the reaction center and associated electron transport components in the chromatophore membrane of *Rps. sphaeroides*. Q, UQ; B, Bchl *a*. Reproduced with permission from Dutton et al. [3].

phospholipid* by dry weight and has a particle weight of 150-170 kD, which includes lipid and detergent [30]. The maximum protein content is 85 kD [169]. Each reaction center contains ~2 molecules of UQ_{10} and only 0.1 *c*-type heme per P870 [30].

b. Spectral Properties [26]. The absorption spectra of wild-type reaction centers before delipidation are similar to those of R-26 reaction centers (Figs. 3 and 4) with additional bands at 440, 470, and 505 nm reflecting residual carotenoids (sphaeroidene and sphaeroidenone) bound to the reaction center. The ratio of A_{285} to A_{365} ($\simeq$1.0) indicates a protein-to-pigment ratio comparable with that in R-26 reaction centers [26, 30]. The light-induced optical changes are essentially the same as in R-26 reaction centers.

c. Redox Potential [26]. The midpoint potential of P870 in isolated wild-type reaction centers (made with CTAB) is +525 mV at pH 7.5, whereas it is +445 mV in intact chromatophores (see Table 5, Section II.F). This 80-mV difference reflects a change in environment and/or conformation for the reaction center upon removal from the chromatophore membrane.

*Major lipids are phosphatidylethanolamine, phosphatidylglycerol, phosphatidylcholine, and an unidentified component.

B. *Rhodospirillum rubrum*

1. Isolation Procedures

The isolation of the photochemical reaction center from *R. rubrum* parallels that from *Rps. sphaeroides.* Gingras and Jolchine [19] obtained the first preparation from a carotenoidless mutant (G-9) by fractionation of chromotophores dissolved in Triton X-100. Wang and Clayton [40] (see also Oezle and Golecki [170] used LDAO in place of Triton. Sucrose density gradient centrifugation of the LDAO extract, followed by gel filtration of the P870-containing zone from the gradient to remove extraneous proteins, provided a preparation severalfold purer than that made by the original procedure [20]. Okamura et al. [23] obtained a pure reaction-center preparation by centrifugation of LDAO extracts at 250,000 X g for 1 hr followed by ammonium sulfate fractionation of the supernatant (as with *Rps. sphaeroides*). Instability of the isolated *R. rubrum* reaction center preparation has been a problem. Gingras [20] refers to an unpublished method for obtaining a stable product, in which LDAO is used for initially solubilizing chromatophores but is exchanged for Triton X-100 during subsequent purification steps on DEAE-Sephadex and DEAE-cellulose.

With wild-type *R. rubrum,* SDS solubilization of chromatophores gives a solution from which a reaction-center preparation has been obtained by sucrose density gradient centrifugation followed by Sephadex G-100 gel filtration [41]. The presence of a large excess of sodium ascorbate during the preparative procedure was found beneficial, but the product contained extrancous proteins. Noël et al. [42] improved the procedure: following solubilization in buffered sodium phosphate plus 0.25% LDAO, the reaction center was obtained by ammonium sulfate fractionation of the resulting solution, and excess LDAO, which alters the absorption spectrum and the photochemical activity of the preparation, was removed by dialysis of the precipitate. A further improvement was achieved by chromatographing the redissolved precipitate on DEAE-cellulose in the presence of 0.1% LDAO [20]. Removal of excess LDAO by dialysis or gel filtration is essential for stability of the product. The isolated material is considered to be pure since it yields only the three polypeptides (L, M, and H) characteristic of purple bacterial reaction centers (see below) on complete dissociation.

2. Reaction Centers

The reaction center isolated from the carotenoidless mutant (G-9) is also made up of three protein subunits (L = 22 kD, M = 26 kD, and H = 29 kD), slightly larger than their R-26 counterparts [23, 43]. Its optical spectra are nearly identical to those of mutant R-26 (Figs. 3 and 4) except that A_{275}/A_{365} is 0.95 for G-9 and 0.88 for R-26 [23]. The G-9 reaction center has a calculated minimum mass of 85 kD [20], and in the chromatophore it appears to be partially exposed on the cytoplasmic side of the membrane [43].

The wild-type *R. rubrum* reaction center made with LDAO consists of 90% protein and 7% pigment by dry weight and has a minimum mass of 90 kD [20]. The experimental particle mass (including lipid and detergent) is 140 kD [42]. The protein is composed of three subunits (L = 21 kD, M = 24 kD, and H = 32 kD) [42], the only NH_2-terminal residue is alanine [41] as in the L subunit of the R-26 reaction center. The wild-type *R. rubrum* reaction center contains no cytochrome and only 0.3 molecule of UQ_{10} on the average [42], but a more recent determination indicates 1 molecule of firmly bound UQ_{10} (G. Gingras, unpublished result, 1977); therefore, in the *R. rubrum* reaction center UQ_{10} probably functions as the quinone in the primary electron acceptor. The reaction center contains 4 Bchl *a*, 2 Bph *a*, and 1 spirilloxanthin per P870 [171].

Smith [41] and Smith et al. [68] found ~1 molecule of spirilloxanthin in the SDS reaction center. This carotenoid is responsible for the absorption bands at 470, 498, and 534-535 nm [42]. (A small band at 680 nm is ascribed to oxidized Bchl *a*.) The light-induced absorbance changes (Fig. 6) are almost identical to those observed in R-26 reaction centers, including the appearance of the (Bchl $a)_2{}^{\cdot +}$ band at ~1250 nm. (Another weaker band appears at ~1140.)

The midpoint redox potential of the primary electron acceptor X may be as low as -370 mV in the larger (100-kD) reaction-center particle [44].

C. *Rhodopseudomonas capsulata*

Reaction centers have been prepared from the carotenoidless mutant *Ala*$^+$ (= *Ala*, *car*$^-$, *bchl*$^+$, *pho*$^+$) of *Rps. capsulata* by the use of Triton X-100 [45] or LDAO [46, 47]. Triton reaction centers contain three protein subunits (L = 22.5 kD, M = 27.5 kD, and H = 32 kD) which account for 80% of the protein in the preparation [45]. The absorption spectrum indicates that a certain amount of light-harvesting Bchl *a* is associated with the reaction center but is irreversibly bleached by strong light. After illumination the reduced reaction centers' absorption spectrum ($A_{750} = A_{855}$) looks similar to that for R-26 reaction centers (Figs. 3 and 4) with a peak at 750 nm ascribed to Bph *a* and peaks at 798 and 855 nm due to Bchl *a*. Upon further illumination (oxidation) the 855-nm band disappears and the 798-nm peaks shifts to the blue [45].

The LDAO reaction centers are not so contaminated with light-harvesting (antenna) Bchl *a* and appear to be generally cleaner preparations than Triton reaction centers. They are free of cytochrome *c* [46, 47], and all the protein consists of three subunits (L = 20.5 kD, M = 24 kD, and H = 28 kD), the H subunit being unpigmented. Since the subunits exist in a 1:1:1 ratio, the reaction center must therefore have a minimum size of 72.5 kD. The Bchl *a*, Bph *a*, and P870

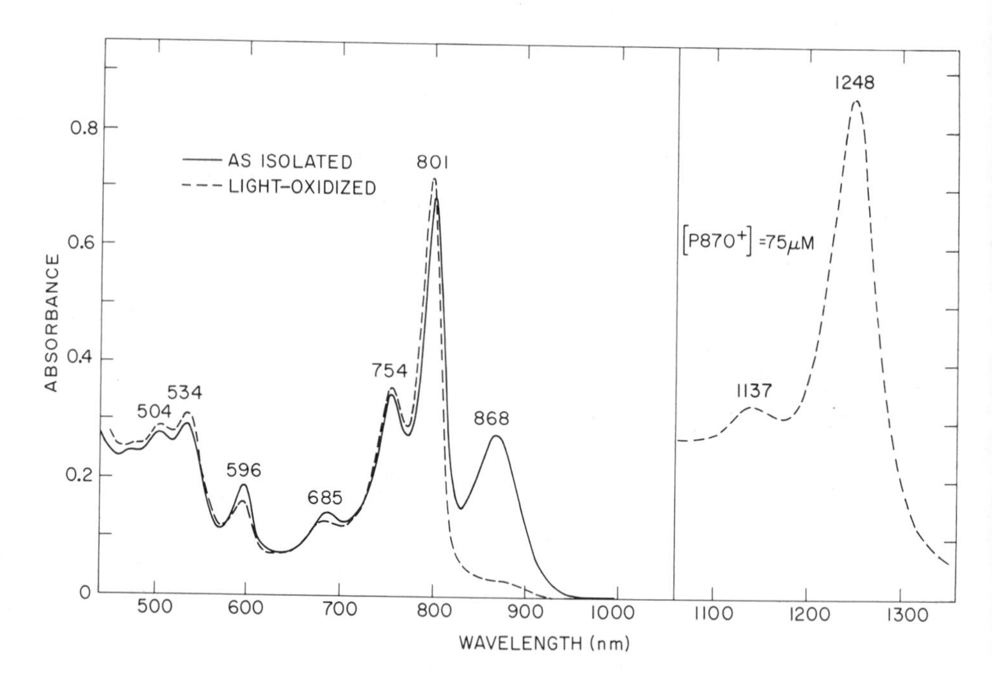

Figure 6 Absorption spectra of an LDAO reaction-center preparation from wild-type *R. rubrum.* The solid curve with labeled maxima is for reaction centers in the dark; the dashed curve on the left side is for reaction centers in actinic light. The dashed curve with labeled maxima on the right is for a 75-μM solution of reaction centers oxidized by ferricyanide.

(850) have levels of 15.5, 11.4, and 7 nmol/mg protein, respectively, and therefore the ratio is 2.2:1.6:1.0. The reaction center appears to contain two LMH trimers associated with 1 P870, 1-2 Bph *a*, ~2 Bchl *a* [46], and ~1.5 UQ_{10} [47]. This corresponds to a minimum of ~150 kD.

Reaction centers made with LDAO have an absorption spectrum (reduced) similar to that of R-26 reaction centers, with peaks at 365 (Bchl *a* + Bph *a*), ~540 (Bph *a*), ~600 (Bchl *a*), ~700 (oxidized Bchl *a*), ~750 (Bph *a*), ~800 (Bchl *a*), and ~850 nm (Bchl *a*) [47]. In the best preparations, A_{280}/A_{802} = 1.25. Maximum bleaching upon illumination occurs at 850 nm instead of at 870 nm [46].

These reaction centers can oxidize mammalian ferrycytochrome *c* and *Rps. capsulata* ferrocytochrome c_2 [47].

Despite the great similarity between the reaction centers from *Rps. sphaeroides* R-26, *R. rubrum,* and *Rps. capsulata,* there is no antigenic cross-reaction between anti R-26 and reaction centers from either *R. rubrum* or *Rps. capsulata* [47].

D. *Rhodopseudomonas gelatinosa*

Reaction centers (~150 kD) prepared from a carotenoidless strain by the use of LDAO have been found to contain only two protein subunits (24 and 34 kD) instead of the usual three [48]. In other respects these reaction centers appear similar to those previously described (see Table 5, Section II.F).

E. *Chromatium vinosum*

1. Isolation Procedures

Methods have been published for isolating reaction centers from *Chr. vinosum* that are similar to these described above [45-47, 49]. Recently a different approach has been used. After a fraction considerably enriched in P883 is obtained by differential centrifugation of Triton X-100 extracts, cold acetone (~200 K) is added to remove the remaining antenna pigments [10, 11, 50, 51]. Preparations made this way, although spectrally pure, are likely to contain extraneous proteins. They are stable for weeks at 0°C and for months at -10°C [10].

2. Reaction Center

a. Composition. The reaction-center particles isolated from *Chr. vinosum* differ from those isolated from *Rps. sphaeroides, Rps. capsulata, Rps. gelatinosa,* and *R. rubrum* in that two *c*-type cytochromes, one of high potential (c_{555}) and the other of low potential (c_{553}), are firmly bound to the reaction center of *Chr. vinosum.* These cytochromes serve as secondary electron donors and give the reaction-center particle the added function of a photoelectron transport particle [52]. At low redox potentials, cytochrome c_{553} (two hemes per primary donor) donates electrons to $P883^+$ at all temperatures [10, 11]; at higher potentials, cytochrome c_{553} is oxidized and cytochrome c_{555} (also two hemes per primary donor) takes over as the secondary donor [53, 54]. In a particle of ~700 kD containing ~10 Bchl *a* per reaction center, the ratio of cytochrome c_{553} to c_{555} is 3:2 [55]. The reaction center also contains Bchl *a,* Bph *a,* one menaquinone [56], and carotenoid (see Table 5, Section II.F).

The ~700-kD particles from *Chr. vinosum* have also been analyzed for protein subunits by SDS-PAGE. Five bands (12, 22, 27, 30, and 45-50 kD) were observed, of which one (45-50 kD) was associated with cytochromes c_{553} and c_{555}. (Five bands were also observed with LDAO reaction centers [49].) The middle three bands (22, 27, and 30 kD) were compared with the L, M, and H bands (22, 23, and 27 kD) of R-26 reaction centers [55].

b. Spectral Properties. LDAO reaction centers from *Chr. vinosum* show a three-peak spectrum (reduced) in the near infrared similar to the spectra of those from nonsulfur bacteria (see Figs. 3, 4, and 6). Peaks at 366 and 386 nm are ascribed to the combined effects of Bchl *a* and Bph *a*. Peaks due to Bchl *a* alone appear at 589, 799, and 875 nm; peaks due to Bph *a* alone appear at 530 and 756 nm. Cytochromes are responsible for the 404-nm peak, and oxidized Bchl *a* for the 689-nm peak. Carotenoids account for the peaks at 445, 480, and 503 nm which are missing in the spectra of reaction centers from green (carotenoid-deficient) cells [49].

c. Function. Upon illumination at redox potentials between -100 and +200 mV the 875-nm band disappears and the 799-nm band shifts to 796 nm to give a difference spectrum (Fig. 7) which is typical for Bchl *a*-containing purple bacteria. Oxidized (Bchl $a)_2$ and oxidized cytochrome c_{553} (at low potentials) or c_{555} (at high potentials) accumulate in the light. Recovery in the dark is rapid in the absence of exogenous mediators or reductants [49].

Shuvalov and Klimov [8] observed that when a B890 complex (B890/P890 = 32) prepared from chromatophores of *Chromatium minutissimum* with Triton X-100 (2.5-3%) is illuminated at redox potentials between -250 and -530 mV and at temperatures between 77 and 293 K, absorption bands disappear at 364 (Bchl *a* + Bph *a*), ~390 (Bchl *a* + Bph *a*), ~543 (Bph *a*), ~595 (Bchl *a*), ~760 (Bph *a*), and ~805 nm (Bchl *a*), and new bands appear at 423, ~600 (broad), and 790 nm (Bchl *a*) (cf. Fig. 8). They made similar observations with chromatophores and B890 complexes of *Thiocapsa roseopersicina* and *Rhodopseudomonas viridis* reaction centers (which also have firmly bound *c*-type cytochromes) but not with chromatophores, B870 complexes, or reaction centers of *Rsp. sphaeroides, R. rub-*

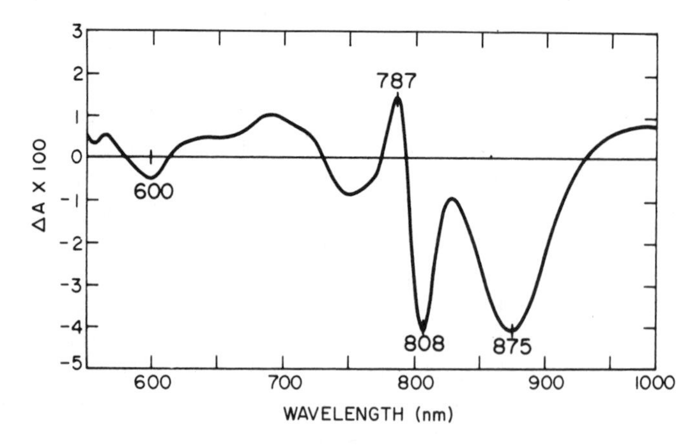

Figure 7 Oxidized (ferricyanide)-minus-reduced (dithionite) difference spectrum of LDAO reaction centers from *Chr. vinosum*. Reproduced with permission from Lin and Thornber [49].

rum, or *Rhodopseudomonas palustris* (which do not have firmly bound cytochromes). (However, Okamura et al. [57] have observed these changes in reaction centers of *Rps. sphaeroides* R-26 to which cytochrome *c* has been added.) Shuvalov and Klimov [8] proposed a light-induced electron transfer from cytochrome *c* to Bph *a* via (Bchl $a)_2$ in the B890 complex under conditions such that the primary acceptor is already reduced in the dark. The midpoint potential of Bph *a* was estimated to be below -620 mV. The ESR signal associated with Bph $a\cdot^-$ had a g value of 2.0025 and a linewidth of 12.5 gauss (see Table 6, which appears at the end of Section II.F).

Dutton et al. [3] and Tiede et al. [10, 11] independently carried out similar experiments at 200 K with Triton-acetone reaction centers from *Chr. vinosum* and essentially confirmed the Russians' observations. To explain the absorbance changes, Tiede et al. [10, 11] proposed that one electron was transferred from cytochrome c_{553} to an intermediate carrier (I) between P883 and X. The carrier obviously involved both Bph *a* (543 and 757 nm) and Bchl *a* (595 and 802 nm). The ESR signal for the reduced I state at 7 K consisted of two superimposed signals at g = 2.003: a narrow signal (ΔH = 15 gauss) and a doublet separated by 60 gauss. Tiede et al. offered two interpretations of the optical and ESR data: (1) that the reduced state of I is shared between one Bph *a* and one Bchl *a* as (Bph $a \cdot$ Bchl $a)\cdot^-$, or (2) that the electron remains on Bph *a* alone with an electrochromic effect on the Bchl *a* (802 → 785 nm). Van Grondelle et al. [51] favored the latter interpretation and showed that around one Bph *a* per reaction center (LDAO-acetone) was reduced in the light. The absorbance changes in the near infrared (see Fig. 8) were explained in terms of a bleaching centered at 758

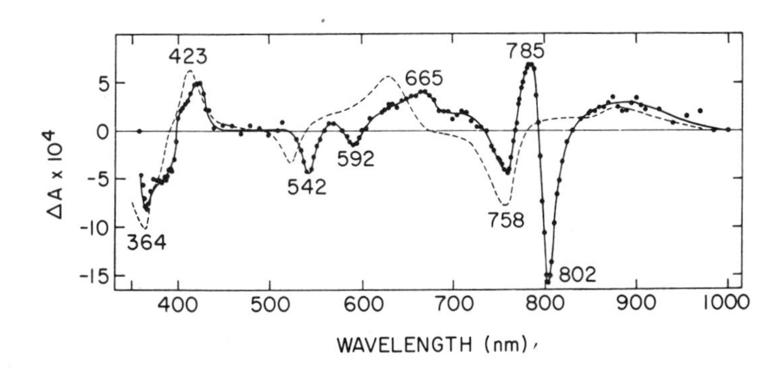

Figure 8 Solid curve: reversible light-induced absorbance changes obtained with chemically reduced reaction centers of *Chr. vinosum* at –15°C. [P870] = 0.16 μM, light path = 1 mm, [Na dithionite] = 10 mM, [TMPD] = 5 μM. Dashed curve: difference spectrum in vitro for reduction of one Bph *a* per P870. Reproduced with permission from van Grondelle et al. [51].

nm and an electrochromic shift of a narrow band at 800 nm to a broader band at 785 nm. Comparison of the optical and ESR data for the I^- state in-vitro data for Bchl *a* and Bph *a* [5] supported this view.

Redox potentials for the various electron carriers in reaction centers, photoactive complexes, and chromatophores are listed in Table 5 (at the end of Section II.F).

F. *Rhodopseudomonas viridis and Thiocapsa pfennigii*

1. Isolation Procedures

The chlorophyll present in the isolated photochemical reaction centers described above is Bchl *a* (+ Bph *a*). Reaction-center preparations have also been made from *Rps. viridis* and *Th. pfennigii* which contain a different chlorophyll, Bchl *b* [5]. This pigment in vitro, and particularly in vivo, absorbs further into the near infrared than any other chlorophyll; therefore studies of some reaction-center characteristics can be made more easily on these two organisms (see below).

The first reaction-center preparation from a carotenoid-containing organism was made from *Rps. viridis* [58]. Photosynthetic lamellae were dissociated by addition of SDS, and the solubilized material was adsorbed to hydroxylapatite. A fraction containing a reaction center (P960) was eluted by a relatively high (0.2 M) concentration of sodium phosphate, and was further purified by ammonium sulfate fractionation [58, 59]. The P960-containing product was contaminated with a 685-nm absorbing pigment arising from chemical alteration of a few antenna Bchl *b* molecules closely associated with P960 which absorb at 1015 nm in intact lamellae (cf. Pucheu et al. [59]). This contaminant was avoided in two ways. Trosper et al. [60] modified the procedure mainly by including sodium dithionite in all P960-containing solutions, after addition of SDS to the lamellae, until the final step. However, substitution of LDAO for SDS provides a surer way of obtaining a spectrally pure reaction-center preparation [61, 62]; furthermore, the primary reaction in preparations made with LDAO functions more like that in whole lamellae. Pucheu et al. [59] were the first to use LDAO to obtain the reaction center of *Rps. viridis.* They fractionated the LDAO extract with ammonium sulfate and further purified it by DEAE-cellulose chromatography and sucrose density gradient centrifugation; alternatively, a dialyzed LDAO extract can be repeatedly chromatographed on DEAE-cellulose. Thornber et al. [62] used a similar procedure: after dilution or dialysis of the LDAO extract, they used chromatography on DEAE-cellulose followed by chromatography on hydroxylapatite for rapid isolation of the reaction center. Cytochromes are so tightly associated with the isolated reaction-center protein that no method has yet been found for dissociating them without destroying the primary photochemical reaction; however, their presence has been used to advantage (see Section II.F.2.c).

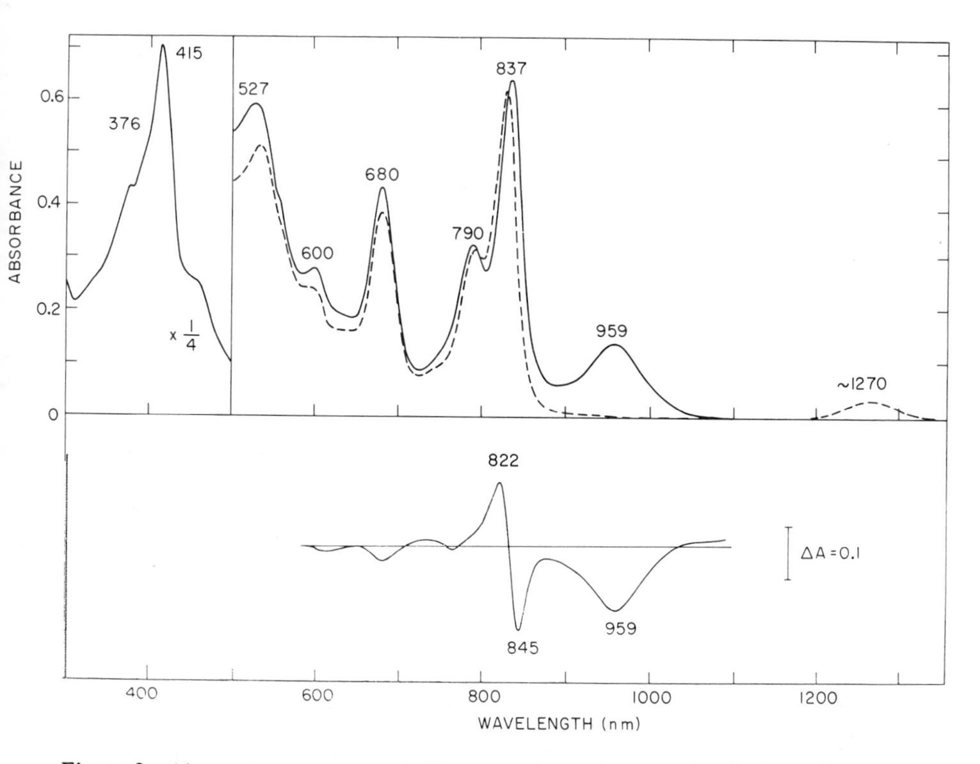

Figure 9 Absorption spectra and difference absorption spectrum of an LDAO reaction-center preparation from *Th. pfennigii* 9111. In the upper part of the figure the solid curve with labeled maxima is for the dark, and the dashed curve with a peak at ~1270 nm is for illumination with white actinic light. The curve in the lower part is the light-minus-dark difference spectrum, obtained by shining blue actinic light (Corning 5-57 filter) on one of two identical samples.

Isolation of the reaction center from *Th. pfennigii* has not progressed as far as that from *Rps. viridis.* A reaction-center preparation has been obtained with LDAO [62], but it still contains the 685-nm absorbing component and also is unstable. The absorption spectrum (Fig. 9) is noticeably different from that for *Rps. viridis* reaction centers (Fig. 10).

2. Reaction Centers

a. Composition. The reaction-center particles isolated from *Rps. viridis* and *Th. pfennigii* are similar to those isolated from *Chr. vinosum* in that they contain two kinds of *c*-type cytochrome firmly bound to the reaction center (see Table 5 on pages 302-309). LDAO and SDS reaction centers of *Rps. viridis* contain

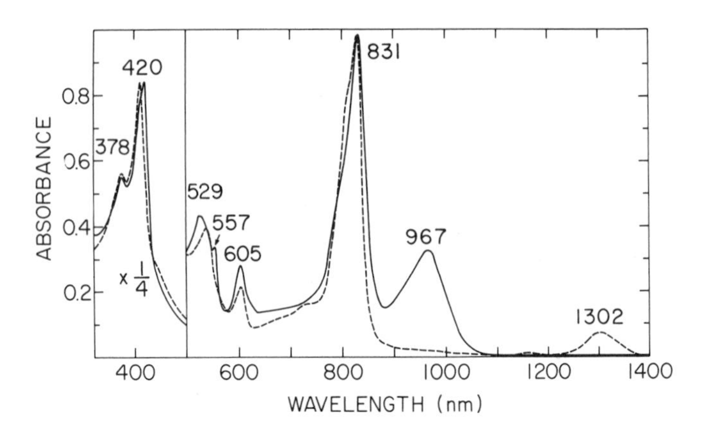

Figure 10 Absorption spectra of an LDAO reaction-center preparation from *Rps. viridis* in 50 mM Tris, pH 8.0. The solid curve is for the preparation reduced with sodium ascorbate, and the dashed curve is for reaction centers illuminated with white actinic light. Absorbance ratios for reduced reaction centers are given in Table 5.

Bchl *b* and Bph *b* in a 2:1 ratio [62], low-potential cytochrome c_{553}, high-potential cytochrome c_{558} [52], and a quinone similar to menaquinone [61]. Ubiquinone is not detectable in the LDAO reaction center even though it is present in the photosynthetic membranes. The reported ratios of components in various reaction-center preparations are listed in Table 4.

The LDAO reaction center appears to lack the L subunit, but it can be broken down into four other protein subunits: 23 ± 1.5 kD, 29 ± 1.5 kD, 37 ± 1.5 kD, and 45 ± 1.5 kD. The subunit stoichiometry seems to be either 2:2:1:2 (minimum 231 kD) or 1:1:1:2 (179 kD) [61]. The experimental particle weight by gel filtration (220-250 kD) agrees rather well with the calculated value of 231 kD, but the composition calculated for a 240-kD particle would include only 0.3 mol of the menaquinone-like substance [61] instead of the 0.6 mol per 6 Bchl *b* shown in Table 4. SDS reaction centers have an experimental particle size found by SDA-PAGE of ~110 kD [58]. The discrepancy between these two values remains to be clarified.

b. Spectral Properties. Absorption spectra of LDAO reaction centers in the presence of ascorbate are shown in Figure 10. In the dark, absorption bands due to Bchl *b* appear at 378, 605, 831, and 967 nm. Upon illumination or chemical oxidation of the reaction center, the 967-nm band disappears, the 831-nm band shifts to 828 nm, and a new band appears at 1302 nm presumably due to the cation radical of (Bchl $b)_2$. At 77 K the 831-nm band is resolved into two peaks

Table 4 Composition of Reaction Centers from *Rps. viridis* (molecules per reaction center)

Component	LDAO[a]	SDS[b]	LDAO or SDS[c]
Bchl *b*	5.9	–	4
Bph *b*	–	–	2
P960	–	1	1
Cytochrome c_{558}	2.1	2	2
Cytochrome c_{553}	0.7	2-3	2.6
MQ-like quinone[d]	0.6	–	–
Dihydrolycopene	–	Present	Present

[a]Data from Pucheu et al. [61].
[b]Data from Trosper et al. [60].
[c]Data from Thornber et al. [62].
[d]MQ, menaquinone.

at 790 and 832 nm and a shoulder at ~850 nm (Fig. 11); the 967-nm band shifts to 987 nm. In the illuminated state the 828-nm band is also resolved at 77 K into a shoulder at 790 nm and two peaks at 808 and 829 nm. Cytochromes (reduced) are responsible for the peaks at the following wavelengths: 420, 522, and 557 nm. Carotenoid peaks are found at 452, 485, and 514 nm [60]. The bands or shoulders at 544 and 790 are ascribed to Bph *b* [60, 62]. (Unfortunately, the peaks at 522, 452, 485, and 514 nm do not appear in Fig. 10.) Although the 790-nm band appears as a shoulder on the 830-nm peak in *Rps. viridis* reaction centers at room temperature, it is clearly separated from the 837-nm peak in *Th. pfennigii* reaction centers (Fig. 9), as is the 750-nm peak clearly separated from the 800-nm peak in Bchl *a*-containing reaction centers.

c. *Function.* When LDAO reaction centers are illuminated and then darkened, both light-on and light-off kinetics are rapid [61]. In contrast, for SDS reaction centers the dark recovery may take up to 30 min unless phenazine methosulfate (20 μM) or sodium ascorbate (50 μM) is added to the medium [52, 58]. At potentials above 0 mV, cytochrome c_{553} (E_{m8} = -20 mV [63]) is oxidized in the dark and cytochrome c_{558} (E_{m8} = +330 mV [63]) serves as the secondary electron donor to P960 (E_{m8} = +330 to 450 mV [63]) in both preparations. There are two c_{558} hemes per reaction center, and each may donate one electron to P960 [6]. Quantum requirements for the oxidation of P960 and of cytochrome c_{558} in SDS reaction centers are about two and three photons per electron respectively [63]. At potentials below 0 mV, both cytochromes are reduced and c_{553} replaces c_{558} as secondary donor (cf. reaction center of *Chr. vinosum* in Section II.E.2.c).

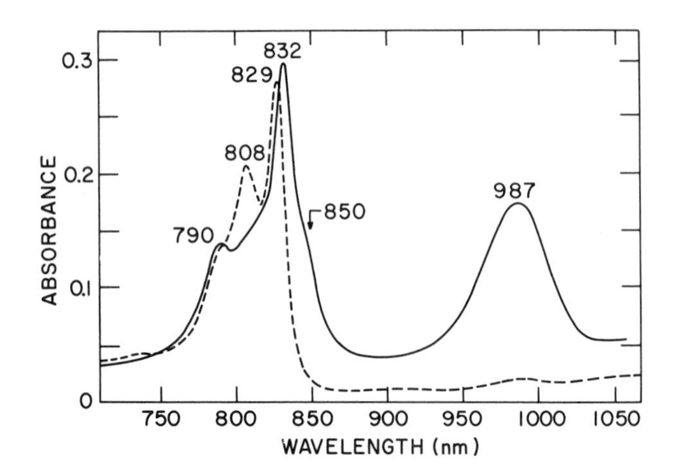

Figure 11 Absorption spectra (77 K) of an SDS reaction-center preparation from *Rps. viridis.* The solid curve is for material reduced with sodium ascorbate in the dark, and the dashed curve is for material exposed to strong actinic light immediately before freezing. Samples were dissolved in 55% glycerol and 22.5 mM Tris, pH 8.0. The optical path was 2 mm. Reproduced with permission from Trosper et al. [60].

The E_m value for P960 in the "best" reaction-center preparations is 50 mV lower than in the membrane (see Table 5). There is no significant difference between the E_m values for cytochrome c_{558} in or out of the chromatophore membrane. For cytochrome c_{553} in reaction center preparations the E_m value varies above and below the value in the membrane.

It has been assumed that the primary electron donor P960 is a Bchl *b* dimer equivalent in structure to the Bchl *a* dimer of P870, P890, etc., or the Chl *a* dimer of P700 [2, 64] (see Section IV.A.2.c). However, the light-induced ESR signal (g = 2.0026) corresponding to $P960^+$ in chromatophores and LDAO reaction centers has a linewidth of ~12 gauss [5, 6, 62], which is halfway between the values for a monomer (14 gauss) and a symmetrical dimer ($1/\sqrt{2} \times 14 = 10$ gauss) [5]. The 12-gauss bandwidth is interpreted to indicate an asymmetric pair of Bchl *b* molecules in P960 [5, 62].

The intermediate electron carrier (I) between P960 and X (= QFe) has been studied by using approaches identical to those used with *Chr. vinosum* preparations. Several groups [6, 9, 62, 65, 66] have examined light-induced optical and ESR changes in chromatophores and reaction centers in the presence of sodium dithionite. The data point to the involvement of a Bph *b* molecule as the immediate acceptor of an electron from P960*, but the electron might possibly be shared between Bph *b* and Bchl *b*. Not only are absorbance decreases at 544 and 790 nm (Bph *b* → Bph $b^{\cdot -}$) observed under reducing conditions, but decreases also occur at 840 and around 600 nm (Bchl *b*) [9, 67] which appear to be equivalent to those at 802-805 nm and 595 nm (Bchl *a*) in *Chr. vinosum*. The ESR. signal associated with the reduced I state at temperatures below 15 K consists of two components (as in *Chr. vinosum*): a narrow signal ($\Delta H \simeq 15$ gauss) and a doublet separated by ~100 gauss [3, 10, 11]. The ESR characteristics (see Table 6) along with ENDOR results show that I is a monomeric tetrapyrrole anion [62]. For the midpoint potential of I in *Rps. viridis*, Prince et al. [6] report a value of -400 ± 25 mV at pH 10.8, whereas Shuvalov et al. [9] and Klimov et al. [67] give a value of -620 ± 20 mV at about pH 11. For the reduction of Bph *b* in dimethylformamide and dichloromethane Fajer et al. [5] find -530 ± 30 mV.

The SDS reaction centers are stable in the dark for weeks at 4°C or 20-30°C. They may also be lyophilized and reconstituted without loss of photochemical activity [52, 58].

III. Green Bacteria

Pure reaction centers have not yet been isolated from green bacteria. The most highly enriched preparations still contain approximately 40 Bchl *a* molecules per reaction center [69]. [This ratio is similar to that occurring in chromatophores of some purple bacteria and in many P700 preparations (see Table 5).] The preparations from green bacteria consist largely of unit-membrane vesicles (Fig. 12) with the reaction centers bound to/in the membrane.

A. *Prosthecochloris aestuarii*

The first reaction-center-enriched preparation was made from *P. aestuarii* 2K, the green member of a consortium known as *Chloropseudomonas ethylica* 2K [70-72]. Fowler et al. [73] isolated a photochemically active subcellular fraction from cells broken in a French pressure cell and separated the extract into two layers on a discontinuous sucrose gradient (see Fig. 13). The upper green layer (density d = 1.12 g cm^{-3}) was enriched in chlorobium chlorophyll (Bchl *c*), which in vivo is contained in chlorophyll bodies (chlorobium vesicles) [74-76]. The lower brown layer (d = 1.18 g cm^{-3}) was enriched in Bchl *a* and in the photochemical activity of the organism, i.e., light-induced bleaching of reaction-center chlorophyll (P840) and oxidation of cytochrome c_{553}, as shown in Figure 14. This type of reaction-center preparation, now called Bchl *a* reaction-center complex I [69, 77] or simply complex I, contains considerable carotenoid in addition to about 80 Bchl *a*, 4 cytochrome c_{553}, and 4 cytochrome b_{563} per P840 [78].

Table 5 Characteristics of Reaction Centers (RC), Photoactive Complexes, and Photosynthetic Membranes

1.	Organism	*Rps. sphaeroides*		
2.	Strain	Carotenoidless (R-26)		
3.	Preparation Weight (including lipid and/or detergent, unless otherwise noted)	RC ~50 kD[a] [31, 23]	RC ~85 kD[a] [31,23]	Membrane – [3]
4.	Spectral characteristics	A_{275}/A_{365} = 0.82 A_{805}/A_{365} = 0.76 A_{805}/A_{853} = 2.4	A_{275}/A_{365} = 0.88 A_{802}/A_{365} = 0.7 A_{802}/A_{865} = 2.3	Omitted
5.	Polypeptides	21 kD 24 kD	21 kD 24 kD 28 kD	Omitted
6.	Light-harvesting Chl per RC	Absent	Absent	Bchl *a*
7.	Carotenoid	Absent	Absent	Absent
8.	Secondary donor E_m (pH)	Absent –	Absent –	Cyt c_2 –
9.	Secondary donor E_m (pH)	Absent –	Absent –	Absent –
10.	Primary donor Reduced wavelength Oxidized wavelength E_m (pH)	(Bchl *a*)$_2$ 853 nm – –	(Bchl *a*)$_2$ 865 nm 1245 nm [34] +450 mV	(Bchl *a*)$_2$ ~870 nm 1245 nm –
11.	Intermediate carrier Oxidized wavelength E_m (pH)	Bph *a* – –	Bph *a* 760 nm -550 mV [130, 125]	Bph *a* – –
12.	Primary acceptor (X) E_m (pH)	$UQ_{10}Fe$ –	$UQ_{10}Fe$ -45 mV (5.7-9.1 [38]	$UQ_{10}Fe$ -180 mV (10.0) [157]

[a]Minimum value without detergent.

Rps. sphaeroides		*R. rubrum*	
Wild-type (Y) [26, 30, 169]		Carotenoidless (G-9)	Wild-type
RC	Membrane	RC	RC
160 ± 10 kD	–	–	140 kD
		[43, 20, 23]	[42]
A_{285}/A_{365} $\simeq 1.0$ A_{802}/A_{365} $\simeq 0.7$ A_{802}/A_{865} $= 2.3$	Omitted	A_{285}/A_{365} $\simeq 1.0$ A_{280}/A_{802} $= 1.25$	A_{275}/A_{365} $\leqslant 1.0$ A_{280}/A_{802} $= 1.22$ A_{802}/A_{365} $\simeq 0.7$ A_{802}/A_{865} $= 2.3$
22 kD 24 kD 27 kD	Omitted	22 Kd 26 kD 29 kD	21 kD 24 kD 32 kD
Absent	Bchl *a*	Absent	Absent
Sphaeroidene	Sphaeroidene	Absent	Spirilloxanthin [41, 171]
Sphaeroidenone	Sphaeroidene	Absent	
Absent	Cyt c_2	Absent	Absent
–	–	–	–
Absent	Absent	Absent	Absent
–	–	–	–
(Bchl *a*)$_2$	(Bchl *a*)$_2$	(Bchl *a*)$_2$	(Bchl *a*)$_2$
870 nm	–	865 nm	865 nm
–	–	–	1248 nm
+525 mV (7.5)	+445 mV	–	–
–	–	–	–
–	–	–	–
–	–	–	–
–	–	–	Present
–	–	–	-370 mV [44]

Table 5 (continued)

1.	*Rps. capsulata*		*Rps. gelatinosa*	
2.	Carotenoidless (*Ala*$^+$)		Carotenoidless strain [48]	
3.	Triton RC – [45]	LDAO RC ~150 kD [46]	1% LDAO Particle ~250 kD	4-8% LDAO RC ~150 kD
4.	–	A_{280}/A_{802} = 1.25	A_{280}/A_{800} = 2.6	A_{280}/A_{800} = 1.85
5.	– 22.5 kD 27.5 kD 32 kD –	– 20.5 kD 24 kD 28 kD –	– – 24 kD 34 kD 43 kD	– – 24 kD 34 kD –
6.	Bchl *a*	Absent	Absent	Absent
7.	Absent	Absent	Absent	Absent
8.	Absent –	Absent –	Cyt c_{552} –	Absent –
9.	Absent –	Absent –	– –	– –
10.	(Bchl $a)_2$ 855 nm – –	(Bchl $a)_2$ ~850 nm – –	(Bchl $a)_2$ 860 nm – –	(Bchl $a)_2$ 860 nm – –
11.	Bph *a* 750 nm –	Bph *a* ~750 nm –	– – –	– – –
12.	– –	UQ_{10} [47] –	– –	– –

Chr. vinosum		
Wild-type (D)		
LDAO or Triton-acetone	SDS-Brij particle	Membrane
RC	~700 kD	–
[3, 10, 11, 49]	[55]	[158]
A_{800}/A_{875} = 2.8	–	Omitted
–	12 kD	Omitted
22.5 kD	22 kD	
26.5 kD	27 kD	
29.5 kD	30 kD	
–	45-50 kD	
Absent	6 Bchl *a*	Bchl *a*
Present	Spirilloxanthin Lycoxanthin Lycophyll	Present
Cyt c_{553}	Cyt c_{553}	Cyt c_{553}
0 mV (8.0)	–	0 mV (7.5)
Cyt c_{555}	Cyt c_{555}	Cyt c_{555}
+340 mV (8.0)	–	+320 mV (7.5)
(Bchl $a)_2$	(Bchl $a)_2$	(Bchl $a)_2$
875 nm	883 nm	883 nm
–	–	–
+490 mV (8.0)	–	+490 mV (7.5)
Bph a^b	–	Bph a^b
756 nm	–	–
–	–	–
MQFe [56]	Present	Present
-160 mV (8.0)	–	-160 mV (8.0) [157]

[b] See text for discussion of the role of Bph *a*.

Table 5 (continued)

1.	*Rps. viridis*			*Th. pfennigii*
2.	Wild-type			Wild-type (9111)
3.	SDS RC 110 kD [8, 52, 58, 63]	LDAO RC 235 ± 15 kD [61]	Membrane – [6]	LDAO RC [62]
4.	A_{280}/A_{830} = 2.5[c] A_{830}/A_{960} = 2.5 [60]	A_{280}/A_{830} = 2.5 A_{830}/A_{965} = 2.3	Omitted	A_{837}/A_{959} = 4.5
5.	– – – –	23 kD 29 kD 37 kD 45 kD	Omitted	– – – –
6.	Absent	Absent	Bchl *b*	Oxidized Bchl *b*
7.	Dihydrolycopene	–	Present	–
8.	Cyt c_{553} –20 mV (8.0)	Cyt c_{553} –	Cyt c_{553} +10 mV (7.0)	Cyt c_{553} –
9.	Cyt c_{558} +330 mV (8.0)	Cyt c_{558} –	Cyt c_{558} +350 mV (7.0)	Cyt c_{556} –
10.	(Bchl $b)_2$ 958 nm 1310 nm +390 to 450 mV (8.0)	(Bchl $b)_2$ 958 nm [62] 1310 nm –	(Bchl $b)_2$ 985 nm – +500 mV (7.8, 10.8) +515 mV [9, 67]	(Bchl $b)_2$ – – –
11.	Bph *b*[d] 790 nm –	Bph *b*[d] 790 nm –	Bph *b*[d] 790 nm –400 ± 25 mV (10.8) –620 ± 20 mV (~11) [9, 67]	Bph *b* 790 nm –
12.	– –	MQ-like –	Present –150 ± mV (7.8)	– –

[c]Estimated from A_{280}/A_{830} = 1.9 [52], A_{830}/A_{960} = 3.3 [52], and A_{830}/A_{960} = 2.5 [60].
[d]See Section II.F.2.c for a discussion of the role of Bph *b*.

Prosthecochloris aestuarii	*Chlorobium limicola* f. sp. *thiosulfatophilum*
Wild-type (2K)	Wild-type (Tassajara)
Membrane [73]	Membrane [69, 81, 91]
–	A_{835}/A_{810} $\simeq 0.4$
–	18 ± 1 kD
–	21 ± 2 kD
–	23 ± 2 kD
–	31 ± 4 kD
–	36 ± 2 kD
–	42 ± 4 kD
~80 Bchl *a*	~80 Bchl *a*
Present	Present
Cyt c_{553}	Cyt c_{553}
+170 mV	+165 mV (6.8)
Absent	Absent
–	–
(Bchl $a)_2$	(Bchl $a)_2$
830, 842 nm	830, 842 nm
–	1160 nm
+240 mV	+250 mV (6.8)
–	–
–	–
–	–
–	Fe-S (?)
–	−540 ± 10 mV (~10)

Table 5 (continued)

1.	*Phormidium luridum*		Spinach chloroplast	
2.	var. *olivacea*		–	
3.	SDS-P700- Chl *a*-Protein ~110 kD [52, 106, 127, 134]	Triton-P700- Chl *a*-Protein – [109]	Digitonin-ether Photosystem I particle – [116]	HP700 particle ~1000 kD per RC [101]
4.	A_{275}/A_{677} = 0.42 A_{437}/A_{677} = 1.3	– –	– A_{434}/A_{673} = 1.9	– –
5.	– 46 kD 48 kD	⩽10 kD 46 kD 48 kD	–	–
6.	40-45 Chl *a*	40-45 Chl *a*	5-9 Chl *a*	24 Chl *a* 6 Chl *b*
7.	β,β-Carotene	β,β-Carotene	Absent	Absent
8.	Absent –	Absent –	Absent –	Cyt *f* –
9.	Absent –	Absent –	Cyt *b* (?) –	Cyt b_6 –
10.	(Chl $a)_2$ ~697 nm ~825 nm –	(Chl $a)_2$ – – –	(Chl $a)_2$ 677, 696 nm – –	(Chl $a)_2$ 698 nm – –
11.	– – –	– – –	– – –	– – –
12.	Absent –	Fe-S center –	– –	– –

Swiss chard chloroplast	Spinach chloroplast		
–	–	–	–
Digitonin-Triton-SDS	Membrane	Triton Photo-	Membrane
Photosystem I	(Photosystem I RC)	system II particle	(Photosystem II RC)
particle [98]		~300 kD	
~140 kD per RC	–	[147,152,153]	
A_{440}/A_{674}	Omitted	A_{436}/A_{673}	Omitted
= 1.6		= 1.4 [101]	
–	Omitted	–	Omitted
–			
–			
70 kD			
40-50 Chl *a*	~200 Chl	~15 Chl *a* [154]	~200 Chl
		~0.5 Chl *b*	
β,β-Carotene (?)	Present	β,β-Carotene	Present
Absent	Plastocyanin	Cyt b_{559} [162]	Cyt b_{559}
–	+340 mV [159]	+58 mV (7.0)	+375 ± 25 mV (7-8) [163]
Absent	Cyt *f*	–	–
–	+380 mV [159]	–	+475 mV [149]
(Chl $a)_2$	(Chl $a)_2$	(Chl $a)_2$	(Chl $a)_2$
697 nm	682, 703 nm [25,141]	680 nm [154]	690 nm [174]
–	815 nm [25, 160]	–	825 nm [145, 160]
+415 mV (8.0) [125]	+375 mV (8.0) [125]	–	$>$+850 mV [149]
	+430 mV (7.0) [14]		
	~500 mV [126]		
–	–	–	–
–	–	–	–
–	–	–	–
Absent [114]	Fe-S center [129, 149]	Plastoquinone	–
–	-530 mV [161]	-325 mV (7.0)	–
		[150]	

Table 6 Characteristics of ESR Signals Associated with Electron Transfer Components of Reaction Centers

Source	g value	ΔH (gauss)	Conditions	References
Bchl a^+	2.0025 ± 0.0001	13	−160°C, $CH_2Cl_2-CH_3OH$	5
Bchl a^-	2.0028 ± 0.0002	12 lines	2-methyltetrahydrofuran	5
Bph a^-	2.0032	~13	1 mW, −160°C, pyridine-water	5
Bchl b^+	2.0025	14	−160°C, $CH_2Cl_2-CH_3OH$	5
Bchl b^-	2.0033	~13	10 μW, −140°C, tetrahydrofuran	5
Bph b^-	2.0033	~13	1 mW, −160°C, pyridine-water	5
Chl a^+	2.0025 ± 0.0001	9	2.6 mW, +23°C, −110°C	175
Chl a^-	2.0027 ± 0.0002	13	−50°C, butyronitrile	5, 164
P680⁺	2.002	7-8	–	144, 146
P700⁺ (average)	2.0025	7.5	–	2
P840⁺ (*C. limicola*)	2.003	9.2 ± 0.5	1 mW, 10 K	69
P870⁺ (*Rps. sphaeroides*)	2.0025 ± 0.0002	9.6 ± 0.2	–	35
P960⁺ (*Rps. viridis*)	2.0026	11.8 ± 0.2	−40°C, −140°C	5
I^- (*Rps. sphaeroides*)	2.0036 ± 0.0002	12.9 ± 0.3	–	31
I^- (*Chr. minutissimum*)	2.0025 ± 0.0005	12.5 ± 0.5	–	8
I^- (*Chr. vinosum*)	2.003	60 (doublet)	50 mW, 4.6 K	3, 10, 11
	2.003	15	50 mW, 15 K	
I^- (*Rps. viridis*)	2.01	100 (doublet)	<15 K	11
	2.003	~15	–	11
	2.0036	12.2 14	10 μW, 110 K 1 mW	Fajer (unpublished)

X^- (*Rps. sphaeroides*)	$g_x = 1.68$	—	20 mW, 10 K	3, 38
	$g_y = 1.82$	Broad		
X^- (*Chr. vinosum*)	$g_x = 1.62$	—	20 mW, 8 K	3, 165
	$g_y = 1.82$	Broad		
X^- (*Rps. viridis*)	$g_x = 1.67$	—	10 mW, 5 K	6
	$g_y = 1.82$	Broad		
Fe-S center (−550 mV) (*C. limicola*)	1.93-4	—	5-10 mW, 10 K	69, 83, 84
Bound ferredoxin	$g_x = 1.86$	—	10 mW, 15-25 K	126, 134
(spinach chloroplasts	$g_y = 1.94$	—		
or P700-Chl *a*-protein)	$g_z = 2.05$	—		
Photosystem I electron	$g_x = 1.78$	—	100 mW, 5-10 K	126, 166
acceptor (spinach or	$g_y = 1.88$	—		
Euglena chloroplasts	$g_z = 2.08$	—		
or cyanobacteria)				

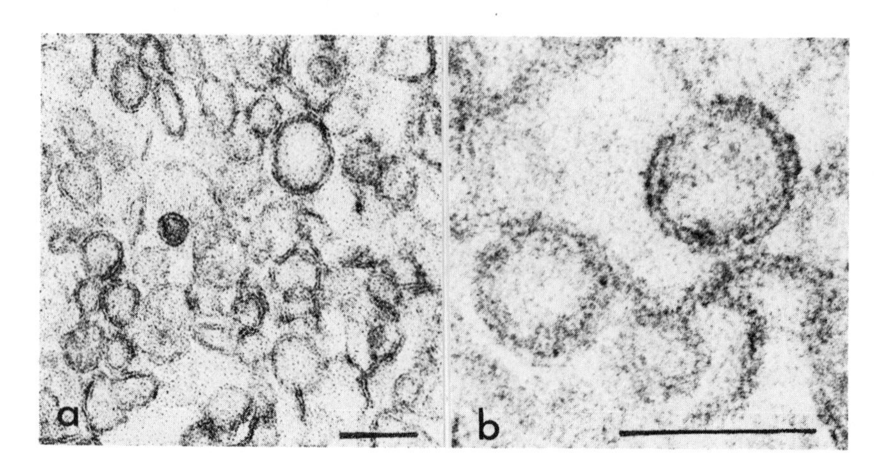

Figure 12 Electron micrographs of a thin section of complex II from *Chlorobium limicola* f. sp. *thiosulfatophilum* strain 6230 (Tassajara). The black bars in a and b each indicate 1,000 Å. Average unit-membrane thickness in this section is 81 ± 10 Å; average vesicle diameter is about 500 Å, varying from about 300 to 900 Å. Complex II was pelleted, fixed with glutaraldehyde, stained with OsO_4, dehydrated, and embedded in Epon. Thin sections (~600 Å) were stained successively with uranyl acetate and lead citrate and then examined in a Philips EM-300 electron microscope. Courtesy of Dr. Myron Ledbetter.

The redox potentials of P840, cytochrome c_{553}, and cytochrome b_{563} are +240, +170, and 0 to -150 mV, respectively [73, 78] (see Table 5). Particles of complex I have molecular weights in excess of 1.5×10^6 and "exhibit a range of sizes consistent with some form of polymeric association of similar subunits" [73].

B. *Chlorobium limicola* f. sp. *thiosulfatophilum*

1. Bchl *a* Reaction-center Complex I

Reaction-center preparations from *C. limicola* f. sp. *thiosulfatophilum* are very similar to those from *P. aestuarii* [73, 79, 80]. The redox potentials of P840 and cytochrome c_{553} are +250 and +165 mV, respectively [81], values which are almost the same as those found in *P. aestuarii.* There are two reactive cytochrome c_{553} hemes per P840. Knaff and Buchanan [82] found a redox potential of -90 ± 20 mV for cytochrome b_{564} in *Chlorobium,* which is also well within the potential range found for cytochrome b_{563} in *P. aestuarii.* The primary electron acceptor (X) has a redox potential of -540 ± 10 mV [69], the same as that reported for an electron acceptor closely associated with P700 in photo-

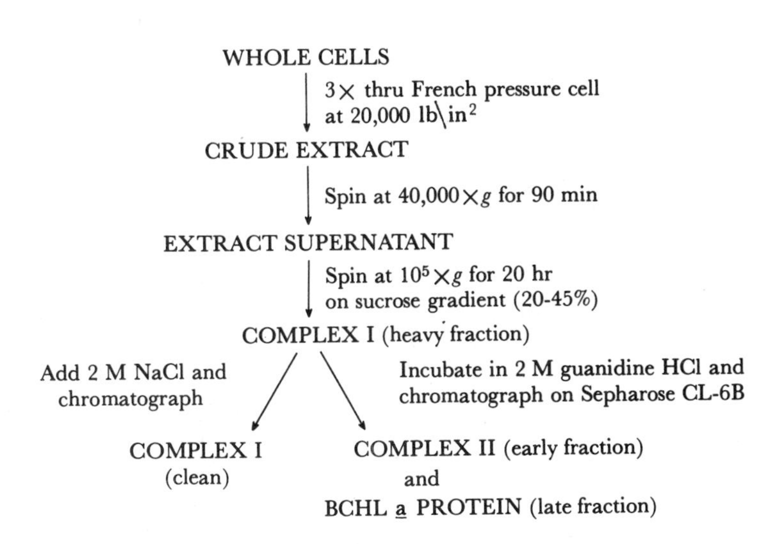

Figure 13 Preparation of Bchl *a* reaction-center complexes I and II from strain Tassajara.

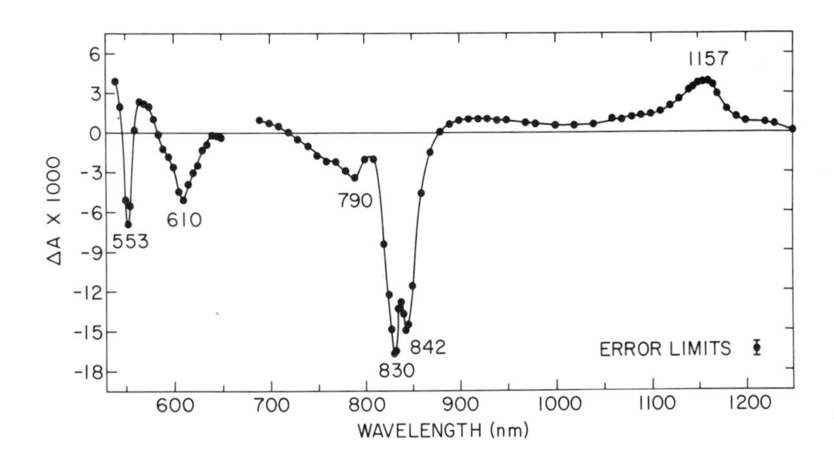

Figure 14 Light-induced change in absorption spectrum in complex I from strain Tassajara (A_{810} = 2.3). Far-red actinic light was used for the region 540 to 650 nm. Orange (605-nm) actinic light was used for the region 690 to 1250 nm. Measurements were made with a Cary 14R spectrophotometer (full scale = 0.1). Reproduced with permission from Olson et al. [69].

system I of cyanobacteria and chloroplasts (see Section IV.A.2.c). ESR studies [81] show the existence in complex I of two iron-sulfur centers (g = 1.90 and g = 1.94). The latter may be related to the primary electron acceptor since there are three iron-sulfur proteins (g = 1.94) with redox potentials of −25, −175, and −550 mV [83]. Light-induced reduction of the lowest-potential iron-sulfur component at 15 K has been reported by Jennings and Evans [84].

The total cytochrome *c* content of complex I is greater than 2 molecules per P840, and the ratio of cytochrome *b* to cytochrome *c* is less than 1 [85], compared with ratios of 4 and 1, respectively, reported for *P. aestuarii* [78].

Electron micrographs showed that preparations of complex I consist of unit-membrane vesicles plus somewhat fewer unit-membrane fragments [69; Ledbetter, Olson, and Shaw, unpublished]. Unit-membrane thicknesses were between 70 and 80 Å, and vesicle diameters ranged from 300 to 1000 Å with the distribution peak at about 500 Å. (For 300- and 500-Å vesicles the anhydrous molecular weights are about 10 and 30×10^6, respectively, assuming d = 1.18 g cm^{-3} [73].) The vesicle images were similar to those shown in Figure 12. Unit-membrane fragments were 200-1000 Å long.

2. Complex II

When complex I is incubated in 2 M guanidine-HCl and then chromatographed on crosslinked dextran or agarose gel (see Fig. 13), two principal components are separated: a larger component with photochemical activity (complex II) and a smaller component without activity (Bchl *a*-protein). This water-soluble Bchl *a*-protein [70, 77] is presumably a trimer similar to the well-characterized protein from *P. aestuarii* [86, 87] in which each subunit contains seven Bchl *a* molecules wrapped in a blanket of protein and the polypeptide has a molecular weight of 39×10^3. The function of this Bchl *a*-protein is to transfer excitation energy from chlorobium chlorophyll (Bchl *c*) in the chlorophyll bodies (chlorobium vesicles) to P840 in the reaction center [88-90].

Comparison of the absorption spectra of complex II and complex I by normalizing absorbance values at 835 nm, as in Figure 15, shows that about half the Bchl *a* is removed from complex I by the 2 M guanidine-HCl. This suggests that the Bchl *a* remaining in complex II may be associated with a protein different from that of the water-soluble Bchl *a*-protein. Complex II still contains (per reaction center) about 40 Bchl *a*, three to four cytochrome *c*, and approximately one cytochrome *b*. There remain two reactive cytochrome c_{553} hemes and considerable carotenoid per P840.

Comparison of the polypeptide compositions of complexes I and II (Table 7) shows that both complexes contain six molecular weight classes of polypeptides ranging from ~18 to ~42 kD [91]. Since the 31-kD polypeptides are the most abundant of the polypeptides in complex II but among the least abundant

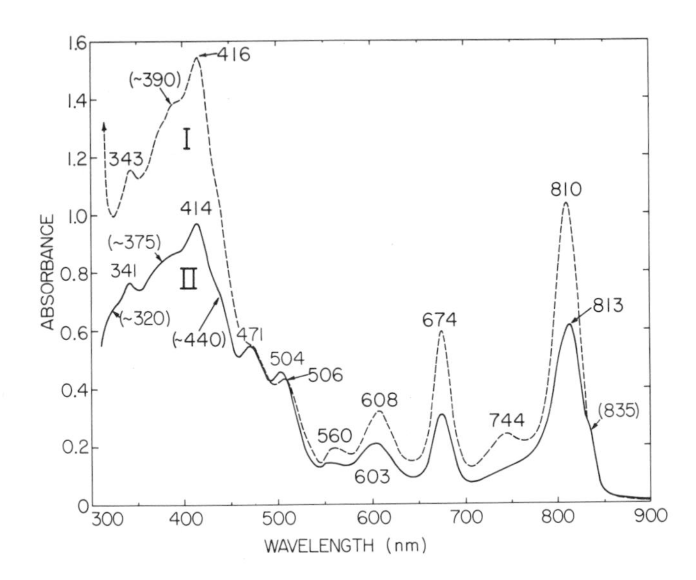

Figure 15 Absorption spectra of complex I (dashed curve) and complex II (solid curve) from strain Tassajara. Complex I is dissolved in 10 mM phosphate (pH 6.8) plus 10 mM ascorbate; complex II in 10 mM phosphate plus 2 M guanidine-HCl. The two preparations were made independently. Reproduced with permission from Olson et al. [69].

Table 7 Molecular Weight and Relative Abundance[a] of Polypeptides in Complexes I and II

Kilodaltons	Complex I	Complex II	Difference (%)
42 ± 4	125 ± 25	15 ± 3	-80 to -90
36 ± 2	50 ± 10	5 ± 1	-85 to -90
31 ± 4	25	25	0
23 ± 2	75 ± 15	10 ± 2	-80 to -90
21 ± 2	25 ± 5	15 ± 3	-10 to -60
18 ± 1	50 ± 10	10 ± 2	-70 to -85
Total	350 ± 65	80 ± 11	-70 to -80

[a]Relative abundance of 31-kD polypeptides is assumed to be the same in both complexes. Electrophoresis in 8 M urea [167] was carried out in 15% polyacrylamide slab gels (acrylamide: bis = 50) by Cecile M. Pickart with the following markers (kD): cytochrome *c* (12.4), lysozyme (14.3), myoglobin (17.2), trypsin (23.3), chymotrypsin (25.7), carbonic anhydrase (29.0), lactic dehydrogenase (36.0), alcohol dehydrogenase (41.0), and ovalbumin (46.0).

Source: Olson et al. [91].

in complex I, it seems reasonable to assume that little or no 31-kD polypeptide is removed from complex I by 2 M guanidine-HCl. With the assumption (used in Table 7) that all the 31-kD polypeptides remain in complex II, it can be estimated that 70-80% of the protein associated with complex I is removed by 2 M guanidine-HCl. Most of the 42-kD polypeptide removed is probably associated with the Bchl *a*-protein. The polypeptides retained in complex II to the greatest extent are the 21- and 31-kD ones, and these may be associated with the reaction center.

Electron micrographs of complex II preparations (Fig. 12) look almost the same as those of complex I. The decrease in protein content apparently does not affect the appearance of the unit-membrane vesicles and segments. Unit-membrane thickness is still 70-80 Å, and vesicle diameters again range from 300 to ~1000 Å. The only noticeable difference between preparations of complex I and complex II is the lower proportion of unit-membrane segments in the latter. With only 20-30% of the protein associated with complex I remaining, complex II retains the integrity of the unit-membrane vesicle, some reaction-center activity, and electron transfer from cytochrome c_{553} to the reaction-center chlorophyll. Electron transfer from the "primary" electron acceptor to cytochrome b_{563} may also occur [91].

Reaction centers have not yet been isolated from the unit membranes of green bacteria. Triton X-100 and LDAO cannot be used because they apparently destroy the reaction centers [91]. The neutral detergent Brij 58 [polyoxyethylene (20) cetyl ether] showed no harmful effects on the reaction centers, but it reacted with the crosslinked dextran used for chromatography. SDS dissociates the unit-membrane vesicles without giving any enrichment of P840 (Thornber, unpublished). The time is now ripe for a serious sustained effort to release intact and functional reaction centers by the use of appropriate detergents on unit membranes of green bacteria.

3. Model of the Photosynthetic Unit

Unit-membrane vesicles have not been seen in electron micrographs of intact green bacteria (see plate 1.4 of Cohen-Bazire [92]); the only unit membrane visible is the 80-Å cytoplasmic membrane [74]. We suppose that upon passage of green bacteria through the French pressure cell, the unit-membrane vesicles and segments are formed from fragments of the cytoplasmic membrane. If this is so, complex I would have the composition and structure of the native cytoplasmic membrane but with the associated chlorophyll bodies removed.

In a recent attempt to formulate a working model [69] (see Fig. 16) of the photosynthetic apparatus, the chlorophyll bodies (~30 megadaltons in size and bounded by a 30-Å membrane [76]) were assumed to be anhydrous with the limiting membrane made up of protein subunits. According to Cruden and

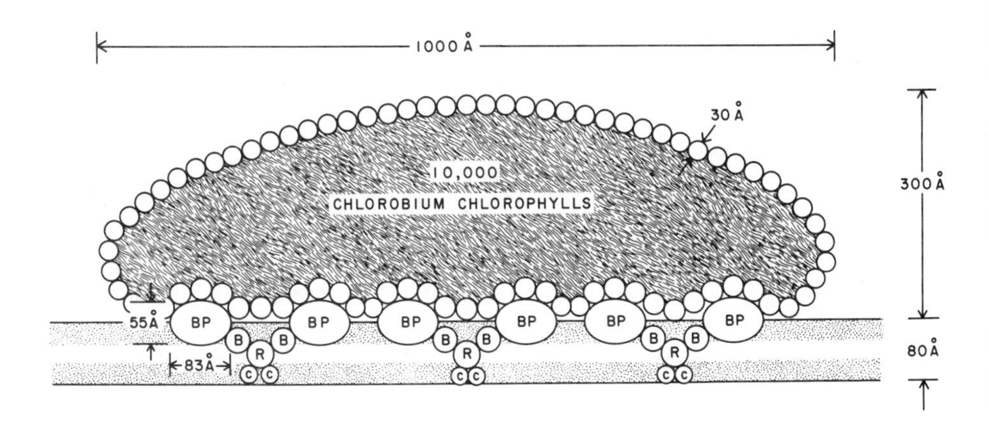

Figure 16 Model of a chlorophyll body (chlorobium vesicle) and associated cytoplasmic membrane. For every 10,000 molecules of chlorobium chlorophyll (Bchl *c*) there are ~800 Bchl *a* molecules (B and BP), ~10 reaction centers (R), and ~20 cytochrome *c* molecules (C). Half the Bchl *a* molecules are inside ~20 Bchl *a*-protein trimers (BP). Membrane components are distributed in two dimensions. Cytochrome *b* and other electron transfer components are not shown. Reproduced with permission from Olson et al. [69].

Stanier [93] the chlorophyll bodies of *C. limicola* f. sp. *thiosulfatophilum* strain 6130 contain 33% protein, 16% lipid, 15% carbohydrate, and 28% chlorobium chlorobium chlorophyll (Bchl *c*). In the model, all the protein is in the 30-Å membrane, and the chlorobium chlorophyll content is equivalent to ~10,000 molecules. Also, each chlorophyll body is appressed to the 80-Å cytoplasmic membrane through ~20 Bchl *a*-protein trimers containing ~400 Bchl *a* molecules, each trimer being half in and half out of the unit membrane. Since the Bchl *a*-protein is water-soluble, the binding to the cytoplastic membrane and also to the chlorophyll body probably involves considerable salt bridging.

The cytoplasmic membrane contains, per chlorophyll body, ~10 reaction centers and ~400 firmly bound Bchl *a* molecules in addition to the 400 Bchl *a* molecules in the Bchl *a*-protein. In the process of preparing complex I,* the chlorophyll bodies are stripped from the cytoplasmic membrane but the Bchl *a*-protein trimers remain attached to it. Presumably the breaking procedure fragments the membrane, and the unit-membrane vesicles and segments of complex I are formed. Upon treatment with 2 M guanidine-HCl, the Bchl *a*-protein and

*For the preparation of complex I from green photosynthetic bacteria (Fig. 13) sodium dithionite should be present at a concentration between 10 and 100 mM to minimize the amount of bacteriopheophytin *c* (chlorobium pheophytin) in the final preparation.

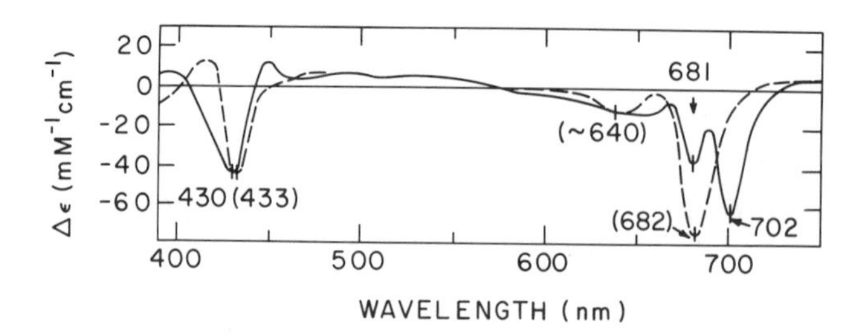

Figure 17 Light-minus-dark difference spectra of P680 (dashed curve) and P700 (solid curve); P680 was measured in spinach subchloroplast fragments prepared with deoxycholate [141], and P700 was measured in digitonin particles (D-114) from spinach [168]. Adapted from van Gorkom et al. [141].

other proteins are released from the vesicle (and segment) membranes leaving the firmly bound Bchl *a* molecules and the reaction centers in complex II.

IV. Cyanobacteria (Blue-Green Algae) and Plant Chloroplasts

The characteristic primary electron donors of photosystem I (PS I) and photosystem II (PS II) are P700 and P680 (or P690), respectively (Fig. 17). Perhaps the most compelling substantiation of the presence of two types of reaction centers in plant photosynthesis (see Fig. 2) came from Boardman and Anderson's observation [94] that the two photosystems could be separated from each other. However, despite this early successful fractionation and much subsequent effort, isolation of either of the two reaction centers (PS I or PS II) totally free of antenna chlorophylls and extraneous proteins has not been achieved.

A. Photosystem I

1. Isolation

Boardman and Anderson's preparation [94] of PS I, which was obtained by differential centrifugation of digitonin-treated chloroplasts, had a Chl/P700 ratio greater than or equal to 120:1. Preparations with slightly improved ratios were obtained subsequently by replacing digitonin with Triton X-100 [95] (see Fig. 18); by DEAE-cellulose chromatography of the PS I particle obtained with digitonin [97]; or by mechanical fractionation of chloroplasts (see Brown [96] for review). The Chl/P700 ratio of such preparations is still greater than 100:1. Greater enrichment in P700 has been achieved by using mixtures of digitonin and Triton X-100 [98]; by using other detergents; or by solvent extraction of PS I particles. Many of the resulting P700 preparations have Chl/P700 ratios around 40:1. The various preparations are listed in Table 8 and described below.

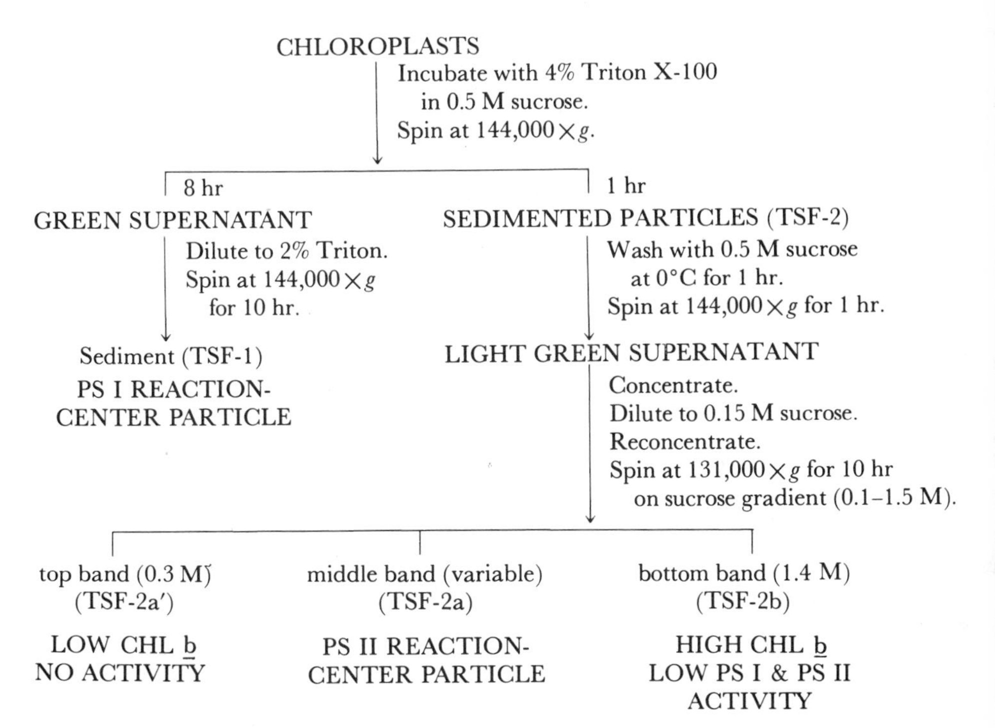

Figure 18 Preparation of PS I and PS II reaction-center particles from spinach chloroplasts with Triton X-100. Reproduced with permission from Vernon et al. [101, 152, 153].

a. HP700 Particle. Working on the basis that absence of carotenoids from membranes of purple bacteria enabled detergents to separate the reaction center from antenna pigments more reacily, Ogawa and Vernon [99] and Yamamoto and Vernon [100] used photosynthetic membranes of cyanobacteria and plant chloroplasts from which carotenoids were missing because of mutation, growth in the presence of inhibitors, or repeated extraction with 15% aceone at –15°C. Carotenoidless membranes were treated with Triton to obtain PS I preparations (TSF-1) having Chl/P700 $\simeq$ 30, Chl a/Chl $b \simeq 4$, and containing cytochromes f and b_6 [101]. These preparations were designated HP700 particles. Later Ke et al. [102] increased the yield of HP700 material and removed the two cytochromes by sucrose gradient and high-speed centrifugation of TSF-1 fractions; they obtained HP700 preparations with Chl/P700 ratios of 35 ± 5.

Table 8 Summary of Attempts to Isolate P700

Chl/P700 ratio in isolated material	Starting material	Detergent used	Fractionation method
⩾100	Photosynthetic lamellae	Digitonin [94] or Triton X-100 [95]	Differential centrifugation
40	Lamellae	SDS [106], LDAO, or Triton X-100 [109]	Hydroxyapatite
	Digitonin PS I particles	LDAO [113] or Triton X-100 [98]	DEAE-cellulose
30	Carotenoid-depleted lamellae	Triton X-100 [100]	Differential centrifugation
20	Lamellae	SDS + LDAO + Triton X-100 [108]	Hydroxyapatite
	French press PS I particles	None	Extraction with acetone (–20°C) [115]
<10	Digitonin PS I particles	None	Extraction with water-saturated ether [116]

b. P700-Chl a-Protein. Meanwhile, Thornber and coworkers [103-105] were studying chlorophyll-proteins isolated from photosynthetic lamellae of cyanobacteria and higher plants dissociated by the anionic detergents SDS or SDBS (sodium dodecyl benzenesulfonate). They later realized [63, 106] that one of these chlorophyll-proteins was highly enriched in P700. This complex, termed P700-Chl *a*-protein [106] was obtained from SDS-solubilized photosynthetic membranes of a cyanobacterium, *Phormidium luridum*, by hydroxylapatite chromatography, with unusually high concentrations (~0.3 M) of sodium phosphate buffer used for its elution. The preparation was further purified by ammonium sulfate fractionation and/or Sephadex gel filtration [103, 106]. It contained essentially only Chl *a*, β,β-carotene, and protein, and it had a Chl/P700 ratio of 40:1. (Because the differential extinction coefficient of P700 was not known, it was assumed to be 100 mM^{-1} cm^{-1} at the maximum bleaching in the red region; this value gave the reported ratio of 70:1 to 80:1. Later, Hiyama and Ke's value [107] of 64 mM^{-1} cm^{-1} was adopted.) Electrophoresis of the isolated complex revealed a single chlorophyll-protein zone (~110 kD).

Subsequently P700 was found to be stable to anionic detergent treatment only in certain cyanobacteria, being destroyed by SDS in other cyanobacteria and in all eukaryotic organisms [108]. Nevertheless a method was developed for preparing the P700-Chl *a*-protein complex from *any* plant or cyanobacterium [109]. Replacement of SDS by Triton X-100 or other nonionic or weakly ionic detergents (see below) and chromatography of the solubilized membranes on hydroxylapatite yields a P700-enriched eluate having the same Chl/P700 ratio (40:1) and absorption spectrum as the P700-Chl *a*-protein prepared from *P. luridum.* However, cytochromes *f* and b_6 are present in the complex isolated from any eukaryotic organism, whereas they are absent from the cyanobacterial preparations regardless of the detergent used. The above method [109] has been applied to a wide variety of plants [110] and the complex is now thought to be ubiquitous in photosynthetically competent plants and to represent 4-30% of their total chlorophyll. Mutants that apparently lack P700 [111, 112] do not contain this chlorophyll-protein. Regardless of the nature of the chlorophylls in the starting organism, the isolated complex contains only Chl *a.*

c. Mixed Detergent Preparations. Malkin [113] obtained a preparation similar to those described above by DEAE-cellulose chromatography of PS I fragments obtained from spinach treated with digitonin and then with LDAO. The isolated material was enriched in P700 (Chl/P700 = 40:1), contained Chl *a,* and probably β,β-carotene and protein, but was devoid of cytochromes. Bengis and Nelson [98], using essentially the same techniques but with Triton X-100 instead of LDAO, obtained a fraction having a Chl/P700 ratio of 100:1 and a Chl *a*/Chl *b* ratio of 40:1 and containing β,β-carotene but no cytochromes and quinones. Treatment of this fraction with 0.5% SDS for 20 min at 0°C followed by sucrose density centrifugation yielded a photochemically active zone having one P700 per 40 to 50 Chl *a* molecules, but apparently lacking bound ferredoxin and the in vivo primary electron acceptor [114]. By using three detergents (Triton X-100, LDAO, and SDS) for the solubilization of photosynthetic membranes, Thornber et al. [108] have routinely been able to obtain, by hydroxylapatite chromatography, fractions having Chl/P700 ratios of 20:1 and generally lacking cytochromes.

d. Preparations Obtained by Organic Solvent Extraction. Kok [14] first demonstrated that a preparation with a higher ratio of P700 to antenna chlorophylls could be obtained by extracting chloroplasts with a critical concentration of acetone. Sane and Park [115] improved the Chl/P700 ratios by starting with PS I particles (stroma lamellae) isolated by differential centrifugation of chloroplasts disrupted in a French pressure cell, and extracting them three times with 100% acetone at -20°C. The final material, resuspended by sonication or homogenization in buffered EDTA, has a Chl *a*/Chl *b* ratio $>$ 10 and an average Chl/P700 ratio of (15 ± 3):1. Recently Ikegami and Katoh [116] used diethyl ether (75-100% saturated with water) to extract PS I particles obtained with digitonin.

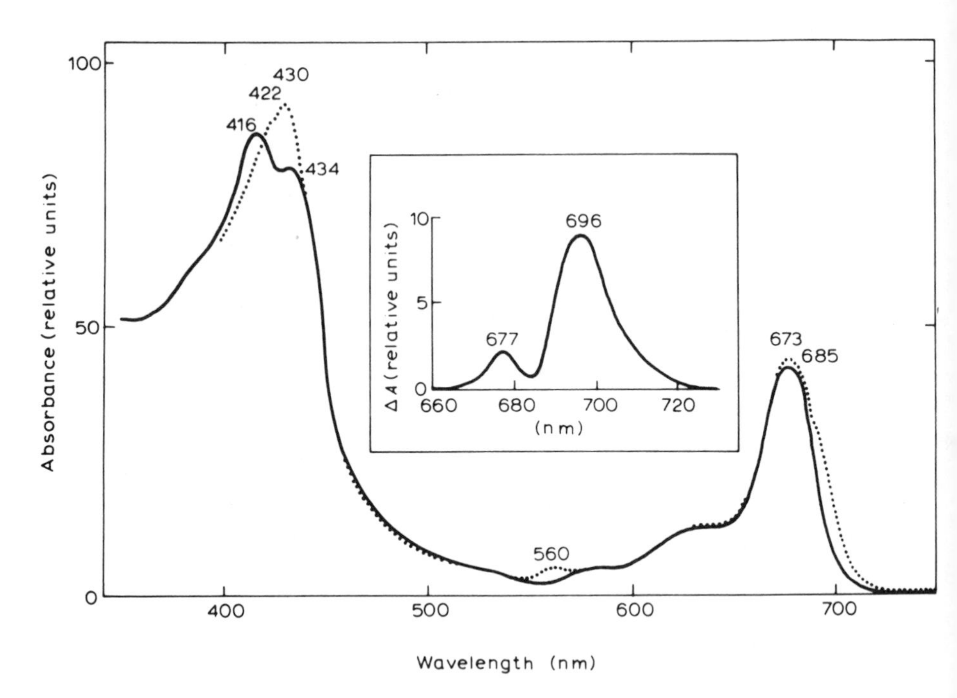

Figure 19 Absorption spectra of ether-extracted PS I particles prepared from spinach with digitonin (Chl/P700 = 6). Dithionite was added to one sample (dotted curve) but not to the other (solid curve). The insert shows the reduced-minus-oxidized difference spectrum for P700. Reproduced with permission from Ikegami and Katoh [116].

The product (Fig. 19) has the highest P700 enrichment currently available, with a Chl/P700 ratio between 5:1 and 9:1. It is free of Chl *b* and carotenoids but contains cytochrome(s).

2. PS I Reaction Centers

The spectral properties and composition have not been unequivocally determined because available P700 preparations are not as pure or as highly enriched in reaction centers as purple bacterial preparations. The best that can be done is to postulate what the characteristics may be by studying the many analyses that have been done on the P700-enriched preparations described above.

a. Absorption Spectrum of P700. The major absorption in the red region of P700 preparations (Chl/P700 $\geqslant$ 40) is due to the absorbance of the antenna chlorophylls and is maximal at 675-677 nm (the 675-676 nm absorbance maximum of the LDAO preparations [R. Malkin, personal communication] was erroneously reported as 673 nm [113]).

In preparations having lower Chl/P700 ratios (<20:1) the red wavelength maximum absorption occurs at about 673 nm; the change in this maximum may be due to removal of antenna chlorophyll molecules of 677-nm spectral form from the preparation by solvent extraction. It is not yet known whether the remaining 673-nm absorbing chlorophylls (see Fig. 19) are components of the reaction center and are analogous to P800 and/or Bph *a* in the purple bacteria, or whether they represent some antenna molecules tenaciously attached to the P700 complex. Thornber et al. [108] reported that one can obtain photochemically active P700 preparations that contain no pheophytin *a,* as evidenced by the lack in many P700 preparations of a minor absorption peak around 540 nm, where the Q_X band of pheophytin *a* would be expected to absorb. Thus it appears that pheophytin *a* cannot serve as an intermediate electron carrier in PS I.

When P700 becomes oxidized either by light or chemically, a slight change is seen in the absorption spectrum of the P700 preparation, which is most easily demonstrated by difference spectrophotometry (see Fig. 17). This change is particularly noticeable around 700 nm, hence the name P700 for the primary electron donor of the PS I reaction center in plants [117]. In this spectral region a loss of absorbance occurs which, in intact chloroplasts and PS I fractions with Chl/P700 ratios greater than 50:1, is maximal at 701-704 nm, and in the more enriched preparations occurs at slightly shorter wavelengths (696-697 nm). This difference in the wavelength of maximum bleaching probably reflects a small change in the environment of the reaction-center chromophores as closely associated antenna chlorophylls and/or lipids are removed. Difference spectra of $P700^+$ – P700 show several other spectral changes correlated with the loss of absorbance at ~700 nm, including decrease in absorbance at about 680-685 nm and in the Soret region at ~435 nm [14, 100, 106, 107] and an increase in the near infrared between 725 and 900 nm [25, 106] (cf. Borg et al. [175]). Difference spectra at 77 K reveal further details of the red spectral changes [107, 108, 118]. There is certainly an increase in absorbance close to 688 nm, but it is not yet clear whether this represents the positive-going peak of a chlorophyll band-shift (similar to that occurring for P800 or P830 upon oxidation of primary donor in the purple bacterial reaction center [99, 100, 119, 120]) or the absorbance of $P700^+$ [82, 121, 122] or a reduced form of Chl *a* acting as an electron acceptor in PS I [123]. ESR data (in particular the linewidth of the g = 2 signal) and spectral data on Chl *a* in vitro and on $P700^+$ indicate that P700 is almost certainly a dimer of Chl *a* molecules [122, 176]. The structures proposed for this special pair of Chl *a* molecules have been described elsewhere [64, 124].

b. Redox Potential of P700. Evans et al. [125] have recently shown that in broken spinach chloroplasts and in Triton PS I particles, the ESR signals usually attributed to P700 (see Table 6) may be due to bulk chlorophyll as well as P700. They found a redox potential of +375 mV at pH 8 for P700 and of +460 to 480 mV for bulk chlorophyll. This new value for P700 is substantially lower than previous values of +430 mV [14] and +500 [126].

c. Composition. Much less is known about the composition of the preparations most highly enriched in P700 [i.e., those obtained by organic solvent extraction (Fig. 19)] than about that of the fractions obtained with detergents (e.g., TSF-1 and P700-Chl *a*-protein). Probably the PS I reaction-center component will be found to contain two different types of chromophores–Chl *a* and a carotene. These two pigments have been found in all P700-containing fractions except those isolated from lamellae from which all carotenoids have been removed by solvent extraction (Fig. 19). Chl *b* occurs only in some P700 preparations made from green plants and, when present, has a much lower concentration than Chl *a*. Since the PS I reaction center is likely to have the same pigment composition in all classes of plants, Chl *b* is unlikely to be a component. Two of the Chl *a* molecules in each moiety are thought to constitute the P700 molecule, while other Chl *a* molecules (almost certainly present, see below) might function as electron acceptors in the primary photochemical event and/or serve the same (as yet unknown) function as P800 or P830 in the purple bacteria. One or two carotene molecules, only a small fraction of the total in most plants, seem to be intimately associated with each P700 [99, 100, 106, 120, 127]. It is interesting that reaction centers in purple bacteria also contain a specific carotenoid [128]. The probable function of the β,β-carotene associated with P700 is to protect the Chl *a* molecules of P700 from photodestruction.

Cytochromes *f* and b_6 are present in most but not all P700 preparations (see above); cytochrome b_{559} is present only in those having relatively high Chl/P700 ratios (>100:1), i.e., the larger subchloroplast particles. The cytochromes participate in the electron transfer reactions between the two photosystems in plants and in the cyclic flow of electrons around PS I. Their association with the components responsible for the primary photochemical event is much less tenacious than that between the membrane-bound cytochromes and the reaction-center components in some purple bacteria (e.g., *Chr. vinosum* and *Rps. viridis*), in which the cytochromes have never been completely removed while the primary photochemical reaction was retained.

Considerable efforts have been made to determine what substances of relatively low molecular weight are present in P700 preparations, since such data might help identify the components participating in the primary photochemical event (in particular the primary electron acceptor) or in secondary electron-transfer reactions around PS I. The nature of the lipids associated with PS I is known [95], but not what lipids might be associated specifically with the P700 moiety. Several possible electron transfer moieties have been ruled out as the primary electron acceptor for P700. For example, pteridines and flavins are not present in one photochemically active P700-Chl *a*-protein preparation [106]. Since one likely participant in the early photochemistry would be a quinone (cf. purple bacteria), it is interesting that fractions such as TSF-1 contain several different quinones, but preparations more enriched in P700 contain a specific quinone in about the same concentration as P700. Shiozawa [120] determined that phylloquinone (vitamin K_1) occurs in an approximately equimolar ratio to P700 in the higher plant P700-Chl *a*-protein, whereas Dietrich and Thornber [106] found an as yet unidentified quinone [probably a substituted naphthoquinone (F. Allen, personal communication)] in the analogous complex in a cyanobacterium [108]. No evidence has yet been found for the participation of quinones in the primary photochemical event.

Participation of membrane-bound Fe-S centers (bound ferredoxins) in the early light-driven events in PS I, and the occurrence of absorption changes at 430 nm associated with reduction of a low-potential electron acceptor, are undisputed [126, 129-133], although just how early in the charge separation process these Fe-S centers participate *is* disputed. On the one hand, Bearden and Malkin [126, 129] believe that one Fe-S center (g = 1.94) is directly and stoichiometrically reduced by an electron from P700*, i.e., it is the primary electron acceptor (cf. Golbeck et al. [131]). Studies on the primary event in isolated P700-Chl *a*-proteins have supported their view. In LDAO or Triton preparations at 15 K, P700 becomes irreversibly oxidized while the Fe-S center becomes irreversibly reduced in the light, whereas in the SDS preparation from cyanobacteria, which was found to contain no Fe-S centers, P700 is not oxidized in the light at the same low temperature (cf. Golbeck et al. [131]). At room temperature the P700 in the SDS preparation does photooxidize, but with a relatively high quantum requirement (8-12). This suggests that P700*, lacking its natural primary electron acceptor in SDS preparations, uses an exogenous acceptor such as oxygen at room temperatures but at 15 K is unable to function [134]. On the other hand, Evans et al. [130] and McIntosh et al. [133] have observed that when the bound Fe-S centers in P700 preparations (except those made by SDS) are reduced by sodium dithionite prior to illumination, P700 still undergoes photooxidation, but reversibly; in addition to the reversibly g = 2 signal of $P700^+$, a reversible ESR signal (g_x = 1.78) is seen. These workers [130, 133 propose that the g = 1.78 signal arises from photoreduction of the true primary electron acceptor and that the g = 1.94 Fe-S center is a secondary acceptor. Further discussion of the controversy regarding the identity of the first component to receive an electron from P700* is not relevant here. Since the Fe-quinone complex of the reaction center of purple bacteria is considered to be the "primary" electron acceptor, even though a bacteriopheophytin molecule receives the electron from the primary donor first, the Fe-S center associated with P700 should be included as a constituent of the PS I reaction center even if the $g_x = 1.78$ component turns out to be the first to receive the electron from P700*. Iron-sulfur centers are in fact present in all P700 preparations made with LDAO or Triton X-100 [24, 108, 134].

An equally important and yet not fully resolved question about P700 is that of its associated polypeptide(s). Boardman et al. [135] have reviewed the extensive data on the polypeptides associated with PS I particles [111, 127, 136, 137]. In brief, P700-containing fractions, when electrophoresed by SDS-PAGE under mild conditions (i.e., without preheating), exhibit a chlorophyll-protein zone of 100-150 kD and, in some fractions, pigment-free polypeptides. (See other reviews [127, 135] for full details.) Note, however, that the electrophoretic mobility of the complex is anomalous [111]; for this and other reasons [127] the size thus determined can be regarded only as an approximation. The S value of the complex (9S) indicates that it is a close approximation. Since SDS destroys P700 activity in all organisms except certain cyanobacteria [108], it is only in these bacteria that P700 can be shown to be present in a ratio of 1 per 40 chlorophylls in the electrophoretic zone [106]; however, all P700-containing organisms contain a P700-Chl *a*-protein homologous to the protein found in cyanobacteris [103, 110].

Not all the protein in the chlorophyll-protein zone of 150 kD is associated with P700. Some, probably the bulk, is conjugated with the antenna chlorophylls present. Complete dissociation of this zone by heating in the presence of SDS, and sometimes also mercaptoethanol, has shown the complex to be composed of subunits. The P700-Chl *a*-protein from *Chlamydomonas reinhardi* y-1 chloroplasts contains two major polypeptides of 64 ± 1 and 49 kD, respectively [138]. Membrane fragments from the cyanobacterium *Anabaena flos-aquae* treated with Triton X-100 and chromatographed on DEAE-cellulose yield a fraction enriched in P700. SDS-PAGE gives reproducible protein bands of 15, 46, 52, and 120 kD [139]. The 46- and 52-kD subunits may correspond to the 46- and 48-kD subunits of the P700-Chl *a*-protein from *P. luridum,* and the 120-kD protein, which is made up of smaller subunits, may correspond to the 110-kD undissociated protein from *P. luridum.* One or two colorless polypeptides of 50-70 kD are widely acknowledged to be present in the PS I reaction center, and P700 in vivo is almost certainly conjugated with a polypeptide of this size [98, 108, 135] which is about equal to the sum of the two essential polypeptides (L + M) of the reaction center of purple bacteria. It appears that a polypeptide(s) of low molecular weight ($\leqslant 10{,}000$) is present also in the Triton, but not in the SDS, P700-Chl *a*-protein [24, 98, 108]. Close association of certain low molecular weight polypeptides with P700 may be inferred also from the studies of Chua et al. [111] on a P700-less mutant of *Chlamydomonas.* This low molecular weight polypeptide(s) may perhaps be derived from the bound ferredoxins [108, 140]. Thornber et al. [108] have proposed a model for the P700-Chl *a*-protein to account for some not easily reconciled data on the complex—an electrophoretic mobility equivalent to that of a protein of 100-150 kD, Chl/P700 ratio of 40:1, a ratio of 7,800 g protein per chlorphyll, and the presence of a major and a minor subunit of 50 ± 2 and 46 kD (cyanobacteria), or 50 and 48 kD (higher plants) [24, 108, 120], or 64 and 49 kD (green algae) [138]. Their model consists of two trimers, one containing two of the larger and one of the smaller subunits, and the other having three of the larger subunits; each subunit has seven chlorophyll molecules, but only the smaller subunit contains P700. (Bengis and Nelson [98] have interpreted their data on their P700 preparation as indicating that one P700 is associated with every two 70-kD polypeptides present.) The model of the P700-Chl *a*-protein is very similar to that of the Bchl *a*-protein of green bacteria [86, 87, 108].

The amino acid composition of the P700-containing polypeptide is not yet known; compositions have been determined only for the P700-Chl *a*-protein (Chl/P700 = 40:1) from a cyanobacterium and from a higher plant (see Table 3). If these reflect the composition of the P700-containing polypeptide, then the latter is similar to the composition of the polypeptides (L + M) of *Rps. sphaeroides* R-26 reaction centers. As expected for membrane-bound components, the analyses (Table 3) show a high proportion of hydrophobic residues.

d. Summary. The extensive data on P700 have been briefly reviewed here; other reviews [12, 14, 108, 126, 127, 129, 132, 135-137] provide more detailed coverage. As yet no P700 preparation is available that is as pure and as highly enriched as the reaction-center preparations from purple bacteria; therefore, virtually all the characteristics attributed to the P700 reaction center must be considered as tentative. The reaction center is envisaged as a pigment-protein complex composed of a polypeptide of 50-70 kD containing a preponderance of hydrophobic amino acid residues. Within this folded polypeptide are at least two Chl *a* molecules forming the primary electron donor, P700. Considering the relatively large size of the polypeptide and the Chl/protein ratios in P700-containing preparations, the number of Chl *a* molecules may be as high as 10; a carotene molecule is very probably associated with the reaction center. The most enriched preparations of P700, obtained by low-temperature solvent extraction, contain 5-9 chlorophylls plus P700 in the minimal entity. The function of the chlorophyll molecules, other than those comprising P700 in the reaction center, is not known. They may participate in the primary photochemical event and/or they may be involved in determining the correct orientations of P700, primary electron acceptor, and polypeptide required for high efficiency of the primary reaction. The identity of the primary acceptor is controversial, but Fe-S centers and a quinone are apparently closely associated with P700. P700 absorbs most strongly at 701-704 nm and 435 nm in the intact membrane [25, 141] and at 696-698 nm and 435 nm in isolated reaction-center preparations [98, 101, 116]. The

~700-nm absorbance maximum is shifted from the red wavelength maximum of Chl *a* in vitro (~663 nm), as would be expected for a Chl *a* dimer. In vivo, cytochromes *f* and b_6 appear to be closely associated with P700. Much remains to be done before the PS I reaction center is fully understood; it is imperative that it be isolated from extraneous material for an unequivocal characterization.

B. *Photosystem II*

1. Separation of PS I and PS II

Andersson et al. [142] and Åkerlund et al. [143] have devised a method for separating membranes from broken, unstacked spinach chloroplasts by counter-counter distribution in an aqueous dextran-polyethylene glycol two-phase system, which separates according to surface properties. PS I "particles" are concentrated in the polyethylene glycol-rich top phase. The material partitioning to the dextran-rich bottom phase consists of large vesicles (membrane thickness ~70 Å) resembling swollen grana disks. The Chl *a*/Chl *b* ratio of the vesicles is about 2.3, and electron transport rates (H_2O to phenyl-*p*-benzoquinone and ascorbate-2,6-dichlorophenolindophenol to $NADP^+$) indicate a PS II/PS I ratio of ~6. There is apparently no enrichment for reaction centers relative to total chlorophyll.

A number of procedures have been developed for separating PS II from PS I and for enriching preparations in PS II, but reaction centers from PS II have not yet been isolated in pure form. Because the reaction center of PS II is harder to detect spectrophotometrically than that of PS I, fewer efforts have been made to isolate the former.

2. PS II Reaction Center

a. Composition. The PS II reaction center consists of a primary electron donor P_{II} (P680, P690), thought to be a Chl *a* dimer [144-146], and a primary electron acceptor A_1 (Q, C-550 X-320) recently identified as a plastoquinone molecule [147, 172]. A 47-kD membrane polypeptide may be associated specifically with the PS II reaction center in *Chlamydomonas reinhardi* chloroplasts [148].

Light-induced absorbance changes due to the photochemical electron transfer from P_{II} to A_1 are shown in Figure 20. The redox potential of P_{II} appears to be greater than +850 mV [149]. A secondary electron donot (E_m = +475 mV) other than cytochrome b_{559} can reduce P_{II}^+ at 77 K [149]. The redox potential of A_1 appears to be -320 mV at pH 7.0 from the redox titration of the variable fluorescence yield of PS II [150]. The electron transfer reactions associated with PS II are summarized in Figure 21.

b. Localization. As shown in Figure 22, the PS II reaction center is oriented in the thylakoid membrane so that P_{II} is close to the inside surface and A_I is near

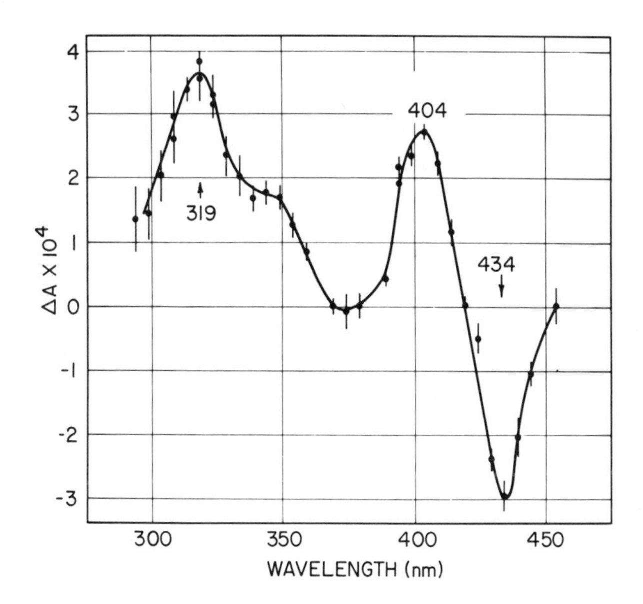

Figure 20 Difference spectrum of the rapidly decaying absorbance changes induced by a single saturating flash in spinach chloroplasts at pH 4.0. The changes between 295 and 360 nm are due mainly to the reduction of the primary acceptor A_1 and those between 400 and 450 nm are due mainly to oxidation of the primary donor P_{II}. Reproduced with permission from Mathis et al. [145].

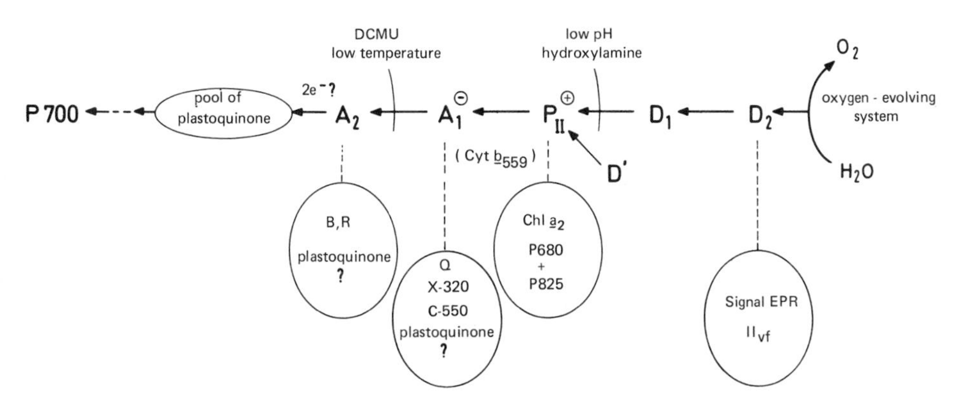

Figure 21 Electron transfer reactions around the PS II reaction center. A_1, A_2, electron acceptors; D_1, D_2, D', electron donors. The inserts attached to electron carriers indicate hypothetical identifications and names commonly found in the literature. Reproduced with permission from Mathis et al. [145].

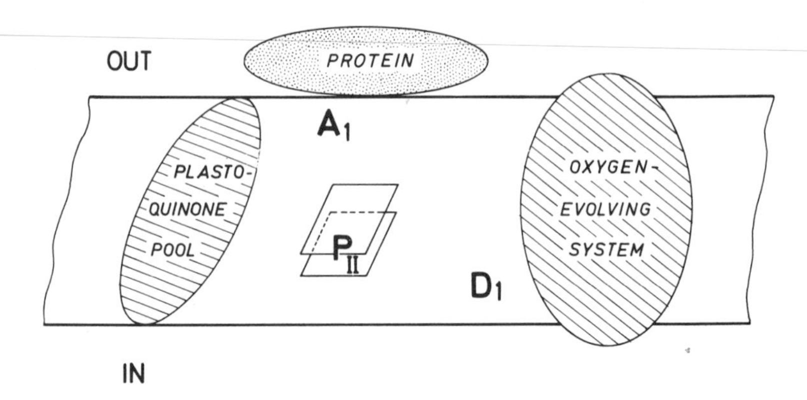

Figure 22 Tentative model for localization of some PS II constituents in the thylakoid membrane. OUT refers to the matrix side of the thylakoid. The stippled area represents a hypothetical protein coat. Reproduced with permission from Mathis et al. [145].

the outside but is not normally accessible to ions [145, 151]. Normally P_{II} is inaccessible to outside electron donors except for tetraphenyl boron or hydroxylamine, but treatment of the membrane with Tris enables secondary donors D_1 and D_2 to accept electrons from external donors and apparently makes P_{II}^+ accessible to hydrophilic ions [145].

The electron donor sites for each PS II reaction center appear to be separate, but the acceptor sites appear to be interconnected. A_I may be connected to the membrane-spanning plastoquinone pool by a specific two-electron carrier [145].

c. Detergent Preparations. Vernon and coworkers [95, 152, 153] and Ke's group [154] have isolated the best characterized PS II reaction-center particle, which is obtained by treating spinach chloroplasts with Triton X-100, as outlined in Figure 18, and fractionating on a sucrose gradient (Fig. 23). The particle size (~10 nm diameter) corresponds to a mass of ~300 kD [154]. The particle contains functional P_{II} (P680) and is 10-times enriched in cytochrome b_{559} and C-550 (A_1) compared with whole chloroplasts. The ratio of total chlorophyll to manganese (heat stable) to β,β-carotene is roughly 15:1.5:1 [101, 154]. The Chl *a*/Chl *b* ratio may vary from 8 (see Fig. 24) to 28 ± 2 [154]. The particles do not evolve oxygen, but will carry out a light-induced electron transfer from 1,5-diphenylcarbazide to 2,6-dichlorophenolindophenol or potassium ferricyanide, a PS II reaction [101, 152]. There is no detectable P_I (P700) or bound Fe-S protein [154].

Wessels et al. [155] treated spinach chloroplasts with digitonin and ob-

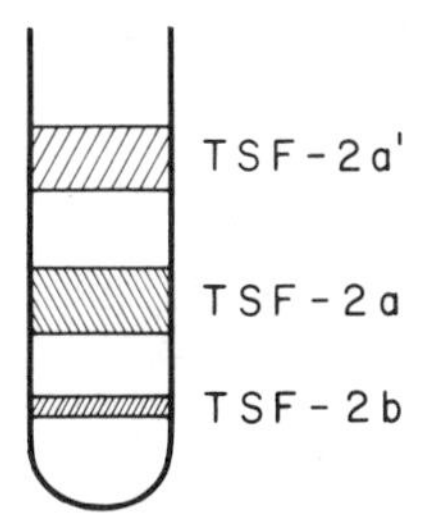

Figure 23 Distribution of fractions from light green supernatant after sucrose density gradient centrifugation (see Fig. 18). Reproduced with permission from Vernon et al. [101].

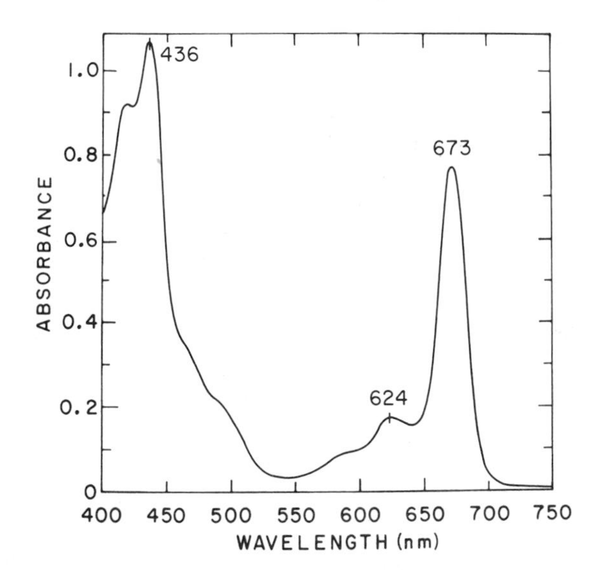

Figure 24 Absorption spectrum at 77 K of PS II reaction-center particle (TSF-2a) prepared from spinach with Triton X-100. Chl a/Chl b = 7.8. Chl/Cyt b_{559} ~15. Reproduced with permission from Vernon et al. [101].

tained PS II particles (size not specified), very similar to those obtained with Triton X-100, that have a high ratio of Chl a/Chl b and are enriched in β,β-carotene and cytochrome b_{559}. At 77 K, light-induced reduction of A_1 and oxidation of cytochrome b_{559} are observed. The ratio of total chlorophyll to A_1 (semiquinone anion in PS II reaction center) ranged from 20 to 70 in various preparations [146]. (In intact chloroplasts this ratio is about 400.) The Chl/reaction-center ratio is slightly higher in the digitonin particle than in the Triton particle.

Treatment of spinach chloroplasts with deoxycholate yields fragments in which PS I and PS II are not well separated. Ratios of C-550 (A_1) to P700 range from <0.2 to ~3 in various fractions. Nonetheless, fractions enriched in PS II have proved useful for photochemical studies of PS II [141, 172].

V. Summary

The best characterized photochemical reaction center has been obtained in a pure state from a carotenoidless mutant (R-26) of a purple nonsulfur bacterium (*Rps. sphaeroides*) by treatment of bacterial membranes with the nonionic detergent Triton X-100 and the weakly zwitterionic detergent LDAO. This reaction center (~85 kD) consists of three polypeptides (~21, ~24, and ~28 kD), four Bchl *a* molecules, two Bph *a* molecules, two UQ_{10} molecules, and one iron atom (Fe^{2+}). The ~28-kD polypeptide and the iron atom are not essential, and an even smaller reaction center (~50 kD) can still carry out the essential photochemistry. A Bchl *a* dimer (P870) serves as the primary electron donor, and the UQ_{10} serves as the primary electron acceptor (X). A single Bph *a* molecule serves as the intermediate electron carrier (I). When a photon is absorbed, there is created an excited state (*) which migrates to P870 ($E_m \simeq 450$ mV) in the reaction center. Within 6 psec, P870* donates an electron to Bph *a,* and within ~200 psec, Bph *a* transfers the electron to UQ_{10} to form ubisemiquinone. The energy of the excited state (1.4 eV) is thereby converted to chemical free energy (~0.6 eV) with an efficiency of ~40%.

Reaction centers (~150 kD) very similar to the R-26 reaction center have been isolated from both carotenoidless mutants and wild types of other purple nonsulfur bacteria (*R. rubrum, Rps. capsulata,* and *Rps. gelatinosa*). Slightly different reaction centers (~100-200 kD) have been isolated from the purple sulfur bacterium *Chr. vinosum* and the Bchl *b*-containing bacteria, *Rps. viridis* and *Th. pfennigii.* These reaction centers, obtained by the use of LDAO and the anionic detergent SDS, contain two types of tightly bound cytochrome *c* which serve as secondary electron donors to the bacteriochlorophyll dimer. (The presence of the cytochromes permits the photochemical trapping of I in the reduced state.) The primary electron acceptor (X) appears to be MQ instead of UQ.

Reaction centers have not yet been isolated from the photosynthetic membranes of green bacteria, but the membranes have been separated without the use of detergents from the chlorophyll bodies which contain ~95% of the light-harvesting chlorophyll. These unit membranes in the form of vesicles (average diameter 500-600 Å (are obtained by breaking the cells and fractioning the crude extract by sucrose density gradient centrifugation. Photochemical activity is associated with the unit-membrane vesicles (complex I) which contain ~80 Bchl *a* molecules per reaction center P840 as well as cytochrome *c*, cytochrome *b*, and carotenoid. Six classes of polypeptide are also found in the vesicles: ~18, ~21, ~23, ~31, ~36, and ~42 kD (Bchl *a*-protein). Upon treatment of complex I with 2 M guanidine-HCl, about half the Bchl *a* is removed from the vesicles along with at least 70% of the protein. The ~21- and ~31-kD polypeptides are least affected by this treatment, and may therefore be associated with the reaction center, since the treated vesicles (complex II) remain photochemically active. The ratio Bchl *a*/P840 is ~40, and the ratio cytochrome c_{553}/P840 is ~2. The nature of I is unknown, but X might be an Fe-S center. The primary donor P840 (E_m = +240 mV) has the proper characteristics for a Bchl *a* dimer.

Cyanobacteria (blue-green algae) and plant chloroplasts contain two kinds of reaction centers in PS I and PS II, respectively, as shown in Figure 2. The PS I reaction center resembles the bacterial reaction centers more closely than does the PS II reaction center. The primary electron donor of PS I is P700, a Chl *a* dimer ($E_m \simeq$ +400 mV), and a bound Fe-S protein may serve as X. There is some evidence for an intermediate electron carrier (I) as yet unidentified. The PS I reaction center has not been isolated free of light-harvesting chlorophyll, but several particles enriched in P700 have been prepared from chloroplast or cyanobacterial membranes by the use of Triton X-100, digitonin, SDS, and/or LDAO. The HP700 particle, prepared from carotenoid-depleted membranes with Triton X-100, has a mass of ~100-1,000 kD, a Chl/P700 ratio of ~30, Chl *a*/Chl *b* ratio of ~4, and also contains cytochromes *f* and b_6. The P700-Chl *a*-protein, prepared with Triton X-100, contains 40-45 Chl *a* per P700 and the Fe-S center thought to be X. When this protein is prepared with SDS, the Fe-S center is missing and photochemistry at 15 K ceases. The SDS protein (~110 kD) contains 46- and 48-kD polypeptides. A PS I particle made from Swiss chard chloroplasts with digitonin, Triton X-100, and SDS has a mass of ~140 kD per P700 and a Chl *a*/P700 ratio of 40-50. The particle contains a 70-kD polypeptide but no Fe-S center to serve as X. Another PS I particle (size unknown) made from spinach with digiton and wet ether contains only five to nine Chl *a* per P700.

The primary electron donor (P680 or P690) of PS II also appears to be a Chl *a* dimer (E_m > +850 mV), and the primary acceptor is a specialized plastoquinone (PQ) molecule (analogous to the UQ and menaquinone molecules in purple bacterial reaction centers). Nothing is known about the possibility of an intermediate electron carrier. PS II particles (~300 kD) have been prepared from spinach with Triton X-100 or digitonin. They contain ~15 Chl *a* molecules and ~0.5 Chl *b* molecules per cytochrome b_{559}, of which one is assumed to occur with each reaction center.

In all photosynthetic reaction centers the primary electron donors (P680 or P690, P700, P840, P870, and P960) seem to be chlorophyll dimers (in spite of some uncertainty about P960). Primary electron acceptors are either quinones (UQ, MQ, or PQ) or Fe-S centers. Intermediate electron carriers (between primary donors and acceptors) are known to exist only in reaction centers of purple bacteria where Bph *a* and Bph *b* function. Three polypeptides are associated with the reaction centers of purple nonsulfur bacteria, but only two (~21 and ~24 kD) are required for photochemistry. In green bacteria, ~21- and ~31-kD polypeptides may be associated with reaction centers. Two polypeptides (~50 and ~60 kD) are associated with a P700-Chl *a*-protein from PS I. The ~50 kD polypeptide, which is thought to be associated with P700, is about equal in molecular weight to the combined molecular weights of the essential polypeptides for bacterial reaction centers. Also a 47-kD polypeptide is associated with PS II in green algae. This data suggests that the minimum protein content may be ~50 kD for all photosynthetic reaction centers.

Recent Developments

Recent work of Davis et al. [177, 178] suggests that P680 in photosystem II may be a ligated Chl *a* monomer. The intermediate electron carrier (I) in PS II appears to be a single molecule of pheophytin *a* [179] analogous to Bph *a* in the reaction centers of purple bacteria. In PS I the intermediate electron carrier is more likely to be a single molecule of Chl *a* [180].

Acknowledgments

J. P. T. gratefully acknowledges the support of the Guggenheim Memorial Foundation during the preparation of this review. We appreciate the helpful comments of Dan Brune, Jack Fajer, Gabriel Gingras, Myron Ledbetter, Paul Mathis, Roger Prince, and J. C. Romijn. We thank Margaret Dienes for major editorial assistance.

References

1. K. Sauer, in *Bioenergetics of Photosynthesis* (Govindjee, ed.), Academic, New York, 1975, pp. 115-181.
2. J. R. Norris, R. A. Uphaus, H. L. Crespi, and J. J. Katz, *Proc. Nat. Acad. Sci. U. S. 68:* 625 (1971).
3. P. L. Dutton, R. C. Prince, D. M. Tiede, K. M. Petty, K. J. Kaufmann, T. L. Netzel, and P. M. Rentzepis, *Brookhaven Symp. Biol. 28:* 213 (1977).
4. J. Fajer, D. C. Brune, M. S. Davis, A. Forman, and L. D. Spaulding, *Proc. Nat. Acad. Sci. U. S. 72:* 4956 (1975).
5. J. Fajer, M. S. Davis, D. C. Brune, L. D. Spaulding, D. C. Borg, and A. Forman, *Brookhaven Symp. Biol. 28:* 74 (1977).

6. R. C. Prince, J. S. Leigh, Jr., and P. L. Dutton, *Biochim. Biophys. Acta 440:* 622 (1976).
7. M. G. Rockley, M. W. Windsor, R. J. Cogdell, and W. W. Parson, *Proc. Nat. Acad. Sci. U. S. 72:* 2251 (1975).
8. V. A. Shuvalov and V. V. Klimov, *Biochim. Biophys. Acta 440:* 587 (1976).
9. V. A. Shuvalov, I. N. Krakhmaleva, and V. V. Klimov, *Biochim. Biophys. Acta 449:* 597 (1976).
10. D. M. Tiede, R. C. Prince, and P. L. Dutton, *Biochim. Biophys. Acta 449:* 447 (1976).
11. D. M. Tiede, R. C. Prince, G. H. Reed, and P. L. Dutton, *FEBS Lett. 65:* 301 (1976).
12. B. Ke, in *Current Topics in Bioenergetics* Vol. 7A (D. R. Sanadi and L. P. Vernon, eds.) Academic, New York, 1978, pp. 75-138.
13. W. W. Parson and R. J. Cogdell, *Biochim. Biophys. Acta 416* 105 (1975).
14. B. Kok, *Biochim. Biophys. Acta 48:* 527 (1961).
15. R. K. Clayton, *Biochim. Biophys. Acta 75:* 312 (1963).
16. I. D. Kuntz, P. A. Loach, and M. Calvin, *Biophys. J. 4:* 227 (1964).
17. P. A. Loach, G. M. Androes, A. F. Maksim, and M. Calvin, *Photochem. Photobiol. 2:* 443 (1963).
18. D. W. Reed and R. K. Clayton, *Biochem. Biophys. Res. Commun. 30:* 471 (1968).
19. G. Gingras and G. Jolchine, *Proc. 1st Int. Cong. Photosyn. Res. 1:* 209 (1969).
20. G. Gingras, in *The Photosynthetic Bacteria* (R. K. Clayton and W. R. Sistron, eds.), Plenum, New York, 1978, pp. 119-131.
21. R. K. Clayton and R. T. Wang, in *Methods in Enzymology* Vol. 23A (A. San Pietro, ed.), Academic, New York, 1971, pp. 696-704.
22. G. Feher, *Photochem. Photobiol. 14:* 373 (1971).
23. M. Y. Okamura, L. A. Steiner, and G. Feher, *Biochemistry 13:* 1394 (1974).
24. F. A. Hunter, Ph.D. Thesis, University of California, Los Angeles, 1976.
25. Y. Inoue, T. Ogawa, and K. Shibata, *Biochim. Biophys. Acta 305:* 483 (1973).
26. R. Reiss-Husson and G. Jolchine, *Biochim. Biophys. Acta 256:* 440 (1972).
27. B. J. Segen and K. D. Gibson, *J. Bacteriol. 105:* 701 (1971).
28. P. A. Loach, D. L. Sekura, R. W. Hadsell, and A. Stemer, *Biochemistry 9:* 724 (1970).
29. L. Slooten, *Biochim. Biophys. Acta 256:* 452 (1972).
30. G. Jolchine and F. Reiss-Husson, *FEBS Lett. 40:* 5 (1974).
31. G. Feher and M. Y. Okamura, *Brookhaven Symp. Biol. 28:* 183 (1977).
32. M. Lutz, J. Kleo, and F. Reiss-Husson, *Brookhaven Symp. Biol. 28:* 367 (1977).
33. D. Rosen, M. Y. Okamura, G. Feher, L. A. Steiner, and J. E. Walker, *Biophys. J. 17:* 67a (1977).
34. D. W. Reed, *J. Biol. Chem. 244:* 4936 (1969).

35. J. R. Bolton, R. K. Clayton, and D. W. Reed, *Photochem. Photobiol. 9:* 209 (1969).
36. S. C. Straley, W. W. Parson, D. C. Mauzerall, and R. K. Clayton, *Biochim. Biophys. Acta 305:* 597 (1973).
37. W. W. Parson and T. G. Monger, *Brookhaven Symp. Biol. 28:* 195 (1977).
38. P. L. Dutton, J. S. Leigh, and C. A. Wraight, *FEBS Lett. 36:* 169 (1973).
39. G. Feher, R. A. Isaacson, M. Y. Okamura, *Biophys. J. 17:* 149a (1977).
40. R. T. Wang and R. K. Clayton, *Photochem. Photobiol. 17:* 57 (1973).
41. W. R. Smith, Jr., Ph.D. Thesis, University of Illinois at Urbana-Champaign, 1972.
42. H. M. Noël, M. van der Rest, and G. Gingras, *Biochim. Biophys. Acta 275:* 219 (1972).
43. R. Bachofen, K. W. Hanselmann, M. Snozzi, H. Zurrer, P. A. Cuendet, and H. Zuber, *Brookhaven Symp. Biol. 28:* 365 (1977).
44. R. Govindjee, W. J. Smith, Jr., and Govindjee, *Photochem. Photobiol. 20:* 191 (1974).
45. K. F. Neith and G. Drews, *Arch. Microbiol. 96:* 161 (1974).
46. K. F. Neith, G. Drews, and R. Feick, *Arch. Microbiol. 105:* 43 (1975).
47. R. C. Prince and A. R. Crofts, *FEBS Lett. 35:* 213 (1973).
48. R. K. Clayton and B. J. Clayton, *Abstracts of the 5th Annual Meeting of the Americal Society of Photobiology,* San Juan, Puerto Rico, 11-15 May, 1977, p. 65.
49. L. Lin and J. P. Thornber, *Photochem. Photobiol. 22:* 37 (1975).
50. J. C. Romijn, *Proceedings of the Bacterial Photosynthesis Conference,* Brussels, Belgium, 6-9 September 1976, Abstr. MB5.
51. R. van Grondelle, J. C. Romijn, and N. G. Holmes, *FEBS Lett. 72:* 187 (1976).
52. J. P. Thornber, in *Methods in Enzymology* Vol. 23A (A. San Pietro, ed.), Academic, New York, 1971, pp. 682-691.
53. G. D. Case, W. W. Parson, and J. P. Thornber, *Biochim. Biophys. Acta 223:* 122 (1970).
54. P. L. Dutton, T. Kihara, J. A. McCray, and J. P. Thornber, *Biochim. Biophys. Acta 226:* 81 (1971).
55. Y. D. Halsey and B. Byers, *Biochim. Biophys. Acta 387:* 349 (1975).
56. M. Y. Okamura, L. C. Akerson, R. A. Isaacson, W. W. Parson, and G. Feher, *Biophys. J. 16:* 223a (1976).
57. M. Y. Okamura, R. A. Isaacson, and G. Feher, *Biophys. J. 17:* 149a (1977).
58. J. P. Thornber, J. M. Olson, D. M. Williams, and M. L. Clayton, *Biochim. Biophys. Acta 172:* 351 (1969).
59. N. L. Pucheu, N. L. Kerber, and A. F. Garcia, *Arch. Microbiol. 101:* 259 (1974).
60. T. L. Trosper, D. L. Benson, and J. P. Thornber, *Biochim. Biophys. Acta 460:* 318 (1977).
61. N. L. Pucheu, N. L. Kerber, and A. F. Garcia, *Arch. Microbiol. 109:* 301 (1976).
62. J. P. Thornber, P. L. Dutton, J. Fajer, A. Forman, D. Holten, J. M. Olson, W. W. Parson, R. C. Prince, D. M. Tiede, and M. W. Winsor, *Proceedings of*

the 4th International Congress on Photosynthesis, Reading, England, 4-9 September 1977, pp. 55-70.
63. J. P. Thornber and J. M. Olson, *Photochem. Photobiol. 14:* 329 (1971).
64. F. K. Fong, *Proc. Nat. Acad. Sci. U. S. 71:* 3692 (1974).
65. T. L. Netzel, P. M. Renzepis, D. M. Tiede, R. C. Prince, and P. L. Dutton, *Biochim. Biophys. Acta 460:* 467 (1977).
66. T. L. Trosper, D. L. Benson, and J. P. Thornber, *Abstracts of the 4th Annual Meeting of the American Society of Photobiology,* Denver, Colorado, 16-20 February 1976, p. 63.
67. V. V. Klimov, V. A. Shuvalov, I. N. Krakhmaleva, A. A. Klevanik, and A. A. Krasnovsky, *Biokhimiya 42:* 519 (1977).
68. W. R. Smith, Jr., C. Sybesma, and K. Dus, *Biochim. Biophys. Acta 267:* 609 (1972).
69. J. M. Olson, R. C. Prince, and D. C. Brune, *Brookhaven Symp. Biol. 28:* 238 (1977).
70. J. M. Olson, in *The Photosynthetic Bacteria* (R. K. Clayton and W. R. Sistrom, eds.), Plenum, New York, 1978, pp. 161-178.
71. N. Pfennig and H. Biebl, *Arch. Microbiol. 110:* 3 (1976).
72. Y. Shioi, K. Takamiya, and M. Nishimura, *J. Biochem.* (Tokyo) *79:* 361 (1976).
73. C. R. Fowler, N. A. Nugent, and R. C. Fuller, *Proc. Nat. Acad. Sci. U. S. 68:* 2278 (1971).
74. G. Cohen-Bazire, N. Pfennig, and R. Kunisawa, *J. Cell Biol. 22:* 207 (1964).
75. S. C. Holt, S. F. Conti, and R. C. Fuller, *J. Bacteriol. 91:* 311 (1966).
76. N. Pfennig and G. Cohen-Bazire, *Arch. Mikrobiol. 59:* 226 (1967).
77. J. M. Olson, E. K. Shaw, and R. M. Englberger, *Biochem. J. 159:* 769 (1976).
78. C. F. Fowler, *Biochim. Biophys. Acta 357:* 327 (1974).
79. C. F. Fowler, B. H. Gray, N. A. Nugent, and R. C. Fuller, *Biochim. Biophys. Acta 292:* 692 (1973).
80. J. M. Olson, K. D. Philipson, and K. Sauer, *Biochim. Biophys. Acta 292:* 206 (1973).
81. R. C. Prince and J. M. Olson, *Biochim. Biophys. Acta 423:* 357 (1976).
82. D. B. Knaff and B. B. Buchanan, *Biochim. Biophys. Acta 376:* 549 (1975).
83. D. B. Knaff and R. Malkin, *Biochim. Biophys. Acta 430:* 244 (1976).
84. J. V. Jennings and M. C. W. Evans, *Proceedings of the Bacterial Photosynthesis Conference,* Brussels, Belgium, 6-9 September 1976, Abstr. MB3.
85. T. Mar and G. Gingras, *Biochim. Biophys. Acta 440:* 609 (1976).
86. R. E. Fenna and B. W. Matthews, *Nature 258:* 573 (1975).
87. R. E. Fenna and B. W. Matthews, *Brookhaven Symp. Biol. 28:* 170 (1977).
88. C. Sybesma and J. M. Olson, *Proc. Nat. Acad. Sci. U. S. 49:* 248 (1963).
89. C. Sybesma and W. J. Vredenberg, *Biochim. Biophys. Acta 75:* 439 (1963).
90. C. Sybesma and W. J. Vredenberg, *Biochim. Biophys. Acta 88:* 205 (1964).
91. J. M. Olson, T. H. Giddings, Jr., and E. K. Shaw, *Biochim. Biophys. Acta 449:* 197 (1976).

92. N. Pfennig and H. G. Trüper, in *Bergy's Manual of Determinative Bacteriology* 8th Ed. (R. E. Buchanan and N. E. Gibbons, eds.), Williams and Wilkins Co., Baltimore, Md., 1974, pp. 24-64.
93. D. L. Cruden and R. Y. Stanier, *Arch. Microbiol. 72:* 115 (1970).
94. N. K. Boardman and J. M. Anderson, *Nature 203:* 166 (1964).
95. L. P. Vernon and E. R. Shaw, in *Methods in Enzymology* Vol 23A (A. San Pietro, ed.), Academic, New York, 1971, pp. 277-292.
96. J. S. Brown, *Photophysiology 8:* 97 (1973).
97. J. S. C. Wessels and M. T. Bronchert, *Proceedings of the 3rd International Congress on Photosynthesis 1:* 474 (1975).
98. C. Bengis and N. Nelson, *J. Biol. Chem. 250:* 2783 (1975).
99. T. Ogawa and L. P. Vernon, *Biochim. Biophys. Acta 180:* 334 (1969).
100. H. Y. Yamamoto and L. P. Vernon, *Biochemistry 8:* 4131 (1969).
101. L. P. Vernon, E. R. Shaw, T. Ogawa, and D. Raveed, *Photochem. Photobiol. 14:* 343 (1971).
102. B. Ke, K. Sugahara, and E. R. Shaw, *Biochim. Biophys. Acta 408:* 12 (1975).
103. J. P. Thornber, *Biochim. Biophys. Acta 172:* 230 (1969).
104. J. P. Thornber, R. P. F. Gregory, C. A. Smith, and J. L. Bailey, *Biochemistry 6:* 391 (1967).
105. J. P. Thornber, J. C. Stewart, M. W. C. Hatton, and J. L. Bailey, *Biochemistry 6:* 2006 (1967).
106. W. E. Dietrich, Jr., and J. P. Thornber, *Biochim. Biophys. Acta 245:* 482 (1971).
107. T. Hiyama and B. Ke, *Biochim. Biophys. Acta 267:* 160 (1972).
108. J. P. Thornber, R. S. Alberte, R. A. Hunter, J. A. Shiozawa, and K.-S. Kan, *Brookhaven Symp. Biol. 28:* 132 (1977).
109. J. A. Shiozawa, R. S. Alberte, and J. P. Thornber, *Arch. Biochem. Biophys. 165:* 385 (1974).
110. J. S. Brown, R. S. Alberte, and J. P. Thornber, *Proc. 3rd Int. Congr. Photosynth. 3:* 1951 (1975).
111. N.-H. Chua, K. Matlin, and P. Bennoun, *J. Cell Biol. 67:* 361 (1975).
112. R. P. F. Gregory, S. Raps, and W. R. Bertsch, *Biochim. Biophys. Acta 234:* 330 (1971).
113. R. Malkin, *Arch. Biochem. Biophys. 169:* 77 (1975).
114. N. Nelson, C. Bengis, B. L. Silver, D. Getz, and M. C. W. Evans, *FEBS Lett. 58:* 363 (1975).
115. P. V. Sane and R. B. Park, *Biochem. Biophys. Res. Commun. 41:* 206 (1970).
116. I. Ikegami and S. Katoh, *Biochim. Biophys. Acta 376:* 588 (1975).
117. T. V. Marsho and B. Kok, in *Methods in Enzymology* Vol 23A (A. San Pietro, ed.), Academic, New York, 1971, pp. 515-522.
118. R. H. Lozier and W. L. Butler, *Biochim. Biophys. Acta 333:* 465 (1974).
119. J. Amesz, D. C. Fork, and W. Nooteboom, *Studia Biophys. 5:* 175 (1967).

120. J. A. Shiozawa, Ph.D. Thesis, University of California, Los Angeles, 1977.
121. J. S. Brown, *Brookhaven Symp. Biol. 28:* 362 (1977).
122. K. Philipson, V. Sato, and K. Sauer, *Biochemistry 11:* 4591 (1972).
123. H. T. Witt, *FEBS Lett. 38:* 112 (1973).
124. L. L. Shipman, T. M. Cotton, J. R. Norris, and J. J. Katz, *Proc. Nat. Acad. Sci. U. S. 73:* 1791 (1976).
125. M. C. W. Evans, C. K. Sihra, and A. R. Slabas, *Biochem. J. 162:* 75 (1977).
126. A. J. Bearden and R. Malkin, *Brookhaven Symp. Biol. 28:* 247 (1977).
127. J. P. Thornber, *Ann. Rev. Plant Physiol. 26:* 127 (1975).
128. R. J. Cogdell, W. W. Parson, and M. A. Kerr, *Biochim. Biophys. Acta 430:* 83 (1976).
129. A. J. Bearden and R. Malkin, *Quart. Rev. Biophys. 7:* 131 (1975).
130. M. C. W. Evans, C. K. Sihra, and R. Cammack, *Biochem. J. 158:* 71 (1976).
131. J. H. Golbeck, S. Lien, and A. San Pietro, *Biochem. Biophys. Res. Commun. 71:* 452 (1976).
132. B. Ke, *Biochim. Biophys. Acta 201:* 1 (1973).
133. A. R. McIntosh, M. Chu, and J. R. Bolton, *Biochim. Biophys. Acta 376:* 308 (1975).
134. R. Malkin, A. J. Bearden, F. A. Hunter, R. S. Alberte, and J. P. Thornber, *Biochim. Biophys. Acta 430:* 389 (1976).
135. N. K. Boardman, J. M. Anderson, and D. J. Goodchild, in *Current Topics in Bioenergetics* Vol. 8 (D. R. Sanadi and L. P. Vernon, eds.) Academic, New York, 1978, pp. 35-109.
136. J. M. Anderson, *Biochim. Biophys. Acta 416:* 191 (1975).
137. L. P. Vernon and S. M. Klein, *Ann. N. Y. Acad. Sci. 244:* 281 (1975).
138. R. Schantz, S. Bar-Nun, and I. Ohad, *Plant Physiol. 59:* 167 (1977).
139. S. M. Klein and L. P. Vernon, *Biochim. Biophys. Acta 459:* 364 (1977).
140. R. Malkin, P. J. Aparicio, and D. I. Arnon, *Proc. Nat. Acad. Sci. U. S. 71:* 2362 (1974).
141. H. J. van Gorkom, J. J. Tamminga, and J. Haveman, *Biochim. Biophys. Acta 347:* 417 (1974).
142. B. Andersson, H.-E. Åkerlund, and P.-A. Albertsson, *Biochim. Biophys. Acta 423:* 122 (1976).
143. H.-E. Åkerlund, B. Andersson, and P.-A. Albertsson, *Biochim. Biophys. Acta 449:* 525 (1976).
144. R. Malkin and A. J. Bearden, *Biochim. Biophys. Acta 396:* 250 (1975).
145. P. Mathis, J. Haveman, and M. Yates, *Brookhaven Symp. Biol. 28:* 267 (1977).
146. H. J. van Gorkom, M. P. J. Pulles, and J. S. C. Wessels, *Biochim. Biophys. Acta 408:* 331 (1975).
147. B. Ke, *Photochem. Photobiol. 20:* 542 (1974).
148. N.-H. Chua and P. Bennoun, *Proc. Nat. Acad. Sci. U. S. 72:* 2175 (1975).
149. A. J. Bearden and R. Malkin, *Biochim. Biophys. Acta 325:* 266 (1973).

150. B. Ke, F. M. Hawkridge, and S. Sahu, *Brookhaven Symp. Biol. 28:* 364 (1977).
151. C. J. Arntzen, C. Vernotte, J. M. Briantais, and P. Armond, *Biochim. Biophys. Acta 368:* 39 (1974).
152. L. P. Vernon and E. R. Shaw, *Biochem. Biophys. Res. Commun. 36:* 878 (1969).
153. L. P. Vernon, E. R. Shaw, and B. Ke, *J. Biol. Chem. 241:* 4101 (1966).
154. B. Ke, S. Sahu, E. R. Shaw, and H. Beinert, *Biochim. Biophys. Acta 347:* 36 (1974).
155. J. S. C. Wessels, O. van Alphe, and G. Voorn, *Biochim. Biophys. Acta 292:* 741 (1973).
156. L. A. Steiner, M. Y. Okamura, A. D. Lopes, E. Moskowitz, and G. Feher, *Biochemistry 13:* 1403 (1974).
157. R. C. Prince and P. L. Dutton, *Arch. Biochem. Biophys. 172:* 329 (1976).
158. M. A. Cusanovich, R. G. Bartsch, and M. D. Kamen, *Biochim. Biophys. Acta 153:* 397 (1968).
159. R. Malkin, D. B. Knaff, and A. J. Bearden, *Biochim. Biophys. Acta 305:* 675 (1973).
160. P. Mathis and A. Vermeglio, *Biochim. Biophys. Acta 369:* 371 (1975).
161. R. H. Lozier and W. L. Butler, *Biochim. Biophys. Acta 333:* 460 (1974).
162. B. Ke, L. P. Vernon, and T. Chaney, *Biochim. Biophys. Acta 256:* 345 (1972).
163. W. A. Cramer and J. Whitmarsh, *Ann. Rev. Plant Physiol. 28:* 133 (1977).
164. D. C. Borg, J. Fajer, A. Forman, R. H. Felton, and D. Dolphin, *Proc. 4th Int. Biophys. Congr., Moscow 2:* 528 (1973).
165. P. L. Dutton and J. S. Leigh, *Biochim. Biophys. Acta 314:* 178 (1973).
166. E. H. Evans, R. Cammack, and M. C. W. Evans, *Biochem. Biophys. Res. Commun. 68:* 1212 (1976).
167. N. W. Downer, N. C. Robinson, and R. A. Capaldi, *Biochemistry 15:* 2930 (1976).
168. B. Ke, *Arch. Biochem. Biophys. 152:* 70 (1972).
169. G. Jolchine and F. Reiss-Husson, *FEBS Lett. 52:* 33 (1975).
170. J. Oleze and J. R. Golecki, *Arch. Microbiol. 102:* 59 (1975).
171. M. van der Rest and G. Gingras, *J. Biol. Chem. 249:* 6446 (1974).
172. H. J. van Gorkom, *Biochim. Biophys. Acta 347:* 439 (1974).
173. D. W. Reed and G. A. Peters, *J. Biol. Chem. 247:* 7148 (1972).
174. M. Gläser, C. Wolff, H.-E. Buchwald, and H. T. Witt, *FEBS Lett. 42:* 81 (1974).
175. D. C. Borg, J. Fajer, D. Dolphin, and R. H. Felton, *Proc. Nat. Acad. Sci. U.S. 67:* 813 (1970).
176. J. J. Katz and J. R. Norris, Jr., in *Current Topics in Bioenergetics* Vol. 5 (D. R. Sanadi and L. Packer, eds.) Academic, New York, 1973, pp. 41-75.
177. M. S. Davis, A. Forman, L. K. Hanson, and J. Fajer, *Biophys. J. 25:* 148a (1979).
178. M. S. Davis, A. Forman, and J. Fajer, *Proc. Nat. Acad. Sci. U.S.*, submitted.
179. V. V. Klimov, A. V. Klevanik, V. A. Shuvalov, A. A. Krasnovsky, *FEBS Lett. 82:* 183 (1977).
180. I. Fujita, M. S. Davis, and J. Fajer, *J. Am. Chem. Soc. 100:* 6280 (1978).

6

Electron Transport and Energy-transducing Systems of *Escherichia coli*

Philip D. Bragg

The University of British Columbia
Vancouver, British Columbia
Canada

I. Introduction

Escherichia coli is a facultative anaerobic organism. It grows on many substrates including acetate and several intermediates of the tricarboxylic acid cycle when oxygen can serve as the terminal electron acceptor. Under anaerobic conditions, growth on many of these substrates will still occur in the presence of nitrate and fumarate which replace oxygen as the terminal electron acceptor. Fermentative growth is possible in the absence of oxygen, fumarate, and nitrate, but only with such substrates as glucose, galactose, or maltose [1-3]. Under fermentative conditions ATP is formed by substrate-level phosphorylation in the Embden-Meyerhof pathway and by phosphoroclastic cleavage of pyruvate to acetate. When the terminal electron acceptors are present, electron transfer through the respiratory chain can provide ATP by oxidative phosphorylation. The nature of the respiratory chains involved in these processes varies with the substrate and terminal electron acceptor, and the extent to which components are common to more than one pathway is not clear at present. Thus, although the NADH oxidase and formate-nitrate reductase systems appear to be separate pathways, it is not known if the NADH-nitrate reductase system involves components of both pathways.

The formation of ATP coupled to substrate oxidation through the respiratory chain involves the generation of an "energized state" of the membrane. It has been proposed that the free energy of oxidation is stored in an electrochemical potential ("protonmotive force") consisting of a proton gradient and/or an electrical potential across the membrane [4-6]. According to this chemiosmotic hypothesis, the components of the respiratory chain are organized in the membrane such that electron transfer through the respiratory chain is coupled to the transfer of protons across the membrane.

The protonmotive force (Δp) of the energized state has two components, the transmembrane pH difference (ΔpH) and the membrane potential ($\Delta\psi$), which are related by the equation

$$\Delta p = \Delta\psi - Z\,\Delta pH$$

where Z is a factor to convert pH to electrical units.

In *E. coli* the energized state can be used to drive such energy-requiring processes as ATP formation (oxidative phosphorylation), active transport, energy-dependent transhydrogenation of $NADP^+$ by NADH, reversal of electron flow in the respiratory chain, and flagella movement [6-8]. The energized state can be generated either by electron transfer through the respiratory chain or by hydrolysis of ATP through a reversal of the reactions of oxidative phosphorylation [4-8]. The latter process involves the membrane-bound ATPase system. The involvement of the ATPase system in the energization of all of the above energy-dependent processes is evident from studies which show that added ATP is ineffective in driving these reactions in *unc A* mutants in which the ATPase is inactive or in cells to which the inhibitor of the ATPase system, N,N′-dicyclohexylcarbodiimide (DCCD), has been added [7, 9, 10] (Table 1). The formation of the energized state by substrate oxidation through the respiratory chain is unaffected by DCCD or the *unc A* mutation. The two pathways for the formation of the energized state and its common use by all of these energy-requiring processes is shown in Figure 1.

Thus, the membrane-bound ATPase system has an important double function being involved in the formation of ATP by oxidative phosphorylation and in the use of ATP to energize the membrane. It can be considered to consist of two parts, F_1 and F_0 [4, 5]. F_1 is an extrinsic protein carrying the active site for ATP hydrolysis. F_0 is the intrinsic membrane protein to which F_1 is attached. It is postulated to be transmembranous and to constitute the channel or pathway for protons which is utilized when the proton gradient is being used to drive the formation of ATP or when it is being generated by hydrolysis of ATP [5].

In this chapter the components, organization, and interaction of the respiratory chain and the ATPase system are discussed.

Table 1 Role of the ATPase System in ATP-dependent Processes in *E. coli* Membranes

Strain	ATPase activity[a]	Oxidative phosphorylation[a]	ATP-dependent NAD^+ reduction by succinate[b]	β-Galactoside or proline transport[a]		Transhydrogenase[a]		Motility[c]	
				Aerobic	Anaerobic	Aerobic	Anaerobic	Aerobic	Anaerobic
Normal	+[d]	+	+	+	+	+	+	+	+
Normal + DCCD	–[d]	–		+	–	+	–	+	–
unc A⁻	–	–	–	+	–	+	–	+	–

[a]Data from Simoni and Postma [7], Gibson and Cox [9], and Cox and Gibson [10].
[b]Data from Poole and Haddock [241].
[c]Data from Larsen et al. [8].
[d]+, present; –, absent.

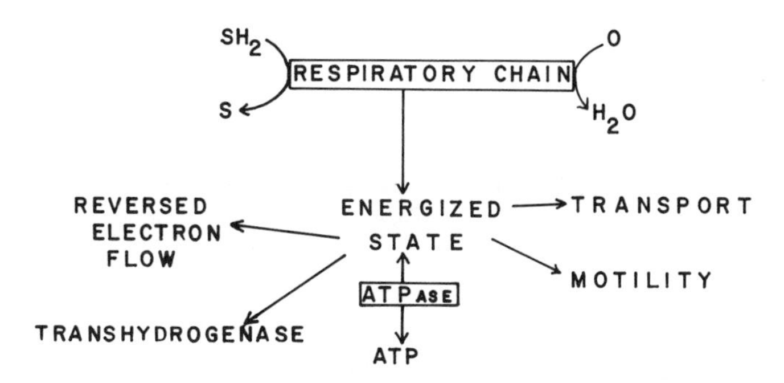

Figure 1 Generation and use of the energized state in the cell membrane of *E. coli.*

II. Subcellular Preparations Used in the Study of Energy-transducing Systems in *E. coli*

Although in many studies intact cells have been used, they are unsuitable for some studies for the following reasons: (1) it is difficult to deplete intact cells of endogenous substrates completely; (2) intact cells are impermeable to a wide variety of biologically important substances such as nucleotides; (3) even if the inner membrane is permeable to certain solutes the outer membrane may prevent their access to the cell; (4) there is often competition by cytoplasmic systems for substrates, and the presence of other than membrane systems can be confusing.

Two types of membrane systems have been developed to overcome these problems. The vesicle system of Kaback generates membrane vesicles from cytoplasmic membranes prepared from whole cells by treatment with lysozyme and EDTA [11]. These membranes are resealed under carefully defined conditions to generate vesicles which probably have the same orientation with respect to the suspending medium as the intact cell. In evidence of this, the vesicles can be energized by substrates such as D-lactate to result in active accumulation of amino acid, certain sugars, sugar acids and phosphates, lactate, pyruvate, deoxycytidine, and Rb^+ or K^+ in the presence of valinomycin [12-14]. Short et al. [15] have shown that almost all of the vesicles are capable of transporting solute. The appearance of the vesicles under freeze-etch microscopy is also consistent with a right-side-out orientation. However, the vesicles have considerable ATPase and NADH oxidase activity. Weiner [16] and Futai [17] found that up to 50% of the glycerol-3-phosphate dehydrogenase, NADH-ferricyanide reductase, and membrane-bound ATPase activities which appeared to be completely inaccessible from the outside of spheroplasts became exposed when the spheroplasts were converted to vesicles. They suggested that the vesicles consisted of a 1:1 mixture of right-side-out and inverted vesicles. Hare et al. [18], using antibody to the ATPase, separated the vesicles into two populations. Those which were not agglutinated by the antibody had low NADH oxidase activity and were able to accumulate proline whereas those which were agglutinated had little ability to accumulate proline and had high NADH oxidase activity. More recently, Futai and Tanaka [19] have re-examined this problem using antibody prepared against D-lactate dehydrogenase which appears normally to be localized on the inner (cytoplasmic) face of the membrane. Their results indicated that 85% of the dehydrogenase remained within the vesicles, suggesting that only 15% of the vesicles were inverted. Measurements of the glycerol-3-phosphate, succinate, and D-lactate-ferricyanide reductase activities of these vesicles suggested that 30-50% of these enzymes were accessible from the outside. This indicated that some rearrangement of the position of these enzymes within the membrane had occurred during the formation of the vesicles. Some rearrangement of the orientation of the ATPase occurred when vesicles were preincubated at pH 8-9 [20]. It can be concluded from these results that membrane vesicles prepared by Kaback's procedure are mainly with the right-side-out orientation but that care must be taken in the course of experiments to avoid rearrangement of the orientation of the enzymes.

In contrast to membrane vesicles prepared by Kaback's procedure, almost all of the membrane vesicles prepared by disruption of cells with a French press or by sonication have an inverted orientation with respect to whole cells [17, 19, 21]. Thus, the active sites of the membrane-bound ATPase and NADH dehydrogenase are fully exposed to the external medium, and proline transport is negligible. In contrast, ATP-driven accumulation of Ca^{2+} by these vesicles occurs readily as would be expected since Ca^{2+} is actively extruded from whole cells [21]. However, reorientation of some of the enzymes cannot be excluded.

In order to avoid confusion as to the orientation of the vesicles, in this review the term "membrane vesicles" will be applied to those vesicles which are prepared by Kaback's procedure and which have the same orientation as the intact cell. "Membrane particles" will be used to refer to membrane preparations which have been obtained by disruption of cells with a French or similar press, or by sonication. Membrane particles are probably sealed vesicles having an inverted orientation compared to intact cells.

III. The Aerobic Respiratory Chain

A. Composition

Reduced-minus-oxidized difference spectra of intact cells of *E. coli* grown aerobically show absorption peaks at 430, 530, 560, 590, and 630 nm [22]. These peaks are attributed to cytochrome b_1 (Soret band), cytochrome b_1 (β band), cytochrome b_1 (α band), cytochrome a_1 (α band), and cytochrome d (α band), respectively. Also present in the spectrum are troughs at about 465 and 650 nm due to the nonheme iron flavoprotein and cytochrome *d*, respectively, and a shoulder at 440 nm on the Soret band of cytochrome b_1 which is due to cytochrome *d*. Hidden by the absorption peaks of cytochrome b_1 are those of cytochrome *o*. This cytochrome reacts in the reduced form with carbon monoxide resulting in a shift in the positions of the absorption maxima. Cytochrome *o* is readily detected in reduced-plus-carbon-monoxide minus reduced difference spectra where absorption peaks at 416, 538, and 567 nm and troughs at 430 and 555 nm are indicative of this cytochrome [23]. Cytochrome *d* also combines with carbon monoxide to show absorption peaks in the difference spectrum at about 440, 640, and possibly 537 nm, and troughs at 443 and 620 nm. Figure 2 shows examples of reduced-minus-oxidized and carbon monoxide difference spectra of membrane particles from cells grown to the exponential and stationary phases of growth on a defined medium with glucose as the substrate for growth.

Besides the cytochromes and flavoprotein dehydrogenases the membrane-bound respiratory chain also contains nonheme iron and ubiquinone and/or menaquinone (vitamin K_2). In Table 2 are shown typical analyses of membrane particles. There is considerable variation in the levels of some components which may be attributed to differences in the respiratory chain produced by the different conditions of growth. Thus, the levels of cytochromes b_1, a_1, and *d* increase significantly as the cell enters the stationary phase of growth and the content of cytochrome *o* decreases (Fig. 2). In exponential phase cells the ratio of cytochrome b_1 to cytochrome *o* is between 2:1 and 5:1 whereas in stationary phase cells it can be 50:1 or greater. Surprisingly, large alterations in the level of the cytochromes are not reflected in major alterations in the oxidase activity. Thus, for the membrane particles giving the spectra in Figure 2 and the analysis shown in Table 2 (last two lines), the NADH oxidase and succinate oxidase activities were 427 and 25, and 362 and 32 ng-atoms oxygen used per min per mg protein for membrane particles from the exponential and stationary phase cells, respectively. More extensive results showing the relative constancy of NADH oxidase activity in the presence of wide variations in the level of the cytochromes have been given by Bragg et al. [29].

B. Factors Affecting the Composition of the Respiratory Chain

The factors controlling the level of the cyrochromes and the oxidase activities in *E. coli* are complex. Since the cells used to obtain the results of Figure 2 were grown in batch culture, there are several factors which are likely to be different between the conditions of stationary and exponential growth. An increasing density of cells as the stationary phase is approached will cause the oxygen concentration of the medium to approach zero [30]. The decrease in the concentration of glucose and the increase in the level of acids in the medium will result both in a decrease in the extent of catabolite repression and in the induction of the synthesis of certain enzymes.

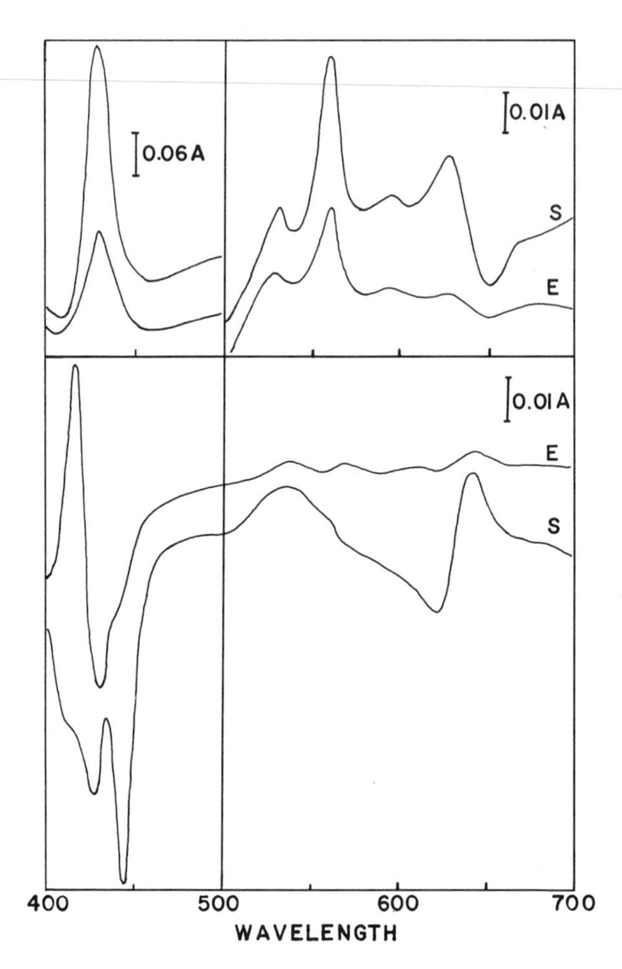

Figure 2 Difference spectra (25°C) of particles prepared from cells grown to the exponential (E) and stationary (S) phases of growth. The two upper curves are dithionite-reduced minus air-oxidized difference spectra. The two lower curves are the dithionite-reduced-plus-carbon-monoxide minus dithionite-reduced difference spectra. The cells were grown on succinate to the exponential or stationary phase of growth. Particles prepared from these cells were suspended to a concentration of 6.0 mg protein/ml and 4.8 mg protein/ml, respectively.

The effect of oxygen concentration on the formation of the cytochromes and other enzymes has been well explored in *E. coli.* Gray et al. [31] and Fujita and Sato [32] found that growth under anaerobic conditions resulted in the formation of cytochromes in addition to those present aerobically. These cytochromes, cytochromes c_{552} and c_{550} were not membrane-bound but were found together with cytochrome b_{562} in the cytoplasmic fraction. Cytochrome c_{552} is involved in the NADH-nitrite reductase system [33] but the function of the

Table 2 Composition[a] of Membrane Particles from Cells Grown Aerobically to the Exponential or Stationary Phase

Strain	Growth substrate and phase	Flavin		MQ[b]	UQ[b]	Fe[b]	ALS[b]	Cytochrome				References
		Total	Acid-soluble					b_1	o	d	a_1	
AN62	Glucose (E)[b]	0.25	–	0.67	4.7	–	–	0.19	0.073	0.027	+[b]	24
NRC482	Glucose (E)	0.11	0.053	–	4.0	3.4	–	0.36	0.087	–	–	25
D1111	Succinate-yeast extract (E)	0.54	–	–	3.9	11.2	–	0.34	0.12	–	–	26
C-1 (S)	Glycerol[c]	0.53	–	–	–	4.3	1.43	0.25	0.10	+	+	27
NRC482	Succinate (E)	–	–	–	–	–	–	0.36	0.066	0.076	0.038	28
NRC482	Succinate (S)[b]	–	–	–	–	–	–	0.96	0.02	0.64	0.25	28

[a]ng-atoms or nmoles per mg protein.
[b]MQ, menaquinone; UQ, ubiquinone; Fe, nonheme iron; ALS, acid-labile sulfide; E, exponential phase of growth; S, stationary phase of growth; +, present but not quantitated.
[c]Cells grown in continuous culture with substrate limiting.

other two cytochromes is unknown. Cytochrome b_{562} is also formed under aerobic conditions so that difference spectra of whole cells may show a distinct absorption peak for this cytochrome [34]. The levels of cytochromes b_1 and d are severalfold higher in aerobically grown cells than in those grown anaerobically [2, 3, 35-37] as also is the activity of NADH oxidase. Thomas et al. [38] using a continuous culture technique found that there were complex changes in the level of these components. Thus, when the partial pressure of oxygen in the medium was below 28 mm of mercury there was a 2.5-fold increase in the level of NADH oxidase and cytochrome d while cytochrome b_1 increased 3.9-fold. However, under strictly anaerobic conditions these levels dropped below those of aerobic cells. Somewhat similar results were obtained by Wimpenny and Necklen [39] who measured the redox potential as a sensitive indicator of the content of dissolved oxygen in the growth medium. Optimum levels of cytochromes b_1 and d were formed when the redox potential of the culture was about 100 mV, lower levels of both cytochromes being formed at more negative (more anaerobic) and more positive (more aerobic) redox potentials. Ishida and Hino [36] suggested that there are several effects of oxygen on cytochrome synthesis. Oxygen increases the level of enzymes in the tricarboxylic acid cycle resulting in a higher level of succinyl-CoA for synthesis of heme. Oxygen enhances the formation of the enzymes involved in heme biosynthesis and is required for two of the steps in heme biosynthesis, although if alternative electron acceptors such as fumarate or nitrate are present, heme biosynthesis can occur under anaerobic conditions [40, 41]. Furthermore, there may be an effect of oxygen on the synthesis of the apoprotein moiety of the cytochromes. Besides the effect on the cytochromes, the level of oxygen affects the synthesis of some of the enzymes feeding in reducing equivalents into the respiratory chain. Thus, succinate dehydrogenase may be up to fourfold higher in aerobically than in anaerobically grown cells [2, 37]. Other enzymes of the tricarboxylic acid cycle show even more marked differences. As proposed by Amarasingham and Davis [42], it is likely that the tricarboxylic acid cycle operates in a cyclic manner under aerobic conditions but under conditions of anaerobic growth it functions as a branched noncyclic pathway. An oxidative branch leading from citrate to α-ketoglutarate serves a purely biosynthetic role. A reductive branch leading from oxaloacetate and malate to succinate via fumarate reductase has a biosynthetic role and provides a means for anaerobic electron transport [43]. The repression of enzymes of the tricarboxylic acid cycle during aerobic growth on glucose is generally considered to be due to the availability of energy from glycolysis which leaves the cycle functioning to generate intermediates for biosynthesis only [1, 2]. The catabolite repression of succinate dehydrogenase synthesis is reversed by cAMP [44].

It is not clear whether NADH oxidase activity undergoes catabolite repression. Gray et al. [3] found that growth on glycerol and pyruvate resulted in a slight reduction in NADH oxidase activity compared to its level in glucose-grown cells. However, Shigeta and Hino [37] found that there was a 30% reduction in the activity of NADH oxidase activity in cells grown aerobically with glucose. With anaerobically grown cells a 70% decrease in activity was seen. Cytochrome b_1 behaved similarly showing a 30% reduction in amount when grown in the presence of glucose under aerobic conditions and a 60% reduction when grown under anaerobic conditions. It is likely that this is due to catabolite repression of protoporphyrinogen oxidase in the heme biosynthetic pathway [45]. Broman and Dobrogosz [46] and Dallas and coworkers [47] have shown that in adenyl cyclase and cAMP receptor protein-defective mutants, the levels of succinate dehydrogenase, NADH oxidase, and cytochromes b_1 and o were lower than in wild-type cells. The levels of these components were increased to those found in wild-type cells when the adenyl cyclase-deficient mutant was grown in the presence of cAMP. The content of cytochrome b_1 increased twofold under these conditions while cytochrome o was elevated by about sevenfold. Daoud and Haddock [48] have found somewhat different results. Using difference spectroscopy at 77 K to resolve the cytochrome b_1 peak into its component cytochromes (Section III. K.1), they found that the level of cytochromes b_{556} and b_{562} were little affected in an adenyl cyclase-cAMP receptor protein double mutant whereas the level of cytochromes b_{558} and d were elevated. Thus, either cAMP exerts a negative effect on the synthesis of cytochromes b_{558} and d or the changes in the level of these cytochromes is a secondary consequence of the mutations. The latter possibility seems more likely since Dallas et al. [47] have shown that several of the membrane systems were affected in these types of mutants.

When *E. coli* is grown anaerobically with nitrate or fumarate as terminal electron acceptors there are further changes observed in the composition of the respiratory chain. These changes are discussed in a later section.

C. Oxidase Activities

Although intact cells or whole cell extracts of *E. coli* can oxidize a variety of different substrates (glucose, galactose, fructose, glycerol, succinate, malate, formate, acetate, pyruvate, glutamate, α-ketoglutarate, citrate, isocitrate, NADH) depending on the conditions of growth, only a limited number of substrates are oxidized by membrane particles or vesicles. Typical activities of membrane particle and vesicle preparations are shown in Table 3. The most active enzyme system is that connected with NADH oxidation. NADPH is not oxidized [53]. NADH is formed as a product by many cytoplasmic enzymes including malate, pyruvate, α-ketoglutarate, glutamate, and 3-phosphoglyceraldehyde dehydrogenases. Thus, these substrates feed reducing equivalents via NADH into the mem-

Table 3 Enzyme Activities of Membrane Particles

		Specific activity[a]							
Growth substrate	Temperature of assay (°C)	NADH oxidase	NADH dehydrogenase[b]	Succinate oxidase	Succinate dehydrogenase[c]	D-lactate oxidase	Formate oxidase	DL-malate oxidase	References
Glucose-nutrient broth (pH 6.5)	30	150	210	80	110	–	1000	–	49
Glucose-nutrient broth	22-25	920	2050	315	435	–	–	–	50
Succinate	30	620	–	115	–	–	–	–	51
Glucose	25	230	–	–	–	68	–	31	24
Glucose	22	718	–	355	–	54	–	–	25
Glucose-amino acids	30	1210	1380	–	–	145	–	–	52
Succinate	25	270[d]	–	540[d]	–	330[d]	–	–	13
Glucose	25	620[d]	–	125[d]	–	300[d]	–	–	13

[a] ng-atoms 0/min and nmol/min per mg protein for oxidase and dehydrogenase activities, respectively.
[b] NADH-ferricyanide reductase.
[c] Succinate-phenazine methosulfate reductase.
[d] Membrane vesicles

brane-bound respiratory chain. Several very active NADH and NADPH dehydrogenases are also present in the cytoplasm or loosely attached to the membrane [53-58]. Since the physiological electron acceptors for these enzymes are unknown, their role in metabolism is unclear. Besides NADH, succinate and D-lactate are actively oxidized by membrane particles and vesicles. The activity of DL-malate oxidase is low and frequently is absent. DL-α-hydroxybutyrate and L-lactate oxidase activities may be present [13]. L-glycerol-3-phosphate oxidase activity is found in membranes from cells grown on glycerol as a result of the induction of a NAD^+-independent flavoprotein dehydrogenase [59, 60]. Membrane-bound formate oxidase activity is found in cells grown anaerobically on nitrate [61] but the highest levels are found if the cells are grown on a nutrient broth-glucose medium at pH 6.5 [49]. A further system funneling reducing equivalents into the respiratory chain at the level of ubiquinone is dihydroorotate dehydrogenase which is involved in the biosynthesis of pyrimidines [62-64]. Besides the pyruvate dehydrogenase complex which yields acetyl-CoA from pyruvate, and the phosphoroclastic system, a third system, pyruvate oxidase, is induced by the accumulation of pyruvate in the medium. This enzyme is a cytoplasmic flavoprotein which converts pyruvate to acetate with the transfer of reducing equivalents to the membrane-bound respiratory chain [65-69].

Thus, besides cytoplasmic enzymes which reduce NAD^+, reducing equivalents can be funneled into the membrane-bound respiratory chain of *E. coli* by a series of flavoprotein dehydrogenases, mainly membrane-bound, which result in the oxidation of NADH, succinate, malate, D- and L-lactate, α-hydroxybutyrate, L-glycerol-3-phosphate, formate, dihydroorotate, and pyruvate. The properties of those dehydrogenases which have been well characterized are discussed in Sections III.D-III.H.

D. NADH Dehydrogenase

It is only recently that this enzyme has been purified from *E. coli* [57, 73] although several attempts had been made previously [50, 53, 54, 70, 71] (see Table 4).

Brodie [70] isolated a NADH-cytochrome *c* reductase from a heat-treated cell-free extract of *E. coli* strain ECFS by fractionation with ammonium sulfate and adsorption on calcium phosphate gel. The enzyme reduced mammalian cytochrome *c*, oxygen, and tetrazolium dyes. The K_m for NADH was 59 μM but NADPH was not oxidized. The prosthetic group of the enzyme was FAD with neither riboflavin or FMN being able to replace it. Sulfhydryl groups were required for enzyme activity since 10^{-5} M *p*-chloromercuribenzoate was inhibitory. It is not clear whether this enzyme is cytoplasmic or whether it was released from the cell membrane by the heat treatment (60°C, 10 min). The latter alternative seems more likely since Bragg and Hou [53, 54] solubilized a similar NADH-

Table 4 Properties of NADH Dehydrogenases of *E. coli*

Name	Source[a]	K_m (μM)		Prosthetic group	Electron acceptors[b]	References
		NADH	NADPH			
NADH-cytochrome *c* reductase	CFE	59	–[c]	FAD	Cyt. *c*, TZ, O_2	70
NADH-quinone reductase	M	300	n.d.[d]	FAD	K_2, K_3	71
NADH oxidase	M	71	–[c]	FAD	Cyt. *c*, TZ, O_2, DCIP	54
Menadione reductase I	M	250	160	FMN, FAD[e]	K_3, DCIP, FC	54
Menadione reductase II	M	14	27		K_3, DCIP, FC	54
NADH dehydrogenase	M	8-12	–	FMN, FAD[e]	K_3, DCIP, FC	50
NADH diaphorase	C	320	–	–	–	50
NAD(P)H menadione reductase	C	12	n.d.[d]	–	K_3, DCIP, FC	72
NADH dehydrogenase	M	30	–[c]	FAD	K_3, DCIP, FC	57, 73

[a]CFE, cell free extract; M, membrane fraction; C, cytoplasmic fraction.
[b]Cyt. *c*, cytochrome *c*; TZ, tetrazolium dyes; K_2, K_3, vitamins K_2 and K_3; DCIP, 2,6-dichlorophenolindophenol; FC, ferricyanide.
[c]Enzyme not able to use NADPH.
[d]n.d., enzyme uses NADPH but the K_m was not determined.
[e]Enzyme activity stimulated by FMN and FAD.

cytochrome *c* reductase activity ("soluble NADH oxidase") from membrane particles of *E. coli* NRC 482 with deoxycholate. This enzyme, which had been partially purified on columns of calcium phosphate and DEAE-cellulose, reduced cytochrome *c,* oxygen, and tetrazolium dyes with a K_m for NADH of 71 μM. NADPH was not oxidized and FAD appeared to be the prosthetic group of the enzyme. This enzyme was inhibited by millimolar concentrations of *o*-phenanthroline and 2,2′-dipyridyl and thus appears to be a nonheme iron flavoprotein.

The solubilization procedure of Bragg and Hou [53] released two other NADH diaphorase enzymes (menadione reductases I and II) from the membrane particles. These enzymes, which were subsequently purified, utilized both NADH and NADPH, and could reduce menadione, ferricyanide, 2,6-dichlorophenolindophenol, and tetrazolium dyes. The prosthetic group of menadione reductase II was not determined. However, menadione reductase I was stimulated by both FMN and FAD. The K_m values for NADH with enzymes I and II were 250 and 14 μM, respectively. Enzyme I was more sensitive than enzyme II to inhibition by dicumarol, pentachlorophenol and *o*-phenanthroline, being more than 60% inhibited at concentrations of 50, 33, and 630 μM, respectively, while enzyme II was essentially unaffected. Menadione reductase I had a molecular weight of 35,000-38,000 [58] compared to 64,000 for enzyme II (Bragg and Hou, unpublished results).

The solubilization procedure of Bragg and Hou [53] was similar to that employed by Kashket and Brodie [71] to prepare a soluble NADH-quinone reductase from membrane particles. The quinone reductase could utilize NADH (K_m = 300 μM) to reduce vitamin K_2 or menadione whereas NADPH reduced menadione only, suggesting that the solubilized fraction contained at least two enzymes, a NADH-vitamin K_2 reductase activity and a NAD(P)H-menadione reductase activity. FAD but not FMN stimulated the NADH-vitamin K_2 reductase activity. This activity was inhibited by 3 mM dicumarol but not by 1.5 mM *p*-chloromercuribenzoate.

NADH dehydrogenase (NADH-ferricyanide reductase) activity has also been extracted with water from lyophilized membrane particles [50]. The enzyme was unstable and was not further purified. NADH acted as substrate for the reduction of ferricyanide, menadione, and 2,6-dichlorophenolindophenol. The activity with NADPH was not reported. Pretreatment of the enzyme as extracted (form I, K_m = 8-12 μM) with substrate (NADH) in the absence of the electron acceptor (ferricyanide) resulted in the transformation of the enzyme to a form (II) with an altered K_m for NADH (43 μM). The amount of NADH required to bring about this conversion was stoichiometrically equivalent to the nonheme iron content of the preparation (14-16 ng-atoms/mg protein) and resulted in reduction of the iron to the ferrous form. Acid-labile sulfide (11-13 ng-atoms/mg protein) and both FMN (0.3 nmol/mg protein) and FAD (0.2

nmol/mg protein) were present in this preparation. The enzyme was inhibited 30% by 0.4 mM dicumarol.

The results above suggest that at least three types of NADH dehydrogenase activities can be extracted from membrane particles of *E. coli:* (1) soluble NADH oxidase or NADH-cytochrome *c* reductase, (2) NADH diaphorase with a K_m for NADH of 8-14 μM, and (3) NADH diaphorase with a K_m for NADH of 250-300 μM. Since the last two enzymes also use NADPH as substrate while the respiratory chain NADH oxidase does not, it is unlikely that either of these enzymes is the respiratory chain dehydrogenase. This is supported by the observation of Bragg and Hou [58] that both enzymes can be released from membrane particles by ethylenediamine tetraacetic acid (EDTA) without affecting the NADH oxidase activity of the particles. Furthermore, it was found that the cytoplasmic fraction of the cell contained several NAD(P)H diaphorase activities. The major two of these enzymes had properties similar to those of the NADH dehydrogenases released from the membrane particles including K_m values for NADH of 12 and 320 μM [50, 58]. It is probable that the cytoplasmic and membrane-bound enzymes are the same. The protein bands released by detergent from the membranes and staining for NADH-tetrazolium reductase activity comigrated with similar bands from the cytoplasmic fraction when examined by polyacrylamide gel electrophoresis using Tris-glycine buffer, pH 9.7 [150], or phosphate buffer, pH 7.1, containing 0.2% deoxycholate [55, 56]. Thus, the enzymes solubilized by detergent from the particles and described above are either loosely bound to the membranes or are cytoplasmic enzymes trapped as contaminants in the sealed vesicles produced by disruption of the cell.

Recently, Dancey et al. [57, 72] have reported the solubilization of a NADH dehydrogenase activity which is probably that of the respiratory chain. This view is based on the following evidence. (1) The enzyme did not oxidize NADPH. (2) It oxidized NADH with a K_m of 30 μM which compares favorably with the K_m values for the membrane-bound NADH oxidase and NADH dehydrogenase of 50 and 33 μM, respectively. (3) The membrane-bound NADH oxidase and NADH dehydrogenase, and the solubilized NADH dehydrogenase were inhibited by 5′-AMP with K_i values of 650, 500, and 500 μM, respectively. The cytoplasmic NADH dehydrogenase activities were not affected by 5′-AMP. (4) Bands staining for NADH-2,6-dichlorophenolindophenol reductase activities in samples of cytoplasm did not comigrate on Triton X-100 containing polyacrylamide gels with those of the solubilized NADH dehydrogenase. (5) Antibody against the solubilized NADH dehydrogenase inhibited membrane-bound NADH oxidase and dehydrogenase activities.

The dehydrogenase solubilized by 5% Triton X-100 from membrane vesicles prepared by the method of Kaback [11] was purified by chromatography on DEAE-cellulose, hydroxylapatite, and DEAE-agarose. Chromatography was carried out in the presence of 1% Triton X-100 in order to prevent reaggregation of the solubilized membrane proteins. The final product showed a 16 to 30-fold purification over the Triton-solubilized membranes. Although the extent of purification appeared to be substantial, the specific activity of the final product was only 0.32, or 0.65 in the presence of cardiolipin, compared to 0.23-0.26 for the membrane-bound enzyme in sonicated vesicles. Polyacrylamide gel electrophoresis of the purified enzyme in the presence of sodium dodecyl sulfate (SDS) gave a protein-staining band of molecular weight 38,000 which in the best preparations accounted for 75% of the stained material on the gel. When the purified enzyme was examined by polyacrylamide gel electrophoresis at pH 9.5 in the presence of 0.1% Triton X-100, a single protein-staining band, also showing enzymic activity, was observed. However, a substantial amount of protein (~50%?) failed to penetrate the gel. Thus, in the best preparations most of the protein in the enzyme preparations can be accounted for by a polypeptide of molecular weight 38,000. If this is the NADH dehydrogenase then the unexpectedly low specific activity of the final product must be due to modification of the enzyme during purification. This seems unlikely since the K_m for NADH and the K_i for AMP of the purified enzyme are identical to those of the membrane-bound dehydrogenase. The molecular weight of 38,000 for the NADH dehydrogenase is low if the beef heart mitochondrial enzyme can be used as a guide. This flavoprotein has a molecular weight of 65,000-70,000 [73, 74]. This fact taken in conjunction with the low specific activity of the enzyme raises the possibility that the 38,000 molecular weight polypeptide could be a contaminant and that the NADH flavoprotein is one of the other bands observed on the SDS gels.

The solubilized enzyme was activated by lipid, cardiolipin being the most effective, and had an absolute requirement for FAD. While NADPH was not a substrate for the enzyme, 2,6-dichlorophenolindophenol (K_m = 5.4 μM), menadione (K_m = 15 μM), and ferricyanide (K_m = 260 μM) would act as electron acceptors with NADH. The purified dehydrogenase was competitively inhibited by AMP (K_i = 0.6 mM), ADP (K_i = 1 mM), and ATP (K_i = 8.5 mM) but the most effective inhibitor was NAD (K_i = 0.02 mM). This compound was a competitive inhibitor which at lower levels of NADH deviated from normal Michaelis-Menten kinetics such that the enzyme became much more inhibited. A large number of other compounds (nucleotides, nucleosides, amino acids, intermediates of glycolysis and the trichloroacetic acid cycle) were without effect on enzyme activity. Thus, the activity of NADH dehydrogenase may be regulated by the product NAD^+. Since the ratio of NAD^+ to NADH is about 3.5 in growing cells, the enzyme would be partly inhibited but would respond cooperatively to increases in the level of NADH.

E. Succinate Dehydrogenase

The only report on a solubilized preparation of succinate dehydrogenase from *E. coli* appears to be that of Kim and Bragg [75]. The enzyme was partially purified by extracting an acetone powder of whole cells with 0.01 M phosphate buffer, pH 7.5, containing 0.025 M succinate. The enzyme could be separated from contaminating cytochrome by chromatography on a column of calcium phosphate gel. The highest specific activity of 0.2 μmol/min per mg protein represented only a twofold purification over that of the acetone powder. The enzyme was eluted from Sephadex G-200 with a molecular weight of about 100,000. A small amount of a higher molecular weight species was also present. Further purification was not attempted due to the lability of the enzyme. In subsequent experiments it was found that succinate dehydrogenase could be readily solubilized from membrane particles using Triton X-100 [26].

Membrane-bound succinate dehydrogenase reacts with succinate with a K_m of 0.28 mM [75]. Kasahara and Anraku [76] obtained a biphasic Lineweaver-Burk plot and calculated K_m values of 0.32 ± 0.11 and 1.2 ± 0.4 mM for the two slopes. Malonate (K_i = 7.3 μM) was a competitive inhibitor of membrane-bound succinate dehydrogenase. Activity was inhibited by *p*-chloromercuribenzoate [75] and Zn^{2+} [76], suggesting that sulfhydryl groups were important for enzyme activity.

The solubilized enzyme, whether extracted from the acetone powder or with Triton X-100, was not further activated by heating with succinate. This contrasts with the behavior of the membrane-bound enzyme in which this treatment increased the activity up to ninefold [75]. To show maximal activation, the membranes had to be prepared in phosphate buffer with or without succinate. If the membranes were prepared in the presence of Tris and succinate then the succinate dehydrogenase was found in its fully activated form. Interestingly, Kasahara and Anraku [76] found that inclusion of potassium cyanide increased the rate at which succinate dehydrogenase was activated by succinate. Houghton et al. [77] found that succinate dehydrogenase activity of membranes was increased at the low concentrations of 2,5-dibromothymoquinone, which caused inhibition of the succinate oxidase pathway. Inhibition of the respiratory chain by cyanide and dibromothymoquinone would increase the level of reduced ubiquinone which is known to cause activation of mitochondrial succinate dehydrogenase. It seems likely that the various factors (substrates, anions, pH, reduced ubiquinone, IDP) which cause activation of the enzyme in mitochondria act by a common mechanism involving the removal of inhibitory oxaloacetic acid from the active site [78, 79]. The same mechanism may apply to the bacterial enzyme. Presumably, the lack of activation of the soluble enzyme was due to the removal of inhibitory oxaloacetate during the preparation of the enzyme.

F. Lactate Dehydrogenases

E. coli contains at least three lactate dehydrogenases [82-85]. From studies with whole cells it appears that the level of the D-lactate dehydrogenase(s) is relatively constant showing little variation under aerobic or anaerobic conditions during growth on various carbon sources. In contrast, the level of the L-lactate dehydrogenase is repressed by growth on glucose, glycerol, pyruvate, and succinate but induced by growth on either D- or L-lactate. The membrane contains both a D- and an L-lactic dehydrogenase which are coupled to the respiratory chain. These enzymes are probably flavoproteins which do not use NAD^+ in their reactions and under physiological conditions are concerned with the oxidation of lactate to pyruvate. However, the reaction in the reverse direction appears to be the role of the D-lactate dehydrogenase of the cytoplasm. This enzyme of molecular weight 115,000 has been purified to homogeneity and extensively characterized by Tarmy and Kaplan [84, 85].

Bennett et al. [83] and Kline and Mahler [82] showed that both L- and D-lactate dehydrogenases could be solubilized from membrane particles by 1% SDS at pH 10 or by treatment with snake venom and deoxycholate, respectively. Further purification of the enzymes was not attempted. Recently, considerable interest has been directed to the membrane-bound D-lactate dehydrogenase since D-lactate has been found to be the best of the physiological electron donors to energize active transport of amino acids and certain sugars in membrane vesicles of *E. coli* [12-14]. There have been two reports of the solubilization and purification to homogeneity of this enzyme. Kohn and Kaback [80] solubilized the dehydrogenase from membrane particles with 0.4 M sodium perchlorate and subsequently purified it by three cycles of chromatography on DEAE-cellulose. During the last two steps 1% Triton X-100 was included in the eluting buffers. Futai [81] started his procedure from spheroplast membranes. The enzyme was extracted from these with 0.63% deoxycholate in the presence of 0.63 M NaCl. The enzyme was reaggregated by dialysis during two subsequent steps and then resolubilized by 0.5% cholate and 1% Triton X-100, respectively. The Triton-resolubilized enzyme was finally purified by chromatography on DEAE-Sephadex and DEAE-cellulose in the presence of this detergent. Both enzymes gave a single band on polyacrylamide gel electrophoresis. A comparison of the two preparations of D-lactate dehydrogenase is made in Table 5. Both preparations had one molecule of noncovalently bound FAD per molecule of enzyme.

The enzyme was not inhibited by sulfhydryl-reacting compounds such as iodoacetate, *p*-chloromercuribenzoate, and N-ethyl maleimide even at concentrations of 0.01 M. Arsenate (0.01 M) caused less than 10% inhibition of the reaction. This contrasts with the NAD^+-dependent D-lactate dehydrogenase isolated from the cytoplasm which was almost completely inhibited by this con-

Table 5 Properties of D-lactate Dehydrogenase

Property	Kohn and Kaback [80]	Futai [81]
Molecular weight	74,000 ± 5,000	71,000 ± 3,000
Prosthetic group	FAD	FAD
Specific activity (μmol/min per mg protein)	75.4	81.5
K_m (D-lactate)	0.9 mM	0.6 mM
K_m (L-lactate)	16 mM	18 mM

centration of arsenate [85]. Oxamate (K_i = 3.4 μM) and oxalate (K_i = 0.9 μM) were competitive inhibitors of the enzyme.

A particularly interesting property of both Kohn and Kaback's and of Futai's preparation was the ability of the purified, solubilized enzyme to rebind to the outer surface of vesicles of a mutant (*E. coli* ML308-225 *dld-3*), in which D-lactate dehydrogenase activity was lacking, to reconstitute D-lactate oxidase activity and D-lactate-dependent transport of amino acids, lactose, and Rb^+ [81, 86, 87].

G. L-Glycerol-3-phosphate Dehydrogenase

This inducible enzyme has been solubilized from spheroplast membranes of a mutant of *E. coli,* strain E27, which is constitutive for the dehydrogenase and where the enzyme appears to constitute about 1% of the membrane protein. The enzyme was solubilized by Weiner and Heppel using 0.1% sodium deoxycholate in the presence of 0.25 M NaCl and subsequently purified by chromatography on DEAE-cellulose and then on phosphocellulose using buffers containing 0.2% Brij 58 [59]. The specific activity of the final product was 5.8 μmol/min per mg protein and represented about a 100-fold purification from the membranes with a recovery of 71%. Futai [81] has solubilized the enzyme from membranes with 0.45 M sodium perchlorate.

The preparation of Weiner and Heppel was nearly homogeneous based on protein-staining following SDS polyacrylamide gel electrophoresis. A molecular weight of 35,000 was determined by electrophoresis. Since gel filtration in the presence of 0.5% cholate gave a molecular weight of 80,000, it is likely that the enzyme consists of two subunits.

The dehydrogenase is a flavoprotein from which the prosthetic group of FAD can be released by boiling. The K_m for DL-glycerol-3-phosphate is 0.8 mM. This is much lower than the value of 25 mM for the membrane-bound enzyme and suggests that some modification of the dehydrogenase had occurred on extraction from the membrane.

When *E. coli* ML308-225 is grown on succinate the enzyme is not induced and the level of L-glycerol-3-phosphate dehydrogenase activity is zero. Using partially purified (ca. 10% pure) enzyme prepared with perchlorate, Futai [81] was able to rebind the dehydrogenase to noninduced membranes of this strain to a level (0.22 units/mg protein) similar to that found in the same strain when the enzyme was fully induced (0.17 units/mg protein). This resulted in the reconstitution of L-glycerol-3-phosphate oxidase activity (400 ng-atoms O per mg protein compared to 230 ng-atoms O per protein in fully induced vesicles) and in amino acid transport coupled to the oxidation of this substrate. This reconstitution is similar to that achieved with purified D-lactate dehydrogenase (see previous section).

H. Pyruvate Dehydrogenase

This enzyme, termed pyruvate oxidase by Hager and coworkers, has been crystallized [65, 66]. The enzyme carries out the following reaction:

$$\text{Pyruvate} + H_2O + \text{enzyme—FAD}$$
$$\rightarrow \text{acetate} + CO_2 + \text{enzyme—FADH}_2$$

The reduced prosthetic group of the pyruvate dehydrogenase is reoxidized via ubiquinone through the respiratory chain [69]. The enzyme is formed in response to increasing levels of pyruvate accumulating in the growth medium and is distinct from the enzymes of the pyruvate dehydrogenase complex and of the phosphoroclastic cleavage of pyruvate [88].

The enzyme was isolated by Williams and Hager [65, 66] from the cytoplasmic fraction of an acetate auxotroph of *E. coli.* The enzyme was purified by fractionation with ammonium sulfate, followed by DEAE-cellulose chromatography, and absorption and desorption from protamine sulfate. It crystallized as rhombohedral crystals. The enzyme is a flavoprotein of molecular weight 265,000, containing four molecules of FAD per molecule of protein. Thiamine pyrophosphate is not present in the enzyme but there are four binding sites for it. Thus, it is probable that the enzyme is a tetramer consisting of four identical subunits each with a binding site for FAD and thiamine pyrophosphate [68]. The reduced flavoprotein is nonautooxidizable but may also be reoxidized by 2,6-dichlorophenolindophenol and ferricyanide. These acceptors can be used in a convenient assay method for the enzyme. Ubiquinone appears to be the immediate electron acceptor for reducing equivalents from the reduced flavoprotein under physiological conditions, since removal of lipids from membrane particles (by treatment with phospholipase C from *Bacillus cereus*) resulted in loss of the pyruvate oxidase activity in a system consisting of membrane particles and purified pyruvate dehydrogenase [68]. Addition of ubiquinone-8 or ubiquinone-6 together with *E. coli* neutral lipid, which had been irradiated to destroy endogenous quinone, resulted in the restoration of oxidase activity. Neutral lipid alone had no effect.

I. Ubiquinone

Ubiquinone is found in the membrane particles of *E. coli* [89, 90] and, as discussed later, has a role in the respiratory chain. Menaquinone (vitamin K_2) may also be present [89, 90]. The relative amounts of these two quinones present depends upon the degree of aeration during growth. Polglase et al. [91] showed that 20-fold more ubiquinone than menaquinone was formed under conditions of vigorous aeration whereas menaquinone greatly predominated under anaerobic conditions. The function of menaquinone in anaerobic electron transport is discussed in a later section. Approximately 85% of the total ubiquinone found in *E. coli* is ubiquinone-8 [92]. Ubiquinones-5, -6, and -7 constitute 1, 2, and 10%, respectively, of the total amount. Ubiquinones-9 and -10 have not been detected but ubiquinones-1, -2, -3, and -4 together account for less than 0.1% of the total ubiquinone. It is likely that the lower isoprenologues of *E. coli* represent biosynthetic precursors of the major ubiquinone, ubiquinone-8. The two major menaquinones are menaquinone-8 and 2-demethylmenaquinone-8 but smaller amounts of menaquinones-6, -7, and -9, and 2-demethylmenaquinone-7, have been detected [93].

J. Nonheme Iron

Insufficient attention has been paid to the presence and function of nonheme iron in the respiratory chain of microorganisms [94]. The information on its role in mitochondrial respiratory pathways indicates that it is of great importance in both electron transfer and energy conservation [94]. Nonheme iron appears to be associated with an equimolar amount of acid-labile sulfide in the mitochondrial respiratory chain proteins.

Nonheme iron can be detected as a signal at $g = 1.94$ in electron paramagnetic resonance (EPR) spectra of membrane particles of *E. coli* which have been reduced by dithionite or substrates [27, 95-97]. As shown in Table 2, the amount of nonheme iron in membrane particles of *E. coli* on a molar basis greatly exceeds that of the cytochromes. However, the amount of nonheme iron found in the membrane particle preparation depends on the availability of iron in the medium. Rainnie and Bragg [30] followed the level of nonheme and heme iron in whole cells of *E. coli* growing in batch culture on succinate during depletion of iron in the medium and during the subsequent recovery on addition of an excess of ferric citrate. When growth was severely limited by iron depletion the amounts of nonheme and heme iron were 69 and 9.2 ng-atoms per gram (wet weight) of cells, respectively. These amounts had risen to 272 and 9.9 ng-atoms of nonheme and heme iron per gram of cells, respectively, within 15 min following the addition of 12 μM ferric citrate, and to 686 and 23.8 ng-atoms per gram of cells in the stationary phase. Thus, the ratio of nonheme iron to heme iron had increased from 7.5:1 in the iron-depleted state to 21:1 in iron-sufficient stationary phase cells. Membrane particles prepared from cells grown on succinate showed nonheme iron/heme iron ratios in the range of 19:1 to 32:1 [26]. The increase in the level of nonheme iron following supplementation of iron-deficient cells with ferric citrate in the experiments of Rainnie and Bragg was associated with an increase in the efficiency of energy conservation as shown by measurement of cell mass and respiratory control ratios.

The presence of acid-labile sulfide, which is characteristic of nonheme iron-containing proteins [94], has been found to be associated with the respiratory chain of *E. coli* [27]. In cells grown in continuous culture under conditions of glycerol limitation, the levels of nonheme iron and acid-labile sulfide were 4.28 ± 0.42 and 1.43 ± 0.27 ng-atoms per mg membrane protein, respectively. Sulfate limitation resulted in a decreased growth yield and the loss of the energy-conserving site associated with the NADH dehydrogenase segment of the respiratory chain. A similar response to sulfate limitation was previously observed with yeast [98].

K. Cytochromes

Reduced-minus-oxidized difference spectra of aerobically grown cells of *E. coli* (measured at room temperature) show absorption bands corresponding to cytochromes b_{562}, b_1, a_1, *o*, and *d*. Cytochrome *o* cannot be distinguished from cytochrome b_1 in reduced-minus-oxidized difference spectra since their absorption bands overlap.

1. *b* Cytochromes

Considerable progress has been made in purifying the *b* cytochromes of *E. coli*. Cytochromes b_1 and b_{562} have been crystallized and extensively characterized [99, 100, 104]. Cytochrome b_{562} was isolated from the cytoplasmic fraction of the cell but a component absorbing at 562 nm has also been detected in reduced-minus-oxidized difference spectra of membranes from *E. coli*. It is not clear whether these two cytochromes are identical and whether the presence of the cytochrome in the membrane is due to contamination by the cytoplasmic fraction. Conversely, the presence of cytochrome b_{562} in the cytoplasmic fraction might be due to its facile release from the membranes.

Cytochrome b_1 was solubilized from the membrane fraction of *E. coli* by prolonged sonication and purified by chromatography on columns of calcium phosphate gel mixed with cellulose. The cytochrome was crystallized from fractions eluted from a second calcium phosphate gel column. The overall yield at this stage was 34% based on the cytochrome b_1 content of the membrane fraction. A less pure preparation (about 50% pure) was obtained by Fujita et al. by solubilization with 0.1% deoxycholate and crude snake venom [101].

The properties of cytochromes b_1 and b_{562} are summarized in Table 6. There are a number of interesting differences between the two cytochromes. The oxidation-reduction potential of purified cytochrome b_1 (E_m = -0.34 V) is much lower than that of cytochrome b_{562} (E_m = +0.113 V), and also much lower than the value of 0 V found for this cytochrome in the membrane-bound form [99]. Deeb and Hager [99] and Fujita et al. [101] found that less pure preparations of the cytochrome had redox potentials of -0.01 to -0.02 V. The former workers attributed the lower redox potential to the removal of a "potential-modifying protein." This was separated from the cytochrome b_1 on the second calcium phosphate gel column. Addition of the potential-modifying protein to the crystalline cytochrome b_1 resulted in an increase in the redox potential from -0.34 V to -0.12 V, half-maximal increase occurring at about 0.02 mg potential-modifying protein per mg crystalline cytochrome b_1. No further information has been published on this interesting protein. An analogous situation has been observed with mitochondrial cytochrome *b* where the low potential (-0.34 V) of the purified cytochrome is increased to about 0 V by addition of a mitochondrial protein fraction [102]. In contrast to cytochrome b_{562}, reduced crystalline cytochrome b_1 is autooxidizable although the membrane-bound form does not have this property. Presumably, the environment about the prosthetic group has undergone some change following release of the cytochrome from the membrane. Also in contrast to cytochrome b_{562}, the purified cytochrome b_1 appears to occur at neutral pH as a multimer of molecular weight 500,000-800,000 [99, 101].

The prosthetic group of both cytochromes b_{562} and b_1 can be removed with acidified acetone. The apoproteins will combine with hemin to reconstitute the cytochromes [101, 103]. Although reconstitution of cytochrome b_{562} proceeds satisfactorily, the reconstituted cytochrome b_1 shows some alterations in the spectrum and, more notably, becomes reactive with carbon monoxide.

Itagaki and Hager [116] have determined the amino acid sequence of cytochrome b_{562}. This, together with other evidence [117, 118], shows that there is considerable similarity both in primary and tertiary structure between cytochrome b_{562} and myoglobin.

Although Deeb and Hager have isolated a single species of *b* cytochrome

Table 6 Properties of Cytochromes b_1 and b_{562}

Property	Cytochrome b_1	Cytochrome b_{562}
Molecular weight	500,000	12,000
Monomer molecular weight	62,000-66,000	12,000
Absorption maxima in reduced-minus-oxidized difference spectrum (ϵ_{mM} in parentheses)[a]		
α	557.5 nm (16)	562 (24.6)
β	527.5 nm (6)	–
γ	427.5 nm (60)	–
Oxidation-reduction potential	-0.34 V	+0.113 V
Prosthetic group	Iron protopor-phyrin IX	Iron protopor-phyrin IX

[a]Measured at room temperature.

Source: Deeb and Hager [99] and Itagaki and Hager [100].

from the membranes of *E. coli*, and in a yield of about 34%, there is evidence that other *b* cytochromes are present in the membrane. While reduced-minus-oxidized difference spectra of membrane particles measured at room temperature showed a single absorption peak at 559-560 nm for the α-band absorption of cytochrome b_1, the kinetics of reduction of this peak suggested that more than one component might be present [25, 26]. Originally the difference spectrum measured at 77 K was also interpreted as showing a single component with a peak at 555-556 nm [105, 106] or 558 nm [107]. However, with a greater number of published spectra, the interpretation of the shoulders seen on peaks recorded at low temperatures is now easier [27, 34, 108-111]. Reduced-minus-oxidized spectra of membrane particles from cells grown aerobically to the exponential and stationary phase are shown in Figures 3 and 4. The α-band region shows distinct peaks at 556 and 558 nm with shoulders at 562 and 548-550 nm. The resolution of the Soret region is not much improved at 77 K over that at room temperature although the absorption maxima are displaced about 3 nm toward lower wavelengths. The *b* cytochromes give a composite peak at 427 nm and cytochrome *d* is responsible for the peak at 437 nm. The α bands at 590 and 628 nm are not further resolved into other components at 77 K. By applying fourth-order finite difference analysis to the reduced-minus-oxidized difference spectra measured at 77 K, Shipp and coworkers [108, 109] were able to obtain a complete separation of the cytochrome peaks in the difference spectrum and so to determine accurately the position of their α absorption maxima. The major peaks in fourth-order difference spectra of the α absorption band region

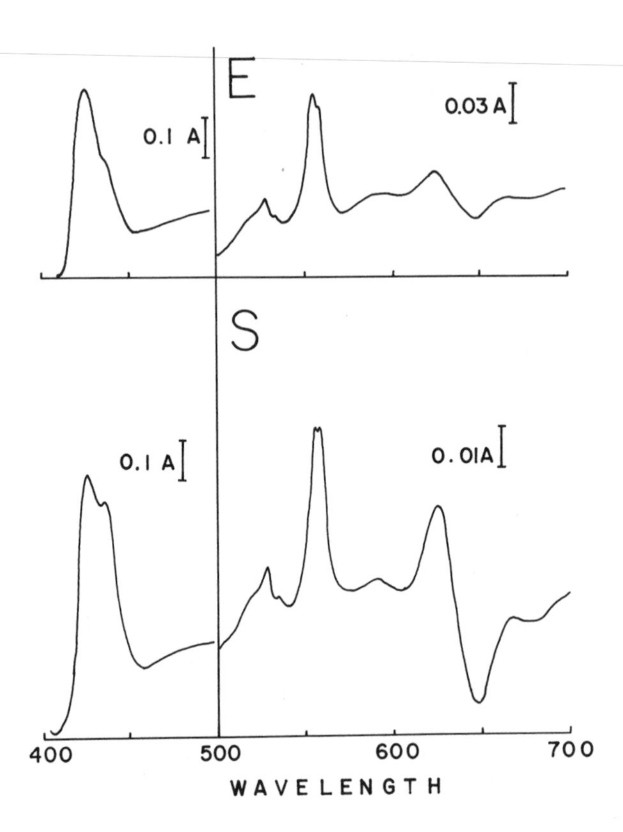

Figure 3 Low-temperature (77 K) reduced-minus-oxidized difference spectra of particles prepared from cells grown to the exponential (E) and stationary (S) phases of growth. The particles were prepared from cells grown on succinate to the required phase of growth and suspended to a concentration of 7.2 mg protein/ml and 5.4 mg protein/ml, respectively. The cytochromes were reduced with NADH.

were at 548-549, 552-553, 556-557.5, 559-562, and 564-566 nm. The first two of these peaks correspond to the shoulder observed at 548-550 nm, and the remaining peaks to those at 556, 558, and 562 nm seen in the difference spectrum. A number of minor peaks between 567 and 575 nm were found in fourth-order difference spectra, particularly of exponential phase cells, but these have not been further characterized [108]. An attempt to relate these absorp-

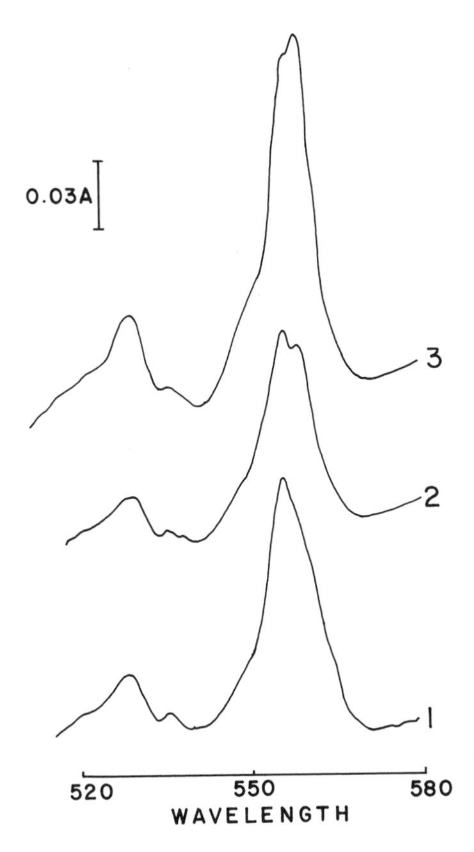

Figure 4 Low-temperature (77 K) dithionite-reduced minus oxidized difference spectra of the cytochrome *b* region of particles grown under different conditions. 1, particles (11.0 mg protein/ml) prepared from cells grown on glucose to the exponential phase; 2, particles (4.1 mg protein/ml) prepared from cells grown on succinate to the stationary phase; 3, particles (5.2 mg protein/ml) prepared from cells grown on complex media to the stationary phase.

tion bands to the cytochromes which have been isolated has not been successful. Thus, the α band of crystalline cytochrome b_1 which is at 557.5 nm at room temperature [99] might be expected to be at about 554.5-555.5 nm at 77 K. No band was observed at this wavelength. Since the properties of the crystalline cytochrome are different from those of the membrane-bound cytochrome, it is possible that cytochrome b_1 is responsible for part of the absorption peak at

556 nm. Haddock et al. [111] suggest that 20-40% of the absorbance at 556 nm is contributed by cytochrome *o*. Cytochrome b_{562} isolated by Itagaki and Hager [100] and Fujita [104] cannot be directly related to a band observed in the difference spectrum at 77 K. It seems unlikely that the peak at 558 nm is due to this cytochrome since this peak is prominent in membrane particles which would have very minor amounts of the cytoplasmic cytochrome b_{562}. No specific α absorption band has been assigned to cytochrome *o* in reduced-minus-oxidized difference spectra. However, on the basis of the shift in the position of the α-band absorption maximum of the reduced cytochrome produced by carbon monoxide, it is possible to tentatively estimate the position of the α-band. Thus, Daniel [112] has suggested that cytochromes *o* may be classified into two groups. In some organisms, including *E. coli,* the absorption maximum is at about 558 nm at room temperature while with others the peak is at about 565 nm. In carbon monoxide difference spectra recorded at 77 K, absorption peaks at 557 and 430 nm were eliminated by reaction with carbon monoxide [27]. If indeed cytochrome *o* has its α-band absorption peak at 557 nm then it is not detectable as a distinct peak at this wavelength in reduced-minus-oxidized difference spectra where it must be hidden by the peaks at 556 and 558 nm. The two *c*-type cytochromes isolated by Fujita [104] from anaerobically grown cells had α-band absorption maxima at 550 and 552 nm at room temperature. At 77 K these bands might be expected to be at 547-549 and 549-550 nm, respectively [108]. Although one of these bands corresponds to that observed in the fourth-order difference spectra, it seems unlikely that it is identical to it since cytochromes c_{550} and c_{552} are found in nonmembranous fractions from anaerobically grown cells. The absorption peaks at 548-549 and 552-553 nm in the fourth-order difference spectra are present in membrane particles from cells grown aerobically. The process of interpretation of the positions of absorption maxima in terms of specific cytochromes is also complicated by the fact that some cytochromes have multiple α absorption bands. However, no obvious relationship suggesting that a cytochrome with a double α absorption is present can be seen between the size of the various peaks under different conditions. At present, the most reasonable interpretation of the spectral data is that the cytochrome b_1 α absorption peak contains contributions from cytochromes c_{549}, c_{552}, b_{556}, b_{558}, b_{562}, and *o* (as identified by the position of the α absorption maximum in the reduced-minus-oxidized difference spectrum).

The relative amounts of the cytochromes present varies with the growth conditions [27, 108, 110]. As can be seen from Figures 3 and 4 the relative amount of cytochrome b_{558} to cytochrome b_{556} increases in the transition from the exponential to the stationary phase of growth. Cytochromes *c* and b_{562} undergo less obvious changes, if any, but it is difficult to asses the contribution made by these cytochromes. In membrane particles from *early* exponential phase cells grown with vigorous aeration in batch culture, cytochrome b_{558} appears to be absent such that direct reduced-minus-oxidized difference spectra show clear resolution of cytochromes b_{556} and b_{562} [113]. However, a reduced amount of cytochrome b_{558} is probably still present since Shipp [108], using fourth-order difference analysis, has shown that in cells grown anaerobically or aerobically, and to different phases of growth, the same group of *b*- and *c*-type cytochromes are present. Dramatic changes in the shape of the α absorption band of cytochrome b_1 simply reflects changes in the relative proportions of these cytochromes.

Cytochromes *d* and a_1 as well as cytochrome b_{558} are at a lower level in early exponential phase compared to stationary phase cells grown in batch culture (Figs. 3 and 4; Table 2) [108, 110]. Cells grown in batch culture in the presence of 150 μM KCN had high levels of cytochromes b_{558}, a_1, and *d* even when harvested in the early exponential phase of growth [113]. If cells growing under conditions of substrate (glycerol) limitation in continuous culture are shifted to conditions of sulfate limitation, then the predominant cytochromes b_{556}, b_{562}, and *o* are now accompanied by cytochromes b_{558}, a_1, and *d* [27]. The factors responsible for increasing the extent of synthesis of cytochromes b_{558}, a_1, and *d* as the stationary phase of growth is approached, or when cell growth is limited by the availability of sulfate or the presence of KCN, are not clear. However, the availability of energy to the cell would have been affected by each of these conditions. Thus, both substrate or oxygen may become limiting toward the stationary phase of growth. KCN inhibits oxidation through the respiratory chain [110] and site I is not functional in sulfate-limited cells [27]. Whether a decline in the supply of energy to the cell, indicated by an increasing [ADP]/[ATP] or [AMP]/[ATP] ratio, induces the formation of these cytochromes remains to be proved. However, the preceding results suggests that in *E. coli* during early exponential growth the respiratory chain consists of cytochromes b_{556}, b_{562}, and *o*. Under the conditions described above, this respiratory pathway can either be integrated with cytochromes b_{558}, a_1, and *d*, or a new respiratory pathway containing these cytochromes can be induced [34].

The presence of multiple *b*-type cytochromes in membrane particles of *E. coli* has also been shown by redox titration [114, 155]. Hendler et al. [114] followed the reduction of cytochrome b_1 at 560 nm in the presence of quinhydrone, phenazine methosulfate, pyocyanine, and 2-hydroxy-1,4-naphthoquinone as redox mediators. A nonlinear plot of potential versus log [oxidized cytochrome]/[reduced cytochrome] was resolved as the contributions of three components of midpoint redox potentials of +220, +110, and −50 mV constituting 39, 33.4, and 27.6%, respectively, of the total cytochrome b_1. The low-potential component was completely removed from the membrane particles with 1% deoxycholate. These potentials were not altered by the addition of

6 mM ATP or 1 mM 2,4-dinitrophenol. No attempt was made by Hendler et al. to correlate the redox potentials with the spectroscopically recognizable cytochromes. Pudek and Bragg [115], in contrast to the results of Hendler et al., found that the midpoint oxidation-reduction potentials of the two major *b*-type cytochromes were about +15 to +34 mV (n = 1) and +165 to 205 mV (n = 1), the exact value depending on the growth phase during which the cells had been harvested. The high- and low-potential components were identified as cytochromes b_{558} and b_{556}, respectively, both from the difference spectra measured at appropriate redox potentials, but also from the relative contributions of the low- and high-potential components to the total absorbance of cytochrome b_1 in membrane particles prepared from exponential and stationary phase cells. The relative contributions of cytochromes b_{556} and b_{558} to the absorbance of the cytochrome b_1 peak in early exponential and stationary phase cells were 73 and 27%, and 60 and 40%, respectively. The midpoint oxidation-reduction potentials of the *c*-type cytochromes and cytochrome b_{562} were not determined due to the relatively small contributions that these components made to the cytochrome b_1 peak. However, the complete reduction of cytochrome b_{562} by ascorbate (in the presence of phenazine methosulfate) suggests that its midpoint oxidation-reduction potential is somewhat more positive than 0 mV [115]. The differences between the results of Hendler et al. [114] and those of Pudek and Bragg [115] have been attributed by the latter workers to be due partly to the incomplete equilibration of the cytochromes with the redox electrode by the mediators used by Hendler et al. Also Hendler et al. used membranes which had been frozen and thawed. The redox potentials of *b*-type cytochromes are very sensitive to alterations in their membrane environment [99, 102]. The midpoint oxidation-reduction potentials of cytochromes b_{556} and b_{558} were not affected by the presence of 14 mM KCN [115]. This may indicate that neither of these cytochromes is cytochrome *o*.

2. *c* Cytochromes

As has been discussed in previous sections, *c*-type cytochromes are present in *E. coli.* The cytoplasmic cytochromes c_{550} and c_{552}, which are formed under anaerobic conditions, will not be further discussed. The membrane-bound cytochromes c_{549} and c_{552} [108] contribute to the cytochrome b_1 α absorption peak to such a minor extent that examination of them is difficult and they are generally considered to be minor components of the membrane. These *c*-type cytochromes are only partially reduced by NADH, succinate, and ascorbate (in the presence of phenazine methosulfate) [115]. It is not known if this is due to the inaccessibility of one or both of the cytochromes to reducing equivalents from these substrates or if they have highly negative midpoint oxidation-reduction potentials.

3. Cytochrome Oxidases

There are possibly three cytochrome oxidases present in *E. coli.* Cytochrome a_1 appears as a minor peak at 594 nm in the difference spectrum of late exponential and stationary phase cells. Cytochrome *d* shows a complex α-band spectrum consisting of a peak at 628 nm and a trough at 650 nm in the reduced-minus-oxidized difference spectrum. As discussed previously, cytochrome *o* is detected by the characteristic spectrum of its compound with carbon monoxide.

Cytochrome a_1 is the sole cytochrome oxidase present in *Acetobacter pasteurianum* and its function as the cytochrome oxidase of this organism has been confirmed by studies of the photodissociation spectrum of its complex with carbon monoxide [119]. In most other species of bacteria, including *E. coli,* cytochrome a_1 is accompanied by other potential cytochrome oxidases so that its function is less clear. In *E. coli* it is most unlikely that it has a significant role as a terminal oxidase. Castor and Chance [119] measured the photodissociation spectrum of the carbon monoxide-cytochrome oxidase complex in exponential phase *E. coli.* Maxima were observed at 416, 535, and 567 nm, which corresponded closely to the absorption peaks at 416, 538, and 567 nm of the carbon monoxide complex of cytochrome *o* [23]. Thus, cytochrome *o* was the major oxidase of exponential phase cells. In stationary phase cells, the photodissociation spectrum indicated that cytochrome *d* as well as cytochrome *o* were the functional oxidases. There was no indication from the photodissociation spectra of the involvement of cytochrome a_1. Confirmation that cytochrome a_1 does not have significant cytochrome oxidase activity in *E. coli* was obtained by Haddock et al. [111]. The kinetics of reoxidation of the reduced cytochromes after the addition of a pulse of oxygen was followed using a stopped-flow dual-wavelength spectrophotometer. Cytochromes *o* and *d* were reoxidized with a half-time of less than 3.3 msec (pseudo first-order rate constant $> 210\ \text{sec}^{-1}$) whereas cytochrome a_1 had a half-time for reoxidation of 25 msec (rate constant = 27 sec^{-1}). Similar results, showing that cytochrome a_1 was reoxidized much less rapidly than cytochromes *o* and *d,* were obtained by Smith et al. [120] for *Hemophilus parainfluenzae.* Cytochrome a_1 in membrane particles of *E. coli* has a midpoint oxidation-reduction potential of +147 mV (n = 1) [115].

Studies of cytochromes *o* and *d* in *E. coli* have been restricted to the membrane-bound cytochromes since they have not as yet been solubilized from membranes of this organism. Cytochrome *o* has been released from membranes of *Bacillus megaterium* and *Mycobacterium phlei* by treatment with pancreatic lipase [121, 122]. Recently cytochrome *o* has been partially purified from Triton X-100 extracts of membranes of *Rhodopseudomonas palustris* [123]. The product migrated as a single band on polyacrylamide gels in the presence of 0.5% Brij 58. It consisted of four polypeptides of molecular weight 30,500, 25,500, 12,200, and 9,500 but it is not clear which of these bands is from cytochrome *o* since the preparation contained much cytochrome *c*. Two types of cytochrome *o* have been purified from *Vitreoscilla* [124]. These cytochromes had unexpectedly low redox potentials of -90 and +100 mV. This might be due to the release of the cytochromes from a membrane environment although they were isolated from the cytoplasmic fraction of the cell. It is possible that they were released during disruption of the cell. These cytochromes have been extensively characterized by Webster and his coworkers [125-128]. The relevance of these results to the firmly bound membrane cytochrome *o* of *E. coli* is not clear.

The oxidation-reduction potential of cytochrome *o* has not been determined. It must have a redox potential greater than 0 mV since it is fully reducible by ascorbate [151]. In the redox titration experiments of Pudek and Bragg [115], the addition of 14 mM KCN did not alter the redox potentials of the *b*-type cytochromes. Since cytochrome *o* is probably a *b*-type cytochrome [129] which reacts with cyanide [110, 130], it is unlikely to be one of the two *b*-type cytochromes identified by redox titration measurements.

The reduced-minus-oxidized difference spectrum of cytochrome *d* shows a peak at 628 nm and a trough at 648 nm. The absorption maximum at 442 nm, seen as a shoulder on the larger cytochrome b_1 peak, probably is the Soret band of reduced cytochrome *d* since it disappears on reaction of this cytochrome with cyanide [110]. The prosthetic group of cytochrome *d* has been identified as an iron-chlorin [131]. The trough at 648 nm observed in difference spectra is seen as an absorption peak in direct spectra of the membrane particles of *E. coli* [110] and *Azotobacter vinelandii* [132] measured in the absence of substrate. This peak is usually considered to be due to the oxidized form of cytochorme *d*. A small absorption peak at 675-680 nm is also present but this does not appear to be related to cytochrome *d* since its apparent level in relationship to cytochrome *d* varies from preparation to preparation [132]. The nature of this chromophore is unknown. The absorption peak at 648 nm disappears on reduction to be replaced by the absorption peak at 628 nm of reduced cytochrome *d*. The conversion of oxidized to reduced cytochrome *d* does not proceed directly but appears to involve an intermediate form, cytochrome *d**, which does not have a detectable absorption band in the 600-700 nm region of the spectrum [110]. Thus, reduction of membrane particles with ascorbate (in the presence of phenazine methosulfate) under anaerobic conditions causes progressive elimination of the 648-nm band without the concomitant appearance of the band at 628 nm. This band appears more slowly. Similar behavior has been noted for cytochrome *d* in *A. vinelandii* [132-134]. The redox potential of cytochrome *d*, determined by measuring the changes in the 628-nm peak, is +260 mV (n = 1) [115]. The absorption band at 648 nm of the oxidized cytochrome *d* does not follow normal redox behavior. When the redox potential of the system is taken from 0 mV to potentials as high as 450 mV, the 648-nm band does not reappear as would have been expected from the value of +260 mV found for the redox changes of the 628-nm band. The 648-nm band does reappear following addition to the anaerobic system of a small pulse of oxygen. These results suggest that the midpoint oxidation-reduction potential of +260 mV applies to the interconversion of reduced (628 nm) cytochrome *d* and cytochrome *d**. Cytochrome *d** may represent another conformation of oxidized cytochrome *d* which is converted to the 648-nm form only in the presence of oxygen [130-134]. This would be somewhat analogous to the behavior of mitochondrial cytochrome oxidase where the oxidized and the so-called "oxygenated" form appear to be conformational forms at the same oxidation state. Both forms can accept four electrons from a reductant, presumably by reduction of the two hemes and two copper atoms present in the mitochondrial cytochrome oxidase molecule [135, 136]. It is not presently known whether cytochrome *d* is a diheme cytochrome and whether copper atoms are involved in the reaction mechanism.

The intermediate form, cytochrome *d**, is present at a very low level in the aerobic steady state. However, there appears to be a temperature-sensitive step in the oxidation-reduction cycle of cytochrome *d* since the steady-state level of cytochrome *d** is increased at lower temperatures. About 28% of cytochrome *d* is present as the intermediate form in the aerobic steady state at 1°C in the presence of ascorbate (with phenazine methosulfate). This amount is increased to 57% at -38°C [137].

Cyanide reacts slowly with cytochrome *d* in membrane particles from stationary phase cells, causing elimination of the 648-nm peak without the appearance of another absorption band. A second-order rate constant of 0.011 M^{-1} sec^{-1} calculated for this reaction appeared to be too low to account for the inhibition of NADH oxidase activity by cyanide for which a second-order rate constant of 0.26 M^{-1} sec^{-1} was obtained [130]. However, in the presence of NADH, the second-order rate constant for the elimination of the 648-nm absorption peak increased to 0.58 M^{-1} sec^{-1}, suggesting that the reaction of cyanide with cytochrome *d* could account for the inhibition of NADH oxidase by this compound in stationary phase cells. The effect of cyanide on the spectrum of cytochrome *d*, and the need for turnover of cytochrome *d* for inhibition to occur, suggests that it interacts with cytochrome *d** and not with the reduced (628 nm) or oxidized form (648 nm) form of cytochrome *d*.

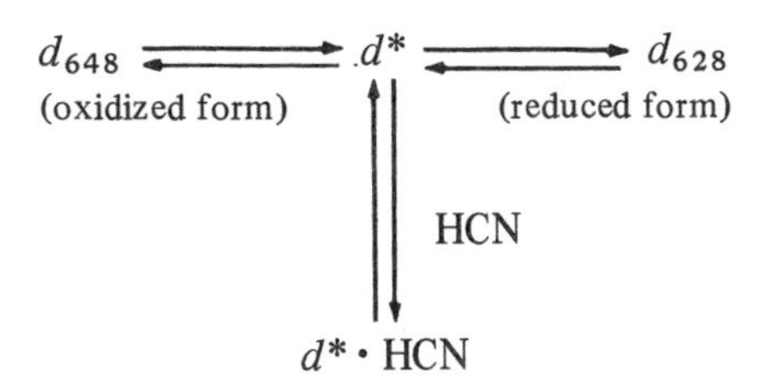

When the oxidation rate of succinate is regulated by using malonate as a competitive inhibitor, the rate constant for the formation of cyanocytochrome *d* is directly proportional to the rate of electron flux through cytochrome *d*. Thus, higher rates of electron flux through the respiratory chain result in higher steady-state levels of cytochrome *d** and so to an increased rate of reaction with cyanide. Kauffman and Van Gelder have come to essentially similar conclusions regarding the reaction of cyanide with cytochrome *d* in *A. vinelandii* [133, 134].

Pudek and Bragg [110] observed that NADH oxidase activity was much more sensitive to inhibition by KCN in exponential than in stationary phase cells. This appeared to correlate with the greater level of cytochrome *o* relative to cytochrome *d* in the exponential phase cells (Table 2). However, the total amount of cytochrome oxidase (cytochromes *o* and *d*) in exponential phase cells is one-fifth of that in stationary phase cells, so that there would be a greater rate of electron flux through the smaller cytochrome oxidase pool in exponential phase cells. This should result in a higher steady-state level of cytochrome *d** and so to an apparently greater sensitivity to inhibition by cyanide. The rate of formation of cyanocytochrome *d* relative to the rate of electron flux through the cytochrome *d* pool was calculated by these workers [130] by dividing the second-order rate constant for the formation of cyanocytochrome *d* by the rate of electron flux through cytochrome *d* (expressed as g-atoms O reduced per sec per mol cytochrome *d*). With stationary phase cells containing only cytochrome *d*, a value of 0.096 mol cytochrome *d* per mol cyanide per g-atom O was obtained. If it was assumed that electrons from the substrates NADH and succinate were partitioned equally between the almost equal pools of cytochrome *o* and of cytochrome *d* in exponential phase cells, a value of 0.085 mol cytochrome *d* per mol cyanide per g-atom O was obtained. This was close to that found for the stationary phase cells and confirmed that the extent of inhibition of respiration by cyanide was dependent not on the apparently different relative abilities of cytochromes *o* and *d* to react with cyanide but upon the relative rate of electron flux through the cytochromes to generate the cyanide-reactive intermediate. Furthermore, these results suggested that both cytochrome *o* and cytochrome *d* functioned as terminal oxidases in cells where both were present, and that electrons from both NADH and succinate passed to oxygen through both of these cytochromes.

L. Respiratory Chain Complexes

Work on the mechanism of oxidative phosphorylation and electron transfer in the mitochondrial respiratory chain has been considerably aided by the isolation of well-defined portions of the respiratory chain [138]. The electron transport chain in beef heart mitochondria has been fractionated into four complexes:

NADH-ubiquinone reductase (complex I), succinate-ubiquinone reductase (complex II), dihydroubiquinone-cytochrome *c* reductase (complex III), and cytochrome *c* oxidase (complex IV). The same approach applied to the respiratory chain of *E. coli* formed aerobically has met with limited success. Treatment of membrane particles with deoxycholate or cholate in the presence of ammonium sulfate solubilized about 50% of the cytochromes and released some of the succinate and D-lactate dehydrogenases, together with three NADH dehydrogenases [25, 53, 54, 139]. Two of the NADH dehydrogenases, which were measured by their ability to reduce menadione, ferricyanide, or dichlorophenolindophenol, resembled similar enzymes found in the cytoplasmic fraction. These enzymes have been discussed in a previous section. The solubilized cytochromes appeared to be associated with a complex of enzymes which was not sedimented by centrifuging at 175,000 $\times$ g for 2 hr. Baillie et al. [25] separated this material ("soluble respiratory complex") on a double column of Sepharose 6B connected in series with Sepharose 4B. It chromatographed as a species of molecular weight 2×10^6. In other experiments, under slightly different conditions, this material appeared to have a molecular weight of about 650,000. It is probable that this complex readily associates to form larger size aggregates. The complex sedimented in a sucrose gradient with a sedimentation constant of 11.3 S. The unexpectedly low sedimentation constant was probably due to the presence of about 30% phospholipid in the preparation. The complex contained cytochromes b_1, a_1, *o*, and *d*, succinate and NADH dehydrogenases, and Ca^{2+}- and Mg^{2+}-activated ATPase activities [25, 139]. Succinate and NADH oxidase activities were restored by addition of exogenous ubiquinone. The composition of the soluble respiratory complex is shown in Table 7. In our best preparation, the concentrations of cytochromes $b_1 + o$ and cytochrome *d* were 2.30 and 2.17 nmol/mg protein. Analysis of the cytochrome b_1 peak in a reduced-minus-oxidized difference spectrum measured at 77 K showed that *c*-type cytochromes, cytochromes b_{556}, b_{558}, and b_{562} were present [28].

Hendler and Burgess [55, 56] have also examined the solubilization of the respiratory chain using detergent. The starting material was a spheroplast membrane preparation which was treated first with 0.4-0.5% and then with 1% deoxycholate. Following chromatography of the first extract on hydroxylapatite and DEAE-cellulose, the following fractions were obtained: (1) two NADH dehydrogenases similar to those purified by Bragg and Hou [53, 54] and also found in the cytoplasmic fraction, (2) D-lactate dehydrogenase, (3) a fraction containing cytochrome b_1 but not cytochromes a_1 and *d*, (4) a fraction containing succinate and NADH dehydrogenases and cytochrome b_1 (this fraction had some succinate oxidase activity). Examination of the spectra presented by Hendler and Burgess suggests that cytochrome *d* was also present. This preparation is probably similar to the soluble respiratory complex of Baillie et al. [25]. The

Table 7 Composition of Soluble Respiratory Complex[a]

Component	Amount (ng-atoms or nmoles per mg protein)
Flavin (total)	0.42
Flavin (acid-soluble)	0.25
Ubiquinone	4.5
Nonheme iron	14.2
Cytochrome b_1	0.92
Cytochrome *o*	0.24
Cytochrome a_1	+[a]
Cytochrome *d*	1.05

[a]+, present but not quantitated.
Source: Data from Baillie et al. [25].

second extract was chromatographed on DEAE-cellulose but no resolution of the components was obtained. Several peaks of material, presumably of different sizes, but each containing cytochromes b_1, a_1, and *d* were obtained. Although dehydrogenase activities did not seem to be present it is possible that they were denatured since a large flavoprotein peak can be seen in difference spectra of these fractions. Although these fractions may be enriched to some extent in the cytochromes relative to the dehydrogenases, it is unlikely that they represent the "cytochrome oxidase" portion of the respiratory chain as claimed by Hendler and Burgess [56].

In conclusion, clear-cut fractionation of the respiratory chain into complexes like those obtained from mitochondria has not been achieved with *E. coli*. Succinate and D-lactate dehydrogenases can be released to leave cytochrome-containing fractions partially depleted of dehydrogenases. The major solubilized product is a complex containing succinate and NADH dehydrogenases, ubiquinone, and cytochromes b_1, *o*, a_1, and *d*, capable of showing NADH and succinate oxidase activities. However, it is not clear at present if this complex is formed by reaggregation of some of its components.

M. Sequence and Organization of the Components of the Respiratory Chain

A number of schemes for the arrangement of the components of the respiratory chain of *E. coli* have been proposed [24, 27, 34, 56, 71, 94, 97, 107, 140]. The essential elements of several of the most important proposals are depicted in Figure 5.

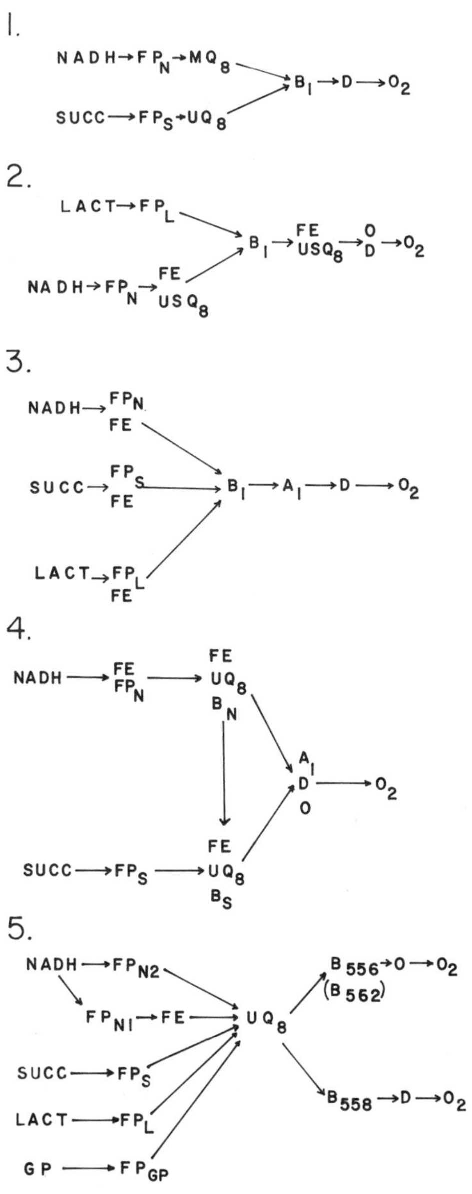

Figure 5 Proposals for the sequence of the components of the respiratory chain of *E. coli.* 1, Kashket and Brodie, 1963 [71]; 2, Cox et al., 1970 [24]; 3, Hendler et al., 1970, 1971, 1974 [107, 97, 56]; 4, Bragg, 1974 [94]; 5, Haddock et al., 1973, 1975 [34, 27]. SUCC, succinate; LACT, D-lactate; GP, L-glycerol-3-phosphate; FP_N, FP_S, FP_L, FP_{GP}, flavoproteins of NADH, succinate, D-lactate, and L-glycerol-3-phosphate dehydrogenases; B_1, D, O, A_1, B_{556}, B_{558}, and B_{562}, cytochromes b_1, *d*, *o*, a_1, b_{556}, b_{558}, and b_{562}; B_N, B_S, cytochromes *b* associated preferentially with the NADH and succinate oxidase pathways, respectively; FE, nonheme iron; UQ_8, ubiquinone-8, MQ_8, menaquinone-8; USQ_8, ubisemiquinone-8.

The earliest proposal for the arrangement of the components of the respiratory chain is that of Kashket and Brodie [71]. In this scheme, reducing equivalents are funneled from succinate and NADH via ubiquinone and menaquinone, respectively, into a common cytochrome chain consisting of cytochromes b_1 and *d* in sequence. Birdsell and Cota-Robles [140] followed this scheme in their proposal but with the introduction of cytochrome *o* which, together with cytochrome *d*, was responsible for carrying electrons from cytochrome b_1 to oxygen.

The scheme of Cox et al. [24] has the flavoprotein dehydrogenases of NADH and D-lactate interacting at the level of cytochrome b_1. Electrons then pass via a complex of nonheme iron protein and ubisemiquinone to the cytochrome oxidases, cytochromes *d* and *o*. A similar complex of ubisemiquinone and nonheme iron protein is suggested to be on the pathway between the NADH flavoprotein and cytochrome b_1.

Hendler and coworkers [56, 97, 107] have proposed a respiratory pathway which differs in three important respects from those proposed previously. The dehydrogenases funnel reducing equivalents into a common cytochrome pathway at the level of cytochrome b_1. However, the dehydrogenases are all considered to be nonheme iron flavoproteins, and nonheme iron is not thought to be located elsewhere in the chain. In contrast to most schemes, cytochrome a_1 is placed in the respiratory chain between cytochromes b_1 and *d*. Lastly, cytochrome *o* is not thought to be a component of the respiratory pathway but to be an optical artifact of the spectrophotometric measurements. However, Castor and Chance [119], using the entirely different technique of photodissociation spectroscopy of the cytochrome *o*-carbon monoxide complex, have shown clearly that cytochrome *o* is a real component of the respiratory chain both in exponential and stationary phase cells.

The schemes proposed by Bragg [94] and by Haddock and coworkers [27, 34] recognize that more than one type or pool of cytochrome *b* may occur in the respiratory chain of *E. coli*. Although Bragg [94] did not specify the identity of the *b*-type cytochrome in the two pools, one of the pools is more closely associated with the NADH pathway and cannot undergo reduction by succinate in the absence of the inhibitor 2-heptyl-4-hydroxyquinoline N-oxide (HOQNO) [26]. Reducing equivalents from NADH can readily reduce both pools of cytochrome b_1. Associated with the cytochrome b_1 in each of the two pools are ubiquinone-8 and a nonheme iron protein. The sequence of electron transfer between the three components in each pool is not indicated. Electrons from the pools are transferred to oxygen by the cytochrome oxidases, cytochromes a_1, *d*, and *o*. The potential relationship of a specific cytochrome oxidase to one of the two pools of cytochrome b_1 is not indicated.

As far as the arrangement of the cytochromes is concerned, by far the most specific scheme is that of Haddock and his coworkers [27, 34]. Reducing equivalents from various substrates are collected at the level of ubiquinone and then funneled into one of two cytochrome-containing terminal respiratory chains. In early exponential cells in batch culture, or in cells growing under substrate-limited conditions in continuous culture, this consists only of cytochromes b_{556} and *o*, although cytochrome b_{562} is almost certainly present. In stationary phase cells another terminal respiratory pathway consisting of cytochromes b_{558} and *d* is also present.

Some of the elements in the schemes proposed above are incorrect. The evidence relating to the sequence of the components of the respiratory chain will now be discussed.

There have been several approaches to determining the role of ubiquinone and menaquinone in the respiratory chain of *E. coli*. Kashket and Brodie [141] separated membrane particles into two populations—large and small particles—by differential centrifugation. The large-particle fraction, containing chiefly succinate oxidase activity, possessed ubiquinone only, whereas the small-particle fraction, having succinate and NADH oxidase activities, contained both ubiquinone and menaquinone. Irradiation of the large-particle fraction with ultraviolet light, which is known to destroy quinones, inactivated succinate oxidase activity but this could be restored by the addition of ubiquinone-10 and menaquinone [71]. Irradiation of the small particles resulted in loss of NADH oxidase activity. Menaquinone but not ubiquinone-10 could restore the lost activity. Kashket and Brodie interpreted these results as indicating that ubiquinone and menaquinone were specifically associated with the succinate and NADH oxidase pathways, respectively (Fig. 5). These results, although suggestive, do not constitute absolute proof that the exogenous quinones are functioning in the same manner as the endogenous quinones. It is possible that the exogenous quinones are generating a new pathway to bypass a damaged region of the respiratory chain. Moreover, irradiation may produce effects on components of the respiratory chain other than quinones. Bragg [142] found that irradiation with near-ultraviolet light resulted in loss of NADH oxidase activity. This loss correlated poorly with loss of ubiquinone but followed closely the kinetics of destruction of cytochrome *d*. Moreover, the increase in the steady-state level of reduction of cytochrome b_1 suggested that the primary lesion was at the level of cytochrome *d*. Lakchaura et al. [143] have pointed out that the ultraviolet radiation source used by Bragg emits much radiation at 410 nm, the absorption of which might account for the observed destruction of cytochrome *d*. Like Bragg, they found that the loss of NADH oxidase activity could not be accounted for solely by the destruction of ubiquinone since a 10-fold greater fluence was required to destroy ubiquinone than NADH oxidase activity [143, 144].

The role of ubiquinone in *E. coli* has also been examined in preparations in which the cytochrome content had been depleted by extraction with detergent. As discussed in a previous section, cholate in the presence of ammonium sulfate solubilized a complex of the respiratory chain of *E. coli* [25]. The soluble respiratory complex contained about 40% of the ubiquinone content of the respiratory particles from which it was derived. Both NADH and succinate oxidase activity were dependent on the presence of exogenous ubiquinone. Although the major ubiquinone, ubiquinone-8, of *E. coli* was effective, the lower isoprenologue ubiquinone-2 was more active possibly because of its greater solubility in aqueous solutions. Addition of ubiquinone to the soluble respiratory complex increased the rate of reduction of cytochrome b_1 by NADH and succinate, and also lowered the aerobic steady-state level of reduction of cytochrome b_1 by succinate and NADH. These results are consistent with a ubiquinone location in the succinate and NADH oxidase chains both before and after cytochrome b_1 in the sequence. This suggestion is similar to that of Cox et al. [24]. The reducibility of ubiquinone by succinate and NADH has been confirmed by direct measurements of its level of reduction in the presence of these substrates [71, 151].

Jones [145] isolated a mutant which did not form ubiquinone unless supplemented with 4-hydroxybenzoic acid. Succinate and malate oxidase activities were very low in cells of the mutant grown without 4-hydroxybenzoic acid. Negligible NADH oxidase activity was present in membrane particles from such cells but addition of ubiquinone restored NADH oxidase activity and increased the rate and extent of reduction of cytochrome b_1 by this substrate. These results are consistent with the involvement of ubiquinone in the NADH and succinate oxidase pathways and with a location between NADH dehydrogenase and cytochrome b_1 in the NADH oxidase pathway. A series of mutants defective in ubiquinone biosynthesis have been isolated by Gibson and Cox [9, 10]. The rate of oxidation of NADH, L-malate, and L- and D-lactate was low in particles prepared from the mutants but could be increased to levels characteristic of the normal strains by addition of exogenous ubiquinone-1 [146, 147]. A comparison of the aerobic steady-state level of reduction by NADH of flavoprotein and cytochromes b_1 and d in the normal and mutant strains showed that the lack of ubiquinone in the mutant caused an increase in the level of reduction of cytochrome b_1 and flavoprotein [24] (Table 8). Addition of ubiquinone-1 decreased the level of reduction to that of the normal strain. This suggests that ubiquinone is located on the oxygen side of cytochrome b_1 and resembles the results with the soluble respiratory complex [25]. However, addition of NaCN at a level causing less inhibition of NADH oxidase activity than was produced by ubiquinone deficiency, resulted in a greater aerobic steady-state level of reduction of cytochrome b_1 and flavoprotein than given by ubiquinone deficiency [24] (Table 8). This indicates that ubiquinone deficiency must also prevent the reduction of cytochrome b_1. Therefore, there is a site of ubiquinone action both before and after cytochrome b_1 in the respiratory chain sequence. Cox

Table 8 Aerobic Steady-state Level of Reduction by NADH of Respiratory Chain Components in Membrane Particles from Strains AN 62 (ubi^+) and AN59 (ubi^-)

Strain	Addition	Time to deplete oxygen in cuvette[a]	Steady-state level of reduction (%)		
			Flavoprotein	Cytochrome b_1	Cytochrome *d*
AN62	–	7	30	10	<5
AN62	0.8 mM NaCN	20	59	43	>50
AN59	–	60	54	23	<5
AN59	40 μM UQ-1[b]	<5	26	14	<5

[a]Inversely proportional to NADH oxidase activity.
[b]UQ-1, ubiquinone-1.
Source: Data taken from Cox et al. [24].

et al. [24] have speculated on the nature of the ubiquinone at these sites. Hamilton et al. [96] found that membrane preparations of normal but not of ubiquinone-deficient strains gave an electron paramagnetic resonance signal at $g = 2.003$ which could be attributed to ubisemiquinone. The signal disappeared on reduction with NADH or D-lactate. This signal accounted for only 2% of the ubiquinone present in the membranes. Cox et al. [24] found that under the same conditions, approximately 50% of the total ubiquinone *on extraction* from the membrane was in the dihydro form. They suggested that ubiquinone *within the membrane* in the absence of substrate under aerobic conditions was completely in the semiquinone form but disproportionated on extraction into the dihydro and oxidized forms. However, because the ubisemiquinone was complexed with nonheme iron protein, the intensity of the signal was only 2% of that which would have been expected. The evidence supporting this hypothesis is very indirect, and more convincing results are required before it can be accepted.

The concentration of ubiquinone may be as much as 25-fold greater than the amount of cytochrome b_1. Newton et al [148] have isolated a mutant (*ubi D*) of *E. coli* which forms 25% of the normal amount of ubiquinone. The NADH oxidase activity in the mutant was only 40% of that of the normal strain and the aerobic steady-state level of reduction of cytochrome b_1 was 18% in the mutant compared to 13% in the normal strain. Thus, in spite of a fourfold molar excess of ubiquinone over cytochrome b_1, membrane particles of this strain show the characteristics of partial ubiquinone deficiency. The large molar excess of ubiquinone over cytochrome b_1 must be required to maintain normal electron transport properties.

Much remains to be learned about the location and nature of nonheme iron in the respiratory chain of *E. coli* [94]. NADH and succinate oxidase activity are inhibited by iron chelating agents [26, 53, 97, 141]. Since succinate dehydrogenase is not inhibited by levels of *o*-phenanthroline and thenoyltrifluoroacetone (TTFA) sufficient to inhibit succinate oxidase activity [75, 97], one of the sites of chelator action must be in the respiratory chain between the dehydrogenase and oxygen. Nonheme iron is probably present in succinate dehydrogenase but is not accessible to this chelating agent. Gutman et al. [50, 149] found equivalent amounts of nonheme iron and acid-labile sulfide in a crude preparation of the NADH dehydrogenase.

Further work on the location of nonheme iron in the respiratory chain of membrane particles has been carried out by Bragg [151], Kim and Bragg [75], and Hendler [97]. Bragg measured nonheme iron reduction in the presence and absence of inhibitors by noting the formation of ferrous *o*-phenanthrolinate following the addition of *o*-phenanthroline to particles reduced by substrates. Ascorbate (in the presence of phenazine methosulfate) and NADH gave the same amount of complex formation as given by dithionite [151]. The ability of ascorbate to fully reduce the species of nonheme iron which could be detected by this method suggested that it must have a more positive redox potential than cytochrome b_1 which is only 70% reduced by ascorbate [151]. Succinate reduced only about one-third of the nonheme iron reacting with *o*-phenanthroline in the presence of NADH, ascorbate, and dithionite [75]. In this preparation the amount of cytochrome b_1 reducible by succinate (55%) was less than that reducible by NADH (90%). Addition of HOQNO increased the level of reduction by succinate of both nonheme iron and cytochrome b_1 to those given by dithionite. These results suggest that there might be more than one species of high-potential nonheme iron possibly associated with a separate pool or species of *b*-type cytochrome. There is insufficient evidence to indicate whether the ascorbate-reducible nonheme iron is located before or after cytochrome b_1 in the sequence. Hendler [97] studied the effect of the iron-chelating agent TTFA on the appearance of the EPR signal of nonheme iron at g = 1.92-1.94 in the presence of substrate. NADH and dithionite produced signals of the same magnitude but succinate gave a signal of about 20-25% of this intensity. These results are superficially similar to those obtained with the *o*-phenanthroline technique. However, they may not be measuring the same species of nonheme iron. Thus, the most intense signal observed at 77 K in pigeon heart mitochondria is that at g = 1.94, due to the low-potential center 1 (E_m = -305 mV) of NADH dehydrogenase [152]. This would not be reduced by ascorbate. Hendler found that TTFA inhibited the formation of the EPR signal and prevented the reduction of cytochromes b_1, a_1, and *d* by NADH, but did not prevent the reoxidation of the cytochromes on aeration. These results suggest that this nonheme iron species is located before cytochrome b_1 in the respiratory chain sequence or is not on the direct pathway of electrons from this cytochrome to oxygen at least in the presence of the chelating agent. Similar conclusions were made by Bragg [151].

In conclusion, it is likely that nonheme iron is present at several regions of the respiratory chain of *E. coli*. It is probable that succinate and NADH dehydrogenases are nonheme iron flavoproteins. Moreover, high-potential nonheme iron is probably associated with the chain at the level of cytochrome b_1 but whether it is before or after this cytochrome in the sequence, or at both positions as suggested by Cox et al. [24], is not clear. More work is required to clarify this problem.

The recognition that there are several species of *b*-type cytochrome in the respiratory chain of *E. coli* has invalidated some of the previous schemes proposed for the sequence of the respiratory chain. As can be seen from Table 9, the extent of reduction of the cytochromes varies with both the substrate and the preparation. Cytochrome b_1 is almost fully reduced by NADH and D-lactate but is reduced to a lesser extent by succinate and ascorbate. This can be attri-

Table 9 Anaerobic Steady-state Levels of Reduction of Cytochromes in Membrane Particles and Vesicles of *E. coli*[a]

Authors	Substrate	Reduction of cytochromes (%)[b]				References
		b_1	a_1	d	o	
Bragg	NADH	100	–	–	100	151
	Ascorbate[c]	60-70	–	–	100	151
Barnes and Kaback, Konings et al.	NADH	84	–	113	–	153, 154
	D-Lactate	97	–	99	–	153, 154
	Succinate	81	–	102	–	153, 154
	Ascorbate[c]	72	–	99	–	153, 154
Hendler and Nanninga	NADH	85	100	100	–	107
	Succinate	65	86	97	–	107
Kim and Bragg	NADH	92	–	–	144	26
	Succinate	60	–	–	71	26
Birdsell and Cota-Robles	NADH	28	–	75	–	140
	Succinate	53	–	94	–	140

[a]All experiments were on membrane particles except those in Refs. 153 and 154 which used membrane vesicles.
[b]Referred to value with dithionite.
[c]In the presence of phenazine methosulfate.

buted to the lower redox potential of these latter substrates. The low extent of reduction of the cytochromes, particularly by NADH, in the preparation of Birdsell and Cota-Robles [140] can be attributed to inhibition by the detergent Brij 58 used in the preparation of the membrane particles. The cytochrome oxidases, cytochromes d and o, are fully reduced by all of the substrates consistent, at least in the case of cytochrome d, with the redox potentials of the cytochromes. There is no indication that reducing equivalents from the different substrates are partitioned to the different cytochrome oxidases.

Examination of the reduction of the components of the cytochrome b_1 peak has been carried out by Pudek and Bragg [28, 115]. NADH and succinate, respectively, reduced 72 and 58% of both cytochrome b_{556} and b_{558} at anaerobiosis. Cytochrome b_{562} was almost completely reduced by both substrates. The kinetics of reoxidation of the reduced cytochromes examined in stopped-flow experiments by Haddock et al. [111] showed that the cytochrome b pool was kinetically heterogeneous in contrast to cytochromes d and a_1. Cytochrome *d* was reoxidized with a half-time of less than 3.3 msec, considerably faster than cytochrome a_1 with a half-time of 25 msec. Cytochrome b_1 oxidation showed at least two phases with half-times of less than 3.3 msec and about 25 msec. These pools of cytochrome b_1 were of about equal size. Haddock et al. have suggested that the fast-oxidizing pool of *b*-type cytochrome is cytochrome *o*. This seems unlikely if cytochrome *o* makes a contribution of only 20-40% to the absorbance at 556 nm as has been proposed [111]. In these experiments the spectral characteristics of the *b*-type cytochromes reacting in the two phases were not determined. Thus, the possibility that the two phases do not represent individual cytochrome species but pools of several *b*-type cytochromes cannot be eliminated.

Cox et al. [24] and Baillie et al. [25] have noted that ubiquinone deficiency or the presence of the inhibitors piericidin A and HOQNO have equivalent effects on the steady-state level of reduction of the cytochromes, suggesting that the site of action of these inhibitors is close to ubiquinone. It is difficult to see why there should be two pools of ubiquinone as well as two sites of action for these inhibitors. A plausible explanation comes from a recent suggestion of Mitchell [156] on the flow of reducing equivalents in the site 2 region of the mitochondrial respiratory chain. This proposal—the protonmotive Q cycle—adapted to the respiratory pathway of *E. coli* is shown in Figure 6. An electron from one of the dehydrogenases together with an electron from a *b*-type cytochrome (possibly cytochrome b_{562}) are transferred to ubiquinone which with the uptake of two protons at the cytoplasmic side of the membrane becomes reduced to ubiquinol. The ubiquinol diffuses to the outer face of the membrane where the protons are discharged into the medium while the electrons are disproportionated between the cytochrome oxidase (cytochrome *o* or *d*) and a *b*-type cytochrome (possibly cytochrome b_{556}). Electrons pass from the *b*-type cytochrome via the other *b*-type cytochrome to reduce ubiquinone on the inner face of the membrane. Ubiquinone diffuses from the external to the inner face of the membrane. It is presumed that the semiquinone cannot diffuse because it is ionized. This would allow sequential two-electron transfer to occur during the reduction and oxidation of the ubiquinone at the two faces of the membrane. The pathway of electrons to oxygen through the cytochrome oxidases would be transmembranous if the reaction site of oxygen is on the cytoplasmic surface of the membrane.

An advantage of this type of scheme is that it accounts for some of the effects of ubiquinone deficiency and of the effects of HOQNO and piericidin A. As discussed above, these effects seem to be manifested by sites of action both before and after the *b* cytochromes. In the scheme of Figure 6 removal of ubiquinone would affect both the reduction and oxidation of these cytochromes. If piericidin A and HOQNO inhibited either the reduction or the reoxidation of

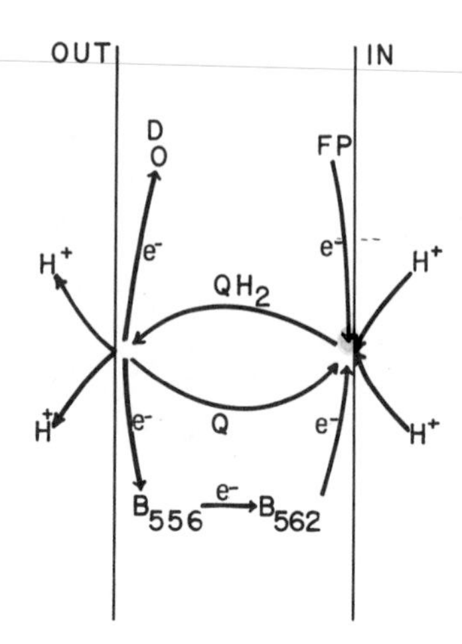

Figure 6 Hypothetical scheme for protonmotive Q cycle in respiratory chain of *E. coli* (see text for description). Q, ubiquinone; QH_2, dihydroubiquinone; FP, flavoprotein dehydrogenases; B_{556}, B_{562}, D, and O, cytochromes b_{556}, b_{562}, *d*, and *o*.

ubiquinone then *both* the reduction and oxidation of the *b* cytochromes would be affected. We have found that HOQNO can inhibit the reoxidation of ubiquinol [151]. A further advantage of this scheme is that it would permit electrons to be transferred from the protonmotive Q cycle to either cytochrome *o* or *d* depending on which was present. The integration of cytochrome b_{558} into this scheme to permit the return of electrons from ubiquinol to ubiquinone by a parallel pathway would also be possible. There is little direct evidence for this scheme. The organization of the cytochromes within the membrane has not been examined although the substrate-binding sites of the D-lactate and NADH and succinate dehydrogenases appear to be on the cytoplasmic face of the membrane [16-20]. One piece of information which is hard to reconcile with this scheme concerns the sites of energy conservation associated with the respiratory chain of *E. coli*. As discussed in a later section, there appears to be two sites of energy conservation in *E. coli*. One site is located in the NADH dehydrogenase region of the chain and the other between ubiquinone and oxygen. If the latter site was associated with the protonmotive Q cycle as suggested by Mitchell then it would not be operative in the ubiquinone-deficient mutant. The results of Singh and Bragg [157] show clearly that there is an energy conservation site in the cytochrome region of the chain in ubiquinone-deficient cells. Thus, either there is a third site of energy conservation or the protonmotive Q cycle is not present as formulated in Figure 6.

IV. Formate-Nitrate Reductase Pathway

A. Composition

E. coli reduces nitrate to nitrite. Thus, under anaerobic conditions nitrate is able to act as an alternate electron acceptor to oxygen. Wimpenny and Cole [158] studied the factors involved in the development of the nitrate reductase system. Cells grown anaerobically with nitrate developed high levels of formate dehydrogenase and nitrate reductase. The level of nitrate in the medium was important. In chemostat studies [158, 159], these workers found that cytochromes c_{552} and *d* and nitrite reductase activities were formed maximally when nitrate in the feed solution was 20 mM. Above 20 mM, *b*-type cytochrome and nitrate reductase activities increased. This was accompanied by an increase in NADH oxidase and NADH-nitrate reductase activities. In contrast, NADH-nitrite reductase activities were maximal at lower concentrations of nitrate. From these data Cole and Wimpenny concluded that there is a specific NADH-nitrite reductase pathway involving cytochrome c_{552} and formate- and NADH-nitrate reductase pathways including a *b*-type cytochrome.

The fact that formate dehydrogenase, *b*-type cytochrome, and nitrate reductase are induced to higher levels in cells grown anaerobically on nitrate than in those grown without this electron acceptor suggests that the formate-nitrate reductase pathway might exist as a specific complex of enzymes in the membrane. The close relationship of these components was confirmed by Ruiz-Herrera et al. [160] who isolated a number of mutants which were unable to reduce nitrate. All of the mutants exhibited reduced or insignificant levels of formate dehydrogenase, nitrate reductase, *b*-type cytochrome, or various combinations of these activities under conditions which produced high levels of these components in wild-type strains.

The relationship of the NADH-nitrate reductase to the formate-nitrate reductase system is still not clear. Since Cole and Wimpenny [158, 159] found that NADH oxidase activity increased at the higher levels of nitrate which also resulted in the optimal formation of the formate-nitrate reductase system, it is possible that the NADH-nitrate reductase system uses the normal aerobic respiratory chain of *E. coli,* with the exception that electrons are ultimately directed to the nitrate reductase. That at least part of the two systems have components in common is suggested by the proton translocation measurements of Garland et al. [161]. With L-malate as electron donor, H^+/NO_3^- and H^+/O values of 3.45 and 3.1, respectively, were observed. Oxidation of D-lactate or glycerol gave H^+/NO_3^- or H^+/O values of about 2 with either substrate. These results suggest that malate oxidation by both nitrate or oxygen involves proton translocation in the NADH dehydrogenase region of the respiratory chain. The second proton translocation site between ubiquinone and oxygen or nitrate could be different depending on the terminal electron acceptor. Further evidence for the divergence of the pathways comes from the work of Kemp et al. [162]. Anaerobic growth of a heme-deficient mutant on nitrate in the absence of 5-aminolevulinic acid resulted in cells lacking cytochrome, NADH oxidase, and NADH-nitrate reductase activities, although nitrate reductase was present. Incubation of the cells in the presence of 5-aminolevulinic acid and chloramphenicol resulted in the appearance of some cytochrome b_1 and in the restoration of NADH oxidase activity. NADH-nitrate reductase activity was not reformed in the presence of chloramphenicol. These results indicate that the apoprotein of cytochrome b_1 of the NADH oxidase is formed in the absence of 5-aminolevulinic acid. However, the formation of the *b*-type cytochrome of the NADH-nitrate reductase pathway requires further protein synthesis. Thus, the NADH oxidase and NADH-nitrate reductase pathways probably diverge at the level of the cytochromes.

B. Formate Dehydrogenase

Two membrane-bound formate dehydrogenases are present in cell envelopes of *E. coli* [2, 3, 158, 159, 163]. Only one of these enzymes is induced by nitrate and so is involved in the formate-nitrate reductase system. The other enzyme is closely associated with hydrogenase in the formate hydrogenlyase complex which is responsible for the conversion of formate to carbon dioxide and hydrogen, and will not be considered here. There have been several attempts to purify formate dehydrogenase. Itagaki et al. [164] solubilized a cytochrome *b*-containing formate-nitrate reductase complex from membrane particles using 0.1% deoxycholate together with crude snake venom. A 10-fold purification of formate dehydrogenase was obtained but the cytochrome and nitrate reductase were purified to the same extent. Linnane and Wrigley [49], using 1 mg deoxycholate per mg envelope protein in the presence of ammonium sulfate, solubilized a preparation of formate dehydrogenase which after a 20-fold purification still contained cytochrome *b* but not nitrate reductase. The specific activity of the preparation was 17 μmol formate oxidized per min per mg protein when ferricyanide was used as electron acceptor. The cytochrome was readily reducible by the formate in fresh preparations of the enzyme. Ubiquinone was not present; therefore, the pathway of electrons from the dehydrogenase to the cytochrome did not involve this quinone.

Recently, Enoch and Lester [165, 166] have been able to obtain formate dehydrogenase in a high state of purity. The enzyme was solubilized from membrane particles by deoxycholate and ammonium sulfate using a procedure similar to that of Linnane and Wrigley [49]. A critical factor at all stages of the purification was the maintenance of strictly anaerobic conditions. The solubilized enzyme was purified by chromatography first on Bio-Gel A-1.5 m and then on

Table 10 Composition of Formate Dehydrogenase

Component	Concentration (nmol/mg protein)	Molar ratio
Heme	6.5	1
Molybdenum	4.8-6.2	0.74-0.95
Selenium	4.4-6.2	0.68-0.95
Nonheme iron	89	14
Acid-labile sulfide	82	13
Flavin	<0.3	0
Ubiquinone	<0.1	0

Source: Enoch and Lester [166].

DEAE-Bio-Gel A in the presence of Triton X-100. At the second chromatography step, formate dehydrogenase was separated from nitrate reductase. The purified formate dehydrogenase was 93-99% pure as judged by polyacrylamide gel electrophoresis in the presence of Triton X-100 or SDS. The specific activity of the final product was 196 and 39 μmol formate oxidized per min per mg protein with phenazine methosulfate and ferricyanide as electron acceptors, respectively. This represented a 150-fold purification over the starting material.

The enzyme as isolated was a detergent-protein complex containing 0.2 mg Triton X-100 per mg protein. The complex had a molecular weight of 590,000 as calculated from the sedimentation constant ($s_{20,w}$ = 18.1 S) and a Stokes radius of 76 Å. The composition of the enzyme is summarized in Table 10. A cytochrome *b* was present with absorption peaks at 428, 531, and 559 nm in the presence of formate. Formate reduced the cytochrome completely in less than 2 sec under anaerobic conditions, suggesting that the cytochrome is a functional part of the dehydrogenase activity. Since flavoprotein and ubiquinone are absent from the enzyme, reduction of the cytochrome is presumably mediated via the molybdo - nonheme iron - acid-labile sulfur portion of the enzyme. At present there is no evidence to show that molybdenum or nonheme iron undergo oxidation and reduction but this seems likely by analogy with the nitrate reductase of *Micrococcus denitrificans* in which molybdenum and nonheme iron undergo redox changes [167].

The composition of formate dehydrogenase suggests that it is a molybdo - nonheme iron - acid-labile sulfide hemoprotein with heme *b,* molybdenum, selenium, nonheme iron, and acid-labile sulfide in the molar ratio of 1:1:1:13-14:13-14, respectively. Using SDS polyacrylamide gel electrophoresis of the purified formate dehydrogenase or gel filtration in the presence of SDS, it was shown that three different types of subunits, α, β, γ, were present, with molecular weights of 110,000-120,000, 32,000-35,000, and 20,000, respectively. The α and β subunits appeared to be present in a 1:1 molar ratio but the molar ratio of α:γ varied from 1:0.2 to 1:1 so that the actual stoichiometry was uncertain. Enoch and Lester favored a structure of $\alpha_4\beta_4\gamma_2$ of molecular weight 608,000 ± 60,000, which compares well with the molecular weight of 615,000 calculated on the basis of four molecules of heme per molecule of enzyme and with the determined value of 590,000. Structures of the type $\alpha_4\beta_4\gamma_4$ and $\alpha_4\beta_4$ (f the γ subunit is an impurity) would give molecular weights of 648,000 and 568,000, respectively. The selenium was present in the α subunit only and was not in an acid-labile form. The role of the selenium is unknown but it may be an essential component of all bacterial formate dehydrogenases. No further information is available about the properties of the other subunits. The heme-carrying subunit has not been identified but on the basis of a heme content of four molecules per 615,000 daltons it is likely to be associated with the α or β subunits.

The purified formate dehydrogenase had a K_m for formate of 0.12 mM. This is about 10-fold higher than that reported for the membrane-bound enzyme by Ruiz-Herrera et al. [168] and may indicate that removal of the enzyme from the membrane results in substantial modification of its properties. Ferricyanide, methylene blue, and 2,6-dichloroindophenol were effective electron acceptors but less so than phenazine methosulfate. Of possible biological significance is the fact that ubiquinone (ubiquinone-6) would also act as an effective electron acceptor giving about 40% of the rate with phenazine methosulfate.

If *E. coli* is grown in the presence of 10 mM tungstate, active formate dehydrogenase is not formed [169]. The band associated with formate dehydrogenase could not be detected on polyacrylamide gels of Triton X-100 extracts of membrane fractions from these cells. A new band was present instead. In contrast to the formate dehydrogenase found when normal cells were used, this material did not contain cytochrome *b*. When tungstate-grown cells were incubated with molybdate in the presence of chloramphenicol to block protein synthesis, formate dehydrogenase activity returned and the band of the enzyme could be detected on the polyacrylamide gels [169]. The formation of the formate dehydrogenase band was associated with the concomitant loss of the band formed in the presence of tungstate. These results suggest that inactive precursors accumulate in the presence of tungstate, presumably due to blocking of the incorporation of molybdenum. Addition of molybdate results in reassociation of cytochrome *b* with the other polypeptides to reconstitute the active formate dehydrogenase.

C. *Nitrate Reductase*

The preparation of this enzyme has been described by Taniguchi and Itagaki [170], Showe and DeMoss [171], Forget [172], MacGregor et al. [173, 174],

Lund and DeMoss [175], Enoch and Lester [165, 166], and Clegg [176]. The work described in the first two papers was not accompanied by analysis of the polypeptide components of the enzyme using polyacrylamide gel electrophoresis; thus, the purity of these preparations is unknown.

The enzyme has been released from the membrane by a variety of techniques. Taniguchi and Itagaki [170] showed that the enzyme was released by heating the membranes at 60°C for 5 min in 5 mM phosphate buffer, pH 8.3. This method was subsequently used by Showe and DeMoss [171], and MacGregor et al. [173]. Lund and DeMoss used Tris buffer at pH 7.3 [175]. Forget [172] used membranes which had been depleted of phospholipids by treatment with cold acetone containing 0.027% NH_4OH. The enzyme was extracted from the treated membranes with 1.5% deoxycholate in 0.1 M phosphate buffer, pH 8.0. Enoch and Lester [165, 166] used deoxycholate (1 mg/mg protein) in the presence of $(NH_4)_2SO_4$ (30% saturation) to solubilize the reductase. MacGregor [174] and Clegg [176] released the nitrate reductase from the membranes by extraction at room temperature with 2% Triton X-100 in 50 mM phosphate buffer, pH 7.2. The subsequent purification steps in most procedures involved chromatography on DEAE-cellulose or DEAE-agarose together with gel filtration. Taniguchi and Itagaki [170] used adsorption on calcium phosphate gel followed by sucrose density gradient centrifugation to purify the nitrate reductase. MacGregor [174] purified the enzyme from Triton extracts by precipitation with antibody raised against enzyme prepared by alkaline heat solubilization [173]. In separations involving the detergent-colubilized enzyme, Triton X-100 was included in the buffer to prevent aggregation. Detergent was not required for the further purification of the enzyme released by heating at pH 8.3.

The properties of the various preparations of purified nitrate reductase are summarized in Table 11. Nitrate reductase is a molybdo - nonheme iron - sulfur protein. Flavin is absent. In the detergent-solubilized preparations, cytochrome *b* is present. This is associated with the presence of the extra subunit of molecular weight 19,000-20,000 which is lacking in the preparations obtained by heating the enzyme. MacGregor [177] has shown that in a *hem A* mutant of *E. coli* in which heme is not synthesized (due to lack of 5-aminolevulinic acid), a form of the nitrate reductase accumulates in the cytoplasm which lacks the 19,000-dalton subunit. Furthermore, Enoch and Lester [165] found that heating the detergent-purified form of the enzyme at pH 8.5 for 5 min at 60°C resulted in loss of this subunit and of the heme.

The enzyme shows association-dissociation behavior which is dependent on the protein concentration in the solution [175, 176]. This applies to both the heat-released and the detergent-solubilized enzyme. Lund and DeMoss [175] found that the heat-released enzyme at high protein concentrations migrated as a 24-S species as determined in the analytical ultracentrifuge. This is in good

Table 11 Properties of Purified Nitrate Reductase

Authors	K_m (NO_3) (mM)	$s_{20,w}$	Molecular weight ($\times 10^{-5}$)		Composition (g-atoms/220,000 g protein)			References
			Intact enzyme	Subunits	Mo	Fe	Sulfide	
Fujita and Itagaki	0.51	25	10	–	0.22	8.8	–	170
Forget	0.49	–	3.2	–	1.0	13.7	12.7	172
MacGregor et al.	–	23	7.2-7.7	1.42; 0.58	0.91	–	–	173
MacGregor	–	–	–	1.42; 0.60; 0.195	–	–	–	174
Enoch and Lester	–	16[a]	4.98[a]	1.55; 0.63; 0.19	–	–	–	166
Lund and DeMoss	–	10-24	8.8	1.55; 0.55	0.94	12.3	12.3	175
Clegg	–	9.9-22.4	9.0	1.5; 0.67; 0.65; 0.2	–	–	–	176

[a]Estimated in the presence of Triton X-100. No correction was made for the presence of bound detergent.

agreement with the values obtained by Taniguchi and Itagaki [170] and MacGregor et al. [173]. At low concentrations of protein a value of 10 S was obtained by sucrose density gradient centrifugation [175]. These two values are close to those determined by Clegg [176] for the detergent-solubilized nitrate reductase and correspond to molecular weights of 200,000-220,000 and 880,000-900,000. These results suggest that nitrate reductase undergoes a reversible monomer-to-tetramer association according to the protein concentration. Tetramers can be seen in negatively stained preparations of the crosslinked enzyme [176]. Assuming that 220,000 is the correct molecular weight, then about 1, 12, and 12 g-atoms of molybdenum, nonheme iron, and acid-labile sulfide, respectively, are present per mol of monomer.

Each monomer probably contains one subunit of about 155,000 daltons and one about 63,000-67,000 daltons (Table 11). MacGregor [174] has shown that the enzyme which is released by heating at pH 8.3 has undergone some proteolysis by a membrane-bound protease and that it is the proteolysis which is responsible for the release of the enzyme from the membrane. Besides the two main subunits, fragments of smaller size produced by proteolysis have been found in preparations of heat-released enzyme [174, 175]. Thus, the estimates of size of the subunits in the detergent-released enzyme would appear to be the most reliable. MacGregor [174] and Clegg [176] have found that the ratio of large (α) to small (β) subunits is close to 1:1 in the detergent solubilized enzyme. Clegg [176] noted that the β subunit yielded two closely spaced bands [of molecular weights 67,000 (β_1) and 65,000 (β_2)] on polyacrylamide gels. The ratio of α: ($\beta_1 + \beta_2$) was 1:1. If these do represent separate polypeptide chains, and not the same chain which has been partially attacked by proteinase, then the high molecular weight form of the enzyme can be represented as $\alpha_4(\beta_1)_2(\beta_2)_2$ which would yield monomers $\alpha\beta_1$ and $\alpha\beta_2$. There is little information on the stoichiometry of the heme-containing subunit (γ) to the α and β subunits. MacGregor [178] found that the ratio γ:α or γ:β was 2:1, which would give a monomer of $\alpha\beta\gamma_2$. Enoch and Lester [165] obtained two forms of nitrate reductase containing 3.2 and 6.7 nmol heme per mg protein. This would be equivalent to 0.7 and 1.5 mol heme per monomer of molecular weight 220,000, and would correspond to monomer structures of $\alpha\beta\gamma$ and $\alpha\beta\gamma_2$ for these two forms of nitrate reductase.

The location of the molybdenum and the nonheme iron in the different subunits has not been investigated yet. However, reduced methyl viologen-nitrate reductase activity only requires the presence of the α subunit [177].

There are a series of mutants in *E. coli* which are referred to as chlorate-resistant (*chl*) mutants. These were isolated by their ability to grow anaerobically on chlorate which is lethal to wild-type strains. The basis for the lethal action of chlorate is its ability to be reduced to a toxic compound by the nitrate reductase system [181]. In the resistant strains, mutations in the nitrate reductase system are present. Seven genetic loci (*chl A-G*) have been mapped [179]. Not all of the mutations occur in the nitrate reductase enzyme. For example, *chl F* is the structural gene for formate dehydrogenase [184]. In contrast to Triton extracts of membranes of mutants *chl C, D* and *E,* in which the subunits of nitrate reductase were not discernible, membranes from *chl A* and *B* mutants appeared to have the normal complement of nitrate reductase subunits [180]. This was confirmed by precipitation of material cross-reacting with antibody against nitrate reductase from Triton extracts of the membranes [181]. The material from *chl A* and *B* mutants contained all three polypeptide subunits (α-γ) as indicated by SDS polyacrylamide gel electrophoresis. However, the amount of the γ subunit was reduced to some extent. A molybdenum-containing factor (Mo-X) is required for the activity of nitrate reductase. This material is absent in *chl A* mutants [182], so it must be presumed that the lack of synthesis of Mo-X is responsible for the defect in the nitrate reductase in this mutant. Mo-X is synthesized in *chl B* mutants but is not incorporated into the membrane-bound nitrate reductase and accumulates in the cytoplasm [183]. The antibody-precipitated material from *chl C* membranes contained small amounts of the α and β subunits of the nitrate reductase together with large amounts of peptide fragments of varied molecular weight. This suggests that the mutant enzyme was extensively degraded by the membrane-bound protease. MacGregor [181] has suggested that it is likely that the α polypeptide is modified in this mutant since modification of the β polypeptide, which appears to be involved in attachment of the nitrate reductase to the membrane [174], would have resulted in release of the modified nitrate reductase into the cytoplasm. It is probable that the γ subunit is also involved in attaching the nitrate reductase to the membrane, presumably through the β subunit. As previously discussed, in *hem A* mutants grown without 5-aminolevulinic acid, the γ subunit is not detectable and the nitrate reductase consisting of α and β subunits only is found in a free form in the cytoplasm. It is probable that *chl E* is the structural gene for the apoprotein of the γ subunit. In contrast to mutants *chl A, B, C,* and *D,* which still make some nitrate-inducible cytochrome *b, chl E* mutants do not form any of this cytochrome [180].

D. Sequence and Organization

Both formate dehydrogenase and nitrate reductase can be isolated as preparations containing a *b*-type cytochrome. The *b*-type cytochromes formed in the presence of nitrate show a distinct peak at 555-556 nm in reduced (with dithionite)-minus-oxidized spectra measured at 77 K [106, 108, 111, 160]. The peak is almost symmetrical although careful examination suggests that cytochromes with absorption maxima at 558 and 562 nm are present in small amounts. Indeed Shipp [108] has found that the fourth-derivative analyses of difference spectra of cells grown anaerobically with nitrate and aerobically were quite similar. This suggests that the normal aerobic respiratory chain of *E. coli* is present but is partially masked in difference spectra by the development of a large amount of cytochrome b_{555} which is specifically associated with the formate-nitrate reductase system.

The presence of *b*-type cytochrome in both formate dehydrogenase and nitrate reductase preparations could be due to the contamination of both by a single species of cytochrome, although this seems unlikely since the two preparations do not share a common subunit [168]. The purified preparations have not been submitted to low-temperature difference spectroscopy so that this point cannot be entirely resolved at present. However, the kinetic experiments of Ruiz-Herrera and DeMoss [106] support the view that two different *b*-type cytochromes, both of which have an absorption maximum at 555 nm (77 K), are present in the electron transfer sequence from formate to nitrate:

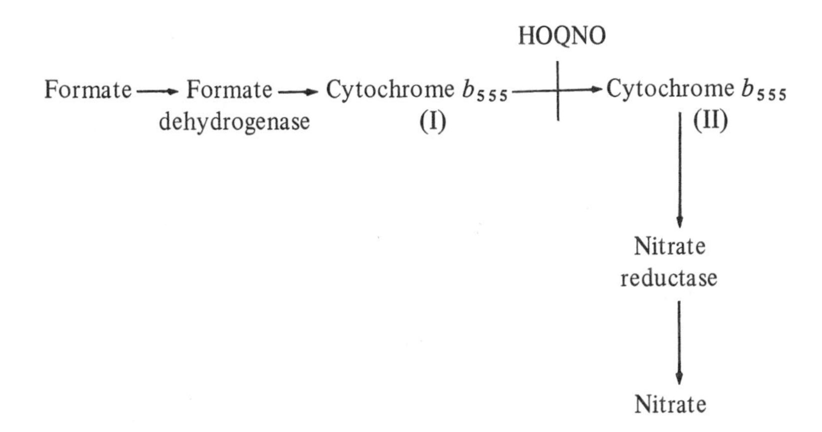

The evidence for this scheme is as follows. There is a lag in the reoxidation by nitrate of about half of the formate-reduced cytochrome b_{555}. The inhibitor of formate-nitrate reductase activity, HOQNO, prevents the reoxidation by nitrate of about half of the cytochrome b_{555} and also inhibits the reduction by formate of half of the cytochrome. The reduction by ascorbate of half of the cytochrome b_{555} is not sensitive to HOQNO and the reduced cytochrome can be completely reoxidized by nitrate.

This scheme accounts for the presence of cytochrome in both the formate dehydrogenase and nitrate reductase preparations. There is no further kinetic evidence for this sequence. Haddock et al. [111] measured the rate of reoxidation of *b*-type cytochrome in a stopped-flow spectrophotometer following the addition of a pulse of nitrate to an anaerobic suspension of cells. The rate of reoxidation of the *b* cytochrome was about one-quarter of the rate of electron transfer to nitrate. They suggested that the most likely explanation for this behavior was that cytochrome *b* was being reduced by substrates at a rate similar to its rate of oxidation by nitrate and nitrate reductase.

Support for this scheme, and for the involvement of ubiquinone in the formate-nitrate reductase pathway, has come from the experiments of Enoch and Lester [165]. Using an acetone-extracted or ultraviolet irradiated membrane particle preparation, Itagaki [185] had shown that restoration of the inactivated formate-nitrate reductase activity required the presence of ubiquinone. Menaquinone was much less effective. Enoch and Lester [165] were able to reconstitute the formate-nitrate reductase pathway from purified formate dehydrogenase, purified nitrate reductase, and ubiquinone-6. Menaquinone was slightly active but phospholipids could not replace the ubiquinone. Under optimal conditions the formate-nitrate reductase activity was 30% of the activity obtained using reduced benzylviologen as an electron donor for the nitrate reductase [165]. Neither the purified formate dehydrogenase nor the nitrate reductase contained ubiquinone. The cytochrome *b* of the purified dehydrogenase was completely reduced by formate but was not reoxidized by nitrate. In contrast, the cytochrome *b* of nitrate reductase was not reduced by formate but was reoxidized by nitrate [165, 166]. Reconstitution of formate-nitrate reductase activity would not occur if cytochrome-deficient nitrate reductase was used, suggesting that at least one of the *b* cytochromes was an obligatory component of this pathway. Thus, the sequence suggested by Ruiz-Herrera and DeMoss is probably correct with cytochromes b_{555} I and b_{555} II being associated with formate dehydrogenase and nitrate reductase, respectively. Ubiquinone is an additional component of this pathway which is probably located between the two cytochromes.

Garland et al. [161] have considered the possible organization of the components of the formate-nitrate reductase pathway. Although formate oxidation by oxygen yields H^+/O ratios of about 4, the H^+/NO_3^- ratio with nitrate as the electron acceptor is 2. Garland et al. [161] believe that the H^+/NO_3^- ratio may have been underestimated. Thus, it is not clear whether the formate-nitrate reductase system is organized in one or two proton-translocating loops across the membrane. The active site of formate dehydrogenase is on the cytoplasmic face of the inner membrane of *E. coli,* with nitrate being reduced at the external surface of the membrane. A possible scheme for proton translocation in the nitrate reductase portion of the system is shown in Figure 7 [161]. Here, ubiquinol is oxidized with discharge of $2H^+$ into the external medium. The electrons are returned to the cytoplasmic face of the membrane by cytochrome b_{555} where together with $2H^+$ they are transferred to nitrate reductase. The reducing equivalents, either 2H or $2(H^+ + e^-)$, are then carried to the external surface of the

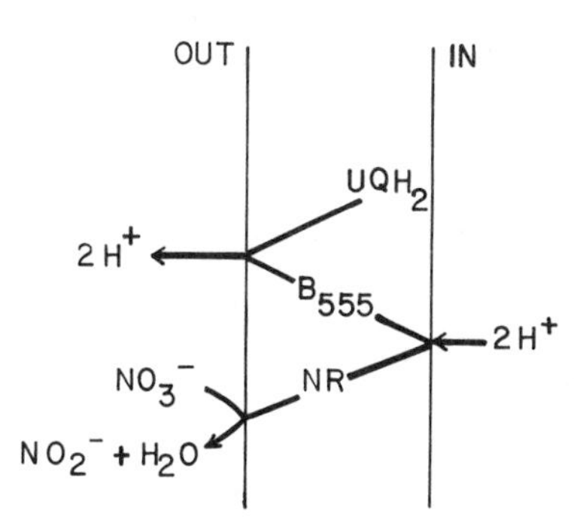

Figure 7 Scheme for proton translocation loop in nitrate reductase (see text for description). UQH_2, dihydroubiquinone; B_{555}, cytochrome b_{555}; NR, nitrate reductase. Based on Garland et al. [161].

membrane by the transmembranous nitrate reductase where they reduce nitrate to nitrite. There is evidence for the transmembranous nature of the nitrate reductase from two types of experiments. Reduced FMN which can serve as a donor of reducing equivalents to nitrate reductase acts from the cytoplasmic face of the membrane [162] whereas nitrate is reduced on the exterior face of the membrane [161]. Lactoperoxidase-catalyzed iodination of membrane-bound nitrate reductase showed that the α and β subunits of nitrate reductase are exposed on the cytoplasmic surface of the membrane and the γ subunit is exposed on the external surface [186].

V. Glycerol-3-Phosphate-Fumarate Reductase Pathway

Growth under anaerobic conditions on glycerol in the presence of fumarate results in the induction of L-glycerol-3-phosphate dehydrogenase and of fumarate reductase [187]. The glycerol-3-phosphate dehydrogenase is different from the membrane-bound enzyme formed under aerobic conditions [187]. Mutants lacking the anaerobic glycerol-3-phosphate dehydrogenase do not grow anaerobically on glycerol with fumarate as the electron acceptor, although growth will occur anaerobically in the presence of nitrate [187]. This suggests that the anaerobic glycerol-3-phosphate dehydrogenase can utilize the respiratory chain terminated by nitrate reductase [188], but the aerobic glycerol-3-phosphate dehydrogenase cannot interact with fumarate reductase. This indicates that there is a specific relationship between the anaerobic glycerol-3-phosphate dehydrogenase and fumarate reductase [189].

The anaerobic glycerol-3-phosphate dehydrogenase has been purified 40-fold from extracts of a strain of *E. coli* with a deletion in the structural gene for the aerobic dehydrogenase [190]. The dehydrogenase can use phenazine methosulfate, menadione and ferricyanide as artificial electron acceptors [190]. It is stimulated three- to fourfold by FAD ($K_m = 10^{-7}$ M) and up to 10-fold by FMN ($K_m = 10^{-4}$ M). The activity of the enzyme is not dependent on the presence of pyridine nucleotides and so it is probably a flavoprotein dehydrogenase. The K_m for substrate in the presence of FAD is 0.1 mM. The dehydrogenase has a molecular weight of 80,000 [190].

It seems likely that the glycerol-3-phosphate dehydrogenase purified above was released from the membranes on disruption of the cell, since about half of the activity can be sedimented by ultracentrifugation [189].

Fumarate reductase has not been purified from *E. coli.* However, it is a membrane-bound enzyme which is distinctly different from the succinate dehydrogenase of the tricarboxylic acid cycle of this organism [191]. Polyacrylamide gels of the membranes of fumarate reductase mutants have suggested that a polypeptide of molecular weight 87,500 could be a subunit of this enzyme [192].

Singh and Bragg [193] have examined the pathway of electrons from glycerol-3-phosphate to fumarate using a strain of *E. coli* unable to form cytochromes unless 5-aminolevulinic acid is present. Glycerol-3-phosphate-fumarate and NADH-fumarate reductase activities were present in the absence of cytochromes. Glycerol-3-phosphate could reduce the endogenous menaquinone of the membranes and the menaquinol so formed could be reoxidized by fumarate. These results suggest that the reaction pathway is

$$\text{Glycerol-3-phosphate} \longrightarrow \text{Dehydrogenase} \longrightarrow \text{Menaquinone} \longrightarrow$$
$$\text{Fumarate reductase} \longrightarrow \text{Fumarate}$$

Reducing equivalents from NADH and dihydroorotate also enter this pathway under anaerobic conditions at the level of menaquinone [193, 194].

Although cytochrome does not appear to be an obligatory component of the glycerol-3-phosphate dehydrogenase-fumarate reductase pathway, the active transport of proline and glutamine is energized under anaerobic conditions by electron transfer between glycerol and fumarate only in cytochrome-containing, not in cytochrome-deficient cells [195]. Konings and Kaback [196] and Boonstra and coworkers [197, 198] have also shown that active transport can be coupled to electron transfer between glycerol-3-phosphate and fumarate in cytochrome-containing cells. Miki and Lin [199] demonstrated that ATP formation from ADP and phosphate can be coupled to these electron transfer reactions.

Thus, cytochrome would appear to be required in this pathway if the process is to lead to membrane energization. This may be due to the requirement of cytochrome to form a proton-translocating loop in the membrane. Boonstra et al [198] have shown that electron transfer by this system in the presence of cytochromes does generate a membrane potential, interior negative, and a transmembrane pH difference is formed in cytochrome-containing but not in cytochrome-deficient cells [193, 200].

VI. Energy-coupled Systems

A. *Oxidative Phosphorylation*

1. Efficiency of Oxidative Phosphorylation

Depending on the growth substrate *E. coli* cells maintain their ATP level at about 2.3-3.6 mM [201, 202]. Holms et al. [201] have calculated the relative contributions made by substrate level and oxidative phosphorylation to the maintenance of this level of ATP. The contribution of substrate level phosphorylation to the ATP pool varied from 0% on acetate to 15% when glucose was the growth substrate. The turnover of the ATP pool was from 250-450 times/min depending on the growth substrate.

There have been several methods used to determine the efficiency of oxidative phosphorylation and the number of phosphorylation sites associated with the respiratory chain. One of these is based on the determination of molar growth yield. It was observed by Bauchop and Elsden [203] that several bacterial species appeared to produce a constant growth yield per mole of ATP available from the growth substrate. A value of 10.5 g dry weight of organism per mole of available ATP was found. Pirt [204] recognized that ATP available for growth had to be distinguished from that required for maintenance of the existing population of cells. He derived the relationship

$$\frac{1}{Y} = \frac{1}{Y_G} + \frac{m}{\mu}$$

where

Y = observed growth yield in grams dry weight of cells per gram of substrate used

Y_G = true growth yield where all energy is used for growth and none for maintenance

m = maintenance coefficient, expressed as grams of glucose used per gram of dry weight of cells per hour

μ = specific growth rate (fractional increase in weight of cells per hour)

This equation can be expressed [205] as

$$\frac{1}{Y_{O_2}} = \frac{1}{Y_{O_2}\text{(true)}} + \frac{M}{N\mu}$$

where

Y_{O_2} = observed growth yield expressed as grams dry weight of cells per mole of oxygen
Y_{O_2} (true) = true growth yield
M = maintenance coefficient expressed as moles of ATP consumed per gram dry weight of cells per hour
N = efficiency of oxidative phosphorylation expressed as moles of ATP formed per mole of oxygen (i.e., P/O ratio × 2)

The Y_{O_2} values are determined over a range of oxygen-limited growth rates. A graph of $1/Y_{O_2}$ versus $1/\mu$ allows the calculation of Y_{O_2} (true). From this the P/O ratio can be obtained using the relationship

$$\frac{P}{O} = \frac{Y_{O_2}\text{(true)}}{2 X Y_{ATP}} = \frac{Y_{O_2}\text{(true)}}{21.2}$$

where Y_{ATP}, the molar growth yield per mole of ATP, is expressed as grams dry weight of cells per mole of ATP. Bauchop and Elsden [203] have found Y_{ATP} to be 10.6

Meyer and Jones [205] have used this method to calculate P/O ratios with intact cells of various species of bacteria. *E. coli* gave a P/O ratio of about 3 under conditions where cytochrome *o* was the predominant oxidase. A ratio of 1-2 was obtained with cells grown under conditions of oxygen limitation where higher levels of cytochromes a_1 and *d* were present. Meyer and Jones [205] suggested that the induction of the cytochrome *d*-terminated pathway bypasses an energy conservation site associated with the cytochrome *o*-terminated pathway.

The calculation of the P/O value by Meyer and Jones [205] depended on the assumption that $Y_{ATP} = 10.6$. Stouthamer and Bettenhaussen [206] have criticized the use of this value to determine P/O ratios since it was originally measured under anaerobic conditions at low growth rates. Experiments designed to measure P/O ratios generally employ higher growth rates under aerobic conditions. Recently, Farmer and Jones [207] have measured Y_{ATP} for *E. coli* growing in a chemostat under a variety of conditions. With glycerol as substrate, Y_{ATP}, which varied between 12.6 and 14.3, was essentially independent of the nature of the growth-limiting nutrient (glycerol, oxygen, sulfate, or ammonium ions). Under substrate-limiting conditions, there was a marked difference in Y_{ATP} depending on the substrate, varying from 7.1 for acetate to 13.9 for glucose. It can be concluded from this that previous estimates of P/O ratios from molar growth yields are probably too high.

In the technique of Hempfling [208], endogenous substrates are allowed to reduce endogenous pyridine nucleotides under anaerobic conditions. A pulse of oxygen is introduced, followed within a short period (up to 10 sec) by a quenching agent (1.5 M $HClO_4$ or ethanolic KOH) to stop the reaction. The amount of phosphate esterified is calculated from changes in the concentrations of the adenine nucleotides. From this value and the extent of oxidation of the endogenous NADH, a value for the $P/2e^-$ (i.e., P/O) ratio can be calculated. Hempfling [208] found the $P/2e^-$ ratio for NADH oxidation to be 3.5 ± 0.3. This value was corrected to 2.6-3.0 when oxidation of the reduced components of the respiratory chain other than NADH was considered. A number of objections can be raised concerning this method. In the "standard" procedure [209] the reaction was quenched 5 sec after exposure of the system to oxygen. Van der Beek and Stouthamer [210] have found that NADH oxidation is complete within 1 sec whereas changes in the levels of the adenine nucleotides continued for 2-3 sec. Moreover, Hempfling's results showed that phosphate esterification also continued for the longer period of time [208]. These results suggest that the ATP pool may be turning over, as indicated by the results of Holms et al. [201], so that the increase in the concentration of ATP may merely represent a change from one steady-state level to another. This objection might also apply to the measurement of the changes in the level of NADH. Furthermore, NADH may not be the only substrate oxidized during the oxygen pulse. An underestimation of the extent of NADH turnover and the possibility that another substrate, not measured, might be oxidized would result in the $P/2e^-$ ratio being too high. In contrast, turnover of the adenine nucleotide pool could result in an underestimate of this ratio. Although there are some uncertainties about the quantitative aspects of this method, it is a valid indication of *oxidative* phosphorylation since mutants lacking the ATPase gave $P/2e^-$ ratios of 0.2 [211].

A third method which has been employed to measure the efficiency of energy conservation in intact cells of *E. coli* measures the stoichiometry of H^+ extrusion from the cells in the presence of the permeant thiocyanate ion following the addition of a pulse of oxygen to an anaerobic suspension. West and Mitchell [212] obtained a H^+/O ratio of 3-4 for the oxidation of endogenous substrates.

Lawford and Haddock [213] observed that the stoichiometries were about 4 with malate and about 2 with succinate, D-lactate, and glycerol as substrates. On the basis of these results, Lawford and Haddock [213] suggested that there are two energy-conserving sites associated with the respiratory chain of *E. coli*. One of the sites would be in the NADH dehydrogenase region of the chain and equivalent to site I in mitochondria. The second site would be associated with the respiratory chain between oxygen and the site of introduction of reducing equivalents from succinate, D-lactate, and glycerol-3-phosphate. This is probably ubiquinone. The existence of an energy conservation site between NADH and ubiquinol has been confirmed by Haddock et al. [214]. Addition of pulses of ubiquinone-1 to anaerobic cell suspensions gave a $H^+/2e^-$ ratio of between 1 and 2. Jones et al. [215] and Farmer and Jones [207] using cells grown under a variety of conditions (substrate, oxygen, sulfate, and ammonium ion limitation) in a chemostat have confirmed that the H^+/O ratio is 4 for oxidation of endogenous substrates. They have attempted to correlate the H^+/O ratio with the type of respiratory chain present in the organism. Only in organisms having cytochrome *c* in the respiratory chain are there two energy conservation sites between ubiquinol and oxygen. These organisms give H^+/O ratios of 6.

This method of estimation of the efficiency of oxidative phosphorylation is not free from potential errors. Thus, because the proton gradient decays rapidly following injection of the oxygen pulse, it is necessary to extrapolate back to zero time to estimate the original size of the pH gradient. This procedure could not detect the presence of a very rapid initial phase in the decay of the proton gradient. This method also depends on the extent of the proton extrusion being fully expressed by the pH gradient. Thus, the pH change must not be buffered in any way or diminished by compensatory movements of other ions. Finally, the method depends on the prediction by Mitchell [4] that two protons are transferred across the membrane during the passage of $2e^-$ through each site of energy conservation. Recent results by Brand and coworkers with mitochondria [216], have indicated that compensatory movements of phosphate have led to an underestimate of the stoichiometry of proton transfer per site. The ratio may be 3-4 protons per site. However, even allowing that the H^+/O values may have been underestimated, the observation that the ratios with NADH as substrate are twice those with succinate, D-lactate and glycerol-3-phosphate indicates that there must be at least two sites of energy conservation associated with the respiratory chain of *E. coli*.

If there are at least two coupling sites in the respiratory chain of *E. coli*, then a P/O ratio of 2 would be the upper value obtainable for the oxidation of NADH by subcellular preparations. This value has not been achieved so far. Kashket and Brodie [141] obtained P/O ratios of 1.3, 1.1, 1.0, 0.5, and 0.3 for the oxidation of pyruvate, malate, α-ketoglutarate, succinate, and NADH, respectively, by unwashed membrane particles which had been supplemented with a preparation precipitated from the cytoplasmic fraction with ammonium sulfate. In the absence of the supernatant, ATP formation was negligible. It is probable that substrate-level phosphorylation contributed to the high values observed by Kashket and Brodie [141]. Thus, succinate, malate, and α-ketoglutarate are readily converted to pyruvate via the malate enzyme [150]. The phosphoroclastic system would then convert pyruvate to acetyl phosphate with the uptake of phosphate from the medium. The phosphate can be transferred from acetyl phosphate to ADP with the formation of ATP. Kashket and Brodie [141] found that KCN or anaerobiosis prevented uptake of phosphate. However, oxidation of NADH is needed to regenerate the NAD^+ required for the substrate-level phosphorylation reactions. The cytoplasmic fraction would provide some of the enzymes required for the conversion of substrate to acetate. These objections do not apply to the oxidation of NADH. The P/O ratio of 0.3 with this substrate [141] is probably the best measurement of the capacity of their system to carry out oxidative phosphorylation.

Because of these problems, Bragg and Hou [217] measured P/O ratios using NADH as the only substrate and correcting the observed values for uptake of phosphate in the absence of substrate. The membrane particles alone gave P/O ratios of 0.1-0.2 but addition of a cytoplasmic fraction increased this ratio to 0.77. Kashket and Brodie [141] had not examined the effect on their system of uncoupling agents other than 2,4-dinitrophenol, which was ineffective. Bragg and Hou found that 0.33 mM pentachlorophenol decreased the P/O ratio by 90%. Generally, membrane particles alone have given P/O ratios in the 0.1-0.2 range [218-223]. However, Turnock et al. [224] and Hertzberg and Hinkle [225] have observed ratios of 0.64 and 0.73, respectively, for the oxidation of NADH by membrane particles. These results do not appear to be artifactually high since Turnock et al. [224] found that the ratio dropped to zero in a mutant with a defective ATPase, and the uncoupler carbonylcyanide-*m*-chlorophenylhydrazone (83 μM) prevented ATP formation in the system of Hertzberg and Hinkle. Hertzberg and Hinkle [225] obtained P/O ratios of 0.38, 0.45, and 0.55 for the oxidation of glycerol-3-phosphate, D-lactate, and succinate, respectively. These results are consistent with the presence of the two sites of energy conservation between NADH and ubiquinone, and ubiquinone and oxygen, suggested by the experiments with intact cells.

The apparent reason for the high P/O ratios obtained by Hertzberg and Hinkle [225] compared to other workers is that they prepared their membrane particles using a very mild disruptive procedure such that only 25% of the cells were broken. Tsuchiya and Rosen [226] have found that these conditions are also optimal for the preparation of inverted membrane vesicles for experiments on Ca^{2+} transport. More severe disruptive conditions resulted in a considerable decline in the transport ability of the vesicles. It is not clear if the low P/O ratios are entirely due to the dislocation of the system which occurs on cell breakage. Respiratory chains in which substrate oxidation is not coupled to ATP formation might be present. Respiratory control has not been demonstrated with respiratory particles of *E. coli* although stimulation of the respiratory rate both by ADP with phosphate and by uncouplers has been shown with particles from *A. vinelandii* [227] and *Paracoccus denitrificans* [228]. The uncoupler 2,4-dibromophenol stimulated respiration in intact cells of *E. coli* [30, 209]. It is also possible that the apparent extent of formation of ATP could be low due to the high ATPase activity shown by membrane particles of *E. coli* [219]. In *P. denitrificans* [229], *A. vinelandii* [230], and *M. lysodeikticus* [231] the rate of ATP hydrolysis is low. The ATPase appears to be held in a form which favors the formation of ATP. Drastic treatment such as hydrolysis with trypsin is required to reveal ATPase activity and to permit ATP to be used as an energy source for calcium transport and membrane energization [230]. In *E. coli,* ATP can be used to energize various energy-dependent reactions in membrane particles without the need for further treatment (see Sections VI.B-VI.D).

Although measurements of P/O ratios have usually been made with membrane particles which have an inverted configuration relative to the intact cell, and in which the active sites of the ATPase and NADH dehydrogenase are in contact with the external medium, oxidative phosphorylation can be demonstrated in right-side-out membrane vesicles provided they are preloaded with ADP during preparation [232]. Uncoupler-sensitive ATP formation coupled to oxidation of D-lactate, succinate, and NADH has been observed. The maximum observed P/O ratio of 0.02 was given by D-lactate.

2. Factors Affecting the Efficiency of Oxidative Phosphorylation

There are some indications that the efficiency of oxidative phosphorylation can be affected by the conditions of growth of the organism. Cavari et al. [233] induced changes in the respiratory system of anaerobically grown *E. coli* by exposing the cells in the growth medium to oxygen. NADH and succinate oxidase activities and the content of cytochromes then increased to levels found in aerobically grown cells. The almost twofold increase in the P/O ratio occurred more slowly than the development of the respiratory chain activities. The development of these systems required concomitant protein and RNA synthesis [233]. As well as the apparent repression, or lack of full induction, of the oxidative phosphorylation system in the absence of oxygen, this system undergoes catabolite repression. In studies of oxidative phosphorylation in whole cells, Hempfling [209] found that during growth on glucose-containing media, the apparent $P/2e^-$ ratio of 0.1-0.3 found during exponential growth increased to 3.9 following exhaustion of the glucose. Growth on acetate, lactate, glycerol, succinate, malate, and glutamate did not result in catabolite repression and $P/2e^-$ ratios of >3 were observed. Subsequently, it was shown that adenosine-3′,5′-monophosphate could reverse the catabolite repression by glucose [234]. The protein(s) subject to catabolite repression have not been identified. The activity of the respiratory chain and the level of the ATPase appear to be unchanged in catabolite repressed and derepressed cells [219].

Iron and sulfate limitation during growth results in impairment of the energy conservation process [27, 30]. Poole and Haddock have investigated the effect of sulfate deprivation in some detail [27]. They found that nonheme iron and acid-labile sulfide concentrations were twofold lower in membrane particles from sulfate-limited cells concomitant with a decrease in the size of the EPR signal for nonheme iron at g = 1.94. In cells grown with limiting sulfate, the H^+/O ratios for L-malate, succinate, and glycerol were 2.21, 2.17, and 2.23, respectively, compared to 3.58, 2.21, and 2.21, respectively, for normal cells. These results are consistent with the loss of a protein containing nonheme iron and acid-labile sulfide, which is associated with energy conservation in the NADH dehydrogenase segment of the respiratory chain.

B. Reversed Electron Flow

Energy-linked reduction of NAD^+ by succinate, first observed by Chance and Hollunger in rat liver mitochondria [235], has subsequently been extensively studied in various systems [236-239]. The reaction involves the transfer of reducing equivalents from succinate, via succinate dehydrogenase and NADH dehydrogenase, to NAD^+ by an energy-requiring process requiring the reversal of the normal direction of electron transport at site 1. In mitochondria, energy can be supplied either by hydrolysis of ATP or by the generation of an energized state at sites 2 or 3. Light energy can be used to drive the reduction of NAD^+ by succinate in chromatophores of *Rhodospirillum rubrum* [238, 239].

The first report of the occurrence of this reaction in *E. coli* was that of Kashket and Brodie [71] who found that NAD^+ was reduced at a low rate (8 nmol/min per mg protein) when membrane particles were incubated under anaerobic conditions with succinate. The rate of formation of NADH was increased fourfold by the inclusion of ATP. Sweetman and Griffiths [240] subsequently investigated this reaction in more detail. Reduction of NAD^+ could be energized by hydrolysis of ATP and to a lesser extent by ITP. GTP, CTP, and ITP were ineffective. The ability of the nucleoside triphosphates to serve as an energy source was related to their ability to act as substrates for the membrane-bound ATPase [240]. Several analogs of NAD^+ could replace NAD^+ as the electron acceptor in the reaction. The hydrolysis of one to two molecules of ATP were required for the reduction of one molecule of NAD^+. The reaction was inhibited by malonate and TTFA, inhibitors of succinate dehydrogenase

[75], and by the uncoupling agents tetrachlorotrifluorobenzimidazole, dicumarol, and pentabromophenol. The maximum activity observed in these experiments was 10.4 nmol NAD^+ reduced/min per mg protein. This is rather low and may account for the difficulty that other workers have had in demonstrating the presence of this reaction in membrane particles. Poole and Haddock [241] have shown that the addition of ammonium sulfate-precipitated fractions from the cytoplasm or from a "low-ionic-strength wash" of particles from glycerol-grown cells resulted in a substantial increase in the rate of the reaction in the presence of ATP. Besides succinate, DL-α-glycerophosphate and D-lactate could provide reducing equivalents for the reduction of NAD^+. The maximum specific activity observed with succinate, DL-α-glycerophosphate, and D-lactate in the presence of the ammonium sulfate-precipitated fraction from wild-type cells was 11.0, 45.2, and 8.4 nmol NAD^+ reduced/min per mg protein. The function of the ammonium sulfate-precipitated fraction is not clear. It contains ATPase, succinate, DL-α-glycerophosphate, and D-lactate dehydrogenase activities [241]. The particles may have been partially depleted of these enzymes during the disruption of the cells by sonication. The stimulating effects of piericidin A and HOQNO on NAD^+ reduction [240], although small, suggest that the pathway of reducing equivalents from succinate to NAD^+ must bypass the site of action of these inhibitors on the respiratory chain. This appears to be close to, although not necessarily involving, cytochrome b_1. Poole and Haddock [241] have shown unambiguously by use of a heme-deficient mutant that cytochromes are not involved in the pathway. In contrast, ubiquinone is required. The energy-linked reduction of NAD^+ by succinate, DL-α-glycerophosphate, or D-lactate could not be shown with membrane particles of a ubiquinone-deficient strain of *E. coli* in the absence of exogenous ubiquinone-1 [241]. It is probable that ubiquinone is the locus at which the reducing equivalents enter the respiratory chain from the several flavin-linked dehydrogenases.

Energy-dependent reduction of NAD^+ energized by the energized state generated by substrate oxidation has not yet been demonstrated in *E. coli* or other bacteria.

C. Energy-dependent Transhydrogenation

The energy-dependent transhydrogenase of mitochondria catalyzes the transhydrogenation of $NADP^+$ by NADH. Input of energy results both in an enhancement in the rate of the energy-independent reaction as well as an increase in the apparent equilibrium constant from 1 to about 500 [242]. Energy can be supplied as an energized state formed either by oxidation of substrate or from ATP by the reversal of the reactions of oxidative phosphorylation. The mechanism of the reaction is unknown. It has been suggested that energization may cause a conformational change in the transhydrogenase with the formation of an active from an inactive form [242]. Energy-dependent transhydrogenation has been less well studied in bacterial systems compared with mitochondria. However, this reaction is more easily demonstrated in *E. coli* than is the ATP-dependent reduction of NAD^+ by succinate. This may be due to the higher level of energization required by the latter reaction as has been shown in chromatophores of *R. rubrum* [239]. The ATP-dependent and respiration-dependent reduction of $NADP^+$ by NADH in *E. coli* membrane particles was first demonstrated by Murthy and Brodie [243] and by Fisher et al. [244]. A convenient assay method based on the procedure of Fisher and Sanadi [245] has been described by Bragg et al. [29]. With membrane particles from cells grown on a defined medium containing glucose the specific activities of both the ATP- and the respiration-dependent reactions are about 100 nmol $NADP^+$ reduced/min per mg protein. Twofold higher specific activities have been obtained with ammonia-washed particles [246]. In the assay method of Bragg et al. [29], NADH in the cuvette is used both as a source of reducing equivalents to reduce $NADP^+$ ans as a substrate which on oxidation through the respiratory chain generates the energized state. The generation of the energized state under the conditions of the assay appears to be limited by the rate of oxidation of NADH since addition of succinate can result in a twofold increase in the rate of transhydrogenation [247]. If NADH oxidation is blocked by treatment of membrane particles with 0.4 N ammonium hydroxide, the transhydrogenation can be energized by oxidation of succinate, D-lactate, or ascorbate (in the presence of phenazine methosulfate [246]. This indicates that energy-dependent transhydrogenation can be coupled to the second site of energy conservation in the cytochrome region of the respiratory chain. Coupling of transhydrogenation through site 1 in the NADH dehydrogenase segment of the respiratory chain was found in cytochrome-deficient membrane particles of a heme mutant of *E. coli* [248]. As an energy donor ATP can be replaced by ITP and to a lesser degree by GTP [249]. The effectiveness of the donor is related to its ability to act as a substrate for the membrane-bound ATPase [249]. The hydrolysis of one to two molecules of ATP are required to energize the reduction of one molecule of $NADP^+$ [249]. As will be described in later sections, the involvement of the ATPase in the ATP-dependent transhydrogenase is clearly shown by the use of mutants and by reconstitution studies. This is also shown by the inhibition of the ATP-dependent reaction by the ATPase inhibitor DCCD [245, 250]. This compound stimulates the aerobic-driven transhydrogenase reaction [245, 250] apparently by decreasing the permeability of the membrane to protons with consequent stabilization of the energized state [250, 251]. Uncouplers such as carbonylcyanide-*m*-chlorophenylhydrazone, pentachlorophenol, and 4,5,6,7-tetrachloro-2-trifluorobenzimidazole inhibit both respiration- and ATP-dependent reactions [217, 245, 249, 252].

The properties of the transhydrogenase enzyme can be examined separately from those of the energization systems by measuring the energy-independent reaction. This can be conveniently measured by following the oxidation of NADPH by NAD^+ in the presence of an NADH reoxidation system (e.g., the respiratory chain of the membrane particles [217]) or by following the reduction of 3-acetylpyridine-NAD^+ by NADPH [252]. Specific activities of up to about 1.2 μmol NADPH oxidized/min per mg protein have been recorded [247].

Hoek et al. [253] have examined the stereospecificity of hydrogen transfer in *E. coli* membrane particles. The transhydrogenase removes the 4A hydrogen atom of NADH. This is similar to the behavior of the enzyme in submitochondrial particles where the 4A hydrogen atom of NADH and the 4B hydrogen atom of NADPH are involved in both energy-independent and energy-dependent reactions [242]. This probably also applies to *E. coli,* and supports the evidence from the inhibition of both of these reactions by 3,3′,5-triiodothyronine [217] that the same enzyme is involved in both energy-dependent and energy-independent transhydrogenase reactions. The transhydrogenase is clearly distinct from the respiratory-chain NADH dehydrogenase in both *E. coli* and submitochondrial particles, although both can oxidize NADH, since the latter enzyme removes the 4B hydrogen of NADH [253].

The specificity of the energy-independent transhydrogenase for NAD^+ is relatively broad, several analogs of NAD^+ being capable of supporting at least 50% of the control rate. The energy-linked reaction is specific for NAD^+ although 3-acetylpyridine-NAD^+ can react at about 10% of the control rate [249]. Singh and Bragg [252] have obtained similar results with *Salmonella typhimurium.*

Houghton et al. [247] have studied the kinetics of the energy-independent transhydrogenase is some detail. The kinetic constants are given in Table 12. Like the mitochondrial enzyme [242], the bacterial transhydrogenase followed the kinetics of a reaction mechanism involving a short-lived ternary complex (Theorell-Chance mechanism). The order in which the nucleotides bind to the enzyme could not be determined. Hoek et al. [253] had found that palmitoyl-CoA was an effective inhibotor of the transhydrogenase of *E. coli,* and Singh and Bragg [252] had observed that 5′-AMP, but not 2′- or 3′-AMP, inhibited the same enzyme in *S. typhimurium.* Houghton et al. examined the kinetics of inhibition by these compounds. $NADP^+$ and palmityl-CoA were competitive inhibitors of the NADPH site and noncompetitive inhibitors of the NAD^+ site. In contrast, 3-acetylpyridine-NADH and 5′-AMP were noncompetitive inhibitors of the NADPH site and competitive inhibitors of the NAD^+ site. This is similar to the behavior of the mitochondrial transhydrogenase [242] and further demonstrates the similarity between the two enzymes. A further point of similarity is that the transhydrogenase from both sources is very sensitive to inactivation by trypsin [253, 254].

Table 12 Kinetic Parameters of Energy-independent Transhydrogenase of Glucose-salts-grown Cells

Parameter	Constant (μM)
K_m (NADPH)	62
K_m ($AcPyNAD^+$)[a]	143
K_{DISS} (NADPH)	10
K_{DISS} ($AcPyNAD^+$)	22
V_{max}	2175[b]
K_i ($NADP^+$)	133
K_i (AcPyNADH)	81
K_i (Palmityl CoA)	0.1
K_i (2′-AMP)	4550
K_i (5′-AMP)	284

[a] $AcPyNAD^+$, 3-acetylpyridine-NAD^+.
[b] V_{max} is expressed as nmol/min per mg protein.
Source: Houghton et al. [247].

The transhydrogenase has not been extensively purified although it has been solubilized with deoxycholate in the presence of 1 M KCl [255] and with Triton X-100 (Homyk and Bragg, unpublished). The reaction sites for NADPH and NAD^+ of the membrane-bound enzyme are on the cytoplasmic face of the inner membrane of *E. coli* [256].

The transhydrogenase activity of membrane particles of *E. coli* is dependent on the composition of the medium in which the cells are grown [29, 245]. Highest activities are found in cells grown on a glucose-salts medium. In the presence of yeast extract or protein hydrolysates the activities of the energy-dependent and energy-independent transhydrogenases are strongly repressed [29, 247, 252] and this has been attributed to the presence of amino acids. The enzyme is not subject to catabolite repression. On the basis of this evidence Bragg et al. [29] have suggested that the role of the transhydrogenase in the cell is to generate NADPH for biosynthetic reactions. If cells containing the repressed level of the transhydrogenase are transferred to a glucose-salts medium the induction of the enzyme can be followed. Induction results in an increase in the levels of energy-dependent and energy-independent transhydrogenase activities [247]. De novo protein synthesis is required. There is little change in the kinetic constants of the enzyme except that the V_{max} increases sixfold. Houghton et al. [247] concluded that induction of transhydrogenase activity results in a severalfold increase in the amount of the enzyme in the membrane without a change in its kinetic properties.

The involvement of lipid in the activities of the energy-dependent and energy-independent transhydrogenase have been investigated by Singh and Bragg [254] using an unsaturated fatty acid-requiring strain of *E. coli.* Transition temperatures due to phase changes in the membrane lipids could not be correlated with changes in the activity of the transhydrogenase with alterations in temperature. Thus, bulk phase lipids do not influence the transhydrogenase activity. However, energy-dependent and energy-independent activities were lost on treatment of the particles with phospholipase A, suggesting that the activity of the transhydrogenase may be dependent on phospholipid molecules in immediate contact with the enzyme [254]. Support for this hypothesis comes from the phospholipid dependency of the activity of the detergent-solubilized transhydrogenase [255].

D. Active Transport

It is usual to limit the term "active transport" to energy-dependent transport in which solute is accumulated in the cytoplasm in an unmodified form. "Group translocation" refers to energy-dependent transport in which the solute is accumulated in a derivatized form, as for example the sugar phosphates which are formed during the transport of glucose, mannose, fructose, glucosamine, hexitols, and β-glucosides by the phosphotransferase systems of *E. coli* and other enteric bacteria [257]. Only active transport will be discussed here.

A wide variety of solutes are taken up by active transport in *E. coli:* galactose, arabinose, ribose, xylose, glycerol, lactose and other β-galactosides, maltose, melibiose, gluconate, glucuronate, glucose-6-phosphate, mannose-6-phosphate, fructose-6-phosphate, α-glycerophosphate, succinate, fumarate, lactate, pyruvate, most amino acids, adenosine, deoxycytidine, Rb^+, and K^+. Pavlasova and Harold [258] observed that uncouplers would prevent the accumulation by intact cells of *E. coli* of thiomethyl-β-D-galactoside under anaerobic conditions. This observation suggested that the transport of the β-galactoside was coupled to the energized state of the membrane generated by hydrolysis of ATP. Klein and Boyer [259] examined this hypothesis in more detail using membrane vesicles and intact cells. Transport of proline by membrane vesicles was supported by oxidation of D-lactate and was resistant to arsenate. In intact cells, high concentrations of arsenate drastically lowered the intracellular level of ATP but had no effect on the transport of proline under aerobic conditions. Under anaerobic conditions, transport of proline was much reduced in the presence of arsenate. Similar results were obtained for the transport of leucine, thiomethyl-β-D-galactoside, and Rb^+ [259]. These results suggested that active transport of the solutes could be energized by an energized state of the membrane generated either from oxidation of substrate through the respiratory chain or by hydrolysis of ATP. Uncouplers of oxidative phosphorylation blocked the use for transport of energy derived both from oxidation and from ATP, suggesting that a similar or identical energized membrane state was formed from either energy source [259]. This conclusion has been substantiated by the use of membrane vesicles and of mutants [7].

In an extensive series of experiments Kaback and his coworkers [11-14] have clearly shown that active transport of solute into membrane vesicles can be driven by substrate oxidation in the absence of ATP hydrolysis by a process which is sensitive to uncouplers. The membrane vesicles lack the components of the cytoplasm including ATP. Uptake of solutes by the vesicles is most effectively supported by oxidation of D-lactate and ascorbate (in the presence of phenazine methosulfate). NADH supports solute accumulation poorly, although it is oxidized by the membrane vesicles. Kaback has claimed a special role for D-lactate in the coupling of oxidation to transport. This view is probably incorrect. Futai [260] has prepared vesicles preloaded with NAD^+ and alcohol dehydrogenase. In the presence of ethanol NADH is generated within the vesicles and on oxidation will now support transport effectively. Presumably oxidation of external NADH bypasses a coupling site in the respiratory chain. The role of ATP in the energization of active transport has been shown in experiments in which mutants unable to form cytochromes were used [195, 248, 261]. The uptake of lactose, phenylalanine, proline, and glutamine in cytochrome-deficient cells lacking the complete respiratory chain occurred in a reaction which was sensitive to uncouplers and to inhibitors of the membrane-bound ATPase such as DCCD, pyrophosphate, and azide. A direct demonstration of the role of ATP in supporting transport has been provided by the experiments of Rosen and his coworkers [262-264]. Calcium ion is extruded from *E. coli* by active transport from the cytoplasm to the medium surrounding the cells [265]. Thus, the direction of active transport is opposite to that found for most of the solutes usually studied. Rosen and McClees [262] found that membrane particles, having an inverted orientation relative to the intact cells or the membrane vesicles usually used in transport studies, accumulated Ca^{2+} in the presence of ATP. The reaction was sensitive to inhibitors of the ATPase. Moreover, removal of the ATPase from the membrane particles, or use of a mutant with a defective ATPase, resulted in loss of transport activity [263]. Transport of Ca^{2+} was also driven by substrate oxidation. Oxidation of NADH was more effective than that of D-lactate in supporting transport, so confirming the results of Futai [260] which indicated that D-lactate did not have a unique role in the energization of active transport.

As discussed elsewhere in this article, the role of ATP and the membrane-bound ATPase in the energization of active transport has been clearly shown by mutants in which the membrane-bound ATPase is inactive or in which the hydrolysis of ATP by the ATPase has been uncoupled from membrane energization [266-276]. The role of the ATPase in the active transport of *all* solutes which use ATP as an energy source has been questioned by Berger [277], Berger and Heppel [278], and Kobayashi et al. [279]. Berger and Heppel [278] have noted that the amino acid transport systems in *E. coli* can be divided into two classes on the basis of the presence or absence of a periplasmic solute-binding protein. The systems can be readily distinguished since transport in the binding-protein class is readily disrupted by osmotic shock of the cells, a process which results in release of the binding protein. The systems not requiring a binding protein are resistant to osmotic shock. The transport systems for glutamine, diaminopimelic acid, arginine, histidine, and ornithine are all strongly impaired by osmotic shock whereas those for proline, serine, phenylalanine, glycine, and cysteine are resistant [278]. The transport of glycylglycine, ribose, and galactose [280-282] also falls into the shock-sensitive category. Berger and Heppel [278] found that shock-sensitive transport systems, in contrast to the resistant systems, could not use energy supplied by oxidation of D-lactate or reduced phenazine methosulfate to drive transport in mutants in which the membrane-bound ATPase was inactive. If a source of glycolytic ATP was supplied, then the shock-sensitive systems were active in the ATPase mutant and were relatively resistant to anaerobiosis and the uncoupler 2,4-dinitrophenol [278]. Shock-resistant systems were inhibited by uncouplers. Arsenate abolished transport by the shock-sensitive but not by the shock-resistant systems [278]. These results are consistent with the use of ATP by the shock-sensitive systems to drive transport by a mechanism not involving the ATPase or the energized state of the membrane. Transport in the shock-resistant systems would be driven by the energized state generated by substrate oxidation or by ATP mediated through the ATPase.

This hypothesis may not be entirely correct. Thus, Plate et al. [283] found that colicin K, which appears to act by de-energization of the energized state of the membrane of *E. coli,* inhibited the transport of both glutamine and proline in an ATPase mutant. This suggests that glutamine and proline transport share a common element sensitive to colicin K. Lieberman and Hong [284] have isolated a temperature-sensitive mutant in the common element of shock-sensitive and shock-resistant systems. Singh and Bragg [195] found that the transport of both glutamine and proline in cytochrome-deficient cells of *E. coli* was sensitive to inhibitors of the ATPase and to uncouplers. It is difficult to reconcile these results with Berger and Heppel's hypothesis [278]. Furthermore, it is not clear how ATP can energize the transport of the shock-sensitive systems. Presumably a mechanism involving phosphorylation and dephosphorylation of the carrier would be involved [278] but no evidence has yet been presented for this mechanism. Lieberman and Hong [284] have suggested that ATP might function as a regulatory effector which could direct the use of the energized state to the transport of solutes belonging to the shock-sensitive systems. This would explain the apparent requirement of these systems for ATP. This very interesting hypothesis remains to be tested.

VII. The Energized State

As discussed in an earlier section the energized state, generated either by oxidation of substrate through the respiratory chain or by hydrolysis of ATP by the membrane-bound ATPase, appears to be used by the oxidative phosphorylation, energy-dependent transhydrogenase and active transport systems, and for flagella movement in *E. coli*. Evidence from mutants indicates that the same form of the energized state is used by all of these processes.

A. Generation of the Protonmotive Force

The chemiosmotic hypothesis of oxidative phosphorylation and of energy coupling to transport of solutes has been particularly useful in the interpretation of the results with bacterial systems [6, 7]. In this hypothesis the energized state is seen as a gradient of protons across the plasma membrane of the bacterial cell. This gradient has two components–the pH difference and the membrane potential. There is evidence that a pH difference and a membrane potential can be generated by *E. coli* cells and vesicles.

As discussed previously, the addition of pulses of oxygen to anaerobic suspensions of *E. coli* in the presence of substrate results in the extrusion of protons from the cells [207, 212-215]. The stoichiometry of proton extrusion to uptake of oxygen yields H^+/O ratios of about 2 or 4 depending on whether one or two sites of energy conservation are traversed by the reducing equivalents from substrate on the way to oxygen [213]. There have been several attempts to measure the magnitude of the protonmotive force generated during oxidation of substrate. Griniuviene et al. [285, 286] measured the uptake of the permeant cations dibenzyldimethyl ammonium (DDA^+) and triphenylmethyl phosphonium ($TPMP^+$) ions by respiring cells which had been made more permeable to these ions by treatment with EDTA in the presence of Tris. At pH 7.2, a membrane potential of about 140 mV, negative inside, was generated by the metabolism of endogenous substrates. That the setting up of this potential was largely due to respiration and not to ATP hydrolysis was shown by the insensitivity of the membrane potential to the ATPase inhibitor DCCD, and to the inhibition of its generation by KCN. Hirata and coworkers [287, 288] have used the distribution of DDA^+ to show that right-side-out membrane vesicles oxidizing D-lactate generate a membrane potential of 100-125 mV, interior negative. More complete studies of the magnitude of the protonmotive force generated in respiring cells, spheroplasts, and vesicles have been made by Padan et al. [289], Collins and Hamilton [290], and Ramos et al. [291]. The results are summarized in Table 13. *E. coli* can generate a proton-

Table 13 Protonmotive Force in Respiring *E. coli*

Preparation	Substrate	Buffer		pH difference	Membrane potential (mV)	Protonmotive force (mV)	References
		pH	KCl (mM)				
Tris/EDTA-treated cells	Endogenous	7	0.01	0.13	122	129	289
		7	1	0.62	84	120	–
		7	150	0.89	16.2	68	–
Spheroplasts	Endogenous	6.45-6.75	0.1	1.65 ± 0.2	132 ± 10	230 ± 15	290
Vesicles	Ascorbate (+phenazine methosulfate)	5.5	50[a]	1.95[b]	74	189	291
	D-lactate	5.5	50[a]	1.7[b]	70	172	–
	Succinate	5.5	50[a]	0	64	64	–
	NADH	5.5	50[a]	0	0	0	–

[a]Potassium phosphate.
[b]Calculated from the results of Ramos et al. [291].

motive force (Δp) of up to 230 mV across the cell membrane by proton extrusion into the medium. The contribution of the pH difference (ΔpH) and the membrane potential ($\Delta\psi$) to the protonmotive force depends on the pH and concentration of K^+ of the buffer in which the preparation is suspended. In Tris/EDTA-treated cells at 0.01 mM KCl the major contribution to the protonmotive force was made by the membrane potential. This contribution was diminished at higher levels of KCl since K^+ ions were able to move into the cells in the presence of valinomycin which was added in these experiments [289]. The contribution of ΔpH to the protonmotive force was higher at more acidic pH values. With both intact cells and membrane vesicles the maximum ΔpH of about 2 pH units was observed at an external pH of 5.5-6.0 [289, 291]. At increasing pH values the contribution by ΔpH to the protonmotive force progressively decreased to become zero at about pH 7.5-8.0 [291]. Above this pH the protonmotive force was entirely composed of the contribution by the membrane potential. The membrane potential changed relatively little over the pH range of 5.5 to 8.5 so that increasing the pH of the medium resulted in a progressive decrease in the magnitude of the protonmotive force from about 190 mV at pH 5.5 to about 70 mV at pH 7.5 [291]. The presence of the uncoupling agents carbonylcyanide-*m*-chlorophenhydrazone and carbonylcyanide-*p*-trifluoromethoxyphenylhydrazone decreased the contribution of both ΔpH and the membrane potential to the protonmotive force, as would be expected from the ability of these compounds to conduct protons across the membrane [6]. With membrane vesicles, valinomycin and nigericin reduced the contributions of the membrane potential and of ΔpH, respectively, to the protonmotive force [291]. This is consistent with their known mode of action. Valinomycin provides a pathway for K^+ to cross the membrane in response to the membrane potential whereas nigericin causes electroneutral K^+/H^+ exchange [6].

The generation of an uncoupler-sensitive membrane potential has been observed with membrane particles (inverted membrane vesicles). Griniuviene et al. [286] found that hydrolysis of ATP or oxidation of NADH resulted in the uptake of the permeant anion phenyl dicarbaundecaborane ion (PCB^-). Although the membrane potential was not calculated from the extent of uptake of PCB^-, it is clear that a membrane potential, positive inside, was generated. The formation of an uncoupler-sensitive pH difference across the membrane of inverted vesicles has been shown by Singh and Bragg [292, 293] using 9-aminoacridine. Hydrolysis of ATP or oxidation of glycerol-3-phosphate, succinate, D-lactate, and NADH generated a pH difference of 3.3-3.7 pH units, with the interior of the vesicle becoming more acidic than the medium. By use of inhibitors of the ATPase and of the respiratory chain it was shown that the size and rate of formation of the pH difference was dependent on the rate of energy supply to the membrane by substrate oxidation and ATP hydrolysis [292, 293]. The formation of the pH difference was associated with at least one of the two regions of the respiratory chain found by other techniques to be involved in energy conservation [213]. Thus, the pH difference could be generated by oxidation of ascorbate in a mutant lacking ubiquinone but not in a cytochrome deficient mutant [294].

B. *Use of Protonmotive Force to Drive ATP Formation and Active Transport*

Substrate oxidation through the respiratory chain and hydrolysis of ATP can generate a protonmotive force across the cell membrane of *E. coli.* If the protonmotive force is equivalent to the energized state then it should be possible to show that the protonmotive force will drive the energy-dependent reactions associated with the cell membrane. At present only ATP formation and active transport have been shown to be driven by the presence of an artificially imposed protonmotive force.

Maloney et al. [295] first showed that the application of a membrane potential could drive ATP formation in *E. coli* and *Streptococcus lactis.* The membrane potential was generated by placing cells with a normal complement of K^+ into a K^+-free buffer. The addition of valinomycin resulted in the diffusion of K^+ from the cells with the generation of a potential, interior negative, across the membrane. In further experiments [296] it was shown that imposition of a membrane potential increased the ATP level of starved cells from about 0.1 mM to about 1.6 mM. Decreasing the size of the membrane potential by diluting the cells into buffer containing 1-3 mM K^+ resulted in a lower yield of ATP. Synthesis of ATP in the absence of a membrane potential was shown by diluting cells originally at pH 8 into a medium at pH 3 which contained valinomycin in the presence of 100 mM KCl to prevent the formation of a membrane potential [296]. Intracellular ATP levels rose from 0.1 mM to 2.1 mM. If the pH of the external medium was 5.3 or greater, little ATP synthesis occurred. There was no net synthesis of ATP on the application of either a pH difference or a membrane potential in the presence of the ATPase inhibitor DCCD or in a mutant in which the ATPase was inactive. Clearly, the membrane-bound ATPase is an essential component of the system which uses the protonmotive force to form ATP. The reverse of this process has been shown by West and Mitchell [297] and by Singh and Bragg [293] who have found that hydrolysis of ATP by the membrane-bound ATPase will generate a pH gradient.

The net synthesis of ATP by the imposition of a pH difference or a membrane potential across the membrane of intact cells has also been shown by Grinius et al. [298]. Similar results have been obtained using right-side-out membrane vesicles [299, 300]. An interesting result to emerge from these experiments is that the formation of ATP had a specific requirement for Mg^{2+} although in hydrolysis of ATP by the same enzyme both Mg^{2+} and Ca^{2+} ions were effective [300].

In their experiments with intact cells Wilson et al. [296] observed that significant synthesis of ATP occurred only when the total protonmotive force attained a value of 200 mV. This compares favorably with the value of 210 mV, calculated by Mitchell [4], which would be required to maintain the ATP/ADP ratio at 1.

The probable involvement of the proton gradient in transport in *E. coli* was first suggested by the experiments of West [301] and West and Mitchell [302]. Uptake of lactose by intact cells under anaerobic conditions in the presence of iodoacetate to poison glycolysis was accompanied by uptake of H^+. A permeant anion such as thiocyanate was required to compensate for the influx of protons. A stoichiometry of one proton consumed per molecule of lactose taken up was established [310]. The cells used in the experiments were not carrying out active transport so that the results of West indicate only that a proton/lactose symport mechanism of the type predicted by West and Mitchell [302] is feasible. Symport of protons with solute has been demonstrated for the uptake into *E. coli* of thiomethyl-β-D-galactoside [273], glucose-6-phosphate [303], galactose and arabinose [304], lactate and alanine [305], malate, aspartate, fumarate, and succinate [306]. With the exception of the dicarboxylic acids, the entry of which is accompanied by two protons per molecule, the uptake of protons and solute probably occurs with a 1:1 molar stoichiometry [303, 305, 306].

It is likely that the component of the protonmotive force responsible for the transport of the solute will not be the same for all solutes. Niven and Hamilton [307] using *Staphylococcus aureus* have observed that cationic amino acids such as lysine respond primarily to the membrane potential, whereas the neutral amino acid isoleucine is transported together with a proton as a cationic species and so requires both ΔpH and the membrane potential for uptake. The anionic amino acid glutamate is taken up as a neutral species during symport with a proton in response to the ΔpH component of the protonmotive force only. The behavior with *E. coli* is consistent with these findings. Thus, the uptake of lactose, thiomethyl-β-D-galactoside, galactose, arabinose, and alanine by proton symport would require both components of the protonmotive force, whereas the transport of glucose-6-phosphate, lactate, malate, succinate, and fumarate would respond to ΔpH only. The only exception appears to be aspartate which in *E. coli* is likely to be taken up in the cationic form by a proton symport mechanism in response to both the membrane potential and ΔpH [306].

The uptake of protons under anaerobic conditions together with the solute being transported is not clear evidence that transport is driven by the protonmotive force. More direct evidence for its involvement have come from experiments in which a membrane potential, interior negative, was created by allowing K^+ to diffuse out of K^+-loaded vesicles in a K^+-deficient medium containing valinomycin [287, 288]. Under these conditions thiomethyl-β-D-galactoside, proline, alanine, serine, lysine, glycine, and threonine, but not glutamate, were rapidly transported in a reaction which was not sensitive to inhibitors of the respiratory chain or of the ATPase [287, 288]. The reaction was inhibited by uncouplers which dissipated the membrane potential. The involvement of the physiological transport systems in the membrane potential-induced transport of proline and thiomethyl-β-D-galactoside was shown by the lack of transport of these solutes in vesicles prepared from mutants having a defect in the specific transport system [288].

Calcium ion is taken up by inverted membrane vesicles (membrane particles) in a reaction which can be energized by ATP or respiration. Tsuchiya and Rosen [264] observed that Ca^{2+} was taken up when membrane particles prepared at pH 5.6 were placed in a buffer at pH 8.5. No uptake was observed if the buffer was at pH 7.3 or if nigericin was added. Nigericin, which catalyzes an electroneutral exchange of protons and K^+, would dissipate ΔpH. Valinomycin, which catalyzes the electrogenic movement of K^+, had no effect. These results indicate that Ca^{2+} was moving in response to the pH difference and not to the membrane potential. Since the interior of the vesicle was more acidic than the medium it is presumed that Ca^{2+} enters by a Ca^{2+}/proton antiport mechanism.

The results of Ramos et al. [291] and Schuldiner and Kaback [308], support the involvement of the membrane potential in the transport of some solutes. The accumulation of $TPMP^+$ was used as a measure of the membrane potential in respiring vesicles. It was found that the efficiency of various substrates in generating the membrane potential by oxidation through the respiratory chain was in the order ascorbate (in the presence of phenazine methosulfate) > D-lactate > L-lactate > succinate > NADH [308]. The efficiency of these substrates in energizing the uptake of lactose was in the same order. Moreover, when the size of the membrane potential was varied by the use of different substrates with various levels of inhibitors, a linear relationship between the size of the membrane potential and the extent of accumulation was observed [308]. Somewhat similar results were obtained by Kashket and Wilson [309] with *Streptococcus lactis,* where the extent of uptake of thiomethyl-β-D-galactoside correlated with the size of the protonmotive force (membrane potential + ΔpH).

A further line of evidence supporting the involvement of the proton gradient comes from work with ATPase mutants. Two of these mutants, NR70 and DL-54, lack ATPase activity and are characterized by an increased permeability of the cell membrane to protons and by an inability to maintain a membrane potential [251, 273]. Associated with this is a lowered ability to accumulate thiomethyl-β-D-galactoside and proline energized by respiration or by the imposition of an artificially-generated membrane potential. The addition of DCCD restores the proton impermeability of the membrane and the ability to maintain a membrane potential and to carry out transport of these solutes [251, 273].

The results discussed above support the view that the movement of solute is coupled to the movement of protons (or of hydroxyl ions in the opposite direction). Recently, it has been found that the transport of some amino acids in *Halobacterium halobium* is coupled to a Na^+ gradient set up by a proton/Na^+ antiport during illumination [311]. West and Mitchell [312] have shown that a proton/Na^+ antiport system is present in *E. coli.* However, the involvement of this in transport has yet to be demonstrated.

VIII. The ATPase Complex

A. Role in Energy Transduction

The formation of ATP coupled to respiration, and the generation from ATP of the energized state, involves the membrane-bound Ca^{2+}- or Mg^{2+}-activated ATPase system. In terms of Mitchell's chemiosomotic hypothesis the ATPase system consists of two parts [4, 5]. The component F_1, which can be readily detached from the membrane, catalyzes the hydrolysis of ATP, and can be purified as a Ca^{2+}- or Mg^{2+}-activated ATPase (see below). The ATPase interacts with the F_0 complex which is an integral protein of the membrane and which does not have ATPase activity. The F_0 complex is suggested to have a translocation pathway through it permitting the access of protons and water to the F_1 component [5]. As discussed later, the properties of ATP hydrolysis by the membrane-bound F_1F_0 complex are somewhat different from those of the solubilized F_1-ATPase. Removal of the F_1-ATPase from the membrane should expose the proton translocation pathway such that the membrane now becomes more permeable to protons. Experiments with F_1-depleted submitochondrial particles by Hinkle and Horstman [313] have confirmed that rebinding the F_1-ATPase results in a decrease in the permeability of the depleted membranes to protons. Oligomycin or DCCD also could bind to the depleted membranes to restore normal proton impermeability.

Experiments with *E. coli* suggest that there exists a parallel system to that found in mitochondria. Three mutants of *E. coli*–NR70, DL-54, and strain 72– in which the F_1-ATPase in inactive or absent appear to have an increased permeability of the membrane to protons [250, 273, 296]. This defect can be observed even in intact cells of NR70 and strain 72 [273, 296]. The increased permeability of these membranes is reflected as a decreased ability of the cells or vesicles to carry out active transport and energy-dependent transhydrogenation [250, 272]. The defect probably results from an increased rate of movement of protons through the F_0 complex. That the defect is located here and is not due to a nonspecific increase in the permeability of the membrane to protons is suggested by the nature of the mutation, and also by the ability of DCCD, an inhibitor of the F_1F_0 complex but not of F_1, to restore both the impermeability of the membrane to protons and respiration-driven active transport [250, 251].

The involvement of the F_1F_0-ATPase system in the formation and use of the energized state of the membrane has received support from reconstitution studies and from the use of mutants.

B. Evidence for Role in Energy Transduction

1. Reconstitution Studies

Although there had been previous reports of the stimulation of oxidative phosphorylation and energy-dependent transhydrogenation in *E. coli* membrane particles by soluble "factors," these components had not been purified and characterized [141, 217]. The first clear evidence from reconstitution studies for the involvement of the F_1-ATPase in energy-dependent reactions in *E. coli* was the demonstration that a homogeneous preparation of the F_1-ATPase solubilized from membrane particles would on addition to F_1-depleted particles restore energy-dependent transhydrogenation of $NADP^+$ by NADH [354]. Removal of the F_1 ATPase from the membrane particles had resulted in the loss of both ATP-driven and respiration-driven transhydrogenase activities. Addition of the F_1 ATPase resulted in the restoration of both transhydrogenase activities to levels found in untreated membrane particles. Restoration of respiration-driven transhydrogenation only was obtained when DCCD was added [250]. These results support the view that the F_1-ATPase is responsible not only for energization of the membrane by catalyzing hydrolysis of ATP, but also for maintaining the impermeability of the membrane to protons by interacting with F_0. In the F_1-depleted particles, the energized state generated by respiration cannot build up due to the permeability of the membrane. It is interesting that the hydrolytically-inactive F_1-ATPase from the *unc A* mutant *E. coli* AN120 can still interact with the F_0 complex in the depleted particles so that respiration-dependent transhydrogenation can function [314]. The reconstitution procedure has been applied to show the involvement of the F_1-ATPase in oxidative phosphorylation [221, 223], energy dependent fluorescence changes [223, 274, 275, 315, 316]. $^{32}P_i$-ATP exchange [223], and Ca^{2+} transport in membrane particles [263, 316]. As discussed in the next section, the reconstitution technique has been of great value in understanding the nature of the defect in energy-uncoupled mutants [221, 223, 250, 274, 275, 315, 317-319].

2. Mutants

A large number of mutants having defects in the ATPase system have now been isolated [7, 9, 10, 320]. The first such mutants were reported by Butlin et al.

[218] and were designated *unc A* (for *unc*oupled). These mutants were incapable of using the energy from substrate oxidation through the respiratory chain to drive oxidative phosphorylation. They lacked membrane-bound ATPase activity and were subsequently found to be unable to carry out ATP-dependent transhydrogenation of $NADP^+$ by NADH [321]. Gibson, Cox, and coworkers have also isolated a second class of "uncoupled" mutants, designated *unc B,* which are characterized by the presence of the normal level of membrane-bound ATPase activity but like *unc A* mutants are unable to form ATP by oxidative phosphorylation or to perform ATP-dependent transhydrogenation [322].

The general phenotype for all *unc* mutants presently reported is as follows: (1) *unc* mutants are able to grow on substrates such as glucose and glycerol where ATP can be provided by glycolysis but they cannot grow on substrates such as succinate, malate, etc., which provide ATP primarily by oxidative phosphorylation; (2) the mutants give low aerobic growth yields when grown on limiting glucose; (3) the respiratory chain activity is normal; (4) oxidative phosphorylation is negligible; (5) ATP-dependent transhydrogenase activity is absent; (6) the *unc* mutations map at 73.5 minutes on the *E. coli* chromosome [7, 9, 10].

The *unc A* mutants differ from *unc B* mutants in lacking membrane-bound ATPase activity. They do not constitute a homogeneous class, however. Mutants AN120, AN249, and AN296, which show the original *unc A401* phenotype, cannot carry out ATP-dependent functions such as ATP-dependent transhydrogenation and ATP-dependent (anaerobic) transport although these reactions are normal when energy is generated by substrate oxidation through the respiratory chain [218, 270, 314]. The inactive F_1-ATPase can be readily removed from membranes of this type of mutant by washing with buffer of low ionic strength [323, 314]. Addition of F_1-ATPase from the parent or wild-type strain to the stripped membrane particles restores the normal level of ATP-dependent transhydrogenase activity to the membrane particles of the mutant. Mutants MDA_1 [319], DG20/1, and BH212 [223] appear to have the same defect as that described above. Using these mutants it was shown that ATP-dependent transport of thiomethyl-β-D-galactoside and ATP-dependent quenching of the fluorescence of 9-amino-6-chloro-2-methoxyacridine (ACMA), a measure of membrane energization, could be added to the list of defective processes in these strains. In several *unc A* mutants [NR70, DL-54, AN285 (*unc 405*), MDA_2, AS12/25] [223, 250, 251, 267, 272, 316, 319, 323] membrane particles lacking a functional ATPase are able to react directly with the F_1-ATPase from wild-type strains to reconstitute normal ATP-dependent transhydrogenation and ACMA fluorescence quenching, and oxidative phosphorylation. Active transport of amino acids and β-galactosides into intact cells of strain NR70 and membrane vesicles of strain DL54 energized by substrate oxidation through the respiratory chain is lower than in the parent strains unless DCCD is added [250, 251]. These results suggest that the mutation which resulted in loss of the ATPase activity of the F_1-ATPase has given rise to impaired binding of the enzyme to the membrane such that protons can leak through the F_0 complex. This defect must be particularly severe in strain NR70 since transport defects can be demonstrated even with intact cells [250]. In the other strains, the F_1-ATPase may dissociate from the membrane during disruption of the cells. It is unlikely that a defect in the F_0 complex is responsible for the lack of binding of the ATPase since normal energy-dependent activities can be reconstituted with wild-type or parental F_1 ATPase [250]. The intact enzyme may not be released into the cytoplasm in the case of strains NR70 and AN285 (*unc 405*) since there was either no reaction of the cytoplasmic fraction with antibodies against the F_1-ATPase or there was no material migrating with a molecular weight equivalent to that of the intact F_1-ATPase in this fraction [272, 316, 323]. Strain NR70 appears to be a deletion mutant [323, 324]. However, in AN 285 it is possible that the inactive F_1-ATPase is dissociated into its subunits when the enzyme is not bound to the membrane. Rosen and Adler [274, 325] have found three strains in which defective F_1-ATPase on removal from the membrane by washing dissociates into subunits in a manner similar to that brought about by cold denaturation.

A third class of mutants with an *unc A* phenotype have been isolated. These are strains N_{I44} [211, 269, 315, 326], MDA_3 [319], BH273, and AS69/1 [223]. Although the evidence is incomplete, these strains appear to lack functional ATPase activity and ATP-dependent processes. However respiration-dependent active transport, quenching of ACMA fluorescence, and transhydrogenation is normal. These strains can be distinguished from other *unc A* strains since addition of wild-type or parental F_1-ATPase to membrane particles of the mutant, either before or after washing treatments designed to remove the inactive ATPase, does not result in reconstitution of ATP-dependent transhydrogenation or fluorescence quenching [223, 250, 315, 319]. This appears to be due to the inability of the normal F_1-ATPase to bind to the membrane. Two explanations for this behavior seems possible. The mutation might have resulted in modification of the binding site for the F_1-ATPase in the F_0 complex as well causing a modification of the hydrolytic site on the F_1-ATPase itself. An alternative explanation is that the modified inactive F_1-ATPase is more strongly bound than usual to the membrane and cannot be released by washing procedures. This would account for the retention of the respiration-driven processes since the membrane would still be impermeable to protons. However, Kanner et al. [275] were not able to detect subunits of the ATPase in 0.5% SDS extracts of the membrane particles using anti-α, anti-β, and anti-γ subunit sera. Whether this indicates that the ATPase was absent or was not solubilized from the particles by the detergent is not clear.

In contrast to *unc A* mutants, *unc B* mutants have an active membrane-bound ATPase. However, ATP hydrolysis is not coupled to generation of the energized state since *unc B* mutants are unable to carry out ATP-dependent active transport and transhydrogenation, or oxidative phosphorylation. These processes are unimpaired if energy is supplied by respiration [211, 223, 268, 269, 275, 276, 315, 322]. As with the *unc A* mutants, there appears to be several distinct types of *unc B* mutants. The "classical" mutants which carry the *unc B401* allele are AN232 and AN283 [322]. Mutants which appear to be of this type are K_{I1}, A_{I44} [275, 315], DG 15/10, and DG 26/4 [223]. They are characterized by a membrane-bound ATPase which can be readily washed from the membrane. This F_1-ATPase is active in reconstituting ATP-dependent quenching of ACMA fluorescence and transhydrogenation with ATPase-depleted wild-type membrane particles. Addition of wild-type F_1-ATPase to depleted membrane particles of the *unc B* mutant does not result in reconstitution of the ATP-dependent reactions [221, 223, 275]. These results indicate that a component of the membrane-located F_0 complex is defective since the presence of bound active ATPase is not sufficient to permit the membrane to be energized by ATP. The modification of the F_0 complex is reflected in the resistance of the membrane-bound ATPase activity to DCCD, a compound which is an effective inhibitor in wild-type strains [327]. F_1-ATPase is only sensitive to this inhibitor when associated with a normal F_0 complex. A second type of *unc B* mutant is exemplified by B_{V4} [211, 269, 275, 315], MDB [319], DG 7/10, and DG 31/3 [223]. In these mutants also, the ATP-dependent reactions cannot be reconstituted in ATP-depleted membrane particles by addition of wild-type F_1-ATPase. However, in contrast to the other type of *unc B* mutant the ATPase stripped from the membrane particles of the mutant, although active hydrolytically, cannot reconstitute these ATP-dependent functions in depleted particles from wild-type strains [223, 275, 319]. Between 40 and 90% of the total ATPase activity is found in the cytoplasmic fraction in these strains [223, 320]. This suggests that the ATPase binds weakly to the membrane which may explain its inability to reconstitute with wild-type membrane particles. Schairer et al. [223] have found that F_1-ATPase functional in reconstitution may be obtained from membranes of strains DG 7/10 and DG 31/3 if ATPase-depleted membrane particles from an *unc A* strain are added to the cell extracts before separation of the membranes and release of the F_1-ATPase. They suggest that since binding is affected in these mutant strains, subunits of the F_1-ATPase necessary for energy coupling might be accessible to proteolysis, and that this might be prevented by binding the F_1-ATPase to the *unc A* membranes during the isolation procedure.

A third type of *unc B* mutant is DG 25/9 [223]. In contrast to the other types of *unc B* mutants, the F_0 complex is not completely inactive. Thus, ATP-dependent reactions are reduced but not absent. The F_1-ATPase of this strain is reconstitutively active but less effective than in the wild-type strain. In some respects this behavior is similar to that of *etc 15* isolated by Hong and Kaback [328]. Bragg et al. [317] found that the membrane particles of this mutant were less effective than those of the wild-type in carrying out respiration- and ATP-dependent transhydrogenation even when supplemented with wild-type F_1-ATPase. The F_1-ATPase of *etc 15* was about half as effective in the reconstitution of transhydrogenation as the wild-type enzyme. This may be due to the modification of the γ-subunit in the mutant ATPase. An apparent difference between strain DG 25/9 and the *etc* mutants is the ability of the former to carry out active transport [223, 328]. Even in intact cells respiration-dependent active transport is impaired in the *etc* strains whereas it is unaffected in DG 25/9. ATP-dependent transport in DG 25/9 is inhibited about 40% compared to a wild-type strain. However, it is difficult to arrive at a firm conclusion since the solutes used in the transport studies were not the same for each mutant.

An uncoupled mutant (*unc D*) which appears to be related to *unc A* mutants has been isolated by Thipayathasana [329]. Like *unc A* mutants it maps near the *ilv* operon in the *E. coli* chromosome, shows reduced growth yields on glucose, has normal respiratory activity and respiration-dependent transhydrogenation, and cannot carry out oxidative phosphorylation. However, it shows a low level of ATP-dependent transhydrogenase activity if Ca^{2+} but not Mg^{2+} is present [329]. This is due to the marked specificity of the ATPase for Ca^{2+} as the divalent metal required for ATP hydrolysis. The total ATPase activity is about 20% of that found in the wild-type strains. The defect in this mutant appears to be in the F_1-ATPase and to result in an alteration in the specificity of this enzyme for divalent cations. Because the defect is in the F_1 and not the F_0 complex of the ATPase system, this mutant should perhaps be considered to be an *unc A* mutant.

C. Kinetic Properties of Membrane-bound and Solubilized ATPase

The properties of the membrane-bound and solubilized (F_1) ATPase have been extensively studied. The solubilization, purification, and molecular properties of the F_1-ATPase are discussed in VIII.D.

The activities of both the membrane-bound and the solubilized enzyme show an absolute dependency on the presence of divalent cations [327, 330-340]. Besides Mg^{2+} and Ca^{2+}, cations Mn^{2+}, Ni^{2+}, Zn^{2+}, and Co^{2+} can activate the enzyme. Roisin and Kepes [327] noted that the effect of cations was dependent both on the cation/ATP ratio and on pH. At Mg^{2+}/ATP ratios of 1:2 and 1:1 optimal activity was shown at pH 7.5-8.0, whereas at a ratio of 2:1 there was little change in activity from pH 7 to pH 9. In contrast, Ca^{2+} did not stimulate ATPase activity between pH 6 and 7 while at pH 7.5 there was only about 20% of the activity given by Mg^{2+} at the same cation/ATP ratios. However, a sharp increase in the activity of the enzyme occurred in the range of pH 8 to 9 to reach that given by Mg^{2+}. On the basis of these results Roisin and Kepes suggested that the ATPase should be considered to be activated only by Mg^{2+} in the physiological pH range [327]. The behavior of the solubilized enzyme to the presence of Ca^{2+} and Mg^{2+} generally is similar to that of the membrane-bound ATPase [332, 333, 335-337]. The optimum molar ratios of Mg^{2+}/ATP and Ca^{2+}/ATP were 0.4:1 and 0.5:1, respectively [333, 335, 341].

Ahlers et al. [342] have also examined the effect of pH on both the membrane-bound and solubilized ATPase activity in the presence of Mg^{2+} and Ca^{2+}. They were able to separate the effect of pH on the dissociation of the CaATP and MgATP complexes from the effect of pH on the dissociation of groups on the enzyme. The pK values of 6.5-6.7 for the dissociation of the MgATP complex, determined with both membrane-bound and solubilized enzymes, were close to the theoretical value of 6.85. These results suggest that the substrate of the ATPase is the MgATP complex. Similar results were obtained for the CaATP complex [342] which indicate that the CaATP complex has a lower affinity than the MgATP complex for the enzyme.

Attempts to determine the K_m for substrate of the ATPase has led to widely different values. Roisin and Kepes [334] found that the K_m value of 0.5 mM for ATP of the membrane-bound ATPase did not vary significantly with pH or with the molar concentration ratio of Mg^{2+}/ATP between 1:2 and 2:1. Hanson and Kennedy [336] obtained K_m values of 0.23 mM and 0.29 mM for the membrane-bound and soluble enzymes, respectively, from experiments in which the concentration of ATP was varied between 1 and 20 mM at a constant concentration of 10 mM Mg^{2+}. In the experiments of Carreira and Mũnoz [337] a K_m value of 2.5 mM was given by the membrane-bound enzyme when the molar concentration ratio of Mg^{2+}/ATP was held at 1:2. However, if ATP concentration was varied over the range of 0.1 to 8 mM at a constant concentration of 4 mM Mg^{2+}, both the membrane-bound and the soluble enzyme gave biphasic Lineweaver-Burk plots from which K_m values of 0.38 and 17 mM ATP, and 1.4 and 20 mM ATP, were calculated for the two forms of the enzyme, respectively [337]. Ahlers and Günther [339] suggest that the reason for these complexities is that the determination of the kinetic constants where the concentration of MgATP or CaATP was varied at a constant molar ratio of cation to ATP resulted in the total cation concentration varying during the assay. Since free Mg^{2+} or Ca^{2+} are competitive inhibitors of the ATPase activity with respect to the substrates MgATP or CaATP, respectively [339], this would result in a variable degree of inhibition occurring during the determinations. In a similar way, variation of the concentration of ATP at a constant level of divalent cation would result in the presence of different concentrations of the inhibitory free cation.

When these factors were controlled by carrying out the appropriate kinetic analyses of the membrane-bound ATPase, Ahlers and Günther [339] obtained K_m values of 20 and 30 μM for MgATP at pH 7.5 and pH 9.1, and 47 and 90 μM for CaATP at these two pH values.

Both the membrane-bound and the solubilized enzyme are stimulated by anions [338, 340]. Chloride was the most effective of the anions examined. Half-maximal activation occurred at a concentration of 10^{-2} M chloride with the membrane-bound enzyme. The activation by chloride was accompanied by a decrease in the K_m for ATP. Ahlers and Günther [340] have speculated on the possible physiological role of anion activation of the ATPase. They suggest that extrusion of H^+ during oxidative phosphorylation would result in the reaction of OH^- with CO_2 to form HCO_3^-. This anion, either directly or following exchange with external chloride ions, might activate ATPase activity. This effect would only be observed at low concentrations of MgATP since the effect of anions would be much less at the higher concentrations of MgATP. Thus, the ATPase activity would be coupled under these conditions to the rate of H^+ extrusion and so to the rate of respiration.

Both membrane-bound and F_1-ATPases hydrolyse GTP, CTP, and UTP [333-336]. ADP, UDP, CDP, GDP, and AMP are not substrates for these enzymes. ADP and phosphate inhibit enzyme activity [333, 335-337].

The effect of other inhibitors on membrane-bound and soluble ATPase activity has been investigated by a number of workers [327, 333-340, 342, 344]. The enzyme activity of both soluble and membrane-bound ATPase is relatively unaffected by oligomycin, carbonylcyanide-*m*-chlorophenylhydrazone (0.5 mM), 2,4-dinitrophenol (1 mM), KCN (1 mM), ouabain (0.25 mM), arsenate (10 mM), N-ethylmaleimide (2 mM), and iodoacetate (1 mM). These results clearly distinguish the *E. coli* ATPase from the enzyme of the mammalian mitochondrion which is inhibited by oligomycin and stimulated by uncouplers such as 2,4-dinitrophenol [345, 346]. The insensitivity of the enzyme to ouabain even in the presence of Na^+, K^+, or $Na^+ + K^+$ shows that the ATPase is different from the ($Na^+ + K^+$)-dependent ATPase of higher organisms [386]. The absence of effect of N-ethylmaleimide and iodoacetate suggest that sulfhydryl groups are not involved at the active site of the enzyme [335, 342]. Compounds reacting with amino groups and histidinyl residues such as 2,4,6-trinitrobenzenesulfonic acid, O-methylisourea, and maleic anhydride were effective inhibitors of both soluble and membrane-bound enzyme activities [338, 342]. Phenylglyoxal and 2,3-butanedione inhibited both enzymes, implicating the involvement of arginine at the active site of the enzyme [342] as has been found with the ATPase of beef heart mitochondria [345]. Another inhibitor of the mitochondrial ATPase [347], 7-chloro-4-nitrobenzo-2-oxa-1,3-diazole, inhibits ATPase activity of the solubilized enzyme through reaction with a tyrosine residue [314] on a β subunit of the enzyme [348].

Sun and Crane [344] have presented evidence for the involvement of zinc at the active site of the ATPase. The reagent zincon reacted with the F_1-ATPase to give the characteristic absorption spectrum of the zincon-zinc chelate. The extent of chelate formation paralleled the loss of ATPase activity until the amount of chelate formed (13 nmol/mg protein) equaled the zinc content of the enzyme preparation (11.5 nmol/mg protein). Addition of Zn^{2+} to the zincon-treated enzyme resulted in the restoration of enzyme activity ot the level seen in the untreated ATPase.

Besides the inhibitors discussed above, which have a clearly defined site of action, there are several others which have been investigated. Sodium azide inhibits the activity of both the membrane-bound and F_1-ATPases [327, 331, 334-338], almost complete inhibition being obtained at a concentration of 1 mM. The antibiotic Dio-9 [314, 327, 334] and DCCD are effective inhibitors of the membrane-bound ATPase but have comparatively little effect on the solubilized (F_1) ATPase [314, 327, 334, 337]. The inhibition of the membrane-bound enzyme is less marked at pH 7.5 than at pH 9.0 [327, 338]. The change in the behavior of the ATPase on solubilization, a phenomenon termed allotopy [387], has been noted with several of the properties of the enzyme. There is shift toward a more alkaline pH for optimum activity on solubilization of the ATPase [327]. The soluble enzyme shows sigmoidal velocity-versus-substrate concentration plots compared to the hyperbolic curves given by the membrane-bound ATPase [337]. Dixon plots of the inhibition of ATPase activity by ADP and by phosphate are linear for the membrane-bound enzyme but nonlinear for the soluble ATPase [337]. Hill plots of the inhibition of ATPase activity by NaCl are different for the two forms of the enzyme [349]. A further distinction between the soluble and membrane-bound enzymes is that the former rapidly loses activity ($t_{1/2}$ = 4 hr) at 0°C in the absence of stabilizing agents such as glycerol or methanol whereas the latter is stable [333, 335]. The soluble enzyme is much more stable if kept at 22°C. This allotopic behavior indicates that the membrane in some manner modulates the properties of the ATPase. The involvement of membrane protein in this process is clearly seen from the action of DCCD. The solubilized ATPase is less sensitive to inhibition by DCCD than is the membrane-bound enzyme [327]. Rebinding of the solubilized ATPase to membranes depleted of the enzyme results in the restoration of increased sensitivity of DCCD [327]. A similar result is obtained with Dio-9 [327]. Various mutants have been obtained in which the membrane-bound ATPase is resistant to DCCD and in which the resistance is associated with modification of a membrane protein [9, 10, 223, 315, 320]. This protein has been purified to homogeneity by Fillingame [350, 351] and by Altendorf and Zitzmann [352]. In mutants in which the membrane-bound ATPase is resistant to DCCD this protein reacts less readily with the inhibitor [350-353]. Clearly, this membrane-bound protein can in some manner influence the active site of the ATPase.

D. *Purification and Structure of F_1-ATPase*

There have been a number of reports on the solubilization and purification of the ATPase [58, 316-318, 327, 332, 333, 335, 336, 341, 348, 354-359]. The first report on the solubilization of this enzyme was that of Bragg and Hou [58] who released it from a membrane particle fraction by treatment at pH 7.4 with 0.2% SDS at 37°C. Evans [332] subsequently modified this procedure using 0.04% SDS at pH 9.0 to release the enzyme. The enzyme, following purification by adsorption on protamine sulfate and gel filtration on agarose, had specific actitities of 74 and 39 μmol/min per mg protein in the presence of Mg^{2+} and Ca^{2+}, respectively. The molecular weight of this preparation estimated by gel filtration was 100,000 [332]. This is substantially below the value of 385,000 for the enzyme prepared by more gentle procedures [341]. It is not clear how the ATPase prepared by Evan's procedure relates to the enzyme with the larger molecular weight. Hanson and Kennedy [336] have also used a detergent, Triton X-100, to release the enzyme from the membrane. The enzyme was purified almost to homogeneity but had an unusually low specific activity at 30°C of 5 μmol/min per mg protein. The reason for this is not clear but it is possible that Triton is inhibiting the activity of the enzyme.

The ATPase is usually released from the membrane by a washing procedure which involves incubation at 4°C or room temperature with a buffer of low ionic strength without Mg^{2+}. EDTA may or may not be present, and in some procedures an alkaline pH (pH 9.0) is used. The stripped membranes are subsequently sedimented by ultracentrifugation and the ATPase in the solubilized fraction is then purified by chromatography on DEAE-cellulose, gel filtration on agarose, sucrose density gradient centrifugation, or by a combination of these procedures [316, 333, 341, 348, 354, 356, 360]. Vogel and Steinhart [359] have developed a large-scale purification procedure involving fractionation with poly(ethylene-glycol) of different molecular sizes. Table 14 summarizes the results of various purification procedures which yield an enzyme preparation showing a single protein band on polyacrylamide gel electrophoresis under nondissociating conditions.

The molecular weight of the ATPase has been variously determined as 300,000 [316, 355], 350,000-390,000 [336, 354, 358], and 400,000-600,000 [335]. On the basis of the molecular weights and stoichiometry of the subunits, the molecular weight of the ATPase is probably about 385,000 [341]. The ATPase generally contains five different polypeptide subunits (α-ϵ) with molecular weights of 54,000-58,000, 51,800-52,000, 31,000-33,000, 20,000-21,000, and 11,000-13,200 [341, 356, 359, 360], present in a stoichiometry of $\alpha_3\beta_3\gamma\delta\epsilon$ [341]. ATPase preparations have been obtained in which the δ subunit is absent. Thus, the preparations of Nelson et al. [348], Kobayashi and Anraku [360], and Hanson and Kennedy [336] lack this subunit. The δ subunit can be removed

Table 14 Specific Activity of Homogeneous ATPase Preparations

Authors	Activating ion	Specific activity[a] (μmol/min per mg protein)	References
Bragg and Hou	Ca^{2+}	43.8 ± 4.6	341, 354
Kobayashi and Anraku	Mg^{2+}	49.7 - 63.4	335, 360
Nelson et al.	Mg^{2+}	100 - 150	348
Futai et al	Mg^{2+}	90 - 120	356
Tsuchiya and Rosen	Mg^{2+}	85 - 110	316
Vogel and Steinhart	Mg^{2+}	39.5[b]	359
Hanson and Kennedy	Mg^{2+}	5.0[c]	336

[a]At 37°C.
[b]At 25°C.
[c]At 30°C.

from the five-subunit enzymes to yield the δ-deficient ATPase by pH values of 9.0 or greater [317, 361]. The δ-deficient ATPase cannot rebind to ATPase-depleted membranes as can the five-subunit enzymes [317, 356]. This suggests that the δ subunit is involved in attaching the ATPase to the membrane. The δ subunit has been purified by Smith and coworkers [361, 362]. The ATPase was resolved into an insoluble fraction containing α, β, and γ subunits, and a soluble fraction composed of the δ and ε subunits by treatment with 50% pyridine [361]. The δ and ε subunits could then be separated from one another by chromatography on DEAE-Sephadex G-25 at pH 6.2. Addition of the purified δ subunit to the δ-deficient ATPase restored its capacity to rebind to deficient membranes. Besides rebinding, the reconstituted ATPase could also catalyze the formation of ATP by oxidative phosphorylation [362] and the ATP-dependent transhydrogenation of $NADP^+$ by NADH [361]. These functions had been lost on removal of the ATPase from the membranes. This suggests that the native structure of the ATPase had been reformed on rebinding the δ subunit.

Pyridine treatment of the ATPase also solubilizes the ε subunit of the enzyme [361]. This subunit has now been purified almost to homogeneity [362]. Addition of the ε subunit to either the five-subunit enzyme or to the δ-deficient ATPase results in substantial (80%) inhibition of ATPase activity [362]. Since these preparations of ATPase already contain bound ε subunit, the significance of this effect needs further examination. The activity of both membrane-bound [355, 363] and purified ATPase [341] is stimulated up to 100% by brief treatment with trypsin or TPCK-trypsin. Examination of purified enzyme after this treatment revealed that the δ- and ε-subunits had been destroyed. Some minor modification of other subunits may also have occurred since the treated enzyme was no longer inhibited by excess Ca^{2+} in contrast to the behavior of the untreated enzyme [341]. Addition of preparations containing the ϵ subunit to the trypsin-treated ATPase resulted in inhibition of ATPase activity [363]. These results suggest that the ϵ subunit might have a role in controlling the hydrolytic activity of the enzyme. Asami et al. [364] and van de Stadt et al. [365] have suggested that the mitochondrial ATPase inhibitor protein, which has many similarities to the ϵ subunit of the *E. coli* ATPase, is involved in determining whether the ATPase is operating in the direction of formation or use of ATP. Substrate oxidation or a low molar ratio of ATP/ADP diminishes the interaction between the inhibitor and the mitochondrial ATPase. Whether similar controls are operating in *E. coli* is not known although it is clear that the ATPase of *E. coli* is capable of acting in both directions. Thus, it is involved in ATP formation by oxidative phosphorylation and in ATP-dependent transport and transhydrogenation. This contrasts with the behavior of the ATPases of *Paracoccus denitrificans, Micrococcus lysodeikticus,* and *Azotobacter vinelandii* in which the rate of ATP synthesis is up to one hundredfold greater than the rate of ATP hydrolysis [229-231]. Moreover, the emmergence of ATPase activity and ATP-dependent reactions occurs only after treatment of the particles with trypsin [230, 231], or in the case of *A. vinelandii,* oxidation of substrate through the respiratory chain in the presence of ADP and phosphate [230].

Less information is available on the function of the α, β, and γ subunits. In the ATPase of *E. coli etc 15* the γ subunit is of slightly smaller molecular weight than in the enzyme from the wild-type strain [314]. The specific activity of the homogeneous enzyme of the mutant is about half that of the wild-type strain [314]. Whether this is due to a direct effect on the active site of the enzyme or to modified binding of the ϵ subunit such as to increase its inhibitory effect is not known. The active site for ATP hydrolysis appears to be associated with the α and/or β subunits. There are several points of evidence supporting this view. Nelson et al. [348] treated the soluble ATPase with trypsin for an extended period of time and then reisolated the enzyme by gel filtration on an agarose column. The reisolated enzyme retained full ATPase activity but consisted only of the α and β subunits. Inhibition of ATPase activity by 7-chloro-4-nitrobenzo-2-oxa-1,3-diazole is due to reaction of a tyrosine residue on the β subunit [317, 348]. Vogel and Steinhart [359] have found that freezing the *E. coli* ATPase results in loss of activity and the splitting of the molecule into three fragments. The three fragments are the monomeric β subunit, an aggregate of α, γ, and ϵ subunits (molecular weight, 100,000), and an aggregate of α, γ, δ, and ϵ subunits. These fragments are inactive but are able to reassociate under carefully defined conditions with full restoration of ATPase activity. This suggests that both α and β subunits need to be present for activity. Recently, Vogel et al.

[366] have examined the reassociation of subunits from several mutant *unc A* ATPases with those of the parent strain. The *unc A* ATPase has no hydrolytic activity but is indistinguishable from the parent enzyme in subunit composition and stoichiometry, bound nucleotides, and the possession of a 7-chloro-4-nitrobenzo-2-oxa-1,3-diazole-reactive tyrosine residue [314]. Vogel et al. [366] have shown that ATPase activity can be reconstituted when the β subunit and the aggregate of α, γ, δ, and ϵ subunits from the mutant are mixed with the aggregate containing α, γ, and ϵ subunits from the wild-type strain. This suggests that the lesion responsible for the lack of ATPase activity in the *unc A* ATPase is due to a modification of an α subunit. Thus, loss of ATPase activity can occur on modification of groups on the α or β subunits, suggesting that both subunits participate at the active site. Since there are three α and three β subunits in the ATPase molecule there is the possibility of at least three active sites being present. In *E. coli* there is no information relating to this but Ferguson et al. [347] have shown that the modification of a single tyrosine residue on one of the three β subunits of the beef heart mitochondrial ATPase by 7-chloro-4-nitrobenzo-2-oxa-1,3-diazole is sufficient to inhibit enzyme activity completely. Thus, either a single active site is present on the ATPase molecule or there are a complex series of cooperative subunit interactions that are an integral part of the catalytic mechanism such that inhibition at one site will effect the others.

Noncovalently-bound nucleotides are present in addition to the polypeptide components of the ATPase. Maeda et al. [367] found that 2.2 molecules of ATP, 0.4 molecules of ADP and 0.1 molecules of inorganic phosphate were present per molecule of the ATPase from *E. coli* W3092. No bound AMP was detected. Bragg and Hou [314] found 1.13 ± 0.09 molecules of ATP and 1.79 molecules of ADP per molecule of the ATPase from *E. coli* AN180. The total number of molecules of bound adenine nucleotide per molecule of ATPase is about the same (2.6-2.9) in these two cases but the ratios of ATP to ADP are different. It is not clear if this is due to the different techniques used for the analysis, or if ATP or ADP can be bound to the same site, or if bound ATP can be dephosphorylated to give bound ADP. Interestingly, both groups found that the nucleotide content of an *unc A* mutant *E. coli* AN120 was identical to that of its parent strain AN180. Bound nucleotides have also been found in the ATPases of mitochondria, chloroplasts, and *Streptococcus faecalis* [368-371]. Their function is unknown but they could function in oxidative or photophosphorylation [370, 372, 373] or in the regulation of ATPase activity [379, 380].

Besides the tightly bound nucleotides the purified ATPase of *E. coli* could bind an additional 0.63 molecules of ADP per molecule of enzyme with a dissociation constant of 9.3 μM for the ADP-ATPase complex [314]. In the mitochondrial ATPase two binding sites can be detected [374]. The dissociation constant for the low-affinity site appears to be similar to the K_i for competitive inhibition of ATP hydrolysis by ADP, whereas the high-affinity site has a similar dissociation constant to the K_m for ADP in oxidative phosphorylation. The K_m for ADP in oxidative phosphorylation has not been measured in *E. coli* but it is likely that it is the high-affinity site which has been measured since the K_i for competitive inhibition by ADP is 0.3-0.75 mM [335, 337]. The weaker binding site would not have been detectable in the experiments of Bragg and Hou [314].

The stoichiometry of the subunits in the ATPase has been determined by measuring the radioactivity associated with each subunit on SDS polyacrylamide gel electrophoresis of the ^{14}C-labeled ATPase. A stoichiometry of $\alpha_3\beta_3\gamma\delta\epsilon$ was found [341]. Recently Vogel and Steinhart [359] have suggested that the stoichiometry of the subunits is $\alpha_2\beta_2\gamma_2\delta_2\epsilon_2$. This was proposed from reconstitution experiments using the two major complexes isolated after cold dissociation of the enzyme. On a molar basis equal amounts of the β subunit (molecular weight = 52,000) and of the subunit aggregate (molecular weight = 100,000) which contained α, γ, and ϵ subunits were required for reconstitution. Two molecules of this aggregate and two of the β subunit would give a molecular weight of 300,000. With the addition of two molecules of the δ subunit, presumed to have been lost during the separation of the subunits, a molecular weight of about 360,000 would be obtained. This is close to the value obtained by gel filtration. The difference between the stoichiometries determined by Vogel and Steinhart [359] and Bragg and Hou [341] has not been resolved. Recently, Kagawa et al. [375], using a method similar to that of the latter workers, have found that the stoichiometry of the subunits in the ATPase of a thermophilic bacterium is $\alpha_3\beta_3\gamma\delta\epsilon$.

Electron micrographs of the purified ATPase negatively stained with uranyl acetate suggest that the subunits are arranged as a planar hexagon around a central subunit (P. D. Bragg, unpublished results). The ATPases of mitochondria and *Micrococcus lysodeikticus* [345, 346, 378] have a similar appearance to that of *E. coli*. The arrangement of the subunits in the ATPase molecule has been explored by Bragg and Hou [341, 376, 377] using crosslinking reagents. The most useful reagent was dithiobis(succinimidyl propionate) since crosslinked material could be readily cleaved by reduction with β-mercaptoethanol to facilitate identification of the crosslinked subunits. Crosslinking of α to β, α to δ, β to γ, β to ϵ, and γ to ϵ subunits was observed. No $\alpha\alpha$, $\beta\delta$, $\gamma\delta$, or $\delta\epsilon$ crosslinked products were found. Although it is difficult to propose a unique arrangement for the subunits from this data some conclusions can be drawn. The formation of $\alpha\beta$ but not $\alpha\alpha$ or $\beta\beta$ crosslinked products suggests that if the α and β subunits do constitute the six peripheral subunits of the planar hexagon seen in electron micrographs as proposed by Bragg and Hou [341], then it is likely that the α and β subunits alternate (Fig. 8). However, the absence of crosslinking does not mean that the subunits are not in proximity. Suitably placed amino groups may not be avail-

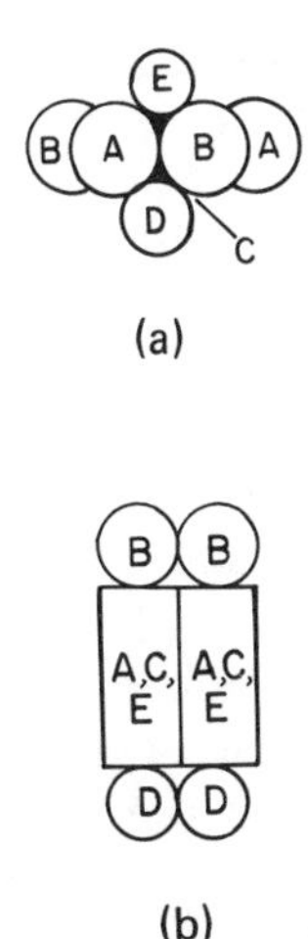

Figure 8 Models for arrangement of subunits in the F_1-ATPase of *E. coli.* (a) Model of Bragg and Hou [341]. View of planar hexagon from the side showing one possible arrangement of the δ and ϵ subunits. The γ subunit in the center of the planar hexagon is shaded. (b) Model of Vogel and Steinhart [359] (redrawn). The arrangement of subunits A, C, and E relative to one another is not specified. A-E indicates subunits α–ϵ.

able on adjacent subunits for reaction with the reagent. Interestingly the β subunit appears to be the most reactive since it crosslinks readily to all of the other subunits with the exception of the δ subunit. The model of Steinhart and Vogel [359] (Fig. 8) also can account for all of the crosslinked products formed but $\alpha\alpha$ and $\beta\beta$ products would also be expected. The absence of these, as indicated above, could be due to the lack of suitably disposed amino groups on the subunits. Thus, the crosslinking data does not distinguish between the model of Vogel and Steinhart and that of Bragg and Hou.

E. The F_0 Complex

A comparison of the properties of the solubilized ATPase with those of the membrane bound enzyme clearly show that the membrane influences the properties of the enzyme. That this interaction is of importance physiologically is shown by the *unc B* class of uncoupled mutants. In these mutants the solubilized ATPase appears to be identical in all respects to the wild-type ATPase and to be able to replace it in reconstitution experiments (Section VIII.B.2). However, in these mutants, or in reconstitution experiments in which wild-type ATPase was re-bound to mutant membranes depleted of ATPase, ATP was not able to act as an energy source [10]. This suggests that there is a lesion in the membrane segment, F_0, of the ATPase system. As discussed previously, F_0 may function as a pathway for protons in the membrane permitting the generation of a transmembrane proton gradient by ATP hydrolysis or the use of such a gradient for the generation of ATP by oxidative phosphorylation. The solubilized ATPase is less sensitive to inhibition by DCCD than is the membrane-bound enzyme [327]. Furthermore, in *unc B* mutants the membrane-bound ATPase may be resistant to inhibition by DCCD [10]. Since DCCD is known to react readily with carboxyl, sulfhydryl, hydroxyl, and amino groups it is likely that the effects of DCCD above can be ascribed to reaction with, or mutation in, a specific polypeptide component of F_0. The DCCD-reactive protein has been isolated and characterized by Fillingame [350, 351] and by Altendorf and Zitzmann [352]. The protein is most effecttively isolated by extracting membrane particles with a chloroform-methanol (2:1) mixture. The DCCD-reactive protein can then be precipitated by the addition of diethyl ether [351, 352]. The material which is precipitated is not entirely homogeneous but may be further purified by chromatography on DEAE-cellulose using an ammonium acetate gradient run in chloroform-methanol-water (3:3:1) solvent, followed by chromatography on a column of the hydroxypropyl derivative of Sephadex G-50 with a solvent of 0.02 M ammonium acetate in chloroform-methanol (2:1) [351, 352]. The material purified in this way has been shown to be that which is labeled when membranes are incubated with [^{14}C] DCCD. Since DCCD is very reactive, other proteins are also labeled but it is only this protein which is not labeled in membranes in which the ATPase is DCCD-resistant [350]. Altendorf and Zitzmann found that this protein did not react with DCCD in an *unc B* mutant *E. coli* K12, DG 7/1 [352].

The DCCD-reactive protein has an unusually high content of nonpolar amino acids [351]. Only 16% of the amino acids are polar. Of the nonpolar amino acids, methionine (10%), glycine (13%), alanine (17%), and leucine (16%) predominate. Histidine, serine, tryptophan, and cysteine are absent. The content of lysine and threonine (1 mol/mol protein) allows a molecular weight of 8,400 to be calculated for the protein. This agrees well with the value of 9,000 obtained by polyacrylamide gel electrophoresis [350]. The value of 12,000-13,000 obtained by Altendorf and Zitzmann [352] is probably too high. Molecular weights in this range are difficult to determine accurately by the technique of polyacrylamide gel electrophoresis, which they used.

Only one-third of the DCCD-reactive protein was converted to the DCCD-modified form when membranes were treated with sufficient DCCD to cause maximal inhibition of ATPase activity [351]. Two feasible explanations for this effect are (1) the ATPase is associated with only one-third of the available DCCD-reactive protein molecules, and (2) the DCCD-reactive protein exists as

Table 15 Properties of F_1F_0 ATPase Complex of *E. coli*

Property	Preparation	
	Hare [382]	Bragg and Hou [383]
Fold-purification	21	17
Specific activity[a]	5.5	2.8
Sedimentation constant	14.7 S	13.8 S
Activators	PC, PS, PE[b]	LL, PC, PG, CL[b]
Concentration of DCCD giving 50% inhibition (μM)	10[c]	15[d]
Number of different polypeptide present	12	16

[a]Expressed as μmol/min per mg protein.
[b]PC, phosphatidylcholine; PS, phosphatidylserine; PE, phosphatidylethanolamine; LL, lysolecithin; PG, phosphatidylglycerol; CL, cardiolipin.
[c]Tested in the presence of phosphatidylcholine.
[d]Tested in the presence of lysolecithin.

an oligomer of perhaps three molecules in the membranes; one F_1-ATPase molecular is associated with each oligomer such that reaction with DCCD of one molecule in the oligomer will preduce inhibition of ATPase activity.

Another approach to the study of F_0 has been to isolate the entire F_1F_0-ATPase system as a DCCD-sensitive ATPase complex [381-383] (Table 15). Nieuwenhuis et al. [381] used deoxycholate in the presence of 1 M KCl to solubilize ATPase activity from membrane particles. The solubilized material was not further purified and still retained 80% of the cytochromes of the particles; however, its ATPase activity was stimulated sixfold by the addition of soybean phospholipids and was strongly inhibited by 60 μM DCCD. Hare [382] employed similar conditions to solubilize the F_1F_0-ATPase complex. The solubilized material was further fractionated on a sucrose gradient run in the presence of 0.18% deoxycholate. The ATPase migrated with a sedimentation constant of 14.7 S. It was stimulated about twofold by *E. coli* phospholipids and, in the presence of phospholipid, the activity was inhibited 84% by 20 μM DCCD. Bragg and Hou [383] have used a somewhat different method to prepare the F_1F_0-ATPase complex. ATPase activity was solubilized with cholate in the presence of ammonium sulfate (0.3 saturation) and 1 M KCl. The extract was fractionated on columns of Sepharose 6B and 4B. ATPase activity was found to be eluted as a soluble complex ("soluble respiratory complex") [25] with components of the respiratory chain. The ATPase complex was resolved from the components of the respiratory chain following treatment with 1.5% Triton X-100 and centrifugation through a sucrose gradient. The ATPase complex migrated with a sedimentation constant of

13.8 S. ATPase activity was stimulated twofold by soybean phospholipids. As in Hare's preparation, ATPase activity was stimulated by phosphatidylcholine. However, in contrast to this preparation the ATPase complex was not affected by phosphatidylethanolamine and was stimulated by cardiolipin. The reason for these discrepancies is not clear. Possibly, different phospholipids are associated with different regions of the complex and the different detergents used in the two preparations may have resulted in the phospholipids being extracted selectively. The ATPase activity of the F_1F_0 complex of Bragg and Hou was inhibited by DCCD and by Dio-9.

Hare's ATPase complex contained at least 12 different polypeptides as shown by gel electrophoresis in the presence of SDS [382]. Five of the polypeptides were subunits of the F_1-ATPase and one was the DCCD-binding protein. In addition to the F_1 and DCCD-binding polypeptides the preparation of Bragg and Hou [383] contained 10 other polypeptides. It is probable that several at least of these polypeptides were not components of the F_0 complex but were contaminatns. Hare [382] attempted to resolve this problem by precipitating the F_1F_0-ATPase complex with antibody against the F_1-ATPase. The immunoprecipitate contained at least six polypeptides in addition to those of F_1 and the DCCD-binding protein. Hare considered that five of these polypeptides were contaminants and suggested that F_0 was composed of the DCCD-binding protein and a polypeptide of molecular weight 29,000. The reasons for excluding some of these other polypeptides are not clear since at least one molecule per molecule of F_1 would appear to have been present. Further work is required to determine the minimum number of different polypeptides in the F_0 complex.

The possible role of the DCCD-binding protein as at least part of the proton channel of F_0 has not been tested directly yet. Such a role is suggested by the experiments of Rosen [273] and Altendorf et al. [251] in which reaction of the membrane with DCCD in certain mutants results in a decrease in the permeability of the membrane to protons. Patel et al. [384] have found that the permeability of the membranes of normal cells can be increased by extraction with chaotropic agents such as guanidine hydrochloride. This is accompanied by a loss of the ability to generate a potential across the membrane and of the capacity to carry out active transport. These effects could be reversed by exposure of the treated membranes to lipophilic or water-soluble carbodiimides [353]. However, with treated membrane vesicles of a DCCD-resistant mutant of *E. coli* only the water-soluble carbodiimides were effective. The membrane-bound ATPase activity was inhibited by both lipophilic and water-soluble carbodimmides in wild-type membranes but only the latter were inhibitory with mutant membranes [353]. Besides confirming the close relationship of the F_1ATPase to the carbodiimide-binding protein in the membrane, these results also suggest that the increase in proton permeability in the membrane on extraction with chaotropic agents is specifically due to an effect on the carbodiimide-binding protein. Thus, alteration in the properties of this polypeptide by mutation directly influences the ability of certain carbodiimides to decrease proton permeability.

At present, only the F_1-ATPase and the carbodiimide-reacting protein have been purified to homogeneity. Other components of the membrane are undoubtedly part of the membrane ATPase system. For example, an *unc B* mutant has a polypeptide of molecular weight 54,000 replaced by one of 25,000 [385]. Is the high molecular weight polypeptide a functional component of F_0? This and other questions will probably be answered by the combined biochemical and genetical approach which has proved so fruitful in studying the molecular biology of *E. coli.*

Acknowledgment

This study was supported by a grant from the Medical Research Council of Canada.

References

1. B. D. Sanwal, *Bacteriol. Rev. 34:* 20 (1970).
2. C. T. Gray, J. W. T. Wimpenny, D. E. Hughes, and M. R. Mossman, *Biochim. Biophys. Acta 117:* 22 (1966).
3. C. T. Gray, J. W. T. Wimpenny, and M. R. Mossman, *Biochim. Biophys. Acta 117:* 33 (1966).
4. P. Mitchell, *Biol. Rev. 41:* 445 (1966).
5. P. Mitchell, *FEBS Lett. 33:* 267 (1973).
6. F. M. Harold, *Bacteriol. Rev. 36:* 172 (1972).
7. R. D. Simoni and P. W. Postma, *Ann. Rev. Biochem. 44:* 524 (1975).
8. S. H. Larsen, J. Adler, J. J. Gargus, and R. W. Hogg, *Proc. Nat. Acad. Sci. U. S. 71:* 1239 (1974).
9. F. Gibson and G. B. Cox, *Essays in Biochemistry 9:* 1 (1973).
10. G. B. Cox and F. Gibson, *Biochim. Biophys. Acta 346:* 1 (1974).
11. H. R. Kaback in *Methods in Enzymology* Vol. 22 (W. B. Jakoby, ed.), Academic, New York, 1971, p. 99.
12. H. R. Kaback, *Biochim. Biophys. Acta 265:* 367 (1972).
13. H. R. Kaback and J. S. Hong, *Crit. Rev. Microbiol. 2:* 333 (1973).
14. H. R. Kaback, *Science 186:* 882 (1974).
15. S. A. Short, H. R. Kaback, G. Kaczorowski, J. Fisher, C. T. Walsh, and S. C. Silverstein, *Proc. Nat. Acad. Sci. U. S. 71:* 5032 (1974).
16. J. H. Weiner, *J. Membr. Biol. 15:* 1 (1974).
17. M. Futai, *J. Membr. Biol. 15:* 15 (1974).
18. J. F. Hare, K. Olden, and E. P. Kennedy, *Proc. Nat. Acad. Sci. U. S. 71:* 4843 (1974).

19. M. Futai and Y. Tanaka, *J. Bacteriol. 124:* 470 (1975).
20. S. A. Short, H. R. Kaback, and L. D. Kohn, *J. Biol. Chem. 250:* 4291 (1975).
21. B. P. Rosen and J. S. McClees, *Proc. Nat. Acad. Sci. U. S. 71:* 5042 (1974).
22. L. Smith, *Bacteriol. Rev. 18:* 106 (1954).
23. B. Revsin and A. F. Brodie, *J. Biol. Chem. 244:* 3101 (1969).
24. G. B. Cox, N. A. Newton, F. Gibson, A. M. Snoswell, and J. A. Hamilton, *Biochem. J. 117:* 551 (1970).
25. R. D. Baillie, C. Hou, and P. D. Bragg, *Biochim. Biophys. Acta 234:* 46 (1971).
26. I. C. Kim and P. D. Bragg, *J. Bacteriol. 107:* 664 (1971).
27. R. K. Poole and B. A. Haddock, *Biochem. J. 152:* 537 (1975).
28. M. R. Pudek and P. D. Bragg, unpublished data.
29. P. D. Bragg, P. L. Davies, and C. Hou, *Biochem. Biophys. Res. Commun. 47:* 1248 (1972).
30. D. J. Rainnie and P. D. Bragg, *J. Gen. Microbiol. 77:* 339 (1973).
31. C. T. Gray, J. W. T. Wimpenny, D. E. Hughes, and M. Ranlett, *Biochim. Biophys. Acta 67:* 157 (1963).
32. T. Fujita and R. Sato, *Biochim. Biophys. Acta 77:* 690 (1963).
33. J. A. Cole and F. B. Ward, *J. Gen. Microbiol. 76:* 21 (1973).
34. B. A. Haddock and H. U. Schairer, *Eur. J. Biochem. 35:* 34 (1973).
35. S. Hino and M. Maeda, *J. Gen. Appl. Microbiol. 12:* 247 (1966).
36. A. Ishida and S. Hino, *J. Gen. Appl. Microbiol. 18:* 225 (1972).
37. K. Shigeta and S. Hino, *J. Gen. Appl. Microbiol. 21:* 149 (1975).
38. A. D. Thomas, H. W. Doelle, A. W. Westwood, and G. L. Gordon, *J. Bacteriol. 112:* 1099 (1972).
39. J. W. T. Wimpenny and D. K. Necklen, *Biochim. Biophys. Acta 253:* 352 (1971).
40. N. J. Jacobs and J. M. Jacobs, *Biochem. Biophys. Res. Commun. 65:* 435 (1975).
41. N. J. Jacobs and J. M. Jacobs, *Biochim. Biophys. Acta 449:* 1 (1976).
42. C. R. Amarasingham and B. D. Davis, *J. Biol. Chem. 240:* 3664 (1965).
43. C. A. Hirsch, M. Rasminsky, B. D. Davis, and E. C. C. Lin, *J. Biol. Chem. 238:* 3770 (1965).
44. Y. Takahashi, *J. Biochem.* (Tokyo) *78:* 1097 (1975).
45. R. Poulson, K. J. Whitlow, and W. J. Polglase, *FEBS Lett. 62:* 351 (1976).
46. R. L. Broman and W. J. Dobrogosz, *Arch. Biochem. Biophys. 162:* 595 (1974).
47. W. S. Dallas, Y. H. Tseng, and W. J. Dobrogosz, *Arch. Biochem. Biophys. 175:* 295 (1976).
48. M. S. Daoud and B. A. Haddock, *Biochem. Soc. Trans. 4:* 711 (1976).
49. A. W. Linnane and C. W. Wrigley, *Biochim. Biophys. Acta 77:* 408 (1963).
50. M. Gutman, A. Schejter, and Y. Avi-Dor, *Biochim. Biophys. Acta 162:* 506 (1968).
51. R. W. Hendler, A. H. Burgess, and R. Scharff, *J. Cell. Biol. 42:* 715 (1969).

52. B. A. Haddock, *Biochem. J. 136:* 877 (1973).
53. P. D. Bragg and C. Hou, *Arch. Biochem. Biophys. 119:* 194 (1967).
54. P. D. Bragg and C. Hou, *Arch. Biochem. Biophys. 119:* 202 (1967).
55. R. W. Hendler and A. H. Burgess, *J. Cell Biol. 55:* 266 (1972).
56. R. W. Hendler and A. H. Burgess, *Biochim. Biophys. Acta 357:* 215 (1974).
57. G. F. Dancey, A. E. Levine, and B. M. Shapiro, *J. Biol. Chem. 251:* 5911 (1976).
58. P. D. Bragg and C. Hou, *Can. J. Biochem. 45:* 1107 (1967).
59. J. H. Weiner and L. A. Heppel, *Biochem. Biophys. Res. Commun. 47:* 1360 (1972).
60. M. Futai, *Biochemistry 13:* 2327 (1974).
61. J. Ruiz-Herrera, A. Alvarez, and I. Figueroa, *Biochim. Biophys. Acta 289:* 254 (1972).
62. C. T. Kerr and R. W. Miller, *J. Biol. Chem. 243:* 2963 (1968).
63. D. Karibian, *Biochim. Biophys. Acta 302:* 205 (1973).
64. D. Karibian and P. Couchoud, *Biochim. Biophys. Acta 364:* 218 (1974).
65. F. R. Williams and L. P. Hager, *J. Biol. Chem. 236:* PC36 (1961).
66. F. R. Williams and L. P. Hager, *Arch. Biochem. Biophys. 116:* 168 (1966).
67. C. C. Cunningham and L. P. Hager, *J. Biol. Chem. 246:* 1575 (1971).
68. C. C. Cunningham and L. P. Hager, *J. Biol. Chem. 246:* 1583 (1971).
69. C. C. Cunningham and L. P. Hager, *J. Biol. Chem. 250:* 7139 (1975).
70. A. F. Brodie, in *Methods in Enzymology* Vol 2 (S. P. Colowick and N. O. Kaplan, eds.), Academic, New York, 1955, p. 693.
71. E. R. Kashket and A. F. Brodie, *J. Biol. Chem. 238:* 2564 (1963).
72. G. F. Dancey and B. M. Shapiro, *J. Biol. Chem. 251:* 5921 (1976).
73. Y. Hatefi and K. E. Stempel, *J. Biol. Chem. 244:* 2350 (1969).
74. R. A. Capaldi, *Arch. Biochem. Biophys. 163:* 99 (1974).
75. I. C. Kim and P. D. Bragg, *Can. J. Biochem. 49:* 1098 (1971).
76. M. Kasahara and Y. Anraku, *J. Biochem. 76:* 959 (1974).
77. R. L. Houghton, R. J. Fisher, and D. R. Sanadi, *FEBS Lett. 68:* 95 (1976).
78. B. A. C. Ackrell, E. B. Kearney, and M. Mayr, *J. Biol. Chem. 249:* 2021 (1974).
79. M. Gutman and N. Silman, *Mol. Cell. Biochem. 7:* 177 (1975).
80. L. D. Kohn and H. R. Kaback, *J. Biol. Chem. 248:* 7012 (1973).
81. M. Futai, *Biochemistry 12:* 2468 (1973).
82. E. S. Kline and H. R. Mahler, *Ann. N. Y. Acad. Sci. 119:* 905 (1965).
83. R. Bennett, D. R. Taylor, and A. Hurst, *Biochim. Biophys. Acta 118:* 512 (1966).
84. E. M. Tarmy and N. O. Kaplan, *J. Biol. Chem. 243:* 2579 (1968).
85. E. M. Tarmy and N. O. Kaplan, *J. Biol. Chem. 243:* 2587 (1968).
86. J. P. Reeves, J. S. Hong, and H. R. Kaback, *Proc. Nat. Acad. Sci. U. S. 70:* 1917 (1973).
87. S. A. Short, H. R. Kaback, and L. D. Kohn, *J. Biol. Chem. 250:* 4291 (1975).

88. A. D. Gounaris and L. P. Hager, *J. Biol. Chem. 236:* 1013 (1961).
89. D. H. L. Bishop, K. P. Pandya, and H. K. King, *Biochem. J. 83:* 606 (1962).
90. D. H. L. Bishop and H. K. King, *Biochem. J. 85:* 550 (1962).
91. W. J. Polglase, W. T. Pun, and J. Withaar, *Biochim. Biophys. Acta 118:* 425 (1966).
92. G. D. Daves, R. F. Muraca, J. S. Whittick, P. Friis, and K. Folkers, *Biochemistry 6:* 2861 (1967).
93. I. M. Campbell and R. Bentley, *Biochemistry 8:* 4651 (1969).
94. P. D. Bragg, in *Microbial Iron Metabolism* (J. B. Neilands, ed.), Academic, New York, 1974, p. 303.
95. D. J. D. Nicholas, P. W. Wilson, W. Heinen, G. Palmer, and H. Beinert, *Nature 196:* 433 (1962).
96. J. A. Hamilton, G. B. Cox, F. D. Looney, and F. Gibson, *Biochem. J. 116:* 319 (1970).
97. R. W. Hendler, *J. Cell Biol. 51:* 664 (1971).
98. B. A. Haddock and P. B. Garland, *Biochem. J. 124:* 155 (1971).
99. S. S. Deeb and L. P. Hager, *J. Biol. Chem. 239:* 1024 (1964).
100. E. Itagaki and L. P. Hager, *J. Biol. Chem. 241:* 3687 (1966).
101. T. Fujita, E. Itagaki, and R. Sato, *J. Biochem.* (Tokyo) *53:* 282 (1963).
102. R. Goldberger, A. Pumphrey, and A. Smith, *Biochim. Biophys. Acta 58:* 307 (1962).
103. E. Itagaki, G. Palmer, and L. P. Hager, *J. Biol. Chem. 242:* 2272 (1967).
104. T. Fujita, *J. Biochem.* (Tokyo) *60:* 329 (1966).
105. J. E. Wilson and L. P. Hager, *Biochim. Biophys. Acta 102:* 317 (1965).
106. J. Ruiz-Herrera and J. A. DeMoss, *J. Bacteriol. 99:* 720 (1969).
107. R. W. Hendler and N. Nanninga, *J. Cell Biol. 46:* 114 (1970).
108. W. S. Shipp, *Arch. Biochem. Biophys. 150:* 459 (1972).
109. W. S. Shipp, M. Piotrowski, and A. E. Friedman, *Arch. Biochem. Biophys. 150:* 473 (1972).
110. M. R. Pudek and P. D. Bragg, *Arch. Biochem. Biophys. 164:* 682 (1974).
111. B. A. Haddock, J. A. Downie, and P. B. Garland, *Biochem. J. 154:* 285 (1976).
112. R. M. Daniel, *Biochim. Biophys. Acta 216:* 328 (1970).
113. J. R. Ashcroft and B. A. Haddock, *Biochem. J. 148:* 349 (1975).
114. R. W. Hendler, D. W. Towne, and R. I. Schrager, *Biochim. Biophys. Acta 376:* 42 (1975).
115. M. R. Pudek and P. D. Bragg, *Arch. Biochem. Biophys. 174:* 546 (1976).
116. E. Itagaki and L. P. Hager, *Biochem. Biophys. Res. Commun. 68:* 1013 (1968).
117. P. K. Warme and L. P. Hager, *Biochemistry 9:* 4237 (1970).
118. P. K. Warme and L. P. Hager, *Biochemistry 9:* 4244 (1970).
119. L. N. Castor and B. Chance, *J. Biol. Chem. 234:* 1587 (1958).
120. L. Smith, D. C. White, P. Sinclair, and B. Chance, *J. Biol. Chem. 245:* 5096 (1970).

121. P. L. Broberg and L. Smith, *Biochim. Biophys. Acta 131:* 479 (1967).
122. B. Revsin, E. D. Marquez, and A. F. Brodie, *Arch. Biochem. Biophys. 139:* 114 (1970).
123. M. T. King and G. Drews, *Eur. J. Biochem. 68:* 5 (1976).
124. D. A. Webster and D. P. Hackett, *J. Biol. Chem. 241:* 3308 (1966).
125. C. Y. Liu and D. A. Webster, *J. Biol. Chem. 249:* 4261 (1974).
126. D. A. Webster and C. Y. Liu, *J. Biol. Chem. 249:* 4257 (1974).
127. C. Y. Liu and D. A. Webster, *J. Biol. Chem. 249:* 4261 (1974).
128. D. A. Webster, *J. Biol. Chem. 250:* 4955 (1975).
129. H. W. Taber and M. Morrison, *Arch. Biochem. Biophys. 105:* 367 (1964).
130. M. R. Pudek and P. D. Bragg, *FEBS Lett. 50:* 111 (1975).
131. J. Barrett, *Biochem. J. 64:* 626 (1956).
132. H. F. Kauffman and B. F. Van Gelder, *Biochim. Biophys. Acta 305:* 260 (1973).
133. H. F. Kauffman and B. F. Van Gelder, *Biochim. Biophys. Acta 314:* 276 (1973).
134. H. F. Kauffman and B. F. Van Gelder, *Biochim. Biophys. Acta 333:* 218 (1974).
135. R. H. Tiesjema, A. O. Muijsers, and B. F. Van Gelder, *Biochim. Biophys. Acta 256:* 32 (1972).
136. J. A. Kornblatt, D. I. C. Kells, and G. R. Williams, *Can. J. Biochem. 53:* 461 (1975).
137. M. R. Pudek and P. D. Bragg, *FEBS Lett. 62:* 330 (1976).
138. Y. Hatefi, W. G. Hanstein, Y. Galante, and D. L. Stiggall, *Fed. Proc. 34:* 1699 (1975).
139. P. D. Bragg and C. Hou, *Arch. Biochem. Biophys. 174:* 553 (1976).
140. D. C. Birdsell and E. H. Cota-Robles, *Biochim. Biophys. Acta 216:* 250 (1970).
141. E. R. Kashket and A. F. Brodie, *Biochim. Biophys. Acta 78:* 52 (1963).
142. P. D. Bragg, *Can. J. Biochem. 49:* 492 (1971).
143. B. D. Lakchaura, T. Fossum, and J. Jagger, *J. Bacteriol. 125:* 111 (1976).
144. H. Werbin, B. D. Lakchaura, and J. Jagger, *Photochem. Photobiol. 19:* 321 (1974).
145. R. G. W. Jones, *Biochem. J. 103:* 714 (1967).
146. G. B. Cox, A. M. Snoswell, and F. Gibson, *Biochim. Biophys. Acta 153:* 1 (1968).
147. A. M. Snoswell and G. B. Cox, *Biochim. Biophys. Acta 162:* 455 (1968).
148. N. A. Newton, G. B. Cox, and F. Gibson, *J. Bacteriol. 109:* 69 (1972).
149. M. Gutman, A. Schechter, and Y. Avi-Dor, *Biochim. Biophys. Acta 172:* 462 (1969).
150. P. D. Bragg, unpublished results.
151. P. D. Bragg, *Can. J. Biochem. 48:* 777 (1970).
152. T. Ohnishi, *Biochim. Biophys. Acta 301:* 105 (1973).
153. E. M. Barnes and H. R. Kaback, *J. Biol. Chem. 246:* 5518 (1971).
154. W. N. Konings, E. M. Barnes, and H. R. Kaback, *J. Biol. Chem. 246:* 5857 (1971).

155. M. T. King and G. Drews, *Arch. Microbiol. 102:* 219 (1975).
156. P. Mitchell, *FEBS Lett. 59:* 137 (1975).
157. A. P. Singh and P. D. Bragg, *Biochem. Biophys. Res. Commun. 72:* 195 (1976).
158. J. W. T. Wimpenny and J. A. Cole, *Biochim. Biophys. Acta 148:* 233 (1967).
159. J. A. Cole and J. W. T. Wimpenny, *Biochim. Biophys. Acta 162:* 39 (1968).
160. J. Ruiz-Herrera, M. K. Showe, and J. A. DeMoss, *J. Bacteriol. 97:* 1291 (1969).
161. P. B. Garland, J. A. Downie, and B. A. Haddock, *Biochem. J. 152:* 547 (1975).
162. M. B. Kemp, B. A. Haddock, and P. B. Garland, *Biochem. J. 148:* 329 (1975).
163. J. Ruiz-Herrera and A. Alvarez, *Ant. V. Leenwenhoek J. Microbiol. Serol. 38:* 479 (1972).
164. E. Itagaki, T. Fujita, and R. Sato, *J. Biochem.* (Tokyo) *52:* 131 (1962).
165. H. G. Enoch and R. L. Lester, *Biochem. Biophys. Res. Commun. 61:* 1234 (1974).
166. H. G. Enoch and R. L. Lester, *J. Biol. Chem. 250:* 6693 (1975).
167. P. Forget and D. V. Dervartanian, *Biochim. Biophys. Acta 256:* 600 (1972).
168. J. Ruiz-Herrera, A. Alvarez, and I. Figueroa, *Biochim. Biophys. Acta 289:* 254 (1972).
169. R. H. Scott and J. A. DeMoss, *J. Bacteriol. 126:* 478 (1976).
170. S. Taniguchi and E. Itagaki, *Biochim. Biophys. Acta 44:* 263 (1960).
171. M. K. Showe and J. A. DeMoss, *J. Bacteriol. 95:* 1305 (1968).
172. P. Forget, *Eur. J. Biochem. 42:* 325 (1974).
173. C. H. MacGregor, C. A. Schnaitman, D. E. Normansell, and M. G. Hodgins, *J. Biol. Chem. 249:* 5321 (1974).
174. C. H. MacGregor, *J. Bacteriol. 121:* 1102 (1975).
175. K. Lund and J. A. DeMoss, *J. Bacteriol. 251:* 2207 (1976).
176. R. A. Clegg, *Biochem. J. 153:* 533 (1976).
177. C. H. MacGregor, *J. Bacteriol. 126:* 122 (1976).
178. C. H. MacGregor, *J. Bacteriol. 121:* 1111 (1975).
179. B. J. Bachmann, K. B. Low, and A. L. Taylor, *Bacteriol. Rev. 40:* 116 (1976).
180. C. H. MacGregor and C. A. Schnaitman, *J. Bacteriol. 108:* 564 (1971).
181. C. H. MacGregor, *J. Bacteriol. 121:* 1117 (1975).
182. C. H. MacGregor and C. A. Schnaitman, *J. Bacteriol. 112:* 388 (1972).
183. C. H. MacGregor and C. A. Schnaitman, *J. Bacteriol. 114:* 1164 (1973).
184. J. H. Glaser and J. A. DeMoss, *Mol. Gen. Genet. 116:* 1 (1972).
185. E. Itagaki, *J. Biochem.* (Tokyo) *55:* 432 (1964).
186. D. H. Boxer and R. A. Clegg, *FEBS Lett. 60:* 54 (1975).
187. W. S. Kistler and E. C. C. Lin, *J. Bacteriol. 108:* 1224 (1971).

188. K. Miki and E. C. C. Lin, *J. Bacteriol. 124:* 1288 (1975).
189. K. Miki and E. C. C. Lin, *J. Bacteriol. 114:* 767 (1973).
190. W. S. Kistler and E. C. C. Lin, *J. Bacteriol. 112:* 539 (1972).
191. C. A. Hirsch, M. Rasminsky, B. D. Davis, and E. C. C. Lin, *J. Biol. Chem. 238:* 3770 (1963).
192. M. E. Spencer and J. R. Guest, *J. Bacteriol. 117:* 954 (1974).
193. A. P. Singh and P. D. Bragg, *Biochim. Biophys. Acta 396:* 229 (1975).
194. N. A. Newton, G. B. Cox, and F. Gibson, *Biochem. J. 244:* 155 (1971).
195. A. P. Singh and P. D. Bragg, *Biochim. Biophys. Acta 423:* 450 (1976).
196. W. N. Konings and H. R. Kaback, *Proc. Nat. Acad. Sci. U. S. 70:* 3376 (1973).
197. J. Boonstra, M. T. Huttenen, W. N. Konings, and H. R. Kaback, *J. Biol. Chem. 250:* 6792 (1975).
198. J. Boonstra, H. J. Sips, and W. N. Konings, *Eur. J. Biochem. 69:* 35 (1976).
199. K. Miki and E. C. C. Lin, *J. Bacteriol. 124:* 1282 (1975).
200. A. P. Singh and P. D. Bragg, unpublished results.
201. W. H. Holms, I. D. Hamilton, and A. G. Robertson, *Arch. Microbiol. 83:* 95 (1972).
202. V. Moses and P. B. Sharp, *J. Gen. Microbiol. 71:* 181 (1972).
203. T. Bauchop and S. R. Elsden, *J. Gen. Microbiol. 23:* 457 (1960).
204. S. J. Pirt, *Proc. Roy. Soc., Ser. B. 163:* 224 (1965).
205. D. J. Meyer and C. W. Jones, *Eur. J. Biochem. 36:* 144 (1973).
206. A. H. Stouthamer and C. Bettenhaussen, *Biochim. Biophys. Acta 301:* 53 (1973).
207. I. S. Farmer and C. W. Jones, *Eur. J. Biochem. 67:* 115 (1976).
208. W. P. Hempfling, *Biochim. Biophys. Acta 205:* 169 (1970).
209. W. P. Hempfling, *Biochem. Biophys. Res. Commun. 41:* 9 (1970).
210. E. G. Van der Beek and A. H. Stouthamer, *Arch. Microbiol. 89:* 327 (1973).
211. D. L. Gutnick, B. I. Kanner, and P. W. Postma, *Biochim. Biophys. Acta 283:* 217 (1972).
212. I. West and P. Mitchell, *J. Bioenergetics 3:* 445 (1972).
213. H. G. Lawford and B. A. Haddock, *Biochem. J. 136:* 217 (1973).
214. B. A. Haddock, A. J. Downie, and H. G. Lawford, *Proc. Soc. Gen. Microbiol. 1:* 50 (1974).
215. C. W. Jones, J. M. Brice, A. J. Downs, and J. W. Drozd, *Eur. J. Biochem. 52:* 265 (1975).
216. M. D. Brand, B. Reynafarje, and A. L. Lehninger, *J. Biol. Chem. 251:* 5670 (1976).
217. P. D. Bragg and C. Hou, *Can. J. Biochem. 46:* 631 (1968).
218. J. D. Butlin, G. B. Cox, and F. Gibson, *Biochem. J. 124:* 75 (1971).
219. P. D. Bragg, P. L. Davies, and C. Hou, *Biochem. Biophys. Res. Commun. 47:* 1248 (1972).
220. M. Watanuki, K. Oishi, K. Aida, and T. Uemura, *J. Gen. Appl. Microbiol. 18:* 29 (1972).

221. G. B. Cox, F. Gibson, and L. McCann, *Biochem. J. 134:* 1015 (1973).
222. M. Mével-Ninio and T. Yamamoto, *Biochim. Biophys. Acta 357:* 63 (1974).
223. H. U. Schairer, P. Friedl, B. I. Schmid, and G. Vogel, *Eur. J. Biochem. 66:* 257 (1976).
224. G. Turnock, S. K. Erickson, B. A. C. Ackrell, and B. Birch, *J. Gen. Microbiol. 70:* 507 (1972).
225. E. L. Hertzberg and P. C. Hinkle, *Biochem. Biophys. Res. Commun. 58:* 170 (1974).
226. T. Tsuchiya and B. P. Rosen, *J. Biol. Chem. 250:* 7687 (1975).
227. C. W. Jones, B. A. C. Ackrell, and S. K. Erickson, *Biochim. Biophys. Acta 245:* 54 (1971).
228. P. John and W. A. Hamilton, *Eur. J. Biochem. 23:* 528 (1971).
229. S. J. Ferguson, P. John, W. J. Lloyd, G. K. Radda, and F. R. Whatley, *FEBS Lett. 62:* 272 (1976).
230. P. Bhattacharyya and E. M. Barnes, *J. Biol. Chem. 251:* 5614 (1976).
231. E. I. Mileykovskaya, G. V. Tikhonova, A. A. Kondrashin, and I. A. Kozlov, *Eur. J. Biochem. 62:* 613 (1976).
232. T. Tsuchiya, *J. Biol. Chem. 251:* 5315 (1976).
233. B. Z. Cavari, Y. Avi-Dor, and N. Grossowicz, *J. Bacteriol. 96:* 751 (1968).
234. W. P. Hempfling and D. K. Beeman, *Biochem. Biophys. Res. Commun. 45:* 924 (1971).
235. B. Chance and G. Hollunger, *Fed. Proc. 16:* 163 (1957).
236. L. Ernster and C. P. Lee, in *Methods in Enzymology* Vol 10 (R. W. Estabrook and M. E. Pullman, eds.), Academic, New York, 1967, p. 729.
237. A. Asano, K. Imai, and R. Sato, *J. Biochem.* (Tokyo) *62:* 210 (1967).
238. D. L. Keister and N. J. Yike, *Arch. Biochem. Biophys. 121:* 415 (1967).
239. D. L. Keister and N. J. Minton, *Biochemistry 8:* 167 (1969).
240. A. J. Sweetman and D. E. Griffiths, *Biochem. J. 121:* 117 (1971).
241. R. K. Poole and B. A. Haddock, *Biochem. J. 144:* 77 (1974).
242. J. Rydström, J. B. Hoek, and L. Ernster, in *The Enzymes* (P. D. Boyer, ed.), Vol. 13C, Academic, New York, 1976, p. 51.
243. P. S. Murthy and A. F. Brodie, *J. Biol. Chem. 239:* 4292 (1964).
244. R. J. Fisher, K. W. Lam, and D. R. Sanadi, *Biochem. Biophys. Res. Commun. 39:* 1021 (1970).
245. R. J. Fisher and D. R. Sanadi, *Biochim. Biophys. Acta 245:* 34 (1971).
246. P. D. Bragg and C. Hou, *Arch. Biochem. Biophys. 163:* 614 (1974).
247. R. L. Houghton, R. J. Fisher, and D. R. Sanadi, *Arch. Biochem. Biophys. 176:* 747 (1976).
248. A. P. Singh and P. D. Bragg, *Biochem. Biophys. Res. Commun. 57:* 1200 (1974).
249. A. J. Sweetman and D. E. Griffiths, *Biochem. J. 121:* 125 (1971).
250. P. D. Bragg and C. Hou, *Biochem. Biophys. Res. Commun. 50:* 729 (1973).
251. K. Altendorf, F. M. Harold, and R. D. Simoni, *J. Biol. Chem. 249:* 4587 (1974).

252. A. P. Singh and P. D. Bragg, *J. Gen. Microbiol. 82:* 237 (1974).
253. J. B. Hoek, J. Rydström, and B. Hojeberg, *Biochim. Biophys. Acta 333:* 237 (1974).
254. A. P. Singh and P. D. Bragg, *J. Bioenergetics 7:* 175 (1975).
255. R. L. Houghton, R. J. Fisher, and D. R. Sanadi, *Biochem. Biophys. Res. Commun. 73:* 751 (1976).
256. R. L. Houghton, R. J. Fisher, and D. R. Sanadi, *Biochim. Biophys. Acta 396:* 17 (1975).
257. P. W. Postma and S. Roseman, *Biochim. Biophys. Acta 457:* 213 (1976).
258. E. Pavlasova and F. M. Harold, *J. Bacteriol. 98:* 198 (1969).
259. W. L. Klein and P. D. Boyer, *J. Biol. Chem. 247:* 7257 (1972).
260. M. Futai, *J. Bacteriol. 120:* 861 (1974).
261. K. A. Devor, H. U. Schairer, D. Renz, and P. Overath, *Eur. J. Biochem. 45:* 451 (1974).
262. B. P. Rosen and J. S. McClees, *Proc. Nat. Acad. Sci. U. S. 71:* 5042 (1974).
263. T. Tsuchiya and B. P. Rosen, *Biochem. Biophys. Res. Commun. 63:* 832 (1975).
264. T. Tsuchiya and B. P. Rosen, *J. Biol. Chem. 251:* 962 (1976).
265. S. Silver, K. Toth, and H. Scribner, *J. Bacteriol. 122:* 880 (1975).
266. H. U. Schairer and B. A. Haddock, *Biochem. Biophys. Res. Commun. 48:* 544 (1972).
267. R. D. Simoni and M. K. Shallenberger, *Proc. Nat. Acad. Sci. U. S. 69:* 2663 (1972).
268. H. U. Schairer and D. Gruber, *Eur. J. Biochem. 37:* 282 (1973).
269. A. Or, B. I. Kanner, and D. L. Gutnick, *FEBS Lett. 35:* 217 (1973).
270. G. Prezioso, J. S. Hong, G. K. Kerwar, and H. R. Kaback, *Arch. Biochem. Biophys. 154:* 575 (1973).
271. T. H. Yamamoto, M. Mével-Ninio, and R. C. Valentine, *Biochim. Biophys. Acta, 314:* 267 (1973).
272. B. P. Rosen, *J. Bacteriol. 116:* 1124 (1973).
273. B. P. Rosen, *Biochem. Biophys. Res. Commun. 53:* 1289 (1973).
274. B. P. Rosen and L. W. Adler, *Biochim. Biophys. Acta 387:* 23 (1975).
275. B. I. Kanner, N. Nelson, and D. L. Gutnick, *Biochim. Biophys. Acta 396:* 347 (1975).
276. H. Rosenberg, G. B. Cox, J. D. Butlin, and S. J. Gutowski, *Biochem. J. 146:* 417 (1975).
277. E. A. Berger, *Proc. Nat. Acad. Sci. U. S. 70:* 1514 (1973).
278. E. A. Berger and L. A. Heppel, *J. Biol. Chem. 249:* 7747 (1974).
279. H. Kobayashi, E. Kin, and Y. Anraku, *J. Biochem. 76:* 251 (1974).
280. J. L. Cowell, *J. Bacteriol. 120:* 139 (1974).
281. S. J. Curtis, *J. Bacteriol. 120:* 295 (1974).
282. D. B. Wilson, *J. Bacteriol. 120:* 866 (1974).
283. C. A. Plate, J. L. Suit, A. M. Jetten, and S. E. Luria, *J. Biol. Chem. 249:* 6138 (1974).

284. M. A. Lieberman and J. S. Hong, *Arch. Biochem. Biophys. 172:* 312 (1976).
285. B. Griniuviene, V. Chmieliauskaite, and L. Grinius, *Biochem. Biophys. Res. Commun. 56:* 206 (1974).
286. B. Griniuviene, V. Chmieliauskaite, V. Melvydas, P. Dzheja, and L. Grinius, *J. Bioenergetics 7:* 17 (1975).
287. H. Hirata, K. Altendorf, and F. M. Harold, *Proc. Nat. Acad. Sci. U. S. 70:* 1804 (1973).
288. H. Hirata, K. Altendorf, and F. M. Harold, *J. Biol. Chem. 249:* 2939 (1974).
289. E. Padan, D. Zilberstein, and H. Rottenberg, *Eur. J. Biochem. 63:* 533 (1976).
290. S. H. Collins and W. H. Hamilton, *J. Bacteriol. 126:* 1224 (1976).
291. S. Ramos, S. Schuldiner, and H. R. Kaback, *Proc. Nat. Acad. Sci. U. S. 73:* 1892 (1976).
292. A. P. Singh and P. D. Bragg, *Eur. J. Biochem. 67:* 177 (1976).
293. A. P. Singh and P. D. Bragg, *Biochim. Biophys. Acta 464:* 562 (1977).
294. A. P. Singh and P. D. Bragg, *Biochem. Biophys. Res. Commun. 72:* 195 (1976).
295. R. C. Maloney, E. R. Kashket, and T. H. Wilson, *Proc. Nat. Acad. Sci. U. S. 71:* 3896 (1974).
296. D. M. Wilson, J. F. Alderete, P. C. Maloney, and T. H. Wilson, *J. Bacteriol. 126:* 327 (1976).
297. I. C. West and P. Mitchell, *FEBS Lett. 40:* 1 (1974).
298. L. Grinius, R. Slusnyte, and B. Griniuviene, *FEBS Lett. 57:* 290 (1975).
299. T. Tsuchiya and B. P. Rosen, *Biochem. Biophys. Res. Commun. 68:* 497 (1976).
300. T. Tsuchiya and B. P. Rosen, *J. Bacteriol. 127:* 154 (1976).
301. I. C. West, *Biochem. Biophys. Res. Commun. 41:* 655 (1970).
302. I. C. West and P. Mitchell, *J. Bioenergetics 3:* 445 (1972).
303. R. C. Essenberg and H. L. Kornberg, *J. Biol. Chem. 250:* 939 (1975).
304. P. J. F. Henderson and A. Skinner, *Biochem. Soc. Trans. 2:* 543 (1974).
305. S. H. Collins, A. W. Jarvis, R. J. Lindsay, and W. A. Hamilton, *J. Bacteriol. 126:* 1232 (1976).
306. S. J. Gutowski and H. Rosenberg, *Biochem. J. 152:* 647 (1975).
307. D. F. Niven and W. A. Hamilton, *Eur. J. Biochem. 44:* 517 (1974).
308. S. Schuldiner and H. R. Kaback, *Biochemistry 14:* 5451 (1975).
309. E. R. Kashket and T. H. Wilson, *Biochem. Biophys. Res. Commun. 59:* 879 (1974).
310. I. C. West and P. Mitchell, *Biochem. J. 132:* 587 (1973).
311. J. K. Lanyi and R. E. MacDonald, *Biochemistry 15:* 4608 (1976).
312. I. C. West and P. Mitchell, *Biochem. J. 144:* 87 (1974).
313. P. C. Hinkle and L. L. Horstman, *J. Biol. Chem. 246:* 6024 (1971).
314. P. D. Bragg and C. Hou, *Arch. Biochem. Biophys. 178:* 486 (1977).
315. F. J. R. M. Nieuwenhuis, B. I. Kanner, D. L. Gutnick, P. W. Postma, and K. van Dam, *Biochim. Biophys. Acta 325:* 62 (1973).

316. T. Tsuchiya and B. P. Rosen, *J. Biol. Chem. 250:* 8409 (1975).
317. P. D. Bragg, P. L. Davies, and C. Hou, *Arch. Biochem. Biophys. 159:* 664 (1973).
318. G. B. Cox, F. Gibson, L. M. McCann, J. D. Butlin, and F. L. Crane, *Biochem. J. 132:* 689 (1973).
319. J. Daniel, M. P. Roisin, C. Burstein, and A. Kepes, *Biochim. Biophys. Acta 376:* 195 (1975).
320. D. L. Gutnick, in *Molecular Aspects of Membrane Phenomena* (H. R. Kaback, H. Neurath, G. K. Radda, R. Schwyzer, and W. R. Wiley, eds.), Springer-Verlag, New York, 1975, p. 76.
321. G. B. Cox, N. A. Newton, J. D. Butlin, and F. Gibson, *Biochem. J. 125:* 489 (1971).
322. J. D. Butlin, G. B. Cox, and F. Gibson, *Biochim. Biophys. Acta 292:* 366 (1973).
323. G. B. Cox, F. Gibson, and L. McCann, *Biochem. J. 138:* 211 (1974).
324. J. Boonstra, D. L. Gutnick, and H. R. Kaback, *J. Bacteriol. 124:* 1248 (1975).
325. L. W. Adler and B. P. Rosen, *J. Bacteriol. 128:* 248 (1976).
326. B. I. Kanner and D. L. Gutnick, *J. Bacteriol. 111:* 287 (1972).
327. M. P. Roisin and A. Kepes, *Biochim. Biophys. Acta 305:* 249 (1973).
328. J. S. Hong and H. R. Kaback, *Proc. Nat. Acad. Sci. U. S. 69:* 3336 (1972).
329. P. Thipayathasana, *Biochim. Biophys. Acta 408:* 47 (1975).
330. T. Günther and F. Dorn, *Z. Naturforsch, 21b:* 1076 (1966).
331. D. J. Evans, *J. Bacteriol. 100:* 914 (1969).
332. D. J. Evans, *J. Bacteriol. 104:* 1203 (1970).
333. P. L. Davies and P. D. Bragg, *Biochim. Biophys. Acta 266:* 273 (1972).
334. M. P. Roisin and A. Kepes, *Biochim. Biophys. Acta 275:* 333 (1972).
335. H. Kobayashi and Y. Anraku, *J. Biochem.* (Tokyo) *71:* 387 (1972).
336. R. L. Hanson and E. P. Kennedy, *J. Bacteriol. 114:* 772 (1973).
337. J. Carreira and E. Muñoz, *Mol. Cell. Biochem. 9:* 85 (1975).
338. T. Günther, W. Pellnitz, and G. Mariss, *Z. Naturforsch. 29c:* 54 (1974).
339. J. Ahlers and T. Günther, *Z. Naturforsch. 30c:* 412 (1975).
340. J. Ahlers and T. Günther, *Arch. Biochem. Biophys. 171:* 163 (1975).
341. P. D. Bragg and C. Hou, *Arch. Biochem. Biophys. 167:* 311 (1975).
342. J. Ahlers, D. Kabisch, and T. Günther, *Can. J. Biochem. 53:* 658 (1975).
343. I. L. Sun, D. C. Phelps, and F. L. Crane, *FEBS Lett. 54:* 253 (1975).
344. I. L. Sun and F. L. Crane, *Biochem. Biophys. Res. Commun. 65:* 1334 (1975).
345. R. B. Beechey, *Biochem. Soc. Spec. Publ. 4:* 41 (1974).
346. W. A. Catterall and P. L. Pedersen, *Biochem. Soc. Spec. Publ. 4:* 63 (1974).
347. S. J. Ferguson, W. J. Lloyd, M. H. Lyons, and G. K. Radda, *Eur. J. Biochem. 54:* 117 (1975).
348. N. Nelson, B. I. Kanner, and D. L. Gutnick, *Proc. Nat. Acad. Sci. U. S. 71:* 2720 (1974).

349. H. Moreno, F. Siñeriz, and R. N. Fariás, *J. Biol. Chem. 249:* 7701 (1974).
350. R. H. Fillingame, *J. Bacteriol. 124:* 870 (1975).
351. R. H. Fillingame, *J. Biol. Chem. 251:* 6630 (1976).
352. K. Altendorf and W. Zitzmann, *FEBS Lett. 59:* 268 (1975).
353. L. Patel and H. R. Kaback, *Biochemistry 15:* 2741 (1976).
354. P. D. Bragg and C. Hou, *FEBS Lett. 28:* 309 (1972).
355. J. Carreira, J. A. Leal, M. Rojas, and E. Muñoz, *Biochim. Biophys. Acta 307:* 541 (1973).
356. M. Futai, P. D. Sternweis, and L. A. Heppel, *Proc. Nat. Acad. Sci. U. S. 71:* 2725 (1974).
357. H. W. Peter and J. Ahlers, *Arch. Biochem. Biophys. 170:* 169 (1975).
358. G. Giordano, C. Riviere, and E. Azoulay, *Biochim. Biophys. Acta 389:* 203 (1975).
359. G. Vogel and R. Steinhart, *Biochemistry 15:* 208 (1976).
360. H. Kobayashi and Y. Anraku, *J. Biochem.* (Tokyo) *76:* 1175 (1974).
361. J. B. Smith, P. C. Sternweis, and L. A. Heppel, *J. Supramol. Structure 3:* 248 (1975).
362. J. B. Smith and P. C. Sternweis, *Biochem. Biophys. Res. Commun. 62:* 764 (1975).
363. F. J. R. M. Nieuwenhuis, J. A. M. van de Drift, A. B. Voet, and K. van Dam, *Biochim, Biophys. Acta 368:* 461 (1974).
364. K. Asami, K. Juntii, and L. Ernster, *Biochim. Biophys. Acta 205:* 307 (1970).
365. R. J. van de Stadt, B. L. de Boer, and K. van Dam, *Biochim. Biophys. Acta 292:* 338 (1973).
366. G. Vogel, H. U. Schairer, and R. Steinhart, *Int. Congress Biochem.,* Abstr. 06-6-199 (1976).
367. M. Maeda, H. Kobayashi, M. Futai, and Y. Anraku, *Biochem. Biophys. Res. Commun. 70:* 228 (1976).
368. D. A. Harris, J. Rosing, R. J. van de Stadt, and E. C. Slater, *Biochim. Biophys. Acta 314:* 149 (1973).
369. N. E. Garrett and H. S. Penefsky, *J. Biol. Chem. 250:* 6640 (1975).
370. D. A. Harris and E. C. Slater, *Biochim. Biophys. Acta 387:* 335 (1975).
371. A. Abrams, E. A. Nolan, C. Jensen, and J. B. Smith, *Biochem. Biophys. Res. Commun. 55:* 22 (1973).
372. P. D. Boyer, R. L. Cross, and W. Momsen, *Proc. Nat. Acad. Sci. U. S. 70:* 2837 (1973).
373. E. C. Slater, *Biochem. Soc. Trans. 2:* 1149 (1974).
374. W. A. Catterall and P. L. Pedersen, *J. Biol. Chem. 247:* 7969 (1972).
375. Y. Kagawa, N. Sone, M. Yoshido, H. Hirata, and H. Okamoto, *J. Biochem.* (Tokyo) *80:* 141 (1976).
376. P. D. Bragg, *J. Supramol. Structure 3:* 297 (1975).
377. P. D. Bragg and C. Hou, *Biochem. Biophys. Res. Commun. 72:* 1042 (1976).
378. E. Muñoz, J. H. Freer, D. J. Ellar, and M. R. J. Salton, *Biochim. Biophys. Acta 150:* 531 (1968).

379. D. A. Hilborn and G. G. Hammes, *Biochemistry 12:* 983 (1973).
380. L. C. Cantley and G. G. Hammes, *Biochemistry 14:* 2968 (1975).
381. F. J. R. M. Nieuwenhuis, A. A. M. Thomas, and K. van Dam, *Biochem. Soc. Trans. 2:* 512 (1974).
382. J. F. Hare, *Biochem. Biophys. Res. Commun. 66:* 1329 (1975).
383. P. D. Bragg and C. Hou, *Arch. Biochem. Biophys. 174:* 553 (1976).
384. L. Patel, S. Schuldiner, and H. R. Kaback, *Proc. Nat. Acad. Sci. U. S. 72:* 3387 (1975).
385. R. D. Simoni and A. Shandell, *J. Biol. Chem. 250:* 9421 (1975).
386. J. C. Skou, *Quart. Rev. Biophys. 7:* 402 (1975).
387. E. Racker, *Fed. Proc. 26:* 1335 (1967).

7

Energy Transduction in *Halobacterium halobium*

Janos K. Lanyi

NASA-Ames Research Center
Moffett Field, California

I. Introduction

The genus *Halobacterium* comprises a number of very similar species, namely *H. halobium, H. cutirubrum, H. salinarium* [1-3], and *H. saccharovorum* [4]. All are rod shaped, about 0.5 μm thick and about 5 μm long, motile, and contain a distinctive red pigment, bacterioruberin [5, 6]. Their unique characteristic is their habitat: highly concentrated or saturated saline solutions, such as exist in the Dead Sea, the Great Salt Lake, and solar evaporative ponds. These organisms have adapted to such an environment by being able to accumulate salt intracellularly, to at least as high a concentration as in the medium [7-10]. Growth is optimal between 3.5 and 5 M and ceases below 3 M salt. Most of the cation in the growth medium must be Na^+, although K^+ is also required [11, 12]. Potassium ion is concentrated by the cells and constitutes the major intracellular cation.

Lowering the NaCl concentration below 3 M is catastrophic for cells, which under these conditions become round and lyse [13], and for various cellular constituents, which lose structural and functional properties. Enzyme activity in most cases is found to be maximal above 2 M salt, and for some enzymes there is an absolute requirement of salt for activity (for a review see Lanyi, [14]). All enzymes appear to be unstable below 1 M salt and some become inactivated in a matter of seconds. A few enzymes are reactivated during slow addition of salt. The structural changes which must accompany such functional alteration have not been extensively investigated. Ribosomes are known to dissociate [15] and the cell envelopes have been found to lose proteins [16-18] and fragment at lowered salt concentrations [17, 19].

Knowledge of the molecular basis of these salt-dependent effects has come from indirect sources. The presence of excess acidic amino acids in proteins from halophiles suggested to Baxter [20], Brown [21], and others [22, 23] that the salt may screen negative charges and is needed to overcome electrostatic repulsion in these structures. Lanyi and Stevenson [24] pointed out that this explanation does not account for the anion specificities observed and for the requirement of high (several molar) salt concentrations, much higher than that which ought to be sufficient for charge screening. On the basis of the effects of various denaturants on NADH-menadione reductase activity (a membrane-bound enzyme), it was proposed that NaCl above 1 M strengthens, by a salting-out effect, those hydrophobic interactions which are necessary for native structure. This hypothesis was later extended to account for results with various other enzymes from halophilic bacteria [14, 25, 26].

Halobacteria lack a rigid cell wall. Instead, on the outside surface of the cytoplasmic membrane there is a hexagonal array of particles, 150-200 Å in diameter, composed mostly of protein [19, 27]. The cytoplasmic membrane of cells grown under aerobic conditions does not appear unusual at first sight. These membranes contain about 30% lipid, and their diverse variety of proteins [28, 29] include the usual assembly of flavoproteins [30, 31] and cytochromes [32-34]. The lipids are unlike those in any other organism, however, all consisting of ether derivatives of dihydrophytol [35-37], a C_{16} chain with four methyl branches, except for a few percent red and yellow carotenoids [5, 38, 39] and squalenes [40, 41]. Some of the effects of these unusual lipids on bilayer structure have been considered [42-44]. The solubilization of the membranes of halophiles in the absence of salt shows many analogies with the inactivation of enzymes from these organisms, and suggests a role for weak hydrophobic interactions in stabilizing the structure of the membranes [17, 44].

When some species of halobacteria, such as *H. halobium,* are grown under restricted aeration, distinct regions, "patches," are formed in the cytoplasmic membrane. These lack carotenoids but contain a purple pigment and appear in freeze-fracture electron micrographs as ordered granular structures [45, 46]. The patches may be isolated and contain only one kind of protein, linked to retinal, which serves as the chromophore [47]. Increasing attention has been focused on this "purple membrane," since through the efforts of Stoeckenius, Oesterhelt and others it is now understood to constitute a unique light-energy-transducing apparatus, which acts as an alternate system to the respiratory chain in generating a H^+ gradient across the cell membrane. This single specialized function for the purple membrane, as a H^+ translocase when illuminated, has lead to conceptual and experimental advantages. For example, the elegance of this system has played an important role in dramatizing the validity of the principle of chemiosmotic energy coupling in bacterial and other membranes.

No obvious relationship between the salt dependence of halobacteria and the appearance of this unique energy-transducing membrane has been found as yet. The physiology of these organisms and their means of coupling the light energy captured to energy-requiring processes is of interest not only because illumination affords an experimental advantage in studying energized processes, but also because energy transduction in these cells must reflect their adaptation to a highly saline growth medium and their reliance on two independent sources of energy. The objective of this review is therefore not only to describe the properties of the purple membrane, but also to explore the interaction of the H^+ gradient generated with other ionic gradients and with various energy-requiring processes, such as amino acid transport and ATP synthesis. Although all the results discussed here have been obtained with *H. halobium* strains R_1 or R_1M_1, recent evidence indicates that other halobacteria also contain purple membrane [6], and the conclusions reached can presumably be extended to some degree to these other species as well.

II. Structure and Function of Purple Membrane

Purple membranes consist of patches, about 0.5 μm in diameter, covering under some conditions as much as half of the surface of *H. halobium* cells. Freeze-fracture electron microscopy reveals that, unlike the red membrane (i.e., that part of the membrane other than purple patches), which contains no in-plane regularity, the purple membrane contains a hexagonal pattern of granularity [46]. When salt is removed the red membrane fragments into small, slowly sedimenting pieces, but the purple membrane can be recovered in the form of sheets [19, 45, 48] of a size corresponding to that of the original patches on the cell membrane. There is no evidence for any irreversible alteration in the purple membrane during exposure to low ionic strengths. This resistance to disruption by the removal of salt provides a convenient method for purifying purple membranes: after extensive dialysis against distilled water only the purple membranes remain intact and they can then be recovered by sedimentation. After sucrose gradient centrifugation a pure preparation is obtained [48, 49].

The purple membrane isolated in this way appears to have a fixed chemical composition—a single kind of protein of molecular weight around 26,000 [47, 50] and specific lipids [38, 39]. The lipids constitute about 25% of the dry weight of the membranes [38, 39]. The lipids in the purple membrane consist almost entirely of the diether analog of phosphatidyl glycerol phosphate (which is also the main lipid in the red membrane) and a glycosulfolipid [38, 39]. Both lipids contain only dihydrophytanyl groups.

In freeze-fracture pictures, the boundary between the main part of the membrane (i.e., red membrane) and the patches (i.e., purple membrane) is clearly visible. Although the two kinds of membranes appear to be contiguous, no structures at their boundary, which might separate them, have been observed. Freeze-fracture treatment presumably cleaves the membranes through their central hydrophobic core. Since in such electron micrographs grains are found almost exclusively on the cytoplasmic half of the purple membranes [46], the pattern of fracturing suggests that the protein in these membranes is asymmetrically located.

The asymmetrical placement of proteins in the purple membrane sheets is also suggested from dried, shadowed electron micrographs of isolated membranes [46], where about 50% of the patches appear smooth (one side up) while the other half exhibits a cracking pattern with 120° angles (other side up). x-Ray diffraction patterns with oriented samples show that the membrane is 49 Å thick and the electron density profile across the width of the membrane is asymmetrical [51, 52]. x-Ray diffraction patterns also reveal that in the plane of the membrane the structure consists of a hexagonal lattice (63 Å repeating distance), with three proteins per unit cell of about 3300 $Å^2$ area [51]. Purple membrane patches should therefore be considered two-dimensional crystals, with approximately 10^5 repeating units per sheet.

Henderson and Unwin [53] used a novel electron diffraction technique to obtain a three-dimensional image of the arrangement of polypeptide chains in the membrane. At 7 Å resolution, seven α-helices were detected; three of these were seen spanning the membranes at right angles, and four were positioned at angles 10-20° inclined from the perpendicular. Although the connections between the α-helices are not yet known, it was suggested that these seven chains belong to one protein molecule. A threefold axis of symmetry in the plane of the membrane is obtained from three proteins facing one another, with the nine right-angle α-helices forming the sides of an equilateral triangle. The lipids are probably arranged into an interrupted bilayer, filling the gaps between the proteins.

The regularity of the structure described above suggests limited motional freedom for its components. Flash-induced dichroism studies [54, 55] indicate that the rotational mobility of the protein is far slower than in other membranes. As expected, stearic acid-type spin labels are highly immobilized in purple membranes [56]. Di-*t*-butyl nitroxide partitions very poorly into these membranes, and once incorporated shows very little temperature dependence for motion (Plachy and Lanyi, unpublished experiments). These observations indicate that the purple membrane does not contain a fluid bilayer. The immobilization of the labels must be the consequence of lipid-protein interaction, but may be caused also by the inherently constrained motion of branched-chain lipid tails [42, 44]. The fact that a difference in lipid composition between red and purple membranes can be observed [38, 39] suggests that very little lateral motion for lipids takes place across the perimeter of the patches in the ctyoplasmic membrane.

The characteristic color of the purple membrane is derived from retinal [47, 57], which is covalently bound to a lysine residue via a protonated Schiff-base linkage [47, 50] and probably interacts noncovalently with neighboring aromatic amino acid(s). The original 380-nm absorption peak of free retinal is thus shifted to about 570 nm [47]. The molar extinction coefficient of the pigment is 63,000 [58]. The lysine bearing the chromophore is a neighbor, once removed, of one of 11 prolines in the molecule, and is therefore probably near a bend between two α-helices [50]. The exact location of the chromophore will become available when amino acid sequencing of the protein is completed. Chemical similarities between the protein-retinal complex in the purple membrane and the visual pigment, rhodopsin, prompted Oesterhelt and Stoeckenius [47] to name the bacterial protein *bacteriorhodopsin.*

Exciton interaction between adjacent chromophores has been proposed on the basis of the observation that native purple membrane exhibits positive and negative circular dichroism peaks near 570 nm, while purple membrane, disorganized by detergent, shows only a positive peak, such as expected for single chromophores [59-61].

Native bacteriorhodopsin is only slightly bleached at normal intensities of light. That a cyclical light reaction takes place was first observed with ether-treated purple membranes, where the reaction time for recovery is slowed down sufficiently for a bleached intermediate to accumulate [58]. This and other photointermediates have also been observed in samples illuminated at low temperatures [62]. The photochemistry of bacteriorhodopsin has been extensively studied using flash spectroscopy [63-67] and modulation excitation spectroscopy [147]. As a first approximation, most of the results fit a single-photon-initiated linear reaction with six intermediates, indicated by their absorption peaks: bR_{570} (ground state), K_{590}, L_{550}, M_{412}, N_{520}, and O_{640} (notation from Refs. 63 and 67). The reaction $bR_{570} \rightarrow K_{590}$ has a half-life of $<$10 psec [68] and it

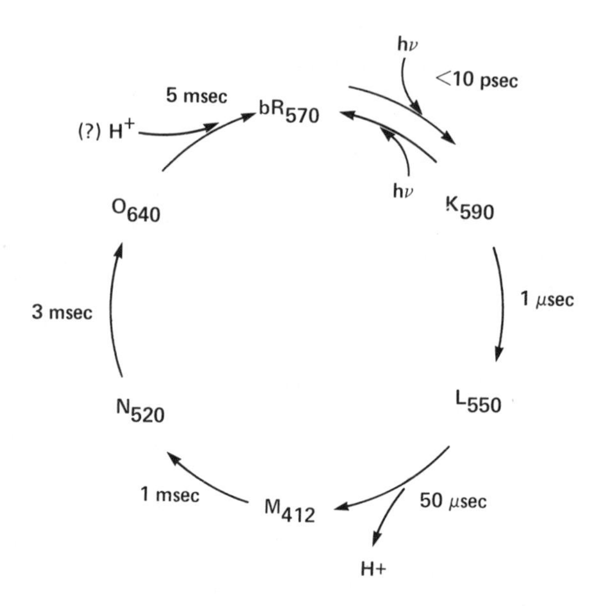

Figure 1 Scheme of the photochemical cycle of bacteriorhodopsin. (Details explained in the text.)

alone requires the absorption of a photon. The reactions resulting in the rise and decay of L_{550}, M_{412}, N_{520}, O_{640}, and finally in bR_{570} will proceed in the dark, with half-lives of 1 μsec, 50 μsec, 1 msec, 3 msec, and 5 msec, respectively, at room temperature [67]. Thus, after the absorption of a photon, bacteriorhodopsin cycles through all its intermediate states within 10 msec. Figure 1 shows the scheme of these interconversions. Since the transition to the first intermediate state, K_{590}, is photoreversible and the spectrum of this intermediate overlaps that of the initial state, bR_{570}, the efficiency of this step is just below 0.5 per photon absorbed [66]. Recently questions have arisen about possible branching reactions involving N_{520} or additional intermediates [67, 69].

Changes in pH during illumination of ether-saturated purple membranes [58, 62] revealed that the photochemical cycle described above is accompanied first by the release and then the uptake of protons. More recently, the H^+ release was tentatively assigned, on the basis of kinetic evidence, to the $L_{550} \rightarrow M_{412}$ transition, and the H^+ uptake to one of the reactions leading from M_{412} to bR_{570} [63, 69-71]. Resonance Raman spectra [72] of bleached purple membranes indicate that the Schiff base is indeed deprotonated in M_{412}. Very little else is known about the chemistry involved in the photon-initiated series of reactions. The formation of retro-retinal apparently does not take place [73]. *Cis-trans* isomerization of retinal has not yet been conclusively ruled out during the photochemical cycle.

First circumstantial and later direct evidence (discussed in the following section) has established that the cyclic deprotonation and protonation of bacteriorhodopsin occurs in a vectorial fashion: the net result of cycling is the translocation of H^+ from one side of the membrane to the other. To understand the molecular details of this process and how it relates to the structure of bacteriorhodopsin is one of the great opportunities and challenges of membrane biology, since this protein is a much simpler ion-translocating apparatus than any others known.

A second kind of light reaction observed with bacteriorhodopsin is the phenomenon of light and dark adaptation. When the purple membrane is kept in the dark, its visible absorption peak slowly shifts from 570 to about 560 nm [49, 74, 75]. Exposure to light results in a shift back to 570 nm. This transition is not accompanied by a gain or loss of protons (Lanyi, unpublished experiments). Since in D_2O light adaptation is accelerated relative to the rate in H_2O (Lanyi, unpublished experiments), while the later reactions in the photochemical cycle are retarded [67, 69], the $bR_{560} \rightarrow bR_{570}$ transition may involve some of the earlier reactions in the cyclical process. Light adaptation may consist of the isomerization of retinal. Light-adapted bacteriorhodopsin was found to contain *trans* retinal, but dark-adapted pigment contained a mixture of *trans* and 13-*cis* retinals [74] or only 13-*cis* retinal [76]. No physiological function for light-dark adaptation has been proposed.

III. Light-induced Development of Protonmotive Force

The chemiosmotic hypothesis [77-81] views energy-transducing membrane proteins as enzymes which selectively take up some of their reactants and/or release some of their products on different sides of the membrane. Since during the course of such a reaction the products accumulate and/or the reactants are depleted from their compartments, the changes in chemical potential in the bulk phases across the membrane (including electrical potential changes if the species involved carry a net charge) enter into the energetics of the overall process. Thus, additional energy is needed whenever gradients are produced, and energy is released whenever gradients are dissipated in the course of the reaction. It follows, therefore, that chemical and electrical gradients can serve as reservoirs of energy when the appropriate molecular machinery is available to first generate and then utilize them. The Mg^{2+}-stimulated ATPase in mitochondria and bacteria is an example of such a proposed machinery; here the hydrolysis and synthesis of ATP is accompanied by apparent movement of H^+ across the membrane. Another type of energy-transducing device does not involve a chemical reaction at all. Rather, such membrane components couple the cross-membrane movement of two or more chemical species. Systems which are thought to function in this manner include those for active transport, the transmembrane movements of amino acids, sugars, etc., all of which are dependent on the coupled movements of H^+ or Na^+ [78, 80-82].

It follows from the concept described above that transmembrane gradients will serve as the means for energy transfer between two transducing components if these are involved in the movement of the same mobile species. Thus, the physiologically important consequence of the chemiosmotic hypothesis is that one membrane process may drive another through the generation and discharge of transmembrane gradients. Exergonic reactions, which take place in membranes (e.g., terminal oxidation), usually serve to generate gradients (e.g., for H^+), while endergonic reactions usually act to dissipate these gradients (e.g., ATP synthesis, active substrate transport). Since the reactions involved in energy transduction are often reversible, the flow of energy can be reversed under certain circumstances.

Two parameters are associated with the gradients developed: the motive force and the capacity of the system. Since the role of H^+ gradients in membrane energetics is widely recognized, the term "protonmotive force" has become familiar. It refers to the difference in chemical potential for protons in the two compartments separated by the membrane, and is given by the relationship

$$\Delta p = \Delta\psi - Z\,\Delta pH \qquad (1)$$

where Δp is the protonmotive force, $\Delta\psi$ is the electrical potential difference across the membrane, Z is a constant of about 58 mV at room temperature, and ΔpH is the pH difference across the membrane. The motive force for a chemical species, i, might be analogously:

$$\Delta i = n\,\Delta\psi + Z \log \frac{a_i^{II}}{a_i^{I}} \qquad (2)$$

where Δi is the motive force, n is the net charge of species i, and a_i^{I} and a_i^{II} are the activities on the two sides of the membrane.

Equations 1 and 2 explicitly state that the electrical and chemical terms can alone or together provide the driving force for the movement of the mobile species.

The capacity of a system is given by the quantity of transported mobile species which yields a given size gradient [77]. For the electrical gradient, the capacity depends on the number of net charges per molecule and the electrical capacitance of the membrane (usually determined by the dielectric properties of the lipid bilayer). For the chemical gradient of H^+, the capacity is given directly by the buffering power of the two phases on either side of the membrane. It is evident that buffering increases the capacity for H^+, since in the presence of strong buffer, even large quantities of H^+ translocated will yield a small ΔpH across the membrane. The capacity of a system is particularly important when the chemical entity transported is a normal intracellular component, such as Na^+ or K^+. The higher the intracellular concentration of these ions, the higher the capacity of these ions and the more energy can be stored in the form of gradients.

Experimental evidence to suggest that the above principle indeed applies to energy conservation in membrane systems has been obtained by a large number of investigators. However, the simplicity and the elegance of the *Halobacterium* system makes it an ideal tool to study these effects.

That purified bacteriorhodopsin responds to illumination by translocating protons across the membrane has been shown by incorporating sheets of purple membranes into planar films (black lipid membrane) and into small, enclosed compartments (liposomes). It was found that when purple membranes were sonicated with soybean lipids the resulting preparations exhibited H^+ uptake during illumination [83, 84]. Electron microscopy confirmed that, for reasons as yet unknown, the purple membrane is incorporated into the liposomes preferentially outside-in, i.e., in the opposite sense than it is oriented in intact cells. In these experiments the quantity of H^+ lost from the medium is up to 20 times the number of bacteriorhodopsin molecules present [85]. It may thus be concluded that the pH change observed is not simply the consequence of ionization changes in bacteriorhodopsin, but must be due to net transport of H^+, corresponding to many photochemical cycles and vectorial deprotonations and protonations for each pigment molecule.

The functional role of bacteriorhodopsin in establishing and maintaining an electrochemical gradient for protons was further demonstrated when Racker and Stoeckenius [83] incorporated mitochondrial ATPase, in addition to purple membrane, into the liposomes and found light-induced phosphorylation of ADP. Purple membrane-containing liposomes have since served in many cases as easily energized systems for studying isolated and reconstituted proteins catalyzing energy-requiring membrane processes [85-87]. Racker and Hinkle [88] and Kayushin and Skulachev [84] later showed that the H^+ uptake by the liposomes is accompanied by the uptake of the membrane-permeant anion tetraphenylboron, a result which can be regarded as evidence for the development of an electrical potential across the liposome membrane (inside positive).

Thin lipid films [89-92] and planar membranes made from a plastic polymer [91], separating two aqueous compartments where electrical measurements can be made, have also been used to study light-induced H^+ translocation by bacteriorhodopsin. When purple membrane is present in such partitions, photopotentials of 100-300 mV have been observed. Asymmetrical orientation of the purple membrane sheets may come from statistical chance [92], since the number of membrane sheets involved is small. Addition of up to 100 mM $CaCl_2$ to one of the compartments causes a large increase in photopotential, consistent with other observations which suggest that some divalent cations inhibit bacteriorhodopsin from one membrane face (the H^+-ejection side) but not the other (Packer, personal communication). Adding La^+ appears to inhibit the H^+-uptake side [90]. An ingenious way to incorporate preferentially oriented purple membranes into planar films involves adding liposomes, prepared as described above, into one of the compartments, together with millimolar concentrations of $CaCl_2$. In a time-dependent manner, presumably due to association of the liposomes carrying purple membranes with the planar membrane, the ability of the system to produce photopotential will increase [89, 90]. Hwang et al. [93] were able to orient purple membrane sheets at air-water interfaces and obtained potentials on illumination.

These experiments demonstrate the proton-translocating function of bacteriorhodopsin and indicate that purified purple membranes not only appear structurally similar to purple membranes in intact cells, but seem to be functionally equivalent, with no obvious impairment of activity resulting from exposure to lowered salt concentration during isolation.

More information about the physiology of *H. halobium* cells can be obtained by studying the behavior of cell envelope vesicles. These contain both red and purple membranes can can be prepared from cells by mechanical breakage [94]. The membrane orientation in the envelope vesicles corresponds largely to that in whole cells [95, 96]. The fraction of inside-in and inside-out vesicles can be estimated by determining NADH-menadione reductase activity in the presence and absence of Triton X-100. Since this enzyme is located on the surface of the red membranes normally facing inside [18], the accessibility of NADH to the enzyme should give a direct percentage of the two vesicle populations [95]. Typically, vesicle preparations contain 85-90% inside-in envelopes. Another indication of the proper orientation of these envelope vesicles comes from pH measurements following short flashes of light [71]. These experiments were carried out with envelope vesicles and with bacteriorhodopsin-containing liposomes in order to ascertain the vectorial nature of deprotonation and protonation during the photochemical cycle. The pH changes occurring after single flashes of 1 msec duration were followed by measuring absorbance changes of a pH-indicator dye. As expected from the direction of membrane orientation and earlier flash spectroscopy, envelope vesicles showed proton *release,* with a time-constant of less than 1 msec, and liposomes showed proton *uptake,* with a time-constant of about 10 msec. These results both confirm the membrane orientation in the two systems and provide additional evidence to the scheme of vectorial H^+ loss and H^+ gain discussed in the previous section.

A critical discussion of the methods used to measure the size of pH gradients is outside the scope of this review. In *H. halobium* envelope vesicles, proton release can be conveniently determined by direct pH measurements in the extravesicle medium, since large pH changes occur during illumination [95, 97-99]. As in liposomes, the number of protons which appear in the medium is much larger than the number of pigment molecules, indicating that there is net translocation of H^+. In addition, heavily buffered vesicles release two to three times more H^+ during illumination [99], suggesting that H^+ release is normally accompanied by a pH rise inside the vesicles. The number of protons translocated at equilibrium is dependent on pH: optimal H^+ release is near pH 3.5-4 and is greatly diminished at and above pH 6 [100]. A similar effect between pH 5.5 and 7.5 is noted in *Escherichia coli* membranes, where proton translocation is caused by substrate oxidation [101]. In *H. halobium* cell envelopes at least, the pH dependence may be caused by direct pH effects on bacteriorhodopsin [102], by slowing down of proton translocation when this process results in high pH (i.e., lowered H^+ concentration) inside the vesicles, or by possible lack of counterion permeability at higher pH, which would allow the development of higher electrical potentials but lower pH gradients. The latter does not seem to be the case, since at higher pH the electrical potential also appears to be less [98], and cation permeability rises rather than falls with pH [103]. Drachev et al. [90] found that in planar membranes, photopotentials from bacteriorhodopsin were diminished when H^+ translocation occurred against a pH gradient.

Electrical potential across the vesicle membrane during illumination has been measured using the fluorescent dye 3,3′-dipentyloxadicarbocyanine(di-O-C_5) [97-99]. As shown earlier by Sims et al. [104], membrane potential (negative inside) will cause this cationic dye to accumulate in the vesicle interior (in the halophilic system mainly on the interior membrane surface), resulting in aggregation-dependent fluorescence quenching. Diffusion potentials for K^+ with added valinomycin were used to calibrate the observed changes: fluorescence decrease was found to be linear with Nernst potential to about -100 mV, with a slope of 0.33% per mV [98]. Both electrical potential, measured in this manner, and pH gradient, determined from exterior pH changes and calculated from interior buffering capacity, are dependent on the presence of K^+. The changes in these two components of the protonmotive force (see Equation 1), are reciprocal. Thus, the pH change is greater (1.5-fold) and $\Delta\psi$ is smaller (threefold) when KCl is the salt present rather than NaCl [98], consistent with the idea that the vesicle membrane is more permeable to K^+ than to Na^+, and the net amount of H^+ displaced at equilibrium, being limited by $\Delta\psi$, is more extensive if a counterion, such as K^+, is available. In NaCl solution the electrical potential reaches -100 to -120 mV, a value confirmed by the accumulation of $[^3H]$ dibenzyldimethyl ammonium ion [99] by the vesicles under these conditions. The pH difference which develops in NaCl solution at an exterior pH of 6.0 is estimated at about 1.8 pH units. The total protonmotive force developed under these conditions is thus somewhat over -200 mV [98].

Measurement of the protonmotive force in whole cells of *H. halobium* during illumination is considerably more difficult than in envelopes, because of preexisting gradients of uncertain size for H^+, K^+, and Na^+, and because of an unexpected kinetic complexity for the light-induced H^+ movements. Both of these questions will be discussed in detail in the following sections. Steady-state values for ΔpH and $\Delta\psi$ have been obtained for cells, however, both in the dark and during illumination [105-108]. These studies used $[^{14}C]$ 5,5′-dimethyl-2,4-oxazolidinedione distribution for pH-gradient determinations and $[^3H]$ triphenylmethyl phosphonium ion ($TPMP^+$) distribution for potential determinations (for a review of these methods, see Maloney et al. [109]). The values obtained by Bakker et al. [107] and Michel and Oesterhelt [108] are, by and large, in agreement: at pH 6.5-6.6 in the dark the cells exhibit pH gradients of 0.45-0.6 units (inside alkaline), and during illumination the gradient increases by 0.12-0.16 units. Wagner and Hope [106] found considerably larger values for the pH gradients. The pH gradient before illumination, and to a smaller extent the light-induced increase, appears to be dependent on the pH, showing considerable decrease between pH 5.5 and 7.5 [107]. Potential gradients are much less pH dependent. Michel and Oesterhelt [108], Bakker et al. [107], and Bogomolni [110] estimate $\Delta\psi$ at -100 to -110 mV in the dark and about -120 to -125 mV wile illuminated. These measurements are subject to some uncertainty because of the possibility that $TPMP^+$ may be bound to the membranes by mechanisms unrelated to electrical potential. The data obtained for *H. halobium* is remarkedly similar to those reported by Ramos et al. [101] for *E. coli* vesicles.

There are various agents which are thought to increase membrane permeability for various cations. The effects of these chemicals on energy coupling in membranes have been of major importance in debating the validity of the chemiosmotic hypothesis [80, 111]. Although these issues are now decided largely on other grounds, the agents (ionophores) still play a useful role in studies of membrane bioenergetics. The uncouplers FCCP, CCCP, etc., which in many systems have been shown to function as proton carriers, discharge the light-induced pH gradients in both whole cells of *H. halobium* [102, 105, 107], and in envelope vesicles [95, 99]. Nigericin, which is thought to facilitate the electrically neutral exchange of K^+ and H^+, has also been shown to discharge ΔpH in envelope vesicles [100]. Valinomycin- and gramicidin-induced cation influx eliminates light-dependent membrane potential in envelope vesicles [98, 99], although gramicidin has no effect on whole cells (Bogomolni, personal communication).

The effects of ionophores on the light-induced gradients in *H. halobium,* are, in sum, similar to those in other systems.

IV. Transmembrane Movements and Gradients of Na^+ and K^+

Freezing-point depression determinations of cell pastes [7] and chemical analyses [8-10] have suggested that the internal salt concentration of extremely halophilic bacteria is very high. Results with what in retrospect appear to be resting (anaerobic, nonmetabolizing, etc.) cells have indicated that the intracellular concentration of K^+ is 3-5 M and that of Na^+ is 0.5-1.2 M. The K^+/Na^+ ratio in actively growing cells may be much higher. An intracellular ratio K^+/Na^+ of >1 is generally found in both prokaryotes and eukaryotes. The gradients for these cations generally have to be achieved with large expenditures of energy, since the capacity of cells for these ions is high, and so the quantity of cations moved across the cell membranes for a given size gradient can be much greater than for H^+. This is even more so for the halobacteria, which contain almost two orders of magnitude more salt than other cells. This means by which Na^+ and K^+ gradients are developed thus become of major importance in the physiology of these bacteria. A related question is the passive permeability of the membranes to Na^+ and K^+, since the movements of these ions across the membranes determine both the capacity for osmotic work in nonenergized cells and the relative size of ΔpH and $\Delta\psi$ during energization.

Several observers [9, 10, 112] have reported that even starved, poisoned, or anaerobic *Halobacterium* cells can retain K^+ for several days against large gradients. Ginzburg and Ginzburg [113, 114] suggested that to account for this phenomenon it must be assumed that intracellular K^+ is tightly bound. Since these authors also claim that the cell membrane is permeable to very large molecules [115], their model of ion movements in these cells is based on cation exchange between the cytoplasm and the external medium due to binding, rather than to membrane transport. This model is probably wrong, since in cell envelopes devoid of cytoplasm large gradients for cations can arise (as discussed below). Furthermore, no binding of K^+ by the cytoplasm of *H. cutirubrum* has been detected by direct measurement [10]. Lanyi and Hilliker [103] recently reported that the slowness of K^+ efflux from *H. halobium* cells can be fully explained by the measured rate of passive K^+/Na^+ exchange in cell envelope vesicles. It appears that one need not postulate intracellular binding for K^+ to account for the ability of cells to retain this ion.

The size of the K^+ gradient in actively growing *Halobacterium* cells is 500-1,000 between inside and outside [11, 12]. A simple model would have K^+ in equilibrium with the membrane potential. In such a model, K^+ influx would be driven directly by $\Delta\psi$, either by unfacilitated or facilitated transport, or both. It is clear, however, that a membrane potential of about –180 mV would be required for a 1000-fold gradient of K^+ (Equation 2), a value much higher than observed in either cells [106-108, 110], or in envelope vesicles [98]. Although one might envision a K^+ accumulation system based directly on ATP hydrolysis, not subject to the above objection, another model will be suggested below, where the entire protonmotive force ($\Delta\psi$ and ΔpH) is utilized for driving K^+ influx.

If the permeability of K^+ in *H. halobium* membranes were much greater than that of other ions, then the K^+ gradient, once established, would represent a large reservoir of energy for the cells. Under such circumstances the chemical gradient of K^+ would maintain a diffusion potential of a size determined by the relative permeabilities of various ions, and described by the Goldman equation [116]. This diffusion potential would be dissipated only by massive cation exchanges across the cell membrane, which, as discussed above, are very slow. The relative permeabilities for various ions in *H. halobium* membranes are not yet known, although some evidence does exist to suggest that K^+ is more permeable than Na^+ [98]. In contrast to other systems, added valinomycin increases K^+ fluxes only two- to threefold [100]. Additionally, electrical potentials of about -100 mV have been reported, on the basis of $TPMP^+$ distribution, in cells kept in the dark, with respiration inhibited [107]. Such measurements, while suggesting the existence of a K^+ diffusion potential, are made inaccurate by an unknown amount of $TPMP^+$ bound nonspecifically to cellular components. Thus, only circumstantial evidence is available at this time to support the idea that the K^+ gradient in halobacteria is in equilibrium with a membrane potential.

Lanyi and coworkers [97, 99] found that when *H. halobium* cell envelope vesicles or cells are illuminated, Na^+ efflux occurs and apparently very high Na^+ gradients (outside $\gg$ inside) are developed. This Na^+ gradient will drive all the active amino acid transport systems, as discussed in Section V. Since under these conditions Na^+ moves across the membrane *against* its electrochemical potential, the energetics of Na^+ transport are very different from those of K^+ transport. Lanyi and MacDonald [99] observed that when Na^+ was included in the envelope vesicles the pH changes measured in the medium during illumination were not as simple as when only K^+ was present in the vesicle interior. Depending on the amount of Na^+ inside the vesicles, there was an initial period of smaller pH change, followed by an increase of ΔpH to the value usually found in the absence of Na^+. The lengths of these initial periods roughly coincided with the time required to deplete the vesicles of ^{22}Na. It appears, therefore, that the efflux of Na^+ is accompanied by an influx of H^+, decreasing the size of the pH gradient. The simplest model to account for this result if H^+/Na^+ exchange or in Mitchell's terminology, *antiport* [78]. When the experiment was performed at increasing initial pH values, between 6 and 7.5, the Na^+-dependent influx of H^+ during illumination became more pronounced. The pH changes under these conditions are quite complicated and are shown in Figure 2: an initial acidification of the external medium is followed by a rise in pH, leading to a transient *reversal* of the pH gradient. This reversal suggests that the H^+/Na^+ antiport cannot proceed with a stoichiometry of 1:1, because such a process would be driven only by ΔpH and thus would be abolished as ΔpH approaches zero. On the other hand, at higher H^+/Na^+ stoichiometries, the exchange would be electrogenic and be driven also by the electrical potential. The potential in such a model could serve as the driving force required to balance the system at a reversed pH gradient. A stoichiometry higher than 1:1 is suggested also by the kinetics of the pH changes in Figure 2, which indicate that H^+ influx is not linearly dependent on ΔpH, but is related by a higher power function.

The electrical potential during the period of H^+/Na^+ exchange was measured with the fluorescent dye di-O-C_5, discussed in the previous section. It was found that Na^+ efflux is accompanied by greatly increased potentials, which decrease as the intravesicle Na^+ is depleted [99]. Since proton translocation by bacteriorhodopsin is almost certainly regulated by the internal pH, the proton ejection should be, in effect, reactivated by the H^+ influx during Na^+ efflux, and the internal H^+ concentration should thereby be kept low. The net result of the various fluxes, according to this model, is the continued removal of Na^+ (and positive charges) from the vesicles, giving rise to the higher membrane potential observed. Ultimately Na^+ would be replaced by K^+, whose influx is energized by the membrane potential. Potential in this model is limited primarily by the passive diffusion of K^+ rather than that of Na^+. Calculations from data of Kanner and Racker [100] indicate that K^+ uptake should continue linearly up to and beyond a gradient of 20 (assuming identical behavior for K^+ and Rb^+). Figure 3 illustrates the model [94, 99] which incorporates these cation movements. An interesting feature of this model is that it predicts positive feedback between active Na^+ efflux and passive Na^+ influx. Oscillations of the electrical potential are indeed observed when small amounts of Na^+ are included in the vesicles [99]. The light-induced efflux of Na^+ from *H. halobium* envelopes and many of the associated effects described above have been recently confirmed by Eisenbach et al. [117]. These authors have also demonstrated ΔpH-induced Na^+ flux and Na^+ gradient-induced H^+ flux in the dark.

The biphasic pH changes, which occur when envelope vesicles containing Na^+ are illunimated, are not seen during a second illumination, unless an interval of several hours is allowed between the two lighting periods [99]. The slow return of the Na^+-dependent effect is consistent with the depletion of the vesicles of Na^+ and the slow exchange of K^+ and Na^+ during the intervening dark time [103].

The net effect of the sustained illumination of envelope vesicles (and pre-

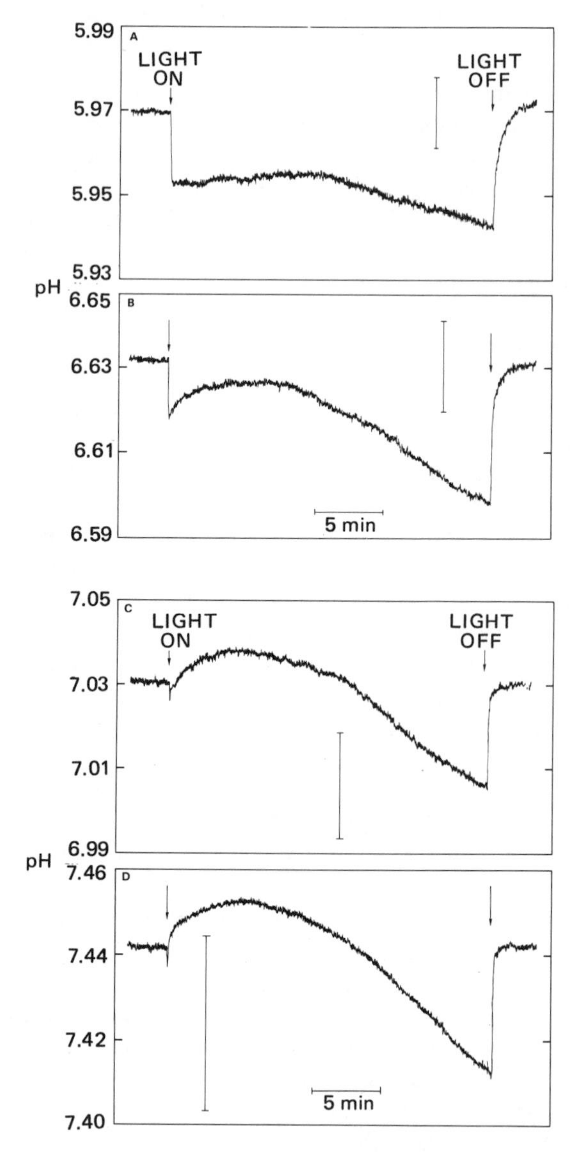

Figure 2 Changes in extravesicle pH during illumination of *H. halobium* cell envelope vesicles. The vesicles were loaded with and suspended in 2.0 M KCl + 1.0 M NaCl, containing 0.2 mM phosphate. After the initial pH had been adjusted, the vesicles were allowed to equilibrate for 30-40 min in the dark. Protein concentration: A, 0.2 mg/ml; B-D, 0.4 mg/ml. Vertical bars: pH changes after addition of 5 nanoequivalents/ml HCl. When the same experiment was carried out in 3.0 KCl, the acidification of the medium was complete in <1 min from the start of illumination, with no further changes until the light was turned off. Reproduced with permission from Lanyi and MacDonald [99].

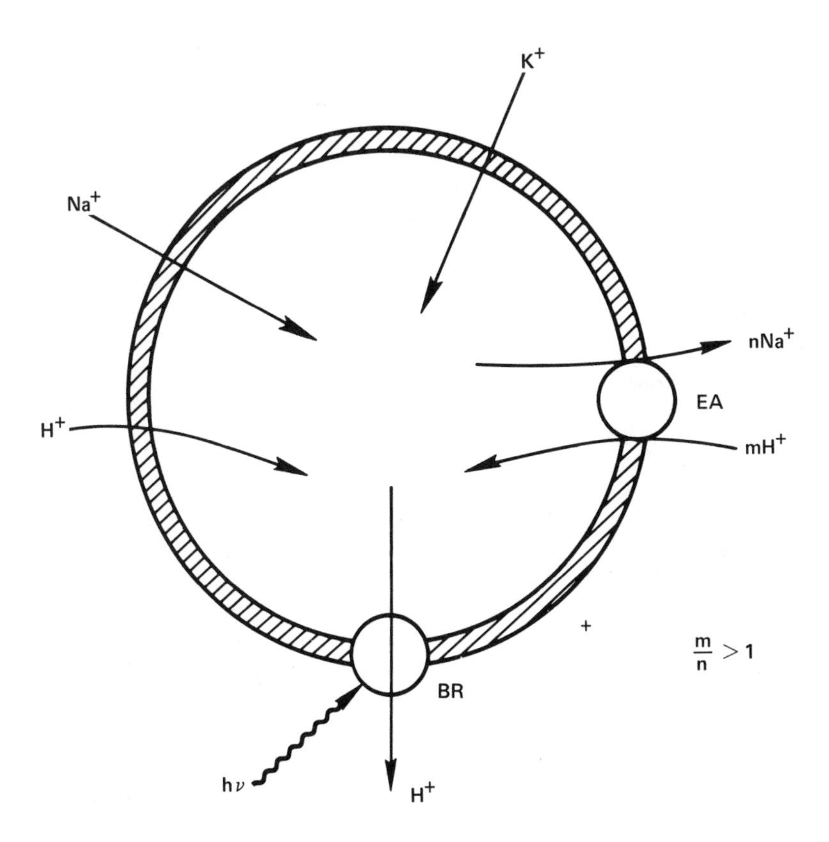

Figure 3 Scheme of cation movements in *H. halobium* cell envelope vesicles. BR, bacteriorhodopsin, EA, electrogenic H^+/Na^+ antiporter. Fluxes of K^+, as well as H^+ and Na^+, other than those facilitated by BR and EA, represent passive ion movements (in the direction of their electrochemical gradients). According to the model, BR captures light energy and gives rise to pH and electrical gradients. These are utilized by EA in affecting Na^+ efflux. The stoichiometry of the ion fluxes through EA are unknown but must be greater than 1. Reproduced with permission from Lanyi and MacDonald [99].

sumably whole cells of *H. halobium* also) is thus seen to be the replacement of internal Na^+ with K^+, resulting under physiological conditions in gradients for both Na^+ and K^+. It is important to realize that according to the model in Figure 3, K^+ influx is driven directly by the electrical potential, arising both from H^+ efflux and from Na^+ efflux. Since Na^+ efflux in this model is energized by ΔpH as well as by $\Delta\psi$, both components of the proton motive force are thereby available to drive K^+ influx. Therefore, the chemical gradient for K^+ should be in equilibrium with the entire magnitude of the protonmotive force, rather than only with its electrical component. Since, in envelopes, the former is near -190 mV while the latter is only about -90 mV, and in whole cells the corresponding values are -150 to -170 mV and -100 to -110 mV, the H^+/Na^+ antiport model can account for nearly a 1000-fold K^+ gradient. However, virtually all that is known about Na^+ and K^+ fluxes during illumination is from experiments with envelope vesicles, and other mechanisms of K^+ accumulation and energy coupling, perhaps involving ATP hydrolysis, are not ruled out in whole cells.

V. Light-energized Amino Acid Transport

In most bacteria the active (concentrative) uptake of amino acids and sugars is energized either (1) by the electron transport system or (2) by the hydrolysis of ATP [80, 118]. For transport of the first category the source of energy has been shown to be the H^+ gradient. In such transport it is generally assumed that the transmembrane movement of the substrates is obligatorily coupled to the movement of H^+ in the same direction, down its electrochemical gradient–i.e., by *symport* [78]. Transport of this type is obtained not only in whole cells but also in cell envelope vesicles, which contain only the cell membrane [119, 120]. In at least one case the protein responsible for active transport has been solubilized with detergent, partially purified, and reconstituted with lipids [121]. The transport of substrates by means in the second category, which involves ATP hydrolysis but not protonmotive force, is more complex. In many of the ATP-dependent transport systems, soluble factors have also been implicated–the "shockable proteins" [118] which have been isolated and show binding of the substrates [122].

Cell envelope preparations from *H. halobium,* which contain purple membrane, transport 19 of the 20 commonly occurring amino acids [123, 124]. Active transport in these vesicles can be energized either by illumination and the resulting chemical and electrical gradients, or by these gradients when they are produced by other means, in the dark. In the envelope vesicles the involvement of ATP synthesis or hydrolysis in energy coupling must be minimal since the envelopes neither contain ATP, or ADP and phosphate, nor produce ATP in measurable quantities when the appropriate substrates are present [97] (Hubbard, personal communication).

As in other bacterial systems (for a review see Ref. 125), transport groups common to several amino acids have been identified in *H. halobium* [123]. Amino acids within such transport groups competitively inhibit the transport of each other, presumably because of the existence of common transport carriers which serve to bind and translocate the substrates. In *H. halobium* the groups are arginine, lysine, histidine; glutamine, asparagine; aspartic acid; glutamic acid; threonine, alanine, serine, glycine; leucine, isoleucine, valine, methionine; tryptophan, phenylalanine, tyrosine; and proline [123]. Kinetic analysis of the components of those transport groups which contain more than one amino acid is too complex to interpret at this time. A survey of maximal transport rates and substrate concentrations at half-maximal velocity (K_m) under various conditions has been made [123].

With the exception of glutamate, the active transport of all the other 18 amino acids can be induced by either a chemical gradient of Na^+ (outside > inside), artificially arranged in the dark by appropriately loading the vesicles or created during illumination as described above, or by an electrical potential, originating from K^+ diffusion in the presence of valinomycin or from illumination [123]. Since both chemical and electrical components of the Na^+ gradient will drive the transport of these amino acids in the absence of other sources of energy, it has been concluded that the fluxes of Na^+ and the amino acids are coupled and thus, in *H. halobium* amino acid gradients are built up at the expense of the Na^+ gradient. The existence of specific transporter proteins, which facilitate the coupling of the two fluxes, is implicit in this model. It is uncertain whether H^+ plays any role in these translocations but the possibility cannot be excluded at present.

Leucine appears to be a representative amino acid, and its transport properties in *H. halobium* have been studied in detail. Figure 4 shows the course of uptake of [^{14}C] L-leucine during illumination. It is apparent that little transport occurs in the dark and illumination causes rapid influx of the amino acid. Leucine concentration gradients of 2-300 are obtained at saturating light intensities ($1\text{-}2 \times 10^6$ ergs cm^{-2} sec^{-1}). The action spectrum for the effect closely resembles the absorption spectrum of bacteriorhodopsin [95]. When the light is turned off, or the membrane potential is abolished with $TPMP^+$, leucine rapidly exits from the vesicles. That this is carrier-mediated transport is indicated by the fact that (1) the process is saturable, with a K_m of about 1×10^{-6} M for leucine, and (2) the transport is stereospecific for L-leucine. Energy coupling involves primarily the electrical potential, since permeant cations (such as $TPMP^+$) completely abolish transport, while buffering the vesicles internally, which diminishes ΔpH but increases $\Delta\psi$ [97], has an enhancing rather than an inhibiting effect. A diffusion potential of K^+, induced by adding valinomycin in the dark, does indeed cause transient leucine transport [95].

As found for the other amino acids [123], active transport of leucine can be obtained in the dark when a Na^+ gradient is arranged such that $Na^+_{out} > Na^+_{in}$ [95]. Since such transport of leucine is much less sensitive to the proton ionophore FCCP than the light-induced uptake, it was concluded that the accumulation of leucine under these conditions is energized directly by the chemical gra-

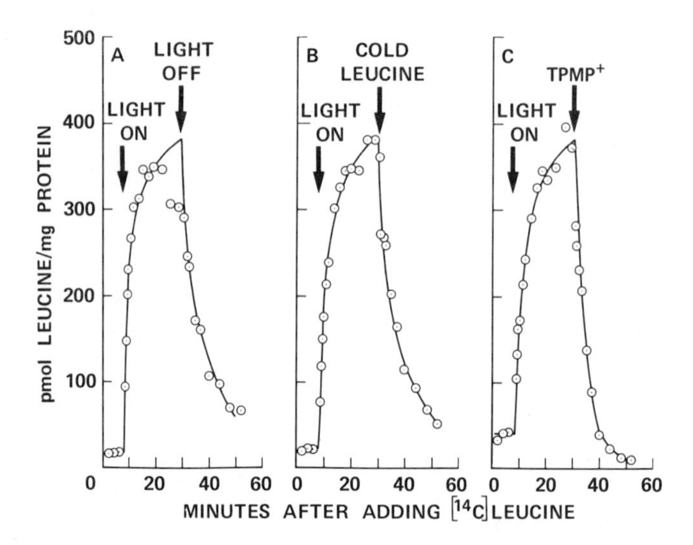

Figure 4 Light-dependent transport of leucine in *H. halobium* cell envelope vesicles. The vesicles were loaded with 3.0 M KCl and suspended in 2.5 M KCl + 0.5 M NaCl. A, no addition; B, when indicated a 100-fold excess of nonradioactive L-leucine was added; C, when indicated $TPMP^+$ was added to 1 mM concentration. Reproduced with permission from MacDonald and Lanyi [95].

dient of Na^+. This point has not yet been tested for other amino acids (except glutamate) and it is possible that for some a secondary gradient of H^+, which could arise through H^+/Na^+ antiport, may be the energizing force under these conditions.

It has been shown that the Na^+ gradient which is created druing illumination does play a role in energizing amino acid transport. Vesicles loaded with significant amounts of NaCl show biphasic uptake for leucine [96], the first phase corresponding to the effect of membrane potential alone, and the second larger phase to the effect of the chemical Na^+ gradient, which is produced only after several minutes. The first phase of transport is completely inhibited by valinomycin but the second, as expected, is only partially inhibited.

Transport of leucine requires the presence of Na^+ on the vesicle exterior [95], as does the transport of all other amino acids [123]. Maximal transport is observed when K^+ is also present in the medium. The K^+ requirement may reflect the fact that the chemical gradient of Na^+ provides a large part of the driving force for transport and this gradient could not arise if an appropriate counterion were not provided.

It is evident from these results that leucine accumulation is energized by both components of the electrochemical potential for Na^+, i.e., $\Delta\psi$ and the chemical gradient of Na^+. The role of $\Delta\psi$, at least, in driving leucine and proline transport in whole cells of *H. halobium* has been confirmed by Hubbard et al. [126].

The study of the efflux of amino acids from de-energized vesicles is complicated by the fact that during illumination the vesicles lose most of their Na^+ and acquire K^+, and these cations do not exchange rapidly in the dark [103]. The collapse of the pH difference and $\Delta\psi$ after the light is turned off does not reflect, therefore, the true state of energization for amino acid transport. The similarity of activation energy (temperature dependence) for influx and efflux of histidine and leucine above 20°C, but not below this temperature, suggests that at higher temperatures efflux may also be carrier mediated [123].

Glutamate transport in *H. halobium* is exceptional in two respects: (1) the transport does not proceed at all unless a chemical gradient for Na^+ is present, regardless of the existence of a membrane potential, and (2) the translocation of this amino acid is predominantly unidirectional [97, 127]. When Na^+ is included in the vesicles, glutamate accumulation shows a lag, roughly proportional to the amount of Na^+, and transport is seen only after the vesicles are depleted of Na^+ [97, 128]. Glutamate transport under these conditions appears coincidentally with the second phase of leucine transport described above [96]. Under lower intensities of light, glutamate transport is not only slower, but shows increased lags, which also suggest the cumulative nature of the driving force for this transport. When the light is turned off, ΔpH and $\Delta\psi$ collapse within 30-40 sec [97-99], but the Na^+ gradient survives for much longer times [98, 103]. As shown in Figure 5, the ability of the vesicles to transport glutamate persists for 15-20 min. The figure also shows that the rate of this postillumination transport depends on the length of previous illumination.

Since glutamate is not transported into vesicles which contain Na^+, even though the membrane potential is maximal (about -100 mV) under these conditions [99], one would suspect that the translocation of this amino acid is electrically neutral. Symport of the glutamic anion with Na^+, at a stoichiometry of 1:1, would fit this model. A diffusion potential of K^+ did not drive glutamate transport (Lanyi, unpublished experiments). When glutamate transport was induced in the dark by a Na^+ gradient, in KCl-loaded vesicles suspended in NaCl, $TPMP^+$ was seen to have little effect on the initial rate of transport [127], even though leucine transport under the same conditions was inhibited. This lack of effect by a $TPMP^+$ diffusion potential (interior positive) also suggests electroneutral translocation for glutamate. On the other hand, when glutamate transport was energized by illumination, i.e., by a Na^+ gradient arising from H^+/Na^+ antiport, both $TPMP^+$ [97] and valinomycin [96] proved to be inhibitory. Similarly, the collapse of the Na^+ gradient and the loss of glutamate-transporting ability of the vesicles were accelerated in the presence of $TPMP^+$ [97]. These

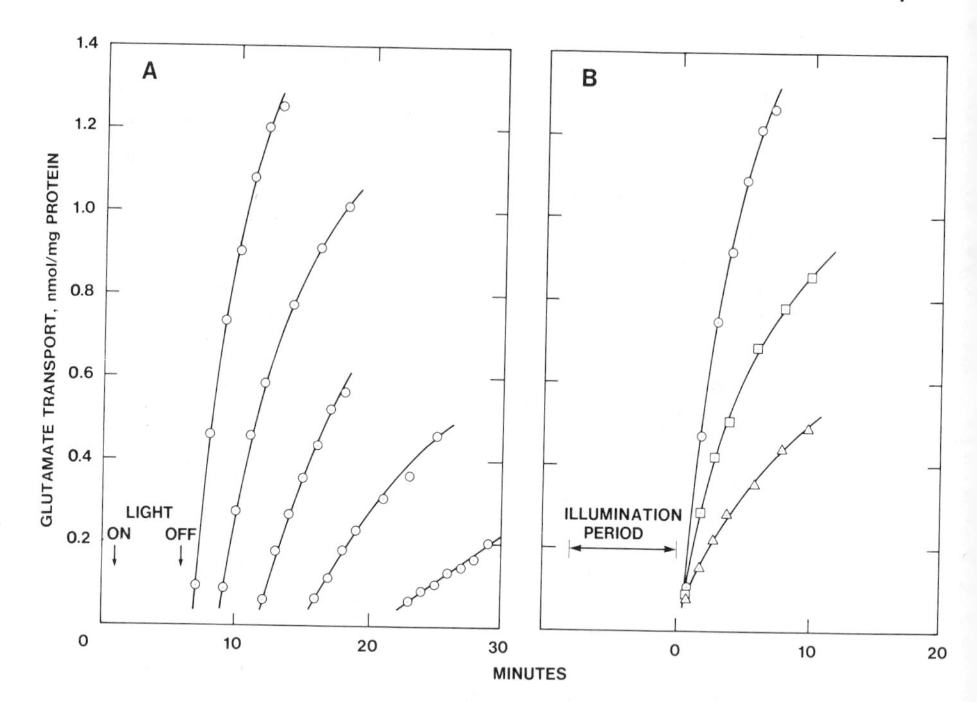

Figure 5 Light-dependent transport of glutamate in *H. halobium* cell envelope vesicles. Vesicles were loaded with approximately 2.9 M KCl + 0.1 M NaCl and suspended in 3.0 M NaCl at zero time. (A) Postillumination transport and the decay of the driving force in the dark. Illumination was for 5 min, and was followed by incubation in the dark of 1, 3, 6, 10, and 17 min, and addition of [^{3}H]glutamate, in separate experiments. (B) Dependence of postillumination transport of glutamate on the length of prior illumination. Illumination was for 0.5 min (△), 1 min (□), and 5 min (○), and was followed by incubation in the dark for 1 min and addition of [^{3}H]glutamate. The results show that maximal glutamate transport is established after several minutes of illumination and that the energized state persists for 15-20 min in the dark. Reproduced with permission from Lanyi et al. [97].

observations suggest that although glutamate transport is indifferent to the presence of membrane potential, its driving force, the chemical Na^+ gradient, arises by an electrogenic process, consistent with the model presented in Figure 3.

If glutamate transport is energized solely by the chemical component of the Na^+ gradient, as suggested above, then glutamate transport rates should provide information about the size of the Na^+ gradient produced during illumination. Lanyi et al. [127] found that a prearranged Na^+ gradient of >500 was required to energize glutamate transport in the dark at rates equivalent to those obtained during light-induced transport. Thus, the Na^+ gradient which arises through H^+/Na^+ antiport may be very large.

Unlike the case of the other amino acids transported, glutamate exit is extremely slow when the light is turned off, even when a large excess of nonradioactive glutamate is added [127]. This irreversibility is not caused by a requirement for Na^+ inside the vesicles for exit, since adding gramicidin, which facilitates exchange of Na^+ and K^+, or preloading the vesicles with Na^+ and glutamate, does not cause efflux either. In fact, glutamate efflux from the envelope vesicles was observed only when ionic conditions opposite to those favoring influx were arranged: Na^+ inside and K^+ outside the vesicles [127]. Thus, glutamate translocation seems to take place only toward the compartment containing K^+ but not Na^+. A result of the low reversibility of the translocation is that glutamate gradients of over 50,000 can be achieved, and as much as 80-85% of the glutamate is removed from the solution by even very small quantities of vesicles. Thus, even though the energy for transport is derived from the Na^+ gradient, the chemical gradient of glutamate is *not* in equilibrium with the gradient of Na^+. Rather, glutamate is trapped in the vesicles while the Na^+ gradient (and glutamate influx) is maintained by the light-initiated active processes. The molecular details of such irreversible transport systems are very difficult to envisage.

Na^+ gradient-dependent substrate transport is widely found in eukaryotic systems (for example, see Refs. 129 and 130), but in prokaryotes, with one exception [131], only H^+ gradient-driven transport has been described. A Na^+ requirement for transport is often attributed to a cofactor role for this cation (for example, see Refs. 132 and 133). It is easy to visualize how in extremely halophilic bacteria and perhaps in some marine microorganisms [134, 135] the presence of NaCl in the growth medium could encourage the evolution of Na^+ symport-based active transport systems. One may speculate, however, that the involvement of Na^+ gradients in energy metabolism may present advantages to other bacteria as well. The capacities of ΔpH and $\Delta\psi$ as energy reservoirs are inherently small, and these gradients collapse with the movement of a relatively small number of protons. A difference in Na^+ concentration across the cell membrane, on the other hand, usually involves a very large quantity of ions. Since substantial Na^+ gradients (outside > inside) seem to exist in several bacteria during active metabolism [136], Na^+-based transport may be a more general phenomenon than generally supposed.

VI. Light-energized Phosphorylation

The formation of ATP from ADP and phosphate is an endergonic process and in mitochondria and many bacteria the energy is supplied by the oxidation of reduced substrates. In such systems, and in others where the energy is provided by light, e.g., chloroplasts, there is evidence for the existence of an energized membrane state, which acts as the means for energy transfer. As described above, the principle of chemiosmotic energy coupling identifies this intermediate state as the electrochemical gradient of H^+ between the phases across the membrane. Powerful support for this concept was supplied by reconstitution experiments of Racker and Stoeckenius [83, 85], Yoshida and coworkers [86], and Ryrie and Blackmore [87] with partially or completely purified ATPase. These authors have shown that liposomes reconstituted with purple membranes and the ATPase phosphorylate when illuminated. From such experiments one is led to expect that light would be a source of energy for phosphorylation in *H. halobium* cells as well.

Photophosphorylation by *H. halobium* cells containing bacteriorhodopsin was first described by Danon and Stoeckenius [137]. These cells maintained ATP levels at a constant value during respiration. When the medium was made anaerobic or respiratory poisons were added, the levels of ATP dropped by about 70%. Illumination under these conditions restored the high intracellular ATP level. The synthesis of ATP, energized either by respiration or by illumination, is sensitive to the usual inhibitors of phosphorylation, e.g., dicyclohexylcarbodiimide, Dio-9, phlorizin [126, 137, 138], as well as to uncouplers [137, 139]. The energy for ATP synthesis in the absence of respiration is clearly provided by the protonmotive force developed during illumination. As in other bacterial systems [140, 141], in *H. halobium* cells ATP synthesis could be induced also by a combination of K^+ diffusion potential and pH gradient in the dark [138].

When intact cells are illuminated the pH changes observed in the medium are complex. The light-induced changes in pH reflect both the expected ejection of H^+ and at least two kinds of transient *influx* of H^+. The first of these reverse fluxes occurs only after long periods (hours) of incubation in the dark between illuminations, and resembles the pH changes found with Na^+-containing envelope vesicles [102, 110]. This type of pH rise during illumination, first reported as an anomalous response [102, 137] is probably due to H^+/Na^+ exchange, which depends on the reaccumulation of Na^+ inside the cells. The second type of light-induced H^+ influx occurs repeatedly, with every illumination, and this effect has been linked to ATP synthesis. The sequence of events under these conditions has been examined closely in a number of laboratories. When whole cells are illuminated the external pH first rises, indicating H^+ influx, then drops below the initial value, indicating subsequent H^+ efflux [102, 106, 138, 139]. At saturating light intensities the duration of the pH rise is less than a minute, although at lower intensities it is longer and does not completely reverse [102]. The pH rise is obtained only at initial pH values above 4.5 [102, 106]. concurrently with the pH rise, ATP is synthesized [102, 138, 139]. The kinetics of the pH change and ATP increase correspond quite well in some cases [102, 110] but poorly in others [142]. Nevertheless, because inhibitors of ATP synthesis, such as dicyclohexylcarbodiimide, abolish the pH rise [102, 138, 139], it has been proposed that the H^+ influx is associated with ATP synthesis. The simplest model was advanced by Oesterhelt [139] who attributed the H^+ uptake to the consumption of H^+ by the ADP + $P_i \rightarrow$ ATP reaction, known to amount to 1 mol at pH > 6.5 in other systems. Since at the high salt concentrations present in *H. halobium* cells the pK of the ionizable groups would be expected to change considerably, it is difficult to evaluate this proposal.

Bogomolni et al. [102] graphically resolved the pH changes during illumination into a rapid alkalinization component and a slower acidification component, and calculated the rate of the H^+ influx component as 2.9 mol H^+ per mol ATP. The light intensity dependence of this H^+ influx was, furthermore, found to be far lower than that of H^+ ejection. The possibility was considered, therefore, that the H^+ influx is not a primary, driven process, but represents the relaxation of a pre-existing gradient for H^+. Such a gradient was indeed demonstrated by Bogomolni et al. [102] who were able to cause its collapse with CCCP or nigericin, resulting in the loss of the light-dependent H^+ influx. Starvation of the cells also caused loss of the H^+ influx, while adding glycerol, an oxidizable substrate, restored it. The existence of a dark-state gradient for H^+ in *H. halobium* cells, maintained by respiration, is supported by measurements of ΔpH and $\Delta\psi$ [107, 108]. Light thus causes further H^+ ejection in these cells, but the H^+ influx is energized primarily by the pre-existing H^+ gradient. According to this model, the H^+ influx represents those protons whose movement down their electrochemical gradient provides the free energy for phosphate bond formation. However, the model is not complete without two added features [102, 110]. Since the main part of the energy for ATP formation is provided by the pre-existing H^+ gradient, poised presumably below a threshold value, the additional protonmotive force during illumination must act in nonlinear manner, as a gating potential. Furthermore, ATP synthesis within the first minute of illumination is seen to continue even though the pH gradient during this time becomes smaller than it had been before the illumination. Measurement of membrane potential under these conditions [110] indicates that the potential initially increases, but by only 10-15 mV. The total protonmotive force during the light-induced synthesis of ATP thus may *decrease* below the dark value. Because of this possibility, it is probably necessary to propose a hysteresis effect as well. Nonlinearity of the ATP/ADP ratio with protonmotive force follows from calculations based on the chemiosmotic hypothesis, and in bacterial cells a threshold value of about -200 mV was indeed found for ATP synthesis [140, 141]. In explaining the hysteresis effect, Bogomolni [110] favors the possibility that the ATPase complex of *H. halobium* contains a regulatory subunit, which would be activated by internal pH or by the increase in protonmotive force.

When illumination is ended the levels of intracellular ATP return to their previous lower value. The external pH changes after the light is turned off are the reverse of those at the beginning of the illumination: first a decrease in pH, then a rise until the preillumination pH is reached. Since this initial decrease in pH is abolished by dicyclohexylcarbodiimide, Bogomolni et al. [102] attribute it to H^+ efflux produced by the hydrolysis of the ATP produced during the illumination period.

VII. Bacteriorhodopsin and the Respiratory Chain as Alternate Sources of Energy

Halobacteria are obligate aerobes [3] and contain cytochromes of the *b* and *c* type, an *o*-type cytochrome oxidase [32-34, 143], and respiratory-chain-linked dehydrogenases [24, 30, 31, 144]. Substrates, which are rapidly oxidized by membrane fragments, include NADH and α-glycerophosphate but not succinate [143].

Since the membranes are not permeable to NADH, and the appropriate dehydrogenase is located on the inside membrane surface [18], inside-in envelope vesicles do not oxidize this substrate. Dimethylphenylene diamine [128], tetramethylphenylene diamine [25, 145], and ascorbate, with and without phenazine methosulfate or ferrocyanide [145], on the other hand, are oxidized rapidly by the vesicles. Inhibition of oxygen uptake and of absorbance change of the phenylene diamines by cyanide and by azide indicates that cytochrome oxidase is involved in the electron transfer.

Bogomolni et al. [102] have shown that the pH changes, which occur when oxygen is introduced into an anaerobic cell suspension, are similar to those occurring during illunination. It seems reasonable, therefore, that protonmotive force could be produced in *H. halobium* either by respiration, as in other systemsm or by bacteriorhodopsin. Cell envelope vesicles transport glutamate when various respiratory substrates are added [128, 145]. The dependence of this transport on extravesicle Na^+, its inhibition by intravesicle Na^+, and other properties are so similar to the light-driven process that Belliveau and Lanyi [128] concluded that the entire sequence of events, leading to the production of the Na^+ gradient, are induced by respiration in much the same way as by the action of light on bacteriorhodopsin.

Oesterhelt and Krippahl [146] described the inhibition of O_2 uptake by *H. halobium* cells when illuminated. This effect is akin to respiratory control observed in other systems but does not necessarily involve ADP or phosphate since nonphosphorylating envelope vesicles show it also (Belliveau, Bogomolni, and Lanyi, unpublished experiments). The photoinhibition of respiration is as much as 30%. It was found that at least 24 quanta of light were absorbed by bacteriorhodopsin when the consumption of one molecule of oxygen is prevented [146]. Since 12 protons would be translocated per molecule of oxygen consumed, it was argued that the quantum yield of the light-induced proton translocation may reach 0.5, in remarkable agreement with later estimates from more direct measurements.

Bogomolni et al. [102] reported that uncouplers diminished the light inhibition of respiration but $TPMP^+$, a membrane-permeant cation which abolishes the electrical potential, did not. Whether respiration is controlled through tight coupling with proton translocation, so that the gradient of H^+ developed during illumination reverses electron transport, or through changes in intracellular pH, is not yet clear.

VIII. Conclusions

The chemiosmotic hypothesis was first formulated by Mitchell about a decade before the discovery of bacteriorhodopsin. By the end of that decade an impressive amount of evidence had already been accumulated in favor of this concept of energy coupling (for a recent review, see Harold [81]). Nevertheless, the simplicity of the light-energy-transducing system in *H. halobium* and its experimental advantages have played a role in removing some objections to the hypothesis and thus must have served to convince many students of membrane energetics of its validity.

From the discussion above, it is evident that energy is conserved in *H. halobium* in various forms: in pigment-protein conformation; in the electrochemical gradient of H^+, Na^+, and probably of K^+; as well as in the gradients of transport substrates and in the phosphate bond of ATP. These have different time-constants and represent energy reservoirs of different size: (1) After absorption of a photon the photochemical states of bacteriorhodopsin store sufficient free energy to translocate a proton within 10 msec. The mechanism of the translocation and the way the energy is transferred to the protons in not yet known, but it is likely that a sequence of binding sites for H^+ are involved, with ordered changes in pK. (2) The translocation of protons results in a gradient for H^+, comprising both chemical and electrical components. The total size of the gradient is at least -180 mV, although its capacity must be relatively small. Thus, when illumination is ended, various coupled processes and passive leaks cause it to collapse within 1 min in envelope vesicles and within a few minutes in intact cells. (3) Antiport of H^+ and Na^+ results in a Na^+ gradient of similar size as the total protonmotive force, poised in the same direction. Complete collapse of the chemical gradient for Na^+ takes several hours and even its initial steep part persists for

15-20 min. (4) During the active efflux of Na^+ a large fraction of the energy stored is shifted to electrical potential, since with active H^+ extrusion the net effect is the removal of Na^+. This potential drives K^+ influx, presumably until the gradients of H^+, Na^+, and K^+ are in equilibrium with one another. Under physiological conditions this results in a large K^+ gradient, poised in the opposite direction from the gradients of H^+ and Na^+. (5) Coupled to the electrical and chemical components of the Na^+ gradient is the active transport of various amino acids. (6) Finally, illumination is accompanied by the synthesis of ATP. The formation of ATP has been linked with H^+ influx, suggesting that the H^+ gradient provides the energy required. In order to account for the results a complex ATP-synthesizing machinery, with gating and hysteresis, has been proposed.

The scheme described above for energy coupling in *H. halobium* admittedly leaves many questions unanswered. It is the hope of many investigators that answers will be easier to obtain with this organism than with others, and that in the process valuable insights into the general problem of energy transduction will be gained.

References

1. H. Larsen in *The Bacteria* (I. C. Gunsalus and R. Y. Stanier, eds.), Vol. 4, Academic, New York, 1963, pp. 297-342.
2. H. Larsen, *Adv. Microbial Physiol. 1:* 97 (1967).
3. R. E. Buchanan and N. E. Gibbons, (eds.), *Bergey's Manual of Determinative Bacteriology* 8th Ed. William and Wilkins, Baltimore, 1974, pp. 270-272.
4. G. A. Tomlinson and L. I. Hochstein, *Can. J. Microbio. 22:* 587 (1976).
5. M. Kelley, S. Norgard, and S. Liaanen-Jensen, *Acta Chem. Scand. 24:* (1970).
6. S. C. Kushwaha, M. B. Gochnauer, D. J. Kushner, and M. Kates, *Can. J. Microbiol. 20:* 241 (1974).
7. J. H. B. Christian and M. Ingram, *J. Gen. Microbiol. 20:* 27 (1959).
8. J. H. B. Christian and J. A. Waltho, *Biochim. Biophys. Acta 65:* 506 (1962).
9. M. Ginzburg, L. Sachs, and B. Z. Ginzburg, *J. Gen. Physiol. 55:* 187 (1970).
10. J. K. Lanyi and M. P. Silverman, *Can. J. Microbiol. 18:* 993 (1972).
11. M. B. Gochnauer and D. J. Kushner, *Can. J. Microbiol. 15:* 1157 (1969).
12. M. B. Gochnauer and D. J. Kushner, *Can. J. Microbiol. 17:* 17 (1971).
13. D. Abram and N. E. Gibbons, *Can. J. Microbiol. 7:* 741 (1961).
14. J. K. Lanyi, *Bacteriol. Rev. 38:* 272 (1974).
15. S. T. Bayley and D. J. Kushner, *J. Mol. Biol. 9:* 654 (1964).
16. H. Onishi and D. J. Kushner, *J. Bacteriol. 91:* 646 (1966).

17. J. K. Lanyi, *J. Biol. Chem. 246:* 4552 (1971).
18. J. K. Lanyi, *J. Biol. Chem. 247:* 3001 (1972).
19. W. Stoeckenius and R. Rowen, *J. Cell Biol. 34:* 345 (1967).
20. R. M. Baxter, *Can. J. Microbiol. 5:* 47 (1959).
21. A. D. Brown, *Bacteriol. Rev. 28:* 296 (1964).
22. D. J. Kushner, *J. Bacteriol. 87:* 1147 (1964).
23. D. J. Kushner and H. Onishi, *J. Bacteriol. 91:* 653 (1966).
24. J. K. Lanyi and J. Stevenson, *J. Biol. Chem. 245:* 4075 (1970).
25. M. M. Lieberman and J. K. Lanyi, *Biochim. Biophys. Acta 245:* 41 (1971).
26. M. M. Lieberman and J. K. Lanyi, *Biochemistry 11:* 211 (1972).
27. C. L. Marshall, A. J. Wicken, and A. D. Brown, *Can. J. Biochem. 47:* 71 (1969).
28. A. D. Brown and R. F. Pearce, *Can. J. Biochem. 47:* 833 (1969).
29. M. F. Mescher, J. L. Strominger, and W. S. Watson, *J. Bacteriol. 120:* 945 (1974).
30. L. I. Hochstein and B. P. Dalton, *Biochim. Biophys. Acta 302:* 216 (1973).
31. L. I. Hochstein, *Biochim. Biophys. Acta 403:* 58 (1975).
32. J. K. Lanyi, *Arch. Biochem. Biophys. 128:* 716 (1968).
33. K. S. Cheah, *Biochim. Biophys. Acta 180:* 320 (1969).
34. K. S. Cheah, *Biochim. Biophys. Acta 216:* 43 (1970).
35. M. Kates, in *Ether Lipids; Chemistry and Biology* (F. Snyder, ed.), Academic, New York, 1972, pp. 351-398.
36. M. Kates, B. Palameta, C. N. Joo, D. J. Kushner, and N. E. Gibbons, *Biochemistry 5:* 4092 (1966).
37. M. Kates, L. S. Yengoyan, and P. S. Sastry, *Biochim. Biophys. Acta 98:* 252 (1965).
38. S. C. Kushwaha, M. Kates, and W. G. Martin, *Can. J. Biochem. 53:* 284 (1975).
39. S. C. Kushwaha, J. K. G. Kramer, and M. Kates, *Biochim. Biophys. Acta 398:* 303 (1975).
40. T. G. Tornabene, M. Kates, E. Gelpi, and J. Oro, *J. Lipid Res. 10:* 294 (1969).
41. J. K. G. Kramer, S. C. Kushwaha, and M. Kates, *Biochim. Biophys. Acta 270:* 103 (1972).
42. W. Z. Plachy, J. K. Lanyi, and M. Kates, *Biochemistry 13:* 4906 (1974).
43. J. K. Lanyi, W. Z. Plachy, and M. Kates, *Biochemistry 13:* 4914 (1974).
44. J. K. Lanyi, in *Extreme Environments: Mechanisms of Microbial Adaptation* (M. R. Heinrich, ed.), Academic, New York, 1975, pp. 295-303.
45. W. Stoeckenius and W. H. Kunau, *J. Cell Biol. 38:* 337 (1968).
46. A. E. Blaurock and W. Stoeckenius, *Nature New Biol. 233:* 152 (1971).
47. D. Oesterhelt and W. Stoeckenius, *Nature New Biol. 233:* 149 (1971).
48. D. Oesterhelt and W. Stoeckenius, in *Methods in Enzymology* Vol. 31 (S. Fleischer and L. Packer, eds.), Academic, New York, 1974, pp. 667-678.
49. B. Becher and J. Y. Cassim, *Biophys. J. 16:* 1183 (1976).

50. J. Bridgen and I. D. Walker, *Biochemistry 15:* 792 (1976).
51. A. E. Blaurock, *J. Mol. Biol. 93:* 139 (1975).
52. R. Henderson, *J. Mol. Biol. 93:* 123 (1975).
53. R. Henderson and P. N. T. Unwin, *Nature 257:* 28 (1975).
54. K. Razi Naqvi, J. Gonzales-Rodrigues, R. J. Cherry, and D. Chapman, *Nature 245:* 249 (1973).
55. W. V. Sherman, M. A. Slifkin, and S. R. Caplan, *Biochim. Biophys. Acta 423:* 238 (1976).
56. C. F. Chignell and D. A. Chignell, *Biochem. Biophys. Res. Commun. 62:* 136 (1975).
57. S. C. Kushwaha and M. Kates, *Biochim. Biophys. Acta 316:* 235 (1973).
58. D. Oesterhelt and B. Hess, *Eur. J. Biochem. 37:* 316 (1973).
59. M. P. Heyn, P. J. Bauer, and N. A. Dencher, *Biochem. Biophys. Res. Commun. 67:* 897 (1975).
60. P. J. Bauer, N. A. Dencher, and M. P. Heyn, *Biophys. Struc. Mech. 2:* 79 (1976).
61. B. Becher and T. G. Ebrey, *Biochem. Biophys. Res. Commun. 69:* 1 (1976).
62. D. Oesterhelt and W. Stoeckenius, *Proc. Nat. Acad. Sci. U. S. 70:* 2853 (1973).
63. R. H. Lozier, R. A. Bogomolni, and W. Stoeckenius, *Biophys. J. 15:* 955 (1975).
64. N. Dencher and M. Wilms, *Biophys. Struc. Mech. 1:* 259 (1975).
65. M. Chu Kung, D. DeVault, B. Hess, and D. Oesterhelt, *Biophys. J. 15:* 907 (1975).
66. C. R. Goldschmidt, M. Ottolenghi, and R. Korenstein, *Biophys. J. 16:* 839 (1976).
67. R. H. Lozier and W. Niederberger, *Proc. Fed. Amer. Soc. Exp. Biol. 36:* 1805 (1977).
68. K. J. Kaufman, P. M. Rentzepis, W. Stoeckenius, and A. Lewis, *Biochem. Biophys. Res. Commun. 68:* 1109 (1976).
69. W. V. Sherman, R. Korenstein, and S. R. Caplan, *Biochim. Biophys. Acta 430:* 454 (1976).
70. B. Chance, M. Porte, B. Hess, and D. Oesterhelt, *Biophys. J. 15:* 913 (1975).
71. R. H. Lozier, W. Niederberger, R. A. Bogomolni, S.-B. Hwang, and W. Stoeckenius, *Biochim. Biophys. Acta 440:* 545 (1976).
72. A. Lewis, J. Spoonhower, R. A. Bogomolni, R. H. Lozier, and W. Stoeckenius, *Proc. Nat. Acad. Sci. U. S. 71:* 4462 (1974).
73. T. Schreckenbach and D. Oesterhelt, *Proc. Fed. Amer. Soc. Exp. Biol. 36:* 1810 (1977).
74. D. Oesterhelt, M. Meentzen, and L. Schuhmann, *Eur. J. Biochem. 40:* 453 (1973).
75. D. Oesterhelt, *Prog. Mol. Subcel. Biol. 4:* 133 (1976).
76. L. Y. Jan, *Vision Res. 15:* 1081 (1975).
77. P. Mitchell, in *Theoretical and Experimental Biophysics* Vol 2 (A. Cole, ed.), Dekker, New York, 1969, pp. 160-216.

78. P. Mitchell, *Symp. Soc. Gen. Microbiol. 20:* 121 (1970).
79. P. Mitchell, *J. Bioenergetics, 3:* 5 (1972).
80. F. M. Harold, *Bacteriol. Rev. 36:* 172 (1972).
81. F. M. Harold, in *The Bacteria* (L. N. Ornston and J. R. Sokatch, eds.), Vol. 6, Academic, New York, 1977.
82. R. D. Simoni and P. W. Postma, *Ann. Rev. Biochem. 44:* 523 (1975).
83. E. Racker and W. Stoeckenius, *J. Biol. Chem. 249:* 662 (1974).
84. L. P. Kayushin and V. P. Skulachev, *FEBS Lett. 39:* 39 (1974).
85. E. Racker, *Biochem. Biophys. Res. Commun. 55:* 224 (1973).
86. M. Yoshida, N. Sone, H. Hirata, Y. Kagawa, Y. Takeuchi, and K. Ohno, *Biochem. Biophys. Res. Commun. 67:* 1295 (1975).
87. I. J. Ryrie and P. F. Blackmore, *Arch. Biochem. Biophys. 176:* 127 (1976).
88. E. Racker and P. C. Hinkle, *J. Membr. Biol. 17:* 181 (1974).
89. L. A. Drachev, A. A. Jasaitis, A. D. Kaulen, A. A. Kondrashin, E. A. Liberman, I. B. Nemecek, S. A. Ostroumov, A. Yu. Semenov, and V. P. Skulachev, *Nature 249:* 321 (1974).
90. L. A. Drachev, V. N. Frolov, A. D. Kaulen, E. A. Liberman, S. A. Ostroumov, V. G. Plakunova, A. Yu. Semenov, and V. P. Skulachev, *J. Biol. Chem. 251:* 7059 (1976).
91. P. Shieh and L. Packer, *Biochem. Biophys. Res. Commun. 71:* 603 (1976).
92. T. R. Hermann and G. W. Rayfield, *Biochim. Biophys. Acta 443:* 623 (1976).
93. S.-B. Hwang and W. Stoeckenius, *J. Memb. Biol. 33:* 325 (1977).
94. J. K. Lanyi and R. E. MacDonald, in *Methods in Enzymology* Vol. 56 (S. Fleischer and L. Packer, eds.), Academic, New York, 1977.
95. R. E. MacDonald and J. K. Lanyi, *Biochemistry 14:* 2882 (1975).
96. J. K. Lanyi and R. E. MacDonald, *Proc. Fed. Amer. Soc. Exp. Biol. 36:* 1824 (1977).
97. J. K. Lanyi, R. Renthal, and R. E. MacDonald, *Biochemistry 15:* 1603 (1976).
98. R. Renthal and J. K. Lanyi, *Biochemistry 15:* 2136 (1976).
99. J. K. Lanyi and R. E. MacDonald, *Biochemistry 15:* 4608 (1976).
100. B. I. Kanner and E. Racker, *Biochem. Biophys. Res. Commun. 64:* 1054 (1975).
101. S. Ramos, S. Schuldiner, and H. R. Kaback, *Proc. Nat. Acad. Sci. U. S. 73:* 1892 (1976).
102. R. A. Bogomolni, R. A. Baker, R. H. Lozier, and W. Stoeckenius, *Biochim. Biophys. Acta 440:* 68 (1976).
103. J. K. Lanyi and K. Hilliker, *Biochim. Biophys. Acta 448:* 181 (1976).
104. P. J. Sims, A. S. Waggoner, C.-H. Wang, and J. F. Hoffman, *Biochemistry 13:* 3315 (1974).
105. T. N. Belyakova, Yu. P. Kadzyauskas, V. P. Skulachev, I. A. Smirnova, L. N. Chekulayeva, and A. A. Jasaytis, *Dokl. Akad. Nauk SSSR 223:* 483 (1975).
106. G. Wagner and A. B. Hope, *Aust. J. Plant Physiol. 3:* 665 (1976).

107. E. P. Bakker, H. Rottenberg, and S. R. Caplan, *Biochim. Biophys. Acta 440:* 557 (1976).
108. H. Michel and D. Oesterhelt, *FEBS Lett. 65:* 175 (1976).
109. P. C. Maloney, E. R. Kashket, and T. H. Wilson, in *Methods in Membrane Biology* Vol. 5 (E. D. Korn, ed.), Plenum, New York, 1976, pp. 1-49.
110. R. A. Bogomolni, *Proc. Fed. Amer. Soc. Exp. Biol. 36:* 1833 (1977).
111. F. M. Harold, *Adv. Microbial Physiol. 4:* 45 (1970).
112. M. Ginzburg, B. Z. Ginzburg, and D. C. Tosteson, *J. Membr. Biol. 6:* 259 (1971).
113. M. Ginzburg and B. Z. Ginzburg, in *Biomembranes* Vol. 7 (H. Eisenberg, E. Katchalsky-Katzir, and L. A. Manson, eds.), Academic, New York, 1975, pp. 219-256.
114. M. Ginzburg and B. Z. Ginzburg, *J. Membr. Biol. 26:* 153 (1976).
115. M. Ginzburg, *Biochim. Biophys. Acta 173:* 370 (1969).
116. G. Szabo, G. Eisenman, and S. M. Ciani, in *Physical Principles in Biological Membranes* (F. Snell, J. Wolken, G. Iverson, and J. Lam, eds.), Gordon and Breach, New York, 1970, pp. 79-135.
117. M. Eisenbach, S. Sprung, H. Garty, R. Johnstone, H. Rottenberg, and S. R. Caplan, *Biochim. Biophys. Acta 465:* 599 (1977).
118. E. A. Berger and L. A. Hepple, *J. Biol. Chem. 249:* 7747 (1974).
119. H. R. Kaback, *Biochim. Biophys. Acta 265:* 367 (1972).
120. H. R. Kaback, *Science 186:* 882 (1974).
121. H. Hirata, N. Sone, M. Yoshida, and Y. Kagawa, *Biochem. Biophys. Res. Commun. 69:* 665 (1976).
122. D. L. Oxender and S. C. Quay, in *Methods in Membrane Biology* Vol. 6 (E. D. Korn, ed.), Plenum, New York, 1977, pp. 183-242.
123. R. E. MacDonald, R. V. Greene, and J. K. Lanyi, *Biochemistry 16:* 3227 (1977).
124. R. E. MacDonald and J. K. Lanyi, *Proc. Fed. Amer. Soc. Exp. Biol. 36:* 1828 (1977).
125. W. Boos, *Ann. Rev. Biochem. 43:* 123 (1974).
126. J. S. Hubbard, C. A. Rinehart, and R. A. Baker, *J. Bacteriol. 125:* 181 (1976).
127. J. K. Lanyi, V. Yearwood-Drayton, and R. E. MacDonald, *Biochemistry 15:* 1595 (1976).
128. J. W. Belliveau and J. K. Lanyi, *Arch. Biochem. Biophys. 178:* 308 (1977).
129. R. K. Crane, *Proc. Fed. Amer. Soc. Exp. Biol. 24:* 1000 (1965).
130. S. G. Schultz and P. F. Curran, *Physiol. Rev. 50:* 637 (1970).
131. J. Stock and S. Roseman, *Biochem. Biophys. Res. Commun. 44:* 132 (1971).
132. S. Kahane, M. Marcus, H. Barash, Y. S. Halpern, and H. R. Kaback, *FEBS Lett. 56:* 235 (1975).
133. K. Ring, H. Ehle, and B. Foit, *Biochim. Biophys. Acta 433:* 615 (1976).
134. J. Thompson and R. A. MacLeod, *J. Biol. Chem. 246:* 4066 (1971).

135. J. Thompson and R. A. MacLeod, *J. Bacteriol. 120:* 598 (1974).
136. F. M. Harold and K. Altendorf, in *Current Topics in Membranes and Transport* Vol. 5 (F. Bronner and A. Kleinzeller, eds.), Academic, New York, 1972, pp. 1-50.
137. A. Danon and W. Stoeckenius, *Proc. Nat. Acad. Sci. U. S. 71:* 1234 (1974).
138. A. Danon and S. R. Caplan, *Biochim. Biophys. Acta 423:* 133 (1976).
139. D. Oesterhelt, *Ciba Fdn. Symp. 31:* 147 (1975).
140. P. C. Maloney, E. R. Kashket, and T. H. Wilson, *Proc. Nat. Acad. Sci. U. S. 71:* 3896 (1974).
141. P. C. Maloney and T. H. Wilson, *J. Membr. Biol. 25:* 285 (1975).
142. D. Oesterhelt, R. Hartman, U. Fischer, H. Michel, and Th. Schreckenbach, *Proceedings of the 10th FEBS Meeting,* North Holland, Amsterdam, 1975, pp. 239-251.
143. J. K. Lanyi, *J. Biol. Chem. 244:* 2864 (1969).
144. J. K. Lanyi, *J. Biol. Chem. 244:* 4168 (1969).
145. K. Andersen, Ph.D. Dissertation, University of Trondheim, Norway, 1975.
146. D. Oesterhelt and G. Krippahl, *FEBS Lett. 36:* 72 (1973).
147. M. A. Slifkin and S. R. Caplan, *Nature 253:* 56 (1975).

Author Index

Numbers in brackets are reference numbers and indicate that an author's work is referred to although his name is not cited in the text. Italic numbers give the page on which the complete reference is listed.

Subject Index